Lazare and Sadi Carnot: A Scientific and Filial Relationship

HISTORY OF MECHANISM AND MACHINE SCIENCE

Volume 19

Series Editor

MARCO CECCARELLI

Aims and Scope of the Series

This book series aims to establish a well defined forum for Monographs and Proceedings on the History of Mechanism and Machine Science (MMS). The series publishes works that give an overviewof the historical developments, from the earliest times up to and including the recent past, of MMS in all its technical aspects.

This technical approach is an essential characteristic of the series. By discussing technical details and formulations and even reformulating those in terms of modern formalisms the possibility is created not only to track the historical technical developments but also to use past experiences in technical teaching and research today. In order to do so, the emphasis must be on technical aspects rather than a purely historical focus, although the latter has its place too.

Furthermore, the series will consider the republication of out-of-print older works with English translation and comments.

The book series is intended to collect technical views on historical developments of the broad field of MMS in a unique frame that can be seen in its totality as an Encyclopaedia of the History of MMS but with the additional purpose of archiving and teaching the History of MMS. Therefore the book series is intended not only for researchers of the History of Engineering but also for professionals and students who are interested in obtaining a clear perspective of the past for their future technical works. The books will be written in general by engineers but not only for engineers.

Prospective authors and editors can contact the series editor, Professor M. Ceccarelli, about future publications within the series at:

LARM: Laboratory of Robotics and Mechatronics
DiMSAT – University of Cassino
Via Di Biasio 43, 03043 Cassino (Fr)
Italy
E-mail: ceccarelli@unicas.it

For further volumes:
http://www.springer.com/series/7481

Charles Coulston Gillispie • Raffaele Pisano

Lazare and Sadi Carnot

A Scientific and Filial Relationship

Second Edition

Springer

Charles Coulston Gillispie
Department of History
Princeton University, New Jersey
USA

Raffaele Pisano
SCité
University of Lille 1, France

ISSN 1875-3442 ISSN 1875-3426 (electronic)
ISBN 978-94-017-7950-0 ISBN 978-94-017-8011-7 (eBook)
DOI 10.1007/978-94-017-8011-7
Springer Dordrecht Heidelberg New York London

Softcover reprint of the hardcover 2nd edition 2014

Printed on acid-free paper

Springer is part of Springer Science+Business Media (www.springer.com)

Foreword

> Les talents écartent l'ennui, chassent le vice, et sèment la vie de fleurs:
> puissance, richesse, vous n'offrez point ces avantages!
> (Legend of a lithograph depicting the spirit of enlightened industry and belonging to a great–great granddaughter of Lazare Carnot)

The first part of this compendium consists of slightly revised excerpts from my monograph, *Lazare Carnot Savant,* published in 1971 (Gillispie 1971). The preface to that book, which is based partly on access to the Carnot family archives, explains how it came to be written, and makes the necessary acknowledgments. It contains a short section on the more famous work of Sadi Carnot, whose *Réflexions sur la puissance du feu et sur les machines propres à développer cette puissance* is generally regarded as the origin of the science of thermodynamics (Carnot 1824). It appeared to me that Sadi Carnot's analysis may be read as an application of his father's invention of the science of machines to heat engines.

I have suggested as much in what follows. My colleague and co–author, Raffaele Pisano, has independently devoted much more intensive study to the work of Sadi Carnot and its influence. It seemed to both of us that it would be well to publish, or in my case to republish, our findings. Both father and son were trained as engineers, Sadi at the newly founded *École polytechnique*. They had no idea that their work would develop into the physics of work and energy. So it came about, however, and that is the reason for including an account of it in a series devoted to the role of engineering mechanics.

The major portion of *Lazare Carnot Savant* is here included. We are not reprinting the appendices, which are reproductions of Carnot's unpublished papers discussed in Chapter 3, Parts A and B, nor the essay on the unpublished mathematical theory of the infinite contributed to my book by my late colleague, A.P. Youshkevitch, which has little to do with engineering mechanics. Princeton University Press has generously granted the necessary permission, for which I am extremely grateful.

Princeton, USA　　　　Charles Coulston Gillispie

Preface

This extraordinary book on two outstanding French engineers and scientists has been written by two authors who combined their efforts to that end in a very fruitful way. The result is based on deep historical, epistemological, and methodological insights that shed completely new light on the scientific and filial relationship between the famous politician, mathematician, and engineer Lazare Nicolas Marguerit Carnot (1753–1823) and his son Nicolas Léonard Sadi Carnot (1796–1832) who was also trained as an engineer at the newly founded *École polytechnique de Paris*.

While Lazare Carnot wrote on machines and adapted the science of mechanics to the science of machines, his son elaborated a general, abstract, thermodynamical theory. Insofar the purpose of Gillispie and Pisano is not to recount the history of mechanics, but to identify the points of entry of Lazare Carnot's engineering science into mechanics (p. 100). The authors are well aware that the analogy between fluid flow and heat flow has often been discussed in the research literature in order to characterize the influence of Lazare's work upon Sadi's (p. 18).

Yet, the authors make a much wider claim for the continuity between the works of the two Carnots. They would like to show that Sadi Carnot's *Réflexions sur le puissance motrice du feu* published after his father's death "may properly be read not only as the foundation of thermodynamics, but also as the culmination of a methodologically and conceptually coherent series" of Lazare Carnot's essays on the science of machines (p. 15).

How can such a strong claim be demonstrated? Gillispie and Pisano study and compare the different steps of Lazare Carnot's thinking about mechanics, of his writings, and compare them with Sadi Carnot's *Réflexions* which can be taken for the foundation stone in the science of thermodynamics (p. 77).

To that end they identify those elements of the arguments that were derived from the work of the father. They amply and clearly explain their methods and methodology as well and include even a whole chapter (Chap. 6) that deals with such general issues like historical methodology, interpretation, and scientific theory.

Their key notion is historical epistemology which is based on the use of logical-historical categories. The authors adopt them in order to investigate Sadi Carnot's scientific thought. They are interested in *effective history* that is history relying on the fundamental choices made by scientists who influenced the interpretation of history by means of crucial choices (p. 156).

Their leading questions read: What is the theoretical organization in the two Carnots? On what principles is it based? In order to answer to these two questions they study the history of science by means of a logical investigation (p. 191). Thus they are able to demonstrate that Sadi Carnot's reasonings with double negative sentences (DNS) are based on non-classical logic. The authors' list in the appendix comprehends 65 such sentences. Their sequence may synthetically express the entire development of Sadi Carnot's scientific thought (p. 205).

Gillispie and Pisano convincingly conclude that Sadi Carnot's "theory has to be qualified as a logical theory because the double negative sentences illustrate for the first time a very detailed structure of Sadi Carnot's arguments, adequately representing Carnot's original scientific thought" (p. 211). But the two authors go even further by explaining that the origin of the idea of the cycle was the analogy with the electric circuit in Alessandro Volta's battery. Hence Lazare Carnot, not Sadi, first had the idea of the cycle of heat machines (p. 234).

The fruitfulness of Gillispie's and Pisano's approach becomes again obvious when they apply the method of historical epistemology to the mathematical footnote in Sadi Carnot's *Réflexions* which combines epistemological and historical approaches to identify significant historical hypotheses (p. 257). The authors claim that Lazare Carnot's synthetic method is present in Sadi Carnot's theory paying particular attention to Sadi Carnot's reasoning process (p. 279). Their hope is certainly justified that their approach can contribute to clarify the birth and development of Sadi Carnot's theory and the historical knowledge of thermodynamics.

Comparing Lazare and Sadi Carnot's theories of the efficiency of a machine, they conclude that "neither theory is based on axioms, but on the program of scientifically resolving a crucial problem that in the minds of the lay people of the time coincided with metaphysics" (p. 370).

Gillispie and Pisano have written a really remarkable book that reveals an impressing knowledge of the huge amount of original and modern publications regarding history and philosophy of science. They always help the reader not to lose track of things by adding summaries, illustrations, and by compiling their arguments or those of their heroes in long lists. In spite of all similarities they have proved between father and son they do not overlook a crucial difference between them. Sadi Carnot's work was deep in a way that his father's was not: It founded the science of thermodynamics (p. 86).

Berlin, Germany Eberhard Knobloch

Acknowledgments

The genesis of such a difficult and lengthy book has deep roots, and the final result has been a long time in the making. When the research and production of a work of this nature is carried out over a significant period of time, many friends and scholars become contributors to both the research and the writing process. I owe gratitude to many such people and will never manage to thank them all appropriately.

My own early research on Lazare and Sadi Carnot began with a lengthy dissertation on Sadi Carnot's logic and mathematics which I wrote while on the faculty of physics at the University of Naples "Federico II", in my native city. Since appearance of this specialized historical and scientific work, many additional papers have been written by me and by others, on Sadi Carnot alone and on the two Carnots jointly, often in collaboration with my adviser Antonino Drago, the first Italian historian since the 1980s–1990s to recognize the importance of studying the two Carnots jointly as a unique program of scientific research in Italy. Therefore, my first acknowledgments are to him.

Of course, this crucial first book to include both of the Carnots would not have been possible, first without Charles Gillispie's approval and vast, indispensable works on Lazare Carnot, and second without all of Robert Fox's historical details and profound research on Sadi Carnot. These are two definitive, worldwide recognized, masters of the history of science and, as well, friends and colleagues of mine. I feel very privileged to have known and to have worked with them.

A special acknowledgment goes out to my friend and historian, Eberhard Knobloch, President of the *International Academy of the History of Science* who generously accepted to write his insightful and very much appreciated *Preface*. I also thank him for his cultural and persuasive and constant, encouragement. A particular acknowledgement and appreciate goes to the *European Society for the History of Science*'s congresses for promoting the history of science in its broadest sense and contacts between scholars and institutions across Europe, and all the world. It gave me the possibility to know and exchange several elements of interest of history of science, mechanics and thermodynamics. Thus I have naturally incurred many debts.

I have to thank the *Centre Françoise Vietè – épistémologies, histoire des sciences et des techniques* and his head, Stéphane Tirard, University of Nantes, where I was welcomed to spend my post–doctoral tenure research for a correlated project on the history of thermodynamics.

I want to thank Ilaria Gaudiello for her kind remarks and readings.

Among the institutions, a special thanks for their precious and kind collaboration in allowing me to study Lazare and Sadi's primary manuscripts. My regular visits to French libraries and archives have also been a source of much pleasure. The courtesy with which I was received in the following archives was exemplary. I express my gratitude to Olivier Azzola of the *Archives de l'École polytechnique de Paris* who offered me very lovely pictures of "X" during Sadi Carnot's time. For their professional and generous availability, I thank Claudine Pouret, of the *Académie des sciences, Institut de France* and Vincenzo de Luise of the *Roberto Stroffolini Archive*, Department of Physics, University of Naples "Federico II". I am particular grateful to Monsieur Gaetan Carnot, member of Carnot's family who was so generous for removing all restrictions and authorized me to publish plates, from the original and special collection of family, portraits both Lazare and Sadi Carnot. The images are conserved in France at *Académie François Bourdon*. Thus deep and warm acknowledgments go to the its director Ivan Kharaba and its archivist, François–Yves Julien, as well.

I intend thank Claude Debru, head of the Cirphles, *École normale supérieure Paris* for his friendly encouragement during this hard job and his interest with Monsieur Jean–François Bach, *Secrétaire Perpétuel* of the French *Académie des sciences.*

Last but not least, I express my warm and friendly gratitude to Marco Ceccarelli, Nathalie Jacobs, Anneke Pot, respectively Springer book Series Editor, Springer Publishing Editor in–chief, and Springer Editorial Assistant for their good job and positive reception of our project on the two Carnots.

I would like to express my sincere gratitude for permission to reproduce materials, readings–improvements, encouragements, collaborations in seeking details on localizations of manuscripts, original handmade or reproduced photographs beneficially cited in the books, and essentially helping with the English translations on my behalf to

Adele Bianchi	*Istituto Lombardo Accademia di Scienze e lettere*
Alessandra Ravelli	*National Archive Club Alpino Italiano*
Annie Chassagne	*Bibliothèque de l'Institut de France*
Fabienne Queyroux	*Bibliothèque de l'Institut de France*
Mireille Pastoureau	*Bibliothèque de l'Institut de France*
Virginia Mateus	*Bibliothèque de l'Institut de France*
Anne Bernard	*Académie des sciences, Institut de France*
Claudine Pouret	*Académie des sciences, Institut de France*
Christiane Pavel	*Académie des sciences, Institut de France*

Anne–Marie Mahe	*Bibliothèque Nationale de France*
Céline Gervaise	*Bibliothèque Nationale de France*
Bendetta Cappi	*Musei Civici di Como. Volta Temple in Como*
Isabella Nobile	*Musei Civici di Como. Volta Temple in Como*
Rachele Viscido	*Musei Civici di Como. Volta Temple in Como*
Dominique Bénard	*Université de Maine*
Ennio Iannucci	*IPSIA Ada Gobetti Marchesini, Torino*
Fausto Barbaglia	*Museo di Storia Naturale, "La Specola", University of Firenze*
Saulo Bambi	*Museo di Storia Naturale, "La Specola", University of Firenze*
Franco di Liberto	*Physics department, University of Naples "Federico II"*
Guillaume Saquet	*Fonds ancien et archives, École nationale des ponts et chaussées. École des ponts ParisTech*
Lavinia Iazzetti	*Archive Accademia delle Scienze, Torino*
Patrizia Ruffo	*Istituto e Museo di Storia della Scienza, Firenze*
Susanna Cimmino	*Istituto e Museo di Storia della Scienza, Firenze*
Roberto Mantovani	*Urbino Museum of Science and Technology, Urbino University*

Paris, France   Raffaele Pisano

Contents

Chapter 1
Biographical Sketch of Lazare Carnot

Known to French history as the "Organizer of Victory" in the wars of the Revolution, and to engineering mechanics for the principle of continuity in the transmission of power, Lazare–Nicolas–Marguerite Carnot (1753–1823) remains one of the very few men of science and of politics whose career in each domain deserves serious attention on its own merits. His father, Claude, lawyer and notary, was among the considerable bourgeois of the small Burgundian town of Nolay, west of Beaune and thus on the opposite side of the ridge from the superb vineyards of the Côte–d'Or. Members of the family still own the ancestral home. Carnot's most notable descendants have been his elder son Nicolas–Léonard Sadi (1796–1832), a principal founder of the science of thermodynamics, and the latter's nephew, also called Sadi, President of the French Republic from 1887 until his assassination in 1894. A minor versifier himself, Lazare named his heir after a Persian poet whose work he admired in translation. His younger son, Hippolyte (1801–1888), wrote the first biography, a source of major importance for the personal details of his father's life.[1] Carnot had his early education in the Oratorian Collège (a school) at Autun. Thereafter his father enrolled him in a tutoring school in Paris, which specialized in preparing candidates for the entrance examinations to the service schools that trained cadets for the Navy, the Artillery, and the Royal Corps of Engineers. Strong in technique and low in prestige, the Corps of Engineers was the only branch of military service in which a commoner might hold a commission. On completing the normal course of 2 years, Carnot graduated from its school at Mézières at the age of 20 in 1773. Gaspard Monge teacher of mathematics and physics, was then at the height of his influence over the cadets.[2]

---

[1]It is an exemplary work of family piety, written for the edification of Carnot's descendants, *Mémoires sur Carnot par son fils* (Carnot 1861–1863). The standard biography is *Le grand Carnot* by Marcel Reinhard (Reinhard). (The 2nd edition, 1994, also includes preface by Charles Coulston Gillispie).

[2]On Mézières, see *L'École royale du génie de Mézières* by René Taton (1964a, pp 559–615).

C.C. Gillispie and R. Pisano, *Lazare and Sadi Carnot: A Scientific and Filial Relationship*, History of Mechanism and Machine Science 19, DOI 10.1007/978-94-017-8011-7_1,

Although Carnot's handling of problems always bore the mark of the engineer he became, it does not appear that he was one of Monge's favorite pupils. There was little mention of the one by the other whereas Monge's disciples were given to fulsome tributes. Nor is it recorded that Monge took the slightest interest in Carnot's writings on mathematics or mechanics or did anything to secure them a hearing. Some further inference may be permitted from their differences in temperament: Monge's that of a pedagogue and mathematician who, when he involved himself in public affairs, proved to be theoretical, emotional, incapable of decision, deficient in judgment, and inattentive to detail; Carnot's that of an engineer, eminently practical, able in conception and execution, and capable of the deeper consistency that inheres in judging of circumstance and practicality rather than asserting moral absolutes. Carnot's approach to mathematics and mechanics was more concerned with operations and fundamentals than that of Monge. Its actual mathematical yield was much less while its significance for the physics of work and energy was much greater.

Carnot's professional life in the 16 years prior to the Revolution did not on the whole prove stimulating. On leaving Mézières with the rank of lieutenant, he entered on a sequence of garrison duties, first at Calais, and then successively at Le Havre, Béthune, Arras, and Aire. Only in Aire did he find cultivated company. There, in the 1780s, he was admitted to the predominantly literary Society of Rosati, a group of beaux esprits famous for having counted Carnot and Maximilien Robespierre among their members rather than for any lasting contribution to letters. Carnot's training did, however, serve him in good stead in several instances. The fortifications of the port of Calais were being rehabilitated during his tour of duty there. He was detached from Le Havre for 3 years of duty on the harbor under construction at Cherbourg, the most considerable and elaborate engineering works undertaken by the military and naval establishment in the latter part of the century. In these and other, lesser assignments he made the reputation of a sound, reliable, and enterprising young officer. His personnel file contains a series of recommendations by his superiors.

Nevertheless, the life was not one to engage the full talent or fulfill the ambition of an able and talented man. In a memoir of 1776, Charles Augustin Coulomb (1736–1806) a fellow engineer and in science a far more famous one, took occasion to testify to its limitations. "After graduating from Mézières a studious young man who would withstand the boredom and monotony of his work had no choice except to cultivate some branch of science or literature that was entirely independent of his professional duty."[3] Later on, Carnot described himself as having been "neglected, lonely, absent–minded, preoccupied, what was called a 'philosophe,' or in other words sort of an odd type."[4] Such was his sense of himself during the years between his graduation from Mézières in 1773 and 1784, when he won a competition set by the Academy of Dijon for an essay on Vauban. In the perspective of his later military

[3]It is quoted by Stewart Gillmor in *Coulomb and the Evolution of Physics and Engineering in Eighteenth-Century France* (Gillmor 1971, Appendix B).

[4]In the Hyppolyte's wording (Carnot 1861–1863, I, p 93).

and political career, the *Éloge de Vauban* (1784) appears, naturally enough given the subject and the discussion that ensued, to be its opening episode.[5]

Although no detailed account of Carnot's studies during the preceding interval is possible, the individuality of his scientific work is such that he must have been occupying it with a self–education in mechanics and mathematics that carried him far beyond the text–book level of the curriculum of Mézières. It must have been in the isolation of garrison life that Carnot set himself to reading and re–reading d'Alembert, Bossut, Bélidor Euler's mechanics, and Daniel Bernoulli's hydrodynamics. His early work bespeaks an auto–didactic familiarity. He had no access to the scientific milieu of Paris. He spent his frequent and lengthy leaves at home in Nolay and in Dijon. While he was at Calais his father sent the younger brother, Claude–Marie–Feulint (1755–1836) to reside with him and to be tutored by Lazare in the mathematics and mechanics for the entrance examinations at Mézières.[6] Feulint succeeded on the first try.

What is more substantial, Carnot's earliest publication was not the *Éloge de Vauban* of 1784, but *Essai sur les machines en général* of 1783. We analyze that work in detail in the next chapter. It contains in principle all that Carnot actually contributed to mechanics, while also constituting the starting point of the French engineering tradition in the science of machines. From documents in the Archives of the Academy of sciences in Paris, in those of the German Academy of Sciences in Berlin, and in those of the Carnot family in their house at Nolay, it is possible to fix with precision the occasions for Carnot's earliest formal composition in the literature of mechanics and mathematics.

Even like the *Éloge de Vauban*, those writings were elicited by competitions set by learned societies. He composed two early drafts of the *Essai sur les machines en général*, the first completed in March 1778 while at Cherbourg and the second in July 1780 at Béthune. The Academy announced the competition in April 1777 for the year 1779 and then adjourned it to 1781 since the judges found none of the initial entries to be worthy of an award. On the second round a memoir by Coulomb won the prize while Carnot's was accorded Honorable Mention. The contents and their conformity to the Academy's desiderata will be discussed in the next chapter, following analysis of the *Essai sur les machines en général* itself.[7] As for Carnot's complementary interests in the foundations of mathematics, his *Réflexions sur la métaphysique du calcul infinitésimal* (1797) grew out of a memoir that he submitted to the *Royal Prussian Academy of Sciences* in 1785.[8]

---

[5]The prize was first offered in 1783, but in the absence of worthy entries, the competition was adjourned to 1784. For the details see both *Le grand Carnot* and *Mémoires sur Carnot par son fils* (Reinhard, I, pp 76–86; Carnot H, pp 99–117).

[6]On that see the arguments presented by Hyppolite (Carnot H, I, pp 94–95).

[7]Below, Chapter 2.

[8]A facsimile of that entry is printed as *Appendix A* in the present author's *Lazare Carnot Savant* (Gillispie 1971). That book also contains an essay *Lazare Carnot and the Competition of the Berlin Academy in 1786 on the Mathematical Theory of the Infinite* (*Ivi*, pp 149–168) by my esteemed

Just prior to his recognition for the *Éloge de Vauban*, Carnot made one more attempt to win the approbation of the Academy of Sciences. On 17 January 1784 he sent in a "Lettre sur les Aérostats" in response to its invitation to interested parties to communicate reflections inspired by the first human flight. On 5 June 1783 Joseph and Étienne Montgolfier had created a sensation by sending a balloon filled with heated air soaring to a height of 6,000 ft above the town of Annonay.[9] Their feat and its elaborate repetitions in Paris raised the engineering problems of locomotion and stability of flight. The studies evoked were by no means all sterile. A memoir by Jean–Baptiste Meusnier de la Place contains the principle that much later governed adjustment of specific gravity of a submarine in accordance with its depth.[10] Carnot addressed himself rather to the problem of a "dirigeable" and proposed a scheme for a propeller to be powered by a motion of systole and diastole created in the balloon by the dissipation and restoration of heat, a sort of jellyfish effect. The notion was visionary enough. What excites interest in retrospect is his discussion of heat, fluids, and the potential awaiting the steam engine. "Notice by the way, gentlemen", he wrote "how much labor will be saved in factories when the mechanism of heat is better understood." And further, "The engine powered by heat provides a very powerful motive principle and the principle can as easily be adapted to moving blades and wheels as to beams and pistons" (Carnot H, I, p 185). Throughout his life, indeed, Carnot maintained a lively interest in the work of mechanical invention, rather in point of principle, criticism, and appreciation, however, than of participation. He does not appear to have been gifted with practical inventiveness himself.

Only with his military writings did Carnot win attention. Biographers and historians interested in Carnot as wartime leader and Member of the Committee of Public Safety have naturally looked in these for clues to what their author would become.[11] In good philosophic vein Carnot invoked the example of Vauban to convey his own enlightened sentiments about natural equality in men of merit and inequities in the social order. The central thrust of the essay was strategic. It celebrated, in the declamatory style thought appropriate for ceremonial occasions, the complements of geometry and enlightenment that impelled Vauban, the founder of professional military engineering, to conceive that the purpose of warfare should be defense of civilization rather than destruction of the enemy.

---

colleague, the late A. P. Youschkevitch of the Moscow Institute for the History of Science and Technology.

[9]On that one can consult *The Montgolfier Brothers and the Invention of Aviation* (Gillispie 1983).

[10]*Mémoire sur l'équilibre des machines aérostatiques* (3 December, 1783). On that see: Meusnier de la Place (1783). Gaston Darboux published this paper in his *Mémories et travaux de Meusnier relatifs à l'aérostation* (Darboux). See also, *The Montgolfier Brothers and the Invention of Aviation* (Gillispie 1983, pp 100–101).

[11]The Academy of Dijon, embarrassed by its experience with its most famous laureate, Jean–Jacques Rousseau, thereafter took care to edit the texts it crowned before publication. As published Carnot's *Éloge de Vauban* was a toned down version of what he had delivered. Its original tenor has been plausibly reconstructed by G. Duthuron (Duthuron, pp 152–165).

Historians have sometimes waxed a little wry at Carnot's expense for having celebrated, when a young Turk, the humanitarian warfare of position, and then, when actually in charge of waging a war, for having put into effect the contrary strategy of mobility, mass, offensive, and conquest. What must be remembered is the difference in circumstances. In the 1780s Carnot was a young engineering officer writing out of the specialist tradition of his corps and upholding its mission in the design and construction of formal works of siege craft and defensive fortification. Doing so required him to contend against the renewed emphasis of the combat arms on gallantry, movement, and command under fire. Chivalric pretensions were the line officers' part in the aristocratic resurgence. In power during the Revolution, however, Carnot did not dispose of the disciplined, technically trained, professional, and careful forces that such a conservative strategy presupposed. His troops were untrained levies, some under arms out of patriotism, and some because of conscription. Patriotic soldiers furnished a commander with dash and conscripted soldiers with mass. Both sorts were altogether different from the armies of Louis XIV and the eighteenth century.

The unexpectedness of what happened to persons swept along in the ineluctability of what happened to society always enhances the historical drama of the French Revolution. There is no need to follow Carnot's political career in detail. It will suffice to evoke it as the context, the disproportionately important context, in which recognition for scientific originality came the way of a no longer young engineering officer suffering increasing frustration over the seemingly dead end of his career as a military engineer and the neglect accorded to his scientific work. The earliest phase, that of the Constituent Assembly, saw that pattern relieved through the new facility of expressing grievances openly and advocating change. At the outset Carnot did not think to go beyond professional matters. That the army must be reformed to secure promotion in accordance with merit rather than noble birth; that equity and dignity must be accorded to the Corps of Engineers; that the Engineers and the Artillery, the two "learned branches (armes savantes)", should be amalgamated into a single, scientifically trained corps, headed by an elective council rather than an authoritarian commander; that warfare itself was to be limited and humanized by a combination of firepower with fortification – these were ideas shared by many of Carnot's colleagues. His advocacy was distinguished in its imperative tone and further in that like many of his countrymen, though few professional colleagues, he went over the heads of his superiors and addressed his reclamations not to the War Office, but directly to the National Assembly.[12]

The way to politics itself opened out of Carnot's reassignment to duty in the North, specifically to Aire in the region of Calais, where he and his younger brother, Feulint, had put down roots among the interests and sympathies of the locality. There he followed for a time in the way smoothed by Feulint's readier congeniality.

---

[12] *Réclamation adressée à l'Assemblée nationale contre le régime oppressif sous lequel est gouverné le Corps Royal du Génie* ... A manuscript copy is preserved in the family archives at Nolay. On these matters, see *Le grand Carnot* by Reinhard (Reinhard, Chapters 11–12, I, pp 148–176).

Lazare Carnot followed Feulint personally and married into the same family. He did so politically and ran alongside his brother as candidates for election as deputies to the Legislative Assembly in September 1791. Feulint had taken part in various councils and deputations to the Constituent Assembly. Better known in the region, he was elected to the first seat on the second ballot. Lazare Carnot for his part had joined the Society of Friends of the Constitution in his garrison town of Aire and had been its president for a time. That was his only political activity prior to his election. He won the ninth seat, also on the second ballot. That so limited a political experience as Feulint's and Lazare's should suffice for election as representatives of the people will be less surprising when it is recalled that the outgoing Constituent Assembly had stipulated, unwisely in the judgment of most historians, that none of its members might be reelected. The French people having had no voice in their own affairs, there were in 1791 no communities in which significant political experience might have been attained. In Reinhard's view, Carnot, now married into a substantial family, entered upon his legislative duties in a much calmer spirit than had been his at the outset of the Revolution in 1789. Already struck down were the aristocracy and the impediments to promotion and dignity that had engendered the angers of his youth. Like the majority of deputies he supposed the Revolution to be over and looked forward to participating in the regime of a constitutional monarchy that would ensue. He can have had no thought of wielding power in posts of the highest responsibility.

He came to do so because the new regime, in which men of his kind, the qualified and able bourgeoisie, could feel at ease was anything but stable. In reality the Revolution had only begun when the Legislative Assembly convened in October 1791. In its further course Carnot's professional military competence in harness with his engineer's ability to improvise arrangements and organize procedures proved to be great advantages, both to him and to the nation. The critical faults in the political structure made his qualities relevant. Throughout the first 6 months of his legislative career the untrustworthiness of the king and many of his ministers dissipated the credibility of combining monarchy with constitutionalism and confirmed Carnot in a kind of latent republicanism. Thereafter, following the declaration of war in April 1792, instigated by the Girondist faction with a view to exhibiting the duplicity of the crown, the unreliability of the leadership of the army and the reverses of the opening campaigns put at a premium the services of any patriotic deputy with military competence.

Lazare Carnot's actual political interventions in the Assembly were maladroit. He was never a notable orator. Still, knowing how he conducted himself in the pressures soon to be upon him, the historian can understand how it was that his traits made themselves felt in the heightened sensibility of those intense and febrile days when the French monarchy was proving its caducity, and no one dared plan openly for what might take its place. Nothing was then observed more narrowly than character. Men took one another's measure, and Carnot's began to come through. His contemporaries felt the intelligence, the probity, and the sense of reality that proved adequate to judging of the rightness of republicanism (for he was never a democrat, and never in rapport with the demagogues and idolaters of humanity); the instinct for concentrating on the job at hand; the fund of toughness about what he could not help (the Terror later and the guillotine) so long as it did not touch him in his own pride; the selflessness about material interests in a man alert in defense of his dignity; the passion to prevail with the right answer; and the unwillingness to compromise his own principles combined with some insensitivity to those of others that might conflict.

His biographer alludes to a mathematical dogmatism in Carnot's spirit.[13] It may not be necessary to read his character that way. On the whole he was a critical thinker about mathematics, but not a notable mathematician. His scientific work was at the opposite pole from dogmatism. It may be preferable to attribute the rigidity in him to a fundamental decency, manifest in the stands he took, in which he appears in retrospect the more sympathetic for a certain maladroitness, his fervor not always suited to the matter, but well suited to the times. His brother's reluctant behavior in contrast gradually revealed a man of the old dispensation insufficiently tempered by an occasional opportunism. Judging of events with a surer instinct, Lazare Carnot moved to the left and the Republic. Originally the less popular of the brothers, only he was returned to the Convention from the Pas–de–Calais in September 1792 following the overthrow of the Monarchy and the dissolution of the Legislative Assembly at the hands of the city of Paris in its rising of 10 August.

Among the hundreds of deputies to the Convention, Lazare Carnot was the one who most effectively improvised the function of the Representative on Mission by means of which the sovereignty of the people was exercised in specific actions throughout the mass of the nation. The substitution of the Republic for the Monarchy was the act in the first instance of the city of Paris. Other municipalities, provincial administrations, civil servants, officers of the army and navy, in short the agencies and agents of government – all that remained in the inertial grip of routine procedures of the Old Regime when the personnel was not, as much of it was and especially so in the armed forces, actively or potentially disposed to Counter–Revolution at home or to treating with the enemy abroad or to both these oppositions now constituting treason. Into such situations of anarchy, confusion, or active disaffection in the winter of 1792 and into the summer of 1793 would come sometimes one, though usually two or three, of the deputies to the Convention, Representatives of the People, dressed in blue waistcoats with brass buttons, red, white, and blue sashes, and felt hats decorated with tricolor feathers. To them was delegated carte–blanche the sovereign authority of the People, and in them was incarnated for the time being its revolutionary will. They might suspend local officials and replace them with others; they might set up special tribunals and sit as judges; they might exercise eminent domain over property, fix prices; and requisition whatever transport, enterprise, livestock, foodstuffs, or supplies were required for military purposes; when on mission to the armed forces, they might supersede commanders and overrule or countermand orders. They might do whatever they thought necessary.

[13]The argument is well treated by Marcel Reinhard in his *Le grand Carnot* (Reinhard, I, p 187).

Thus on the morrow of 10 August 1792 the National Assembly designated commissioners to apprise the several armies that the Republic had supplanted the monarchy and to secure their allegiance to the regime. Lazare Carnot, Claude–Antoine Prieur de la Côte–d'Or and Anne–Pierre Coustard were dispatched to the Army of the Rhine. Carnot was the chief of mission. They arrived in Phalsbourg on the 14th, summoned the acting commander, the former duc de Biron et de Lauzun, together with his staff and immediate subordinates, and required them on the spot to reply one after the other to the question: "Do you purely and simply accept the decrees of the National Assembly, YES or NO?"

All accepted except one, whom they immediately relieved of duty. Thereupon they moved among the rank and file from one unit to the next explaining the portent of events in Paris, staged a triumphal entry into Strasbourg as embodiments of the sovereign Republic, outmaneuvered the mayor, a constitutional monarchist, purged the local administration of persons of his persuasion, replaced a number of suspect officials throughout Alsace, entered into diplomatic relations with the Swiss on problems of the scrambled border, and returned triumphantly to Paris on 4 September to assure the Assembly that the Army of the Rhine was trusty and the eastern Departments loyal to the Republic (Reinhard, Chapter 18, I, pp 254–287).

Between Carnot's election to the Convention a few weeks later and his entry into the Committee of Public Safety in August 1793 he was much absent from Paris in the discharge of further missions: to the Army of the Pyrenees when Spain entered the war; to the Pas–de–Calais and the North when the initial conquest of the Austrian Netherlands (i.e. Belgium) turned into confusion and defeat; to the Army of the North when its commander, Dumouriez, deserted to the Austrians, taking captive with him the Minister of War and several members of the Convention. At that humiliating nadir of military misfortune Carnot overrode Dumouriez's demoralized successor and organized the defense himself.

His actions were no less firm in politics. He voted the execution of Louis XVI. He moved the decree to annex Belgium and Monaco. He proposed the conscription of all men between 20 and 25 ("every citizen is born a soldier"). Those measures were consistent with the principle that he proposed for the basis of the Declaration of Rights: that the "safety of the people is the supreme law." Carnot's was a tough republican will, and his reputation that of a reliable, not to say a ruthless, patriot when in the late summer of alarm, defeat, civil war, and treason he was called upon by the more politically minded men constituting the great Committee of Public Safety to join its membership together with a younger Burgundian and fellow engineer, Prieur de la Côte– d'Or.

The historical vision of the French Revolution tends to be dominated by the climactic events of the Year II of the Republic, which is to say the autumn, winter, spring, and early summer of 1793–1794 prior to the overthrow of Robespierre on 9 Thermidor (27 July, 1794) and termination of the Jacobin Terror as an instrument of government associated with his spirit and dominant influence. Those months were dramatic ones from any standpoint, and not least Carnot's. Barely 2 years previously he had been a captain of engineers relieved that the way to promotion was finally open, settling into family life, and resigned to the imminent approach of middle age. Now he found himself called to membership in the body of 12 men who were ruling France and determining the destiny of the Revolution.

The Committee of Public Safety was like no government that had ever existed. Originating in early 1793 to exercise the Convention's general supervision, first over the War Office, and then over all ministries, it rapidly came to supersede them and to take executive power into its own hands. Everything about it was anomalous except its energy. All powerful in action, it existed only at the will of the Convention, which renewed its mandate monthly and only for a month. Implementing its decrees through the Jacobin Club of Paris and the network of Jacobin clubs throughout France, it both controlled and depended upon what was in effect a one–party political system. By its own definition this government was merely provisional, "revolutionary until the peace". With no precedents, it evolved its own procedures. Its responsibility was collective. Any or several members signed and validated decrees for the whole, working round the clock behind the doors of their council room in the Tuileries, sometimes two or three of them at a time, sometimes five or six, never the whole 12. Some were off on mission to the armed forces or to some point where counter–revolution threatened or revolution needed their firm guidance. Their creation was the levée–en–masse, which mobilized the French people into the first incarnation of a modern nation in arms beating back its enemies, overrunning its frontiers, and destroying the assent of history to the old order in Europe.

At home the Committee destroyed dissent. Its instruments were the Revolutionary Tribunal and the guillotine. No survivor involved in the application of that Terror could ever after escape either the inner necessity to shift the responsibility from his conscience elsewhere or the external necessity to justify himself when public reaction set in after Thermidor. The purposes of the present monograph does not require reaching a moral judgment on Carnot's part in the Terror, whether it was one of application or acquiescence. Several things may be said, however. On the one hand his ruthlessness is not to be gainsaid. He encouraged Louis Turreau, deputy on mission in the rebellious Vendée, to put down counter–revolution and "exterminate the brigands to the last man". He signed the decree ordering that the city of Lyons should be brought into line, "torch in hand and bayonets fixed". Toulon, also in rebellion, was to be burned to the ground by red hot cannon balls.[14]

On the other hand there is justice in the view, originating with Hippolyte if not with Carnot himself, that dissociates him from the zealotry of Robespierre, Saint–Just, and Couthon, who would have made of the Revolution a moral crusade for the regeneration of humanity, and woe in the Republic of Virtue to whoever faltered or fell short. He is to be distinguished too from those other colleagues who appear to

---

[14]On that see *Le grand Carnot* by Reinhard (Reinhard, II, pp 86–87). On the governance of the Committee of Public Safety, see *Twelve Who Ruled: the Committee of Public Safety during the Terror* by Robert R. Palmer (Palmer).

have been adventurers and malcontents rather than crusaders in a great cause: Collot d'Herbois, a paranoid and unsuccessful actor; Billaud–Varenne, an unscrupulous drifter and would–be lawyer; Hérault de Sechelles a renegade aristocrat; Barère the wily politician who ever sought to please. It was not in the company of such as these, fanatics, sadists, or demagogues, that Carnot was with his own kind, but rather in that of Prieur de la Côte–d'Or and Prieur de la Marne, one an engineer and one a lawyer (they were unrelated); Jeanbon Saint–André a Protestant, sometime pastor, and ship's captain; and Robert Lindet careful, industrious, and the senior member in point of age.

Carnot concentrated on supreme direction of the war, with Prieur de la Côte–d'Or's assistance with respect to ordnance, supply, materiel, weaponry, and the revolutionary production of saltpetre and gunpowder. He began by reorganizing the armies and their command in the autumn and winter of 1793–1794. His also was the strategy of a three–pronged offensive into Belgium in the spring of 1794. Thereby the French armies passed over into the attack that they sustained against all of Europe for the better part of 20 years. Robespierre, who had no head for military matters, avoided the responsibility of signing the Committee's decrees concerning prosecution of the war, just as Carnot tended to eschew those concerning matters of police and revolutionary justice (Reinhard, II, p 89). War and Terror were intimately related, however. Military security was invoked at the time, and has often been accepted by historians, as justification for the strong hand of the Terror. What is more to the present point, it was certainly military success in the spring of 1794 that rendered the Terror unnecessary and finally intolerable to a people no longer fearful of foreign invasion and in a position instead to export despotic interference in ordinary life from its territory to that occupied by its armies.

Therein lay the reason that after 9 Thermidor Carnot alone among his erstwhile colleagues remained in office, albeit not continuously, under the Convention and into the regime of the Directory that ensued in 1795. "Carnot organized the victory", cried an anonymous voice from the floor of the Convention when at a tumultuous session of that expiring body on 20 May 1795 he was under attack as an erstwhile terrorist (Reinhard, II, p 158). Fearful of despotism, the framers of the Constitution of the Directory placed the executive power in the collective hands of five men to be elected by the upper house of the legislature upon nomination by the lower. In November 1795 Carnot was elected a Director. During the nearly 2 years he held office, he was the leading figure of that precarious regime, ill fated in the absence of a stable, sturdy center by the irreconcilability of surviving radicalism on the left and reviving royalism on the right.

In the early days the threat from the left seemed the more dangerous. Carnot took charge of putting down with a very heavy hand a rising called Conspiracy of the Equals, led by an obscure journalist, one François–Émile Babeuf. Regrouping the armies after the disarray of Thermidor, and directing them across the Alps, he seemed at first to handle military administration with his old sure touch. The sword he chose turned out to have two edges. His was the decision to vest command of the Army of Italy in the youthful General Napoleon Bonaparte, with whom Carnot conducted a correspondence that for a short while bore the appearance of exchanges between patron and protégé, mentor and pupil, Aristotle and Alexander. This time, however, instead of saving the Republic, the success of its military strategy doomed it.

Carnot was driven from office 2 years before that dénouement. In the spring of 1797, the legislative elections resulted in a royalist resurgence. Unwilling to accept the undoing of the Revolution, the factions of the left prepared to reject the verdict of the polls and to purge the two assemblies of their reactionary members. Their coup occurred on 18 Fructidor (4 September 1797). Carnot was persuaded that the Constitution could not be preserved by violating its provisions and refused to lend himself to that maneuver. Warned that his fellow Directors planned his arrest, he escaped from his quarters in the Luxembourg, went into hiding for a few weeks, and then fled into Switzerland. So ended the political career in which, though never holding power alone, he shared in the supreme power for a longer period, August 1793 until September 1797, than any revolutionary statesman prior to Napoleon Bonaparte and longer too than any minister of State had done throughout the reign of Louis XVI. At few periods in modern French history would four almost uninterrupted years of power seem short.

Thereupon, turning 50 years of age, Carnot resumed the studies in mechanics and mathematics that he had begun with the ambitions of a young engineer in the Old Regime. After the strenuous life of a national leader he reengaged his interest in technical studies with every appearance of enhanced enthusiasm. Mechanics and politics shared nothing in content. Nevertheless, given the historical juncture, the two aspects of Carnot's life were not different in the qualities required of him. The activism animating the nascent profession of engineering was congruent with the activism animating the large political affirmations of the Revolution. In the great world of politics, men acted on the belief that by taking charge of their own affairs they could add to their stature. In the special world of engineering the practices on which men acted in building and working contained problems that were brought into the house of science, where they alimented analytical mechanics and helped turn it into modern physical science.

Carnot had written one major book, *Réflexions sur la métaphysique du calcul infinitésimal* (Carnot 1797), before his entry into government and perhaps prior to the Revolution. Like the *Théorie des machines en général* he developed it out of a memoir he had submitted in 1785 to an academic prize competition, this one set by the Prussian Academy of Sciences, Arts, and Letters in Berlin. The subject was the theory of mathematical infinity. His remaining technical books followed very rapidly on his retirement from government: *Lettre au citoyen Bossut contenant des vues nouvelles sur la trigonométrie* (Carnot 1800a); *De la corrélation des figures de géométrie* (Carnot 1801); the revision of *Essai sur les machines en général* (Carnot [1783] 1786) under the title *Principes de l'équilibre et du mouvement* (1803a); the book he considered his major work, *Géométrie de position* (1803b); and *Mémoire sur la rélation qui existe entre les distances respectives de cinq points quelconques pris dans l'éspace, suivi d'un essai sur la théorie des transversales* (Carnot 1806b).

The newly founded Institut de France constituted the forum in which Carnot led a professional life throughout the Napoleonic regime. Embedded in the very constitution of the Directory, the Institute replaced the *Académie des sciences*, which had been suppressed as a privileged body in August 1793. Like the Academy before it, the First (or scientific) Class of the Institute fulfilled two functions. On the one hand it saw to the advancement of Science by publishing its memoirs and endorsing publication of worthy treatises submitted to the judgment of its committees. On the other hand it advised government departments and the public on technological matters by examining the design and merit of new machines, industrial processes, and agricultural methods submitted by inventors and entrepreneurs in hopes of a patent, premium, or subsidy. Throughout the Eighteenth Century in the great days of the old Academy, its leading members esteemed the former of these functions and regarded the latter with ill–concealed impatience. Something of that differential was no doubt inherent in the value structure of the science of the Enlightenment. It would oversimplify and exaggerate matters to say that the utilitarian and humanitarian emphasis of the Revolution transvalued those values. Nevertheless, the Institute did devote a larger share of its attention to applied science and engineering.

There was, in short, a displacement toward engineering; or perhaps it would be more accurate to say that engineering problems occupied a larger share of the attention and respect of scientists. Such was certainly Carnot's preference. He came into the Institute from that world and was at ease in it as he probably would not have been in the company of the old Academy. The effect of the Revolution was to legitimate and accelerate this shift, not to cause it. Once his own treatises just mentioned were published (which had no immediate effect in science proper whatever their significance for its later course), Carnot devoted his time, energy, and interest to many commissions of the Institute charged with examining the merits of many of the numerous mechanical inventions that attest to the fertility of French technical imagination in the early Nineteenth Century.

These tasks were no mere desultory hobbies of retirement. In an average year Carnot would sit on 12 or 15 such commissions, some on machine technology, some on military technology, some on mathematics. Carnot would frequently be charged with framing the report. A number of these reports amounted to memoirs reviewing the whole state of a subject. In reporting on a notational scheme for descriptive algebra, for example, Carnot summarized the development of "analysis situ" (now called topology) from its invention by Leibniz through the contributions of Euler and d'Alembert to his own applications of algebra to geometry.

Normally, however, Carnot scrutinized inventions of a more practical sort. He was a member of the Commission that Robert Fulton invited to watch his steamboat mount the Seine in the summer of 1804. Even more intriguing was an internal combustion motor called the Pyréolophore by its inventors, the brothers Claude and Joseph Niépce, of whom the latter was known later for his invention of the tintype and association with Daguerre in the development of photography. The Niépce motor, said Carnot in an enthusiastic report (1806), was the first device ever imagined for drawing motive power from the expansive force of heated air. It presented the advantage over the steam engine that all the fuel was employed to produce the expansion that went into the motive force. In 1809 Lazare Carnot reported to the Institute on another heat engine invented by a prolific inventor called Cagniard de Latour. As in the Niépce motor the expansive agent was air but employed in a much subtler manner. The fundamental notion was to pipe atmospherically cool air into the bottom of a bath of heated water and to draw power from its thermal expansion on rising.[15] These devices have in common with the ideal heat engine later imagined by Sadi Carnot that they draw motive power from the expansibility common, not just to steam, but to all gases. That feature renders them more interesting as probable intermediaries between the work of father and son than are certain other, perhaps equally ingenious inventions on which Carnot framed reports, notably a force pump operated by air compression, invented by a certain Lingois; a number of hydraulic machines constructed and operated by a prominent industrialist, Mannoury–Ectot; and a memoir of 1814 by Dupin on the stability of floating bodies.

Such were Carnot's occupations during the decade from 1804 to 1814 when he was in his 50s and his two children were boys. He never abandoned his interests in military strategy nor his role as a defender of the Revolution. He wished to help defend the country amid the crumbling of the Napoleonic system and offered his services to Napoleon at the moment when the catastrophic retreat from Moscow reached the Rhine. In those desperate circumstances, Napoleon appointed him Governor of Antwerp. Carnot commanded the defense. It was the only direct command in his career, and he maintained French control even after Napoleon's first abdication. He then accommodated himself to the initial return of the Bourbons and even addressed a memoir to Louis XVIII counseling him about the aspects of the previous quarter century that the king must needs respect in order to preserve his throne. Nevertheless, Carnot rallied to Napoleon upon the return from Elba and took the Ministry of the Interior during the Hundred Days.

That act, rather than having voted for the execution of Louis XVI, exhibited the consistency of the old revolutionary and was not forgiven by a monarchy that had had to be restored twice. His wife having died and Sadi Carnot having graduated from the *École polytechnique*, Carnot took the younger son, Hippolyte, into exile. Forbidden the Rhineland because of its proximity to France, he and Hippolyte settled in Magdeburg. There Carnot corresponded with German colleagues, published a few volumes of mediocre verse, and brought up Hippolyte, who became a Saint–Simonian.

There on 2 August 1823 Lazare Carnot died.

[15]For detail on these and other devices inspected by Carnot, see *Lazare Carnot Savant* (Gillispie 1971, Chapter 1, pp 27–29, notes 47–54). On the Cagniard engine, see *Sadi Carnot and the Cagniard Engine* by Thomas S. Kuhn (1961, pp 567–574).

Chapter 2
The Science of Machines

The great events that Carnot touched and shaped never caused him to lose sight of the adaptation of the science of mechanics to the science of machines that he thought to initiate when still an obscure young engineer. Fortunately for the historian, the documentation is adequate to permit following his conceptions from their genesis in the entry he prepared for the prize contest set by the Academy of sciences in April 1777 right through their development during his lifetime into the subject matter of a new branch of science in the 1820s. Since his own ideas were expressed in their most individual and unadorned form in his first publication, the *Essai sur les machines en général* of 1783, it will be best to present them through the medium of the detailed analysis of that work that occupies the current chapter. Thereupon it will be informative to look first back and then forward. The stages through which Carnot formulated his approach may be observed in the successive memoirs he submitted to the Academy in 1778 and 1780. Twenty years later, an ostensibly retired statesman, he extended and developed the subject in *Principes fondamentaux de l'équilibre et du mouvement.* Beyond this, it is one of the purposes of the present monograph to exhibit that Sadi Carnot's *Réflexions sur la puissance motrice du feu*, published in 1824, the year after his father's death, may properly be read not only as the foundation of thermodynamics, but also as the culmination of a methodologically and conceptually coherent series of Carnot essays on the science of machines. After surveying the literature of that subject a little more at large, it will be appropriate finally to consider the relevance of Carnot's mathematical writings to his mechanics and his work in science.

2.1 Summary of *Essai sur les machines en général*

There is no difficulty in understanding why the scientific community should have ignored Carnot's *Essai sur les machines en général* in the 1780s. His book does not read like the rational mechanics of the eighteenth century. It had long since

C.C. Gillispie and R. Pisano, *Lazare and Sadi Carnot: A Scientific and Filial Relationship*, History of Mechanism and Machine Science 19, DOI 10.1007/978-94-017-8011-7_2,

become normal to compose treatises of mechanics addressed to a professional public in the language of mathematical analysis; though Carnot reasoned no less rigorously than did contemporary mathematical argument, he conducted the discussion verbally, conceived the mathematical expressions he did employ in a geometric or trigonometric rather than algebraic spirit, and usually went on to explain in words what the formulas contained. The genre was apparently of an altogether lower order than that of d'Alembert and Lagrange or Euler and the Bernoulli family. Judging by the style alone, prolix and naive, a contemporary reader might easily have supposed the book to be among the many negligible writings that retailed merely elementary mechanics under one pretext or another.

Yet, the essay, despite its title, could never have served the purpose of a practical manual for designing or employing actual machinery. Carnot proceeded on the basis of a highly abstract definition of a machine: it was an intermediary body serving to transmit motion between two or more primary bodies that do not act directly one on another. Carnot lodged his complaint against the existing mode in mechanics right at the outset. Since its problems were normally limited to analyzing the interactions among primary bodies, the practice had tacitly arisen of abstracting from the mass of the machine itself as if it were inertia free. While simplifying problems in mechanics, however, that method of treatment, carried over from the geometrization of simple machines in statics, complicated a dynamical study of machines. It was standard procedure to deduce from the laws of mechanics the particular rules of equilibrium and motion in each class of simple machines, i.e., cords, the lever, the crank, the pulley, the wedge, the screw, and the inclined plane. No general principles existed, however, that were capable of containing the conditions of motion and equilibrium in all machines. In order to establish such principles, the mechanist needed to treat machines like other bodies and account for their mass in his analysis. Only then would the science of machines in general be feasible, and then it would come down to resolution of the following problem:

> Given the virtual motion of any system of bodies (i.e., that which each of the bodies would describe if it were free), find the real motion it will assume in the next instant in consequence of the mutual interaction of the bodies considered as they exist in nature, i.e., endowed with the inertia that is common to all the parts of matter.[1]

Already it will appear that, verbally expressed–though the argument was, the *Essai sur les machines en général* was beyond the comprehension of readers without formal education. It presupposed the competence of scientifically literate persons, but was not couched in language that would attract their interest. The essay would have been accessible to trained engineers, in other words, and since the author himself was of a first generation of them, the impediment was that it could have been appreciated only in a profession that barely existed.

A difficulty about the content complements that about the form. The *Essai sur les machines en général* mingled novelty of approach with the most elementary aspects

[1] *Essai sur les machines en général* (Carnot 1786, § X, p 21).

of mechanics in such an unassuming fashion that it is not easy to distinguish in retrospect what was new from what was obvious. Indeed, to treat Carnot's scientific career in point of discovery would be relatively unprofitable. Innovations there were, to be sure, and far from negligible. But they did not occur on the frontiers of unsolved problems in the science of mechanics. They occurred behind the front lines, so to say, and rather than consider Carnot to have been a seeker and finder in the conventional scientific sense, it will be better to approach his work as the critic might that of a writer who should introduce a new idiom into a literature in order to bring existing, largely unnoticed resources to bear on different purposes. What Carnot added to mechanics accrued largely in virtue of what he thought to do with it.

Before entering into his reasoning, a reader needs to be reminded of certain contemporaneous physical conventions about the structure of matter. He will not get beyond the first few pages of the *Essai sur les machines en général* without encountering the distinction between hard bodies and elastic bodies that entered mechanics in the early stages of collision theory in the seventeenth century and disappeared into the theory of elasticity in the nineteenth. Carnot's work came into that development somewhat past its midpoint and inherited as assumptions the positions adopted by Maupertuis.[2] Perfectly hard bodies were held to be indeformable and perfectly elastic bodies to contain forces capable of restoring their initial shape and volume after compression or shock of impact. For completeness there should have been the third category of soft bodies, deformable and incapable of self–restitution. In the seventeenth century, John Wallis had addressed one of the classic papers inaugurating these studies to the problem of the behavior of inelastic bodies in collision.[3] He wrote of hard bodies but, paradoxically enough, soft bodies would behave in the same way in collision since they would not rebound. Resilience or its absence being the significant alternatives, elasticity and hardness sufficed for the idealized conditions. The polarity was unsymmetrical, however, because hardness did not admit of degree whereas the degrees of elasticity could be attributed to a complement of softness.

In modeling nature itself, hard bodies were taken to be fundamental, their properties a manifestation of the impenetrability of matter inhering in the ultimate corpuscles. In actual bodies, solid matter was imagined to. consist in such corpuscles connected to each other by rods or shafts–rigid, inextensible, and incompressible rods in hard bodies, springs in elastic bodies. In order to follow Carnot's thought, it will further be important to appreciate the way in which other physical states related to the model, for that was not in the usual notion of progression from solid through liquid to gas. Instead, liquids were fluids congruent with hardness in mechanical

---

[2]For a helpful account, see *The significance of "hard bodies" in the history of scientific thought* (Scott 1959, pp 199–210).

[3]On that one can see *A Summary Account of the General Laws of Motion,* 26 November 1668 (Wallis).

properties in that they were incompressible though deformable, and gases were fluids mechanically congruent with elasticity in deformability and resilience.

These distinctions created dynamical difficulties for Carnot that might arouse impatience in his modern reader. Historically, however, it would be misleading to consider them false problems, for working through them was precisely what led Carnot in the direction of physics of work, power, and ultimately energy, wherein they did indeed become obsolete. The embarrassment grew out of the famous issue in eighteenth–century dynamics about whether momentum (MV) or "live force" (MV^2) should be admitted to be the fundamental quantity conserved in interchanges of motion. Ignoring (as is legitimate for present purposes) metaphysical aspects, it was a limitation upon the employment of the principle of live force that although seventeenth–century collision theory conserved its quantity in the interaction of perfectly elastic bodies, in the supposedly more fundamental case of hard bodies, MV^2 was conserved only when motion was communicated smoothly ("by insensible degrees" is the usual phrase), and not in impact or collision.

Looking back at Carnot's writing, one can see that a treatise in which the quantity now called work (force · distance) was to be designated as the measure of what machines accomplish was in some sense bound to presuppose its equivalence to kinetic energy ($\frac{1}{2}MV^2$). How Carnot saw in advance that he must move towards such a convertibility can only be conjectured. Dimensionally it derived from the equivalence in the law of fall of "mgh" to "$\frac{1}{2}mv^2$", known since Galileo as the means of equating the velocity that a body would generate in falling a certain distance with the force required to carry it back. Beyond that, the relation had to do with his object in thinking of machines as massive bodies in action and not mere extensions of rigidity serving the geometrical transfer of inertial motion, or in simpler words with his having been an engineer. Certainly, too, his ideas carried over the role of live–force conservation from hydrodynamics into what he would have regarded as the deeper problem of hard–body interactions.

Following the publication of Daniel Bernoulli's famous treatise in 1738,[4] hydrodynamics was the subject in which the principle of live force was basic and indispensable in the solution of engineering problems. There is thematic evidence running through all Carnot's writings on mechanics that the hydraulic application of principles and findings, although ostensibly a special case, actually held a major if not a primary place in his thinking.[5] It would have been likely enough at the outset, though not strictly correct, to think of motion being communicated by liquids in accordance with the principle of continuity and to suppose from the common incompressibility of liquids and hard bodies that the conservation of live force might profitably be taken to be the principle governing the latter types of interaction, in nature the most fundamental.

---

[4] *Hydrodynamica sive de viribus et motibus fluidorum commentarii* (Bernoulli D).

[5] Scholars have often cited the analogy between fluid flow and heat flow in discussing the influence of Lazare's work upon Sadi's, though, as will appear, the present monograph makes a much wider claim for the continuity between the work of father and son. See below Chapter 2, pp 44–45, Chapter 3, pp 77–78, Chapter 5, pp 130–131.

At any rate, whether for those reasons or others, the strategy of Carnot's *Essai sur les machines en général* was decided by the necessity to outflank the discrepancy between continuous and discontinuous change of motion in hard–body interaction and to define the sense in which the principle of live force might equally well be employed in continuous or discontinuous interchange of motion between elastic or inelastic bodies. To that end Carnot tacitly made use of distinctions that, although he never saw so far, later developed into those between potential and kinetic energy, work and power, input and output, scalar and vector quantity, reversible and irreversible process.

In his avowed point of view Carnot robustly admitted to the engineer's inaptitude for metaphysics. There had always been, he acknowledged, two distinct approaches to the science of mechanics, the experiential and the rational. The former, to which he adhered, took its departure from those basic notions that we draw from our gross experience of nature and to which we give the names body, power, equilibrium, and motion. These ideas neither could nor needed to be defined. They were primary, and other conceptions were derivative, such as velocity and the various types of force in terms of which laws of motion were framed. The second approach was one that would make mechanics a purely rational science. Its adherents began with hypotheses, deduced the laws that bodies in motion would exhibit if the hypotheses were correct, and then compared their conclusions with phenomena. What they thus gained in elegance was at the cost of incomparably greater difficulty, for nothing was more embarrassing in exact science, and especially in mechanics, than an effort to formulate definitions entirely free of ambiguity. That was something he would not and need not attempt[6] in his *Essai sur les machines en général* (Carnot 1786, pp 105–106).

Carnot's reluctance with respect to definitions makes the *Essai sur les machines en général* hard to follow for persons schooled in nineteenth–and twentieth–century terminology. Only by the use he made of terms is it borne in on the modern reader what he meant by them. Perhaps Carnot's ambiguity may have been felt at the time, for in reworking the material 20 years later for the *Principes de l'équilibre et du mouvement*,[7] he gave a glossary. Even those terms that are dimensionally

---

[6]Carnot did not pose the choice between with the two schools of mechanics as explicitly Jouguet implies–i.e., that what distinguishes them is that one takes mass to be fundamental and force for a derivative notion, and the other takes force to be fundamental and the laws of motion and impact for derivative. Carnot's one distinction may indeed come down to that, but he did not say so, and Jouguet like other commentators on his mechanics confuses his *Essai sur les machines en général* of 1783 with the *Principes fondamentaux de l'équilibre et du mouvement* of 1803 (Carnot 1803a). On Jouguet see his *Lectures de mécanique. La mécanique enseignée par les auteurs originaux* (Jouguet II, p 72).

[7]Perhaps it will be useful to reproduce the dimensional glossary from Carnot's *Principes fondamentaux de l'équilibre et du mouvement* (Carnot 1803a, § 18, p 13). Denoting mass by m, linear distance by e (espace), and time by t, then

> "Any quantity of the form or reducible to the form $\frac{e}{t}$ is called velocity".
> "Any quantity of this form, $m\frac{e}{t}$ is called quantity of motion".

what might be expected did not carry the full meaning of later usage. For example, the simplest, "vitesse", might better be rendered speed than velocity, for Carnot disposed of no convention for combining intensity and direction of quantity in a single expression, but always specified the sense in which a velocity or force was to be reckoned. The present discussion will follow Carnot's own usage, translating "vitesse" usually as velocity in order to avoid awkwardness, and understanding that by "force" he usually meant quantity–of–motion (momentum) in moving bodies. Given his point of view, only secondarily did he need to allude to Newtonian force, the product of mass times acceleration. When he did, he normally employed the phrase widely used in the eighteenth century, "force–motrice", or "motive–force", dimensionally identical with "dead force" or statical pressure.

From among the basic principles of mechanics proper, Carnot chose two for an introductory criticism that also established his own standpoint. The first stated for the general law of equilibrium in weight–driven machines the condition that the center of gravity of the system be at the lowest point possible. The second, which Carnot called "Descartes' famous law of equilibrium"; will be unfamiliar under that designation both to the student of modern mechanics and the student of Descartes, who in fact made no such statement. The principle was that two forces in equilibrium were in inverse ratio to their "vitesses" (by which Carnot meant the motions they produce) at the instant when one prevailed infinitesimally over the other, thus initiating a "small" motion.

Carnot preferred to rely on the center of gravity principle. True, it was open to certain objections. For one, its applicability appeared to be limited to machines driven by weight. That defect, however, Carnot dismissed as merely apparent. It was always possible to reduce the operation of other forces to that of gravity by replacing their agency in principle with that of a weight acting over a pulley. This imaginary transformation of the system may at first annoy the modern reader as a somewhat sophomoric evasion of the difficulty, but it would be better to see in it the engineer's way of reducing an abstract problem to his own terms. Anyone who has been to engineering school will have resorted to similar devices, although it would be more germane to recall for a moment the analytic role of the experience and manipulation of weight in the mechanics of an Archimedes, a Stevin, or a Galileo. Even Lagrange was not above it.

"Any quantity of this form, $\frac{e}{t^2}$ is called accelerating or retarding force".
"Any quantity of this form, $m\frac{e}{t^2}$ is called motive force".
"Any quantity of this form, $m\frac{e}{t}$ or of this, $m\frac{e}{t^2}$ is called simply force or power." [The ambiguity here reflects the historical situation. Although the concept of power (puissance) as the rate work or energy changes with respect to time was implicit in Carnot's analysis, he could also use the word as a synonym for force, often with the implication of latent force as the ability to exert effort].
"Any quantity of this form, $m\frac{e^2}{t^2}$ is named motive live force, moment of motive force or moment–of–activity".
"Any quantity of this form, $m\frac{e^2}{t}$ is named moment of the quantity of motion, or quantity of action".

The significance for Carnot of the last two quantities will appear in what follows.

A further objection was more serious. There were exceptions to the statement that equilibrium required that the center of gravity be at the lowest possible point, as indeed there were to all assertions involving maxima and minima. There might be provisional or temporary equilibrium in other configurations of the system. In order to obviate that difficulty, Carnot employed a type of argument that his reader soon comes to recognize for characteristic. He restated the principle by appealing to the operation of an ideal machine. The form is arbitrary. No forces are applied other than weights. No motion has been imparted. In the quiescent state, the sum of the resistances of the fixed supports estimated vertically must equal the weight of the system. Suppose now that a "small" motion begins. A portion of the weight must have gone into producing movement, and only with the remainder are the fixtures loaded. The difference between weight of the system and load on the fixed supports will be the force depressing the center of gravity at a rate (here Carnot used the word "vitesse" where we should say acceleration, a term he rarely employed) equal to that difference divided by the mass of the system. It follows that if the center of gravity did not descend, there would be equilibrium:

> To demonstrate that several weights applied to any machine whatever are in equilibrium, it suffices to prove that if the machine be left to itself, the center of gravity of the system will not descend.[8]

What is noteworthy is the mode of reasoning, which combined the ideal with the operational and did so in a negative and restrictive way. In a treatise purporting to rest upon experience, a reader might expect that a demonstration would consist in a generalization of experiments. Not at all–it was based upon the exclusion in an idealized, unrealizable thought experiment of what we recognize from our general experience would not occur in the behavior of real objects. Nothing was new about the logic: Stevin, in obtaining the law of the inclined plane from the exclusion of perpetual motion, had adapted the geometric proof from the absurd to physics. The assertion was that a proposition was correct if its contrary entailed consequences that were physically unthinkable. Nothing was new about the model: the motions of systems of mass–points had long been the object of an analysis that had idealized bodies and turned rational mechanics into a form of mathematics. But, as in other aspects of Carnot's work, what made the difference was the use he repeatedly drew out of such arguments until they became the distinctive idiom of the science of machine motion.

As for the so–called principle of Descartes, Carnot found in it disqualifying flaws. It was less general than the center–of–gravity principle thus transformed, from which it could be deduced by conversion of its forces into weights acting over pulleys. It applied only to systems in which no more than two forces were at work. More seriously, it envisaged only the relative amounts of the forces in equilibrium, whereas in requiring their vertical projections the center–of–gravity principle specified also the direction of those forces. (It is just in these passages that one may begin to appreciate how Carnot's attempts to analyze the manner in

---

[8] Carnot 1786, § II, p 14.

which forces transmitted by shafts, cords, and pulleys would constrain and move points within systems composed of rigid members, clumsy though these constructs seem, they nevertheless belong to the pre–history of vector analysis in exhibiting awareness that the quantity of a force comprises direction as well as intensity).

Nothing in the principle of Descartes even required that two forces in order to be in equilibrium must act in opposite senses, and it could not, therefore, specify in what their opposition consisted. In raising that problem, Carnot anticipated one of the clarifications of which he was always proudest, the distinction between what he called impelling forces ("forces–sollicitantes") and resisting forces ("forces–resistantes"). Not by a metaphysical differentiation between cause and effect would we tell which is which, but merely by the geometry of the system. As shown in the sketch, a force that made an acute angle with the direction wherein the motion occurred would be called impelling; a force that made an obtuse angle would be called resisting (Fig. 2.1).

Fig. 2.1

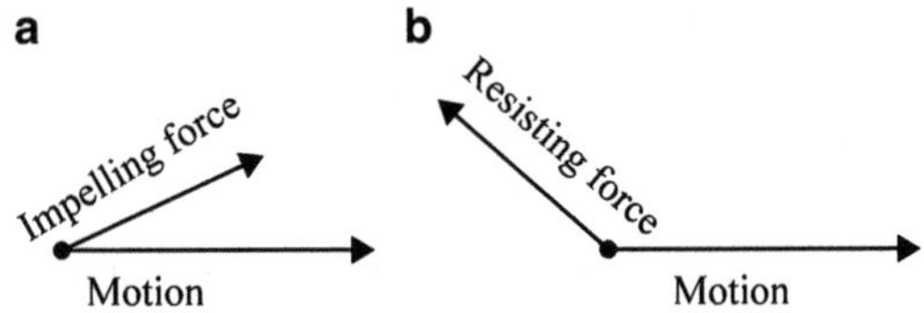

Finally, Carnot lodged another criticism against the principle of Descartes together with all those that invoked a "small"–i.e., infinitesimal–motion arising in the system. He had in view here the principle of virtual velocities, of course, although he did not identify it by name. This approach never specified what determined that infinitesimal motion. If it was necessary to invoke a new mechanical principle for that, then the original one was inadequate. If, on the other hand, the geometry of the system sufficed to determine the nascent motion, how did it do so? The identical objection lodged against analyses that considered two machines in two states infinitely close to each other. In specifying how the machine must move in passing from one to the other, either it was necessary to invoke a new principle, or else the determination was present in the geometry of the system, and in the latter case it was a defect of the principle that it failed to make evident the geometric conditions of the motion. In fact, however, such motions were subject to certain conditions, in definition of which Carnot was going to propose for the whole class the designation of geometric motions. As will appear, his geometric motions differed from virtual velocities in being finite, and amounted to possible or actual displacements in which no internal work was performed or energy consumed within the system.

Both of the foregoing principles applied only to equilibrium, and though Carnot alluded to d'Alembert's principle as justification for extending equilibrium principles to motion, he made no actual use of it, preferring in practice to found the reasoning on the conservation of live force. From among laws of motion and equilibrium themselves, Carnot chose two as axiomatic. The first was the equality in opposite senses of action and reaction, which he did not identify as Newton's third law of motion. It applied to all bodies without exception. The second applied only to hard bodies and asserted that in their interactions, whether by impact or pressure, their relative velocity in the next instant was zero. From these two laws, two corollaries followed: first, that the intensity of interactions between two bodies depended only on their motion relative to each other, and second, that the force they exerted on each other in impact was always along a perpendicular to their common surface at the point of contact.

Carnot began the argument itself by deducing from these fundamental principles an equation stating that in the motion of a system of hard bodies the net effect of mutual interactions among the corpuscles constituting the system was zero. Applied to generalized systems of hard bodies, this equation reduced to an expression equivalent to the conservation of live force. Just there, however, Carnot encountered the inapplicability of that principle to hard–body collision. It was mainly in order to turn this difficulty that Carnot defined for purposes of analysis the class of geometric motions, i.e., displacements depending for their possibility on the geometry of a system and not on rules of dynamics based in physics. For such motions Carnot might ignore the supposed loss of live force in inelastic collision. Invoking them permitted him to transform his fundamental equation into an indeterminate of general application into which arbitrary values determining the motions might be introduced in the solution of particular cases.

In later terms, this analysis amounted to a derivation of conservation of moment–of–momentum or torque from conservation of energy or work. Since what Carnot required for a theory of machines was precisely the license to disregard the internal constraints and interactions of systems of bodies and forces, he made conservation of moment–of–momentum ("moment of the quantity–of–motion", in his words) the fundamental principle in mechanics generally. From it he deduced the Principle of Least Action and restatement of conservation of live force in an altered form expressing that what machines transmit is power rather than motion. Indeed, the central thrust of Carnot's contribution to mechanics was tacitly to transform the analysis of motion into the analysis of power.

The second part of the *Essai sur les machines en général* applied Carnot's adaptation of the principles of mechanics to the operation of machines themselves. The most important definition was of the quantity now called work. Carnot called it "Moment–of–Activity" and identified it clearly as the basis for measuring input against output in machine processes. Recurring to the principle of moment–of–momentum, Carnot deduced from it the most original of his propositions: that, in the case of a machine transmitting motion smoothly, the work done by the impelling forces equals the work done by the resisting forces. The principle, since named after Carnot, follows directly: characteristically, he thought of it and stated it restrictively to the effect that the work done by a system of moving parts equals the work done on it only if percussion or turbulence be eliminated in the transmission of motion (or power). A concluding scholium inveighed against the chimaera of perpetual motion, discussed the factors of mechanical advantage, developed the conditions for efficient design of machinery with particular emphasis on hydraulic power, and made the distinction already outlined between the experiential and rational approaches to the science of mechanics.

Such were the main heads and findings of Carnot's first publication. Readers interested largely in a qualitative summary of what he did may find this outline sufficient and prefer to pass on directly to Chapter 3, which discusses the background of the *Essai sur les machines en général* and the ensuing development of the science of machines. Readers more immediately concerned with the inwardness of eighteenth–century mechanics itself may wish to follow the detailed précis and analysis of the argument that occupies the remainder of the present chapter.

2.2 Geometric Motions

Imagine, Carnot charged his reader, any system of hard bodies in which the virtual motion is modified into some other motion in consequence of the internal constraints and interactions among the bodies. The example that he meant but did not specify was in consequence of their being assembled into a machine. The general problem of mechanics was to find that other motion, and Carnot began his resolution by summing the interaction of contiguous corpuscles, m' and m''. From the two basic axioms (equality of action and reaction and zero relative motion following collision) and designating by

F'	The action of m'' on m', i.e., the "force" (quantity of motion) that m'' imparts to m',
F''	The reaction of m' against m'',
V' and V''	The respective velocities immediately after impact,
q' and q''	The angles between the directions respectively of V' and F' and those of V'' and F'',

Carnot obtained the expression[9]

$$\int F'V'\cos q' + \int F''V''\cos q'' = 0$$

[9]Perhaps it will be helpful to give here Carnot's explanation of this method of representing the projection of a force or any directed quantity upon another in terms of the elementary trigonometry of a right triangle (Fig. 2.2).

Fig. 2.2

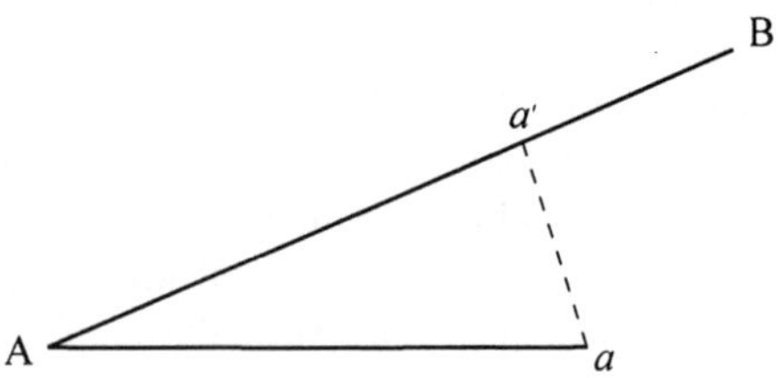

for the interaction of neighboring corpuscles taken two by two; which according to his wont he also stated verbally:

> That is to say, that the sum of the products of the quantities of motion impressed on each other by the corpuscles separated by each of these little inextensible wires or incompressible rods [. . .] multiplied by the velocity of the corpuscle on which it is impressed, evaluated in the direction of that force, is equal to zero.[10]

Carnot next considered the system in motion as a whole in order to analyze the internal interactions of the constituent parts, and adopted designations that he thereafter used quite consistently for the several quantities. He called

The mass of each corpuscle	m
Its virtual velocity	W
Its actual velocity	V
The velocity that it *loses*,[11] in the sense that W is the resultant of V and of this velocity which is	U
The force (quantity of motion) that each contiguous particle imparts to m and from which derives all the motion it receives from the system	F
The angle between the directions of W and V	X
The angle between the directions of W and U	Y
The angle between the directions of V and U	Z
The angle between the directions of V and F	q

Aa represents the initial quantity and AB the direction in which it is to be "estimated". Thus, the projection $\overline{Aa'}$ of the force Aa estimated in the direction $\overline{AB}$ is given by "$Aa \cos aAB$". The explanation is from the *Principes fondamentaux de l'équilibre et du mouvement* (Carnot 1803a, § 26, p. 16). In the *Essai sur les machines en général* he did not stop to explain these elementary devices but simply employed them. (The interest lies mainly in the example they afford of his geometric and trigonometric way of visualizing relations. This comes out initially in the 1778 memoir for the *Académie des sciences*. (See: Gillispie 1971, Appendix B esp., § 52–60)). In the expression given in the text above, the velocity of m' imparted by the force F' estimated in the direction of F' is "$V' \cos q'$". Since F' and F'' denote quantity of motion, what this expression says in the notation of a modern primer would be $\sum M'V'^2 + \sum M''V''^2$. Carnot in this book used the sign "Σ" indifferently for integration and summation.

[10]Carnot 1786, § XV, p 26.

[11]It was one of Carnot's central conventions that when a system is set in motion, the difference between virtual and actual velocity in any component part is "lost to" or as the sign may be "gained from" the mutual interactions or constraints within the system. The usage comes from d'Alembert's mechanics. Carnot used the terms in the *Essai sur les machines en général* of 1783 and explained them in the *Principes fondamentaux de l'équilibre et du mouvement* of 1803 (Carnot 1803a, § pp 41–43; pp. 24–26 and Pl. I, Fig. 4), and did so in the following manner. If a body M tends to move with the virtual velocity $\overline{MW}$, but is constrained instead to take on the velocity $\overline{MV}$, then if a parallelogram of velocities is constructed, the velocity $\overline{MW}$ maybe resolved into two components. of which one, $\overline{MV}$, will be the actual or remaining velocity and the other $\overline{MU}$will be called the velocity "lost" by the body M. Prolonging MW in the opposite direction will give us

Given these designations, Carnot by means of simple algebra obtained from the previous equation an expression that he called his first fundamental equation (E), as follows:

$$\int mVU \cos Z = 0 \text{ (E)} \tag{2.1}$$

Only later in the argument did he point out that this expression was formally identical with conservation of live force. (Since $U \cos Z = V$ the expression reduces to $\int mV^2 = 0$). He could not simply invoke that principle directly, however, given his assumption about the inapplicability of the principle to hard–body collision. He was bound by the generality in which he had set himself the problem to convert the expression into one applicable to interactions of elastic body and of hard body (the more so as the latter were generally taken for the term of comparison in nature) whether motion was communicated by impact or by insensible degrees.

To that end, Carnot introduced at this, the critical juncture of his argument, the notion that he always regarded as his most significant contribution, to mathematics as well as to mechanics: the idea of geometric motions, i.e., displacements that depended for their possibility only on the geometry of a system quite independently of the science of dynamics. Carnot's geometric motions became virtual displacements in later mechanics. The idea was an interesting one, therefore; it played a distinctive part in his own analysis; more signally than any other element it exhibits that a kind of operational economy in Carnot's reasoning was what differentiated it from the classical mode of analytic mechanics.

Imagine, Carnot asked in regard to a generalized system of hard bodies, that just as the system is struck, its actual motion be stilled and it be made instead to describe two successive movements arbitrary in character but subject to the condition that they be equal in velocity and opposite in direction. The configuration of the system being given, it was clear that such an effect could be accomplished in infinitely many ways, and (what was essential) by purely geometric operations.

In the *Essai sur les machines en général* Carnot did not at once achieve a definition of this class of motions that was both simple and adequate to the use he needed to make of them (although about the use itself, as will appear, he was perfectly clear). Thus, he laid down first that motions would be geometric if they

---

the complementary case, a construction for velocity "gained", $\overline{MU'}$in configurations in which the constraints augment initial virtual velocity (Fig. 2.3).

Fig. 2.3

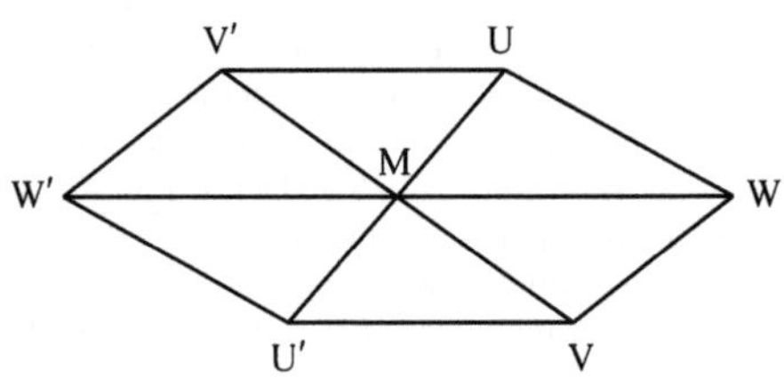

involved the constituent bodies of a system in no displacement relative to each other. The converse would not be true, however, for motions that did involve such displacements might be geometric. For example, in the case of a machine composed of two weights suspended in equilibrium over a wheel and axle, a motion consisting of the descent of one from a height equal to the circumference of the wheel compensated by the raising of the other through a length equaling the circumference of the axle would be geometric since the equal and opposite movement could occur. Moreover, Carnot's purpose required him in a later passage to extend the concept to include motions of a system that meet the condition of reversibility in virtue of eliminating from consideration the components that (in later terminology) do no work. Perhaps, therefore, it will be legitimate for the sake of clarity to give the definition he did achieve in reworking the material in 1803 for the *Principes de l'équilibre et du mouvement*, wherein he suppressed allusions to these imaginary and reversible displacements, and laid down:

> Any motion will be called geometric if, when it is impressed upon a system of bodies, it has no effect on the intensity of the actions that they do or can exert on each other when any other motion is impressed upon them.[12]

In the *Essai sur les machines en général* however, Carnot was thinking primarily in terms of bodies and displacements when it was a question of internal interactions in a system and only derivatively in terms immediately translatable as work and energy. Applying the concept of geometric motion to a generalized system of hard

---

[12]Carnot (1803a), § 136, p 108. It will help clarify movement tangential to the surface both stages in his thought to give a few examples.

(a) From the *Essai sur les machines en général* of 1783 (Carnot 1786):

1. Two globes in contact. An impulse displacing both in the same direction along the line of centers would produce a geometric motion; an impulse separating them along the line of centers would not.
2. Several bodies attached by flexible but inextensible wires to a common center. Any motion in which all remain equidistant from the center is geometric even if they shift among themselves; any motion altering the length of a radius is not.
3. A body moving on a curved surface. A movement tangential to the surface would be geometric; any departure from the tangent would not.

In each instance, the justification is that the equal and opposite motion is possible in the first case and not in the second (*Ivi*, 28–30*n*).

(b) From the *Principes de l'équilibre et du mouvement* of 1803:

1. Two bodies, A and B, are fixed to either extremity of a lever arm; the arm is rotated around a fixed point C. Each body assumes an angular velocity proportional to AC and BC respectively; since neither of these velocities influences the action of one body on the other whether by gravity or in any other manner the rotation of the lever is a geometric motion (Carnot 1803a § 139, p 111).

bodies, he noted that by definition the relative velocity of neighboring corpuscles would be zero during the initial instant of such a motion. Designating by

u The absolute velocity of the corpuscle m in the initial instant of a geometric motion
U As before, the velocity "lost" to the internal interactions,
z The angle between u and U,

then, since the corpuscles of the system would not tend to displace relative to each other in consequence of the velocity "u" alone, the mutual interactions within the system would be the same whether it was considered that "m" was animated solely by the velocity lost "U" or by the combined velocities "u" and "U". But if all the corpuscles were animated by the velocity "U" alone, equilibrium would necessarily obtain. Therefore, the real velocity after interaction would be "u", and by reasoning similar to that which yielded the first fundamental equation,

$$\int mVU \cos Z = 0, \text{ (E)} \tag{2.1}$$

Carnot had his second fundamental equation,

$$\int muU \cos z = 0, \text{ (F)} \tag{2.2}$$

In which he has replaced an actual, physical velocity "V" with an idealized geometric velocity "u".

Let us summarize, then, more explicitly than Carnot in the toils of his argument did himself, what precisely he thought to have gained by transposing the problem into terms of geometric motions. In the first place, that step justified him in extending the principle of conservation of live force to hard–body interactions, whether sudden or smooth, and that is how he interpreted Equation (F) (see Eq. 2.2). As he observed later in the *Principes de l'équilibre et du mouvement*, all bodies are susceptible of geometric motion, whether hard, soft, or elastic, whether solid or fluid, for the reason that by definition these motions had no bearing on the internal interactions of bodies and were entirely independent of these dynamical distinctions (Carnot 1803a, § 144, pp 114–115). Hence, Equation (F) (see Eq. 2.2) extended to hard–body interaction, and, since it contained as a special case Equation (E) (see Eq. 2.1), which was itself a statement of conservation of live force, it authorized Carnot to take that principle as fundamental in the realm of geometric motion. And that, in the second place, is all he needed, since among the geometric motions of which a system was capable would be the real motion that it assumed upon any actual impulse. In the third place, therefore, Equation (F) (see Eq. 2.2) represented the solution to the problem originally posed–i.e., given the virtual motion of any system of hard bodies, to find the actual motion of the system upon communication of external forces.

It will be well to specify what Carnot meant by a solution. Since Equation (F) (see Eq. 2.2) was indeterminate, it would hold whatever the value of "u". Provided that the motion be geometric, particular values and arbitrary directions might be attributed to the indeterminate according to the conditions of the specific problem, and it would then always be possible to formulate equations between the unknowns. For an example, Carnot produced the problem of a system of bodies unchanging in relative positions and containing no attachments to external fixtures. The solution could be drawn from Equation (F) (see Eq. 2.2) by supposing arbitrarily that all points of the system were subjected to a geometric motion such that their velocities "u" were parallel to a given right line. Since "u" was then constant, Equation (F) (see Eq. 2.2) became

$$\int mU \cos z = 0,$$

which stated that the sum of the components of the forces "lost" to the constraints in the arbitrary direction of "u" was zero, and hence the resultant force was the same as if each body had been free. This was a "well–known principle", observed Carnot, not identifying it.

A second example was that of the same system made to rotate around a given axis, so that each of the points described a circle in a plane perpendicular to the axis. The movement being geometric, "R" being the radius of "m" and "A" a constant for all points, it was clear that

$$u = AR$$

and that Equation (F) (see Eq. 2.2) became

$$\int mRU \cos z = 0.$$

That is to say, the sum of the moments relative to any axis of the forces "lost" to the mutual interactions was zero–another "well–known principle" said Carnot, not identifying it in that place for conservation of angular momentum or moment–of–momentum.[13] Of that, to which he recurred in a fundamental way, more needs to be said in the next section. What is characteristic to notice here is that such results were what Carnot meant by solutions to the problem. Contrary to what might have been expected of a young engineer, there were no numerical problems and solutions in the *Essai sur les machines en général* just as there were no formulations not applicable generally to any type of machine.

[13]On that please consult *Essai sur les machines en général* (Carnot 1786, § XVII, p 33).

2.3 Moment–of–Momentum

Having obtained a general solution for his fundamental problem of mechanics, Carnot now needed to state it in a form suited to drawing out the consequences for the theory of machines. To that end he turned to demonstrating its equivalence to the law called in later mechanics that of the moment–of–momentum, which involved much more for him than the example just cited concerning moments of forces relative to different axes of rotation. Carnot named the principle moment of the quantity–of–motion. Adducing it in terms of his geometric motions, he employed it for the same purpose that (according to Truesdell) physicists usually have done, in order "to obviate the need to specify the mutual forces among the particles of a rigid or deformable body."[14]

Referring still to a generalized dynamical system of bodies, wherein "m" is the mass of each body and "V" its velocity, suppose there be impressed upon it a geometric motion of velocity "u" which in its direction makes the angle "y" with the direction of the velocity "V". The quantity

$$muV \cos y$$

would then be the moment of the quantity of motion "mV" with respect to the geometric motion "u" and its sum, $\sum muV \cos y$ would be the moment of the quantity–of–motion of the system with respect to the geometric motion imparted to it. Retaining the notation of the basic problem and expressing it by

$\int muW \cos x$	The moment of the quantity–of–motion of the system before impact,
$\int muV \cos y$	The moment of the quantity–of–motion of the system after impact,
$\int muU \cos z$	The moment of the quantity–of–motion "lost" in impact,

Carnot showed by a simple trigonometric argument that

$$\int muU \cos z = 0$$

and hence

$$\int muW \cos x = \int muV \cos y$$

[14] *Whence the law of moment of momentum?* in his *Essays in the History of Mechanics* (Truesdell 1968b, p 243).

so that he could state a fundamental theorem:

> In the impact of hard bodies, whether the impact be immediate or be transmitted by means of any springless machine [machine quelconque sans ressort], it is always true that with respect to any geometric motion–
>
> 1. The moment of the quantity–of–motion lost by the system as a whole is equal to zero.
> 2. The moment of the quantity–of–motion lost by any portion of the bodies of the system is equal to the moment of the quantity–of–motion gained by the remaining portion.
> 3. The moment of the quantity of real motion of the general system immediately after impact is equal to the moment of the quantity–of–motion of the same system immediately before impact.[15]

These three propositions were identical at bottom, being simply interchangeable ways of stating the solution contained in Carnot's fundamental Equation (F) (see Eq. 2.2). Nevertheless, the third was the most important to Carnot, for it was from that way of looking at the problem that he later in the argument drew out his own injunction to continuity in the transmission of power. Already in this passage, however, the principle appeared to him to be the most valuable in all the science of mechanics. For here was a quantity that did indeed remain unaltered in any impact, whether direct or indirect. The quantity was not what Descartes had thought it to be, the sum of the quantities of motion. That conservation holds only for particular directions and only when the system is free. Neither was it the sum of the live forces. That quantity is conserved only in the gradual transmission of motion. Here was this further quantity, however, that could not be diminished by obstacles interposed in the way of the motions of the system, nor yet by the machines that transmitted the motions, nor finally by percussions that might intervene; and that was the moment of the quantity–of–motion of the system in general with respect to any geometric motion it could perform. That principle (Carnot went on) contained all the laws of equilibrium and motion of hard bodies. He would next show that it might equally well be extended to other bodies whatever their nature or degree of elasticity.

In the inquiry just cited, Truesdell remarks that specialists in classical mechanics consider their subject to be one based on three fundamental laws: "the conservation or balance of force, torque, and work, or, in other terms, of linear momentum, moment of momentum, and energy" (Truesdell 1968a, pp 241–242). It would falsify the way it looked to Carnot himself to say that he saw the science so categorically. Yet it would not falsify what was actually in his reasonings to say that he found the first of little use for his purposes, felt that he could assume the third while requiring to give it more prominence than it was currently receiving, and while not claiming to originate the second, did believe that he had given an original argument for a principle the generality and profundity of which was then far from understood, and which he required in order to draw consequences that were his own. Even so it was not his originality on which he insisted but the conviction the argument gave to a rationality of mechanical practice that he felt should have been possible at any time. For Carnot was proud of his subject itself rather than vain of what he individually contributed to it.

---

[15] Carnot (1786), § XXII, pp 43–44.

Carnot justified his strong statement of conservation of moment of the quantity–of–motion with a series of corollaries and remarks that deduced from it principles he had been assuming and expressed them in a form applicable to the motion of machines. In the first he demonstrated that among the geometric motions of which a system was capable, the one that would actually occur would give a minimum value for the sum of the products of each of the masses by the square of the velocity lost–though what he really meant, he hastened to point out was that the differential would be zero,

$$d\int mU^2 = 0,$$

which could also be true of a maximum or, under certain conditions, of a value that was neither a maximum nor a minimum. (No doubt this possibility was what he had had in mind at the outset in noting the exceptions to which the conventional statement of the lowest center of gravity principle was liable (Carnot 1786, § II, p 12)). Carnot explicitly recognized the analogy of this proposition with Maupertuis' Principle of Least Action. The remark is revealing of the background of his own thinking, for, excepting d'Alembert, Carnot in the *Essai sur les machines en général*, mentioned no other writer on mechanics more recent than Descartes. Departing from d'Alembert's point of view that the fundamental phenomena are those of inertial motion, the quantity of which constitutes force in his thinking, he directed the analysis to determining what forces could do, their action–though for Carnot this was an engineering concept and never a metaphysical one.

In the second corollary, Carnot turned to live force and stated that in hard–body impact the sum of live forces before collision equals the sum after collision plus the sum of the live forces that would obtain if the velocity remaining to each moving part were equal to what it had lost. The proof consists in a simple and obvious trigonometric transformation of Carnot's first fundamental Equation (E) (see Eq. 2.1)

$$\int mVU \cos Z = 0$$

into the expression

$$\int mW^2 = \int mV^2 + \int mU^2.$$

The analogy, Carnot here pointed out, between his Equation (E) (see Eq. 2.1) and the conservation of live force thus amounted virtually to a demonstration. So also did the analogy of that equation with the same conservation in systems of hard bodies of which the motion changed insensibly, for then, "U" becoming infinitesimal, "U^2" became an infinitesimal of the second order. He developed and established these points in the remaining two corollaries.

In the third corollary, Carnot turned attention to a subject that he had not yet discussed and that may be called Newtonian force only if it be disclaimed immediately that he saw it in such guise. For Carnot, the case was the one he had just mentioned, that of a system of hard bodies in which motion was communicated by insensible degrees. It was "motive force" that would have that effect, "force–motrice", and since he had not previously considered it, he required additional notation according to which

The mass of each body is	m
Its velocity is	V
Its motive force is	p
The angle between V and p is	R
Its velocity in any geometric motion of the system is	u
The angle between u and p is	r
The angle between V and u is	y
The element of time is	dt

Then the Equations (E) and (F) were shown to take the respective forms[16]:

$$\int mVpdt \cos R - \int mVdV = 0$$

$$\int mupdt \cos r - \int mud\,(V \cos y) = 0$$

Considering, then, a body freely describing a trajectory in uniformly–accelerated motion, he integrated the former of those equations over time and space and obtained the conservation of live force in the form

$$\int mV^2 = \int mK^2 + \int mV'^2$$

where "V" was final velocity, "K" initial velocity, and "V'" velocity acquired under a constant motive force in an undetermined time.

Finally, Carnot in his fourth corollary derived from the conservation of moment of the quantity–of–motion a proof that his fundamental indeterminate Equation (F) (see Eq. 2.2)

$$\int muU \cos z = 0$$

[16]The terms "$pdt \cos R - dV$" represented the velocity lost by m in the direction of V in consequence of the interactions of the bodies, and were substituted for "$U \cos Z$" in Equation (E) (see Eq. 2.1). Similarly "$pdt \cos r - d\,(V \cos y)$" was the velocity lost by m in the direction of u, and substituted for "$U \cos z$" in Equation (F) (see Eq. 2.2).

governed motion and equilibrium not only in the case of hard body (that he had already proved in deriving it), but that it was a general law holding good for all bodies whatsoever. The proof being a digression from the theory of machines itself, it will not be necessary to follow it in detail. It turned on pointing out that in hard–body problems, Equation (E) (see Eq. 2.1) became one of the determinates of the indeterminate (F) (see Eq. 2.2) since by definition in that case, $u = V$ (i.e., the real motion is geometric). For other types of body, expressions drawn from their nature would be needed in order to establish determinate values. For example, in respect of the force F that bodies exerted on each other, if in other bodies it was n times unity, then it would be possible to put $\frac{U}{n}$ in place of U and express Equation (E) (see Eq. 2.1) in the form

$$\frac{n}{1-n}\int mVU\cos Z = \int mU^2.$$

In the case of perfect elasticity, in which convention would put $n = 2$ this expression reduced to

$$\int mW^2 = \int mV^2,$$

which, of course, was conservation of live force. That conservation was for elastic collision what the Equation (E) (see Eq. 2.1) was for hard bodies, i.e., the characteristic determinant, which was just what he had undertaken to demonstrate at the outset.

Three remarks completed Carnot's formulation of the principles of equilibrium and motion themselves. The first referred the expressions he had derived to Cartesian coordinates, in which they appeared much simpler. Indicating by single, double, and triple primes the components of the quantities referred to three mutually perpendicular axes, and considering the conditions of equilibrium, Equation (E) (see Eq. 2.1) reduced to $0 = 0$; the fundamental Equation (F) (see Eq. 2.2) became

$$\int mu'W' + \int mu''W'' + \int mu'''W''' = 0;$$

and the general indeterminate for equilibrium in the case of motive forces became

$$\int mu'p' + \int mu''p'' + \int mu'''p''' = 0.$$

A second remark was more significant of Carnot's own style of thought. Traditionally mechanics had not only abstracted from the massiness of machines, but had treated obstacles and fixtures like the fulcrum of a lever either as stationary or as outside the system. Strictly speaking, however, no object in nature could be considered stationary, and no body was without mass. Rather than thus limiting the scope of the analysis, Carnot proposed a different mode of visualizing, one involving approximations to the material reality. He would let the fixtures and obstacles affecting the motions of a system be considered bodies of infinite mass and density though mobile in principle and hence susceptible of geometric motions. He would let bodies that merely communicated motion without offering any resistance to changes in the state of the system be considered bodies of infinitesimal mass and density. Of the two notions, the former seems the more significant, not for any practical difference it made in mechanics, but because this mode of accountability for mass bears a striking ancestral resemblance to that of the accountability for heat in Sadi Carnot's analysis of a heat engine drawing upon and discharging into reservoirs and sinks of heat of infinite capacity. By means of the latter notion, on the other hand, the mathematical theory of each of the classic types of machine might be derived from Equation (F) (see Eq. 2.2).

Finally, Carnot began turning the discussion toward the object ultimately in view, the application of the principles of equilibrium and motion in the form just given them to the theory of machines. For that purpose, he found the word "power" (puissance) more natural than "force", and it would appear that Carnot's discussion marks an early stage of the process in which that word began making its transition from the usages of ordinary life into the service of exact science and engineering. Nothing could be more fundamental, of course, than the existence of bodies acting on and modifying the motions of other bodies in accordance with first principles. Cases would arise, however, in which it would prove convenient to abstract from the mass of bodies for the sake of considering only the effort being made by them–the pull of the cord against the load, the thrust of the wind against the sail, the drag of the current upon the boat, the resistance of the body to impulsion, the friction between the bearing and the moving part. When it was a question of effect, the size or nature of the agent made no difference, whether the source of power were man, beast, or machine; wind, water current, or weight. As for machines, they were considered to be assemblages of immaterial obstacles and moving parts capable of transmitting action without internal reaction. Carnot took them to be, in other words, bodies of infinitesimal density and mass, to which were applied forces external to the system–powers in a word–but Carnot was consistent with his own realism. Machines in fact possessed mass, and he would not neglect it. He would simply consider its effect as he did the other forces or powers exerted by agents external to the system. Any force was to be taken as a quantity of motion "lost" by the agent that exerted it, whatever that agent might be. If, then, that force was designated "F" its quantity was evidently the same as that expressed by the product "mU" in all the foregoing discussions; and if "Z" was the angle between the direction of that force and any geometric velocity "u" imparted to the system, then the fundamental Equation (F) (see Eq. 2.2) might be expressed as follows:

$$\int Fu \cos Z = 0 \text{ (AA)}. \tag{2.3}$$

That was the form in which Carnot employed it in the discussion of machines in general that ensued in Part II, wherein when he wrote of force he was thinking of what he also called power, and tacitly taking its quantity over time.

It is clear, therefore, that we have reached the crucial turn that Carnot gave to the arguments of classical mechanics in what was said of machines. Classically machines were taken to be agents for the transmission of motion. The quantity of motion was an expression of force and involved mass. Abstracting from the mass of machines in order to study motions, the formulations of mechanics took no account of the motions of machines themselves. Substituting the notion of power for that of force, Carnot was able to transpose the notion of machines into that of agents for the transmission of power. Instead of neglecting the mass of the machines, he considered it to be one of the powers affecting motions. The problem became, therefore, adaptation of the laws of motion to laws of power and specification of the optimum conditions for transmission of power, and it was in order to prepare the ground that Carnot had brought the concept of motions "lost" and "gained" into the purview of the classical conservation principles and models of the structure of matter.

2.4 Moment–of–Activity–the Concept of Work

In all the foregoing, which constituted Part I of the *Essai sur les machines en général*, Carnot had not proposed to add to the science of mechanics but to state its principles in a form and arrange them in a sequence applicable to the science of "machines properly speaking" that he next inaugurated in Part II. There the writing went more directly to the point, the tone became more businesslike, and the yield of the· successive propositions was more evident in the analysis they afforded of the transmission of power.

The analysis began with Carnot's distinction between forces applied to a machine in motion according to whether their direction made an acute or an obtuse angle with the direction of motion of the point of application. The former he called "moving" or "impelling" forces, the second being the adjective usually employed. The latter he called "resisting" forces. Elsewhere in the *Essai sur les machines en général*, Carnot observed mildly that the definition would avert dispute about whether forces were to be considered causes or effects (Carnot 1786, p iv). Here it sufficed to notice that impelling forces could become resisting and vice versa if the direction of the motion changed, and that in any power system each of the forces would be impelling or resisting with respect to any given geometric motion imparted to the system according to the angle between the respective directions.

Carnot needed those metaphysically indifferent definitions for themselves. He also needed the distinction between the two classes of force for its auxiliary importance in his discussion of what in retrospect appears the most significant recognition in all his writing on mechanics and on mathematics, that of the importance to the science of power of the quantity that is now called "work" and that he called "moment–of–activity". At first glance it might appear that extending as he did the distinction between impelling and resisting force into one between moments–of–activity "consumed" and "produced" by forces applied to a system arrested him on the threshold instead of in the presence of a unified concept indistinguishable in substance from that of work. Such was not the case, however–though a little complex in the phraseology, the distinction was simply that between work done on or by a system.

Let a force "*P*," he laid down, be moving with a velocity "*u*" (for, though he did not say so, such a movement could only be geometric), and let the angle between "*P*" and "*u*" be "*z*" then the quantity "$P \cos zudt$" (where "*dt*" is the differential of time) would be what Carnot defined as the *Moment–of–Activity* consumed by the force "*P*" during the time "*dt*". Characteristically enough, Carnot was thinking about the gravitational instance for his example. Let a weight fall the distance "*H*" in time "*t*". Then, "$dH = u \cos zdt$", and the moment–of–activity consumed during time "*dt*" would be "$P \int dH$" or "*PH*"–the product of force by distance: foot–pounds in the Anglo–American dimensions of elementary physics.

Now, it is very difficult to establish categorical priorities in the history of mechanics. It is equally difficult, however, to find specialists in mechanics prior to Lazare Carnot who explicitly singled out that quantity to be the measure of what forces and powers accomplish and a unit in terms of which the fundamental conservations were to be stated. It is significant, moreover, that he stated it as the sum of the moment–of–activity consumed in successive instants–i.e., as what was accomplished by a process occurring in time. He was thinking about power, and not merely its dimensional equivalent, the product of displacement by motive force.

For it is clear that he did know what he was doing. Considering an entire system of forces applied to a machine in motion, the moment–of–activity consumed by the totality of the forces would be the difference between the sum of moments–of–activity consumed by the impelling forces and the sum of moments–of–activity consumed by the resisting forces. Since impelling forces made an acute angle and resisting forces an obtuse angle with motions, the cosines would be positive in the former case and negative in the latter, and the sign of moments–of–activity of impelling forces would thus be positive and of resisting forces negative.

Turning now to the reciprocal point of view, the moment–of–activity *produced* by a force was to be understood as the product of its magnitude evaluated in a direction *contrary* to its velocity multiplied by the distance that the point of application traversed in an element of time. Obviously, then, the moments–of–activity *consumed* and *produced* were equal quantities of opposite signs, and the difference between them was identical in the terminology of power to that in mechanics proper between moments–of–the–quantity–of–motion gained or lost with respect to a given geometric motion. In a final definition, Carnot combined his two classes of force and with them his two classes of moment–of–activity into a single designation. He called the moment–of–activity *exerted* by a force that which was *consumed* by an impelling force or *produced* by a resisting force. Since he did not know which was which, he would in practice always take the cosine of the smaller of the two supplementary angles between the force and the motion of the point of application, so that the moment–of–activity exerted by a force would always be a positive quantity.

Clearly then, this notion contained what is later to be found in the employment given by the science of physics to the concept of work. Having laid down these definitions for the mechanics of machines, Carnot proceeded to state what he called its fundamental theorem:

> Whatever the state of rest or motion of any system of forces applied to a machine, if a geometric motion be imparted to it without altering the forces, the sum of the product of each by the velocity of the point of application, evaluated at the first instant and in the direction of the force, will be zero.[17]

That statement was, of course, simply the adaptation to a system of forces or powers of the principle of conservation of the moment–of–the–quantity–of–motion for a system of bodies in motion. Expressed in the following form,

$$\int Fu \cos z = 0,$$

it was evidently identical with the Equation (AA) (see Eq. 2.3) into which he had transformed his fundamental indeterminate (F) (See Eq. 2.2) for the purpose of applying it to the transmission of power (Carnot 1786, § XXX, p 63). Indeed, it was obvious (Carnot remarked immediately) that strictly speaking this fundamental theorem was nothing more than the principle of Descartes with its defects repaired–i.e., generalized beyond equilibrated pairs to any system of forces with regard both to their direction and intensity.

Following the fundamental theorem, Carnot set out a series of six corollary propositions. The first three concerned the conditions of equilibrium of systems of machines and weights of various sorts on which geometric motions were impressed, and the fourth asserted the balance of torques around each of three mutually perpendicular axes along which forces in equilibrium could be resolved. Except for redeeming his opening promise to give a rigorous derivation of the original lowest center–of–gravity principle of equilibrium, Carnot did not seem much interested in these propositions, for he gave them in summary fashion and mainly for the sake of completeness.

The fifth was different: "Special law concerning machines of which the motion changes by insensible degrees" he called it, and now at last he had come to what he did certainly recognize to be his own, though still without insisting on his originality. The law ran thus:

> *In a machine of which the motion is changing by insensible degrees*, the moment–of–activity consumed *in a given time by the impelling forces* equals the moment–of–activity exerted *in the same time by the resisting forces.*[18]

That would be true if it could be proved that the moment–of–activity consumed by all the forces of the system during the given time was zero. If "*F*" designated each of the forces, "*V*" its velocity, and "*z*" the angle between them, then the requirement

[17]"THÉORÈME FONDAMENTAL. Principe général de l'équilibre & du mouvement dans les machines. XXXIV" (Carnot 1786, p 68).

[18]Carnot (1786), § XLI, pp 75–76.

was that $\int FV \cos Zdt = 0$; but by the fundamental theorem just stated, it was clear that $\int FV \cos Z = 0$. Hence the corollary followed from the general principle of equilibrium in machines.

It was not, however, in this almost tautological proof that Carnot's interesting contribution consisted, nor yet in the statement of the special law, although he did recognize it to be "the most important in all the theory of the motions of machines properly speaking". For he introduced the principle of continuity in the transmission of power by way of exemplifying an application of this special law, and then developed its consequences in the scholium that concluded the *Essai sur les machines en général* (Carnot 1786, § XLIX–LXIV, pp 81–89).

When working out his own thoughts Carnot usually preferred reasoning on weights and asked the reader to suppose that they be the powers applied to a machine and that

m be the mass of each of the bodies,
M designate the total mass of the system,
g be the force of gravity,
V be the actual velocity of a body m,
K be its initial velocity,
t be the time elapsed since the motion began,
H be the height through which the center of gravity of the system falls in t time, and
W be the velocity acquired in the height H.

Two types of force were involved in the operation of the machine, the pull of gravity and the inertial resistance of the various bodies to change of motion. By definition, the moment–of–activity *consumed* in the whole system by the action of gravity during time t was

$$MgH,$$

which was equivalent to

$$\frac{1}{2}MW^2.$$

Considering now the forces of inertia, the velocity of m being V and becoming in the next instant $V + dV$, its force of inertia in the direction of V would be

$$m\frac{dV}{dt}$$

Reverting again to definitions, the moment–of–activity exerted by the forces of inertia during time dt was

$$m\frac{dV}{dt}Vdt, \text{ or simplifying, } mVdV.$$

The moment–of–activity *consumed*, therefore, during time t by the forces of inertia was

$$\int mVdV$$

Integrating and completing the integral, it became

$$\frac{1}{2}mV^2 - \frac{1}{2}mK^2$$

(Carnot sometimes adopted the device of including the constant of integration in the term for initial velocity). Therefore, the moment–of–activity *consumed* in time t by all the bodies of the system would be

$$\frac{1}{2}\int mV^2 - \frac{1}{2}\int mK^2.$$

Now, by the conditions of the problem, inertia was a resisting and gravity an impelling force. Thus, by the special law just stated in the corollary, it would be true that

$$MW^2 = \int mV^2 - \int mK^2$$

or

$$\int mV^2 = \int mK^2 + MW^2.$$

As usual, Carnot also told verbally what he had just done:

> *In a weight–driven machine of which the motion is changing by insensible degrees, the sum of the live forces of the system after a given time is equal to the sum of the initial live forces plus the sum of the live forces that would obtain if all the bodies of the system were animated by a common velocity equal to that due to the height through which the center of gravity of the system has fallen.*[19]

It followed immediately that (a) in a weight–driven machine in uniform motion, the center of gravity of the system remained at constant height (since then $V = K$, $W^2 = 0$, $H = 0$); (b) No matter in what manner a weight was raised to a certain height, the forces producing that effect must have been such as to consume a moment–of–activity equal to the product of the weight by the height; (c) To produce any movement by insensible degrees in a system of bodies, the forces (Carnot said "powers") to produce that effect must have consumed a moment–of–activity equal to half the quantity by which the sum of the live forces of the system was increased.

[19]Carnot (1786), § XLII, pp 77–78.

From these last two propositions it further followed that in order to raise a weight M from rest to a height H while imparting to it a velocity V, the forces employed to that end must consume a moment–of–activity equal to

$$MgH + \frac{1}{2}MV^2.$$

And it seems worth following Carnot's reader through these elementary steps since they end with this expression dimensionally and notationally identical to an energy statement. Conceptually, of course, the range was extremely restricted, but there the expression was, nonetheless, and it would be a generalization and not an alteration of this type of thinking that would extend that range. So far, indeed, all that was new was the type of thinking, for dimensionally and algebraically the equivalence of those two terms was a relation as old and elementary as Leibniz's early discussions on the true measure of force. What Carnot had contributed was couching the relations in terms of the operation of machines, inertial masses transmitting powers over distances, and thinking of live force as the capacity for consuming moments–of–activity of machines in motion, or in later words, of forces doing work over distances and in time.

Now, however, we come to the consequence that was his own both in substance as well as genre. The previous reasoning assumed that motion in the system changed by insensible degrees. Suppose, however, that some discontinuity, some impact or percussion, intervened. Then let

- h be the height from which the center of gravity has fallen at the moment of percussion,
- X be the sum of the live forces immediately before percussion,
- Y be the sum of the live forces immediately after percussion,
- Q be the moment–of–activity that the live forces must consume throughout the entire movement,
- q be the moment–of–activity that the live forces must consume up to the moment of percussion.

Finally, suppose for the sake of simplicity that the movement of the system began from and ended in rest. It would be evident from the relation just established that

$$q = Mgh + \frac{1}{2}X,$$

and similarly that the moment–of–activity to be consumed following percussion would be

$$Q = Mgh + \frac{1}{2}X - \frac{1}{2}Y.$$

Now Carnot appealed back to his statement of what he had held in Part I of the *Essai sur les machines en général* to be the fundamental law of motion and equilibrium in mechanics proper, the conservation of the moment–of–the–quantity–of–motion (Carnot 1786, § XXII, pp 43–45) and specifically to its first corollary, which found that of all the motions of which a system was capable, that which would actually occur would be the geometric motion such that the sum of the live forces of each of the masses would be a minimum[20] (Carnot 1786, § XXIII, pp 45–48). Therefore, since $X > Y$ the moment–of–activity to be consumed in raising M to the height H was larger than if there had been no percussion, for in the latter case the equation would simply be

$$Q = MgH.$$

It followed that the same impelling forces could raise the same weight to a greater height if percussion were avoided than if it occurred. Carnot deferred an explicit statement of this, his Principle, to the concluding summary. For any machine, the condition of maximum efficiency is that $q = Q$:

> Now, in order to fulfill that condition, I say, first, that any impact or sudden change is to be avoided, for it is easy to apply to all imaginable cases the reasoning developed in that of weight–driven machines; whence it follows that whenever there is impact, there is simultaneously a loss of moment–of–activity on the part of the impelling forces, a loss so real that their effect is necessarily diminished [...]. It is then with good reason that we have proposed that in order to make machines produce the greatest possible effect, they must never change their [state of] motion except by insensible degrees. We must except only those that by their very nature are subject in their operation to various percussions, as are most mills. But even in this case it is clear that all sudden changes should be avoided that are not essential to the constitution of the machine.[21]

Such was Carnot's Principle. It is characteristic that he should have worked it out by example, claimed a generality that he did not actually demonstrate, and then exhibited his awareness that in actual practice the condition of perfect efficiency is an ideal to be approximated so far as the nature of the process allows. To the demonstration he added a final corollary on hydraulic machines, observing merely that since a fluid might be regarded as an infinity of solid but detached corpuscles–and since he had already proved what was usually taken to be merely experimental truth, i.e., the conservation of live force in incompressible fluids wherein motion changes insensibly (that is without splashing or turbulence)–everything that he had demonstrated for systems of hard bodies held equally good for masses composed of incompressible fluids (Carnot 1786, § XLVIII, p 80). But that extension was more important than the afterthought it appears to be in the formal structure, and it will be easier to appreciate its significance and indeed the quality of Carnot's performance in the mechanics of machines if we accompany him into the scholium in which he enlarged more informally, and more naturally, on what he hoped his *Essai sur les machines en général* would accomplish.

[20]This is the principle that Carnot recognized as analogous to Maupertuis' Law of Least Action.

[21]Carnot (1786), § LIX, pp 91–92.

2.5 Practical Conclusions

In the concluding discussion containing the statement of Carnot's principle just quoted, he relaxed his formality and spoke his mind. According to Corollary *V*, the moment–of–activity consumed in a given time by impelling forces equals that exerted in the same time by the resisting forces in a machine that changes gradually in its state of motion. Actually, he observed, that proposition contained nearly all the applicable part of the theory of machines. In practice most working devices were powered by agents–animals, springs, weights–that exerted "dead force" continuously. Most real machines, furthermore, once set in motion soon reached a steady pace of operation such that the forces required to keep the process going balanced the elements of resistance. It now appears, therefore, that this corollary had been the goal of the argument all along, and we would not be forcing his meaning to call it his Work Principle. Indeed, we would not altogether be forcing his language, for in one example of the equivalence between moment–of–activity and effect, he remarked it propos of an arrangement for raising a weight a certain distance, that no machine could be designed by which it would be possible "with the same work [travail] (that is to say the same force and the same velocity)" to lift the object higher in the given time (Carnot 1786, § LIV, p 85).

In turning now to practice, Carnot maintained the generality of his discussion. Designating by

$$Q \quad \left\{ \begin{matrix} \text{the work done} \\ \text{the moment}-\text{of}-\text{activity consumed} \end{matrix} \right\}$$

by impelling forces and by

$$q \text{ that } \quad \left\{ \begin{matrix} \text{done on} \\ \text{exerted by} \end{matrix} \right\}$$

the resisting forces, then Corollary *V* the Work Principle, could be symbolized by

$$Q = q.$$

It followed that there were two related sets of conditions for the most efficient possible operation of machines. The first was that the greatest possible mechanical advantage be secured and the second that no motion be wasted. In order to clarify the former condition, the relation could also be expressed

$$FVt = q,$$

where F was the resultant of all impelling forces and V the resultant of their velocities. Achieving the maximum effect would involve varying those three factors. A fourth variable would be the direction in which the impelling forces acted, but it was obvious (though Carnot labored the point in one of his trigonometric excursions) that for best results the force ought always to be applied in the direction of the velocity. As for the intensity of force, the time of application, and speed of operation, no general rules applied and experience would need to govern. For example, a man might turn a crank one foot in diameter for eight hours a day making an effort equivalent to twenty–five pounds at a rate of one revolution every two seconds. If he were to work faster, his output would suffer for he could no longer maintain the twenty–five pound exertion. If he turned the handle more slowly, he increased F proportionately less than V, and the moment–of–activity would diminish. Every source of power would have its own maximum, of which Carnot could say only that it was determined by its physical constitution and that experience alone would find it.

No hidden resources of power lay in the capacities of machines, therefore, and what theory offered was identification of the factors among which economy would dictate the wise proportions to observe. If time were not important, force might be economized at its expense. But (the reader may feel a bit impatiently) so elaborate a discussion can hardly have been required to exemplify the ancient maxim that what is gained in force is lost in time or speed. Carnot clothed the point in new terminology, but gave no new findings, and what must be appreciated is that he evidently thought that rigorous demonstration would help and to that end expressed himself with an engineer's passion and urgency. These matters, so obvious to him, must need the conclusive proof he was giving, else why the persistence of chimerical schemes for perpetual motion under one pretext or another, why the continuing waste of ingenuity and money? It would falsify Carnot's book to underemphasize that, perhaps its most deeply felt point.

The novel considerations that Carnot brought forward, however, pertained mainly to the second set of requirements for maximum efficiency, the avoidance of waste motions. Instead of generalizing the reasoning that had established his Principle in the example of weight–driven machines, so as to cover all possible cases, Carnot turned to developing its relevance for hydraulic machines, and did so in a series of paragraphs that seem particularly striking when compared to the reasoning in the memoir by his son, *Réflexions sur la puissance motrice du feu*, where heat is treated like a fluid in flow.[22]

Looking at the ordinary waterwheel from an ideal point of view, Carnot observed that its design embodied two faults. First, the water was generally allowed to fall onto the blades and thus to transmit motion by percussion. Second, after striking the blade the water ran off with a velocity that was entirely lost to the process. In order that a hydraulic machine be as efficient as possible, its design should be such that the water would lose all its motion to the mechanism, and should do so without turbulence or splashing. It made no difference what form the machine might take. As always, Carnot was dealing in the analysis with idealized machines. He recognized

---

[22]This analogy has often been noticed, e.g., Charles Brunold in his *L'entropie. Son rôle dans le développement historique de la thermodynamique* (Brunold 1930, pp 37–40).

that descending to actuality might make it desirable to depart from these conditions. For simplicity of construction, nothing was likely to prove better than wheels turned by the impact of water.

Practically, moreover, the closer the design approached to satisfying one of the conditions, that is absorbing all the motion of the water in the wheel, the greater would be the impact and the greater the loss of power to percussion. The lesser the impact, conversely, the less would be the· proportion of the power of the water transferred to the wheel. The shrewd designer would, therefore, regard these conditions not as goals to be attained but as norms to be approached in the degree that circumstances render the one saving or the other relatively more important. Indeed, it was an ultimate condition of efficiency in machines of all types that no motion be produced extraneous to their purpose, and readers of Sadi Carnot's memoir will also recognize this injunction adapted to the heat engine in his requirement that no differential of temperature be admitted that does not measure itself in a change of volume in the gas confined in a cylinder. As his father originated the example, it showed the most efficient pump to be the one that delivered water into the reservoir at velocity zero.

Expanding in his closing remarks on the concept of moment–of–activity, Carnot introduced a final illustration of its utility that carried his perception of its significance beyond the study of machines into generalized physics. Suppose the problem to be one of a system of bodies mutually attracting one another by forces that vary as any power of the distance. (Only at this late stage did he move into a mechanics of which the model was clearly Newtonian, and then only as an object of an analysis developed out of other considerations and other problems than those of central force systems.) Suppose the system was impelled to move from some given configuration to another. No matter, then, what the sequence in which the individual bodies were displaced; no matter, further, what route they took, provided only that no percussion intervened; no matter, finally, what sort of machines effected the transformations–quite independently of all such incidental means to the end, the moment–of–activity that the external agents consumed would always be the same, assuming the system to be at rest in its initial and final states.

Now, then, this certainly might be taken for a model of the kind of analysis that the physics of work and energy has found useful ever since those topics became explicit in the 1820s, '30s, and '40s. Its most recognizable offspring again was the heat cycle of Sadi Carnot, which considered a system in view of what had been done to or by it in shifting from an initial to a final state. The family resemblance was most marked in the abstractedness of the system, in the notion that process consisted in the transition between successive "states", in the restriction that this transition be gradual and continuous, in the requirement that all changes be reversible (which for Lazare Carnot was still to say that all motions be geometric), in the indifference (given those conditions) to the details (rate, route, or order of displacements), in the relevance of this extreme schematization to the actuality of operating tools, engines, and machinery. Clearly, the relationship between the science of machines and thermodynamics was similar to that between Lazare and Sadi Carnot. It was one of parentage.

Chapter 3
The Development of Carnot's Mechanics

It is fortunate that the two memoirs Carnot submitted to the *Academy of sciences* in Paris in 1778 and 1780 have survived in its archives. They give access to the form and content of his thinking about mechanics and the science of machines when he was a young and hopeful officer 5 or 6 years out of engineering school. The present chapter traces the development of his thinking from these, its earliest recorded expressions, through to the publication, in (1803a), of his *Principes fondamentaux de l'équilibre et du mouvement*. The method is to compare these earlier and later versions to the *Essai sur les machines en général*, and to continue to an analysis of Sadi Carnot's *Réflexions sur la puissance motrice du feu* of 1824 in order to bring out its filiation with his father's work. The relevant, theoretical sections of the early two memoirs are reprinted by Charles Gillispie in his *Lazare Carnot Savant* (Gillispie 1971, Appendices B and C, pp 270–340). Readers may wish to refer points in the discussion, or indeed the discussion as a whole, to the original texts.

3.1 Argument of the 1778 Memoir on the Theory of Machines

A notation on the cover page of the manuscript of Carnot's first memoir indicates that he learned of the Academy's announcement of a prize contest from the *Gazette de France* of 18 April 1777. The Academy specified the subject to be:

> The theory of simple machines with regard to friction and the stiffness of cordage, but it [the Academy] requires that the laws of friction and the examination of the effects resulting from stiffness in cordage be determined by new experiments conducted on a large scale. It

C.C. Gillispie and R. Pisano, *Lazare and Sadi Carnot: A Scientific and Filial Relationship*, History of Mechanism and Machine Science 19, DOI 10.1007/978-94-017-8011-7_3,

requires further that these experiments be applicable to machines used in the Navy such as the pulley, the capstan, and the inclined plane.[1]

From a further note it appears that the Academy received the entry Carnot sent on 28 March 1778. Assuming that he had set to work immediately, it took him, therefore, just under a year to design and carry out his experiments and compose the argument. For device he misquoted a line from Lucretius: "Videndum/Qua ratione fiant et qua vi quaeque gerantur"[2] (Gillispie 1971, p 271). In setting the competition the responsible members of the Academy would have had no thought that anything of importance to the science of mechanics itself was to be expected in consequence of the entries it might elicit or on the part of the contestants it might attract. The emphasis was on the naval application, and what the Academy clearly wanted was studies of friction. None of the papers having satisfied the judges, it reset the same subject for the contest of 1781, and then awarded the prize to Coulomb, whose investigation gave them exactly that and clearly deserved to win. His memoir *Théorie des machines simples en ayant égard au frottement de leurs parties et à la roideur du cordage* (Coulomb 1785, pp 161–332) remains one of the cardinal contributions to the knowledge of friction.[3]

---

[1]The translation is from original manuscript of which a photographic reproduction was edited in: Gillispie (1971), Appendix B, pp 270–296. Hereafter we quote Lazare Carnot's *Memoires* (1778, 1780) using «§» as reported in Gillispie (1971), Appendices.

[2]Probably relying on memory while working at Cherbourg, he had in mind line 129 of *de rerum natura*, Book I, but spoiled the scansion by inverting the word order; he also displaced the gerundive from the end of line 131. The entire passage reads

Qua propter bene cum superis de rebus habenda
nobis est ratio, solis lunaeque meatus
qua fiant ratione, et qua vi quaeque gerantur
in terris, tum cum primis ratione sagaci
unde anima atque animi constet natura videndum,
et quae res nobis vigilantibus obvia mentis
terrificet morbo adfectis somnoque sepultis,
cernere uti videamur eos audireque coram,
morte obita quorum tellus amplectitur ossa.

A translation of the words Carnot chose might read: "Let us look to an account of how all things come to pass and of the forces that govern them".

[3]For a discussion of the contest centered upon the winning memoir, and for the importance of Coulomb's work on friction, see Gillmor (1971), chap. IV. The judges were d'Alembert, Bézout, Bossut, Condorcet, and Trudaine de Montigny.

Carnot's own taste and interest, however, responded primarily to the opening phrase, and he submitted his entry as a *Mémoire sur la théorie des machines.* It is divided into two parts, experimental and theoretical. As will appear, the organization of Part II already exhibited the structure within which Carnot's thought about the subject developed throughout his life. Obedient to the Academy's injunction, he did dutifully set to work to determine experimental laws of friction and binding, and Part I of the memoir consists of some 20 folio pages detailing the results. An account of these experiments would be no help in understanding the difference Carnot's work finally made in mechanics, but perhaps the reader will be curious to know what he undertook. Employing the services of an assistant, he spent considerable pains upon two sets of determinations, one concerned with friction and the other with cordage. For the work on friction Carnot constructed the device pictured in the following (see Fig. 3.1).

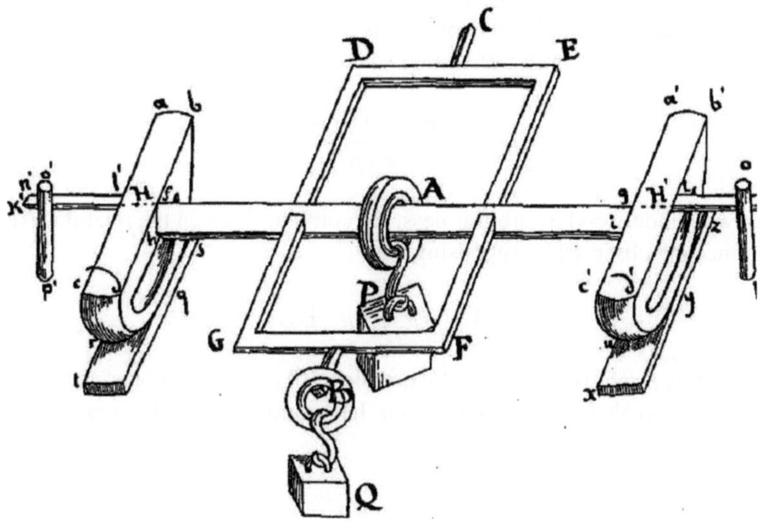

Fig. 3.1 Machine for valuing friction (The illustration is in fact from the 1780 memoir, but it is clear from the context that it is the same device of which the drawing has been separated from the 1778 text.)

The instrument operated in a simple manner for the purpose of determining values for friction between surfaces that (in the so-called first type) slide one across the other and (in the second type) roll one upon the other. The principle of the machine was that the statical moment of a weight Q (see Fig. 3.1) just sufficient to initiate rotation was the measure of the force of friction (with which it was in equilibrium) between the surfaces of the hemispheres and of the iron shoes *srtg* and *zuxy*. The inner circumference of the rings A and B and the upper blade of the axle HH' were knife edges.

With the retaining pins *op* and $o'p'$ in place, the axle was free only to turn. A sufficient weight Q would cause the surfaces of the hemispheres to slide across the supporting shoes, and the friction measured was of the first type. With the pins *op* and $o'p'$ removed, the axle was free to displace. A weight Q would cause the hemispheres to roll on the supports, and the friction measured was that of the second type. Carnot made, too, a determination of the effect of adherence. In measuring sliding friction, the values were always significantly greater if the device was allowed to rest on the supports for an appreciable time than they were when the measurement was made at the instant it was loaded. Adherence appeared to have no appreciable effect, however, upon rolling friction.

Carnot employed increasing weights for P and Q, and tabulated the ratios of the two classes of friction and of adherence to pressure over a range between 100 and 2,000 lb for the total weight of the movable cradle. In each class of measurement, the ratio of friction to pressure decreased with increasing load. In good engineering fashion, he sought to express the results in formulas that turned out no more beautifully than such empirical expressions usually do.

The formula for the ratio x of sliding friction to pressure was

$$x = \frac{A + Bp + Cp^2 + \ldots etc.}{1 + bp + cp^2 + \ldots etc.}$$

Carnot well understood its arbitrariness. He could have given the relation a simpler appearance, he observed, in supposing that

$$x = A + Bp + Cp^2 + \ldots etc.$$

except that as p increased, so would x, which effect was the contrary of what the experiments revealed. On the other hand, he would fall into the opposite error in supposing that

$$x = A + Bp^{-1} + Cp^{-2} + \ldots etc.$$

The best he could say for the formula adopted was that it fell as closely as possible between the two extremes.

Further experimentation to determine the influence of velocity on the two classes of friction led Carnot to produce formulas even more empirical in appearance. For sliding friction,

$$X = \frac{\phi + \mu\pi}{1 + \varpi\pi} \cdot \frac{1 + bu}{1 + cu},$$

and for rolling friction

$$X = \frac{\phi + \mu\pi}{1 + \varpi\pi} + \frac{\phi'}{1 + \varpi'\pi},$$

where x was the ratio of friction to pressure; ϕ, μ, ω, ϕ' and ω' were constants depending on the nature and degree of polish of the surfaces; π was the pressure exerted by a single point; u was velocity; and b and c were constants of velocity determined by a method at once cumbersome and inaccurate.

The experiments on stiffness and binding in cordage were of a similar type, and the results no more applicable. Perhaps it will suffice to reproduce the illustration of the pulley-type instrument that Carnot designed and employed in order to determine how the properties of ropes and wires affected the equilibrium conditions of weights of varying magnitude (Fig. 3.2).

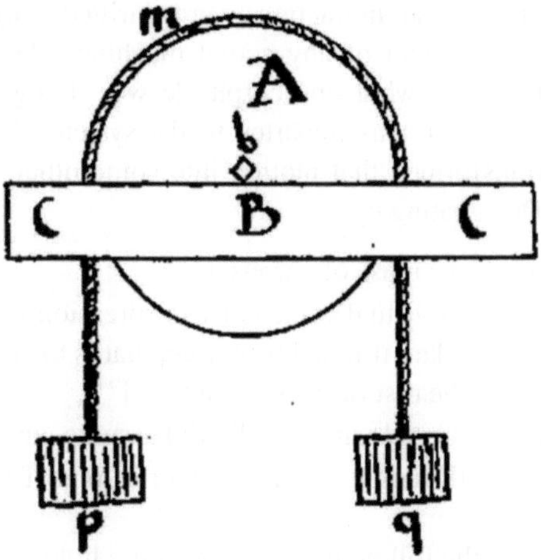

Fig. 3.2 Instrument for determining equilibrium conditions (Gillispie 1971, p 65)

The account of these experiments occupies § 1 through 26, some 16 folia of a manuscript containing 64 folia all told. At the outset Carnot promised that Part II would contain their "application to machines". What Part II actually contains is three sections, the first 16 folia discussing the principle of "machines en général" in § 27 through 50, and the remaining two, in § 51 through 84, applying the principles respectively to equilibrium and to motion in the cases of the seven types of simple machine. It is the first section of Part II that is most interesting to the historian, for it is one of those relatively rare documents that admits him to the genesis of an approach that was later to become distinctive of a whole new way of seeing and doing a science.

That it was an application of the experiments was a pretense on Carnot's part, however. It is difficult, in fact, to see that they had anything to do with the theoretical discussion of machines in general except to serve as an excuse to qualify the memoir for this particular competition. The excuse is the less convincing in that Carnot had not the time or did not take the trouble to fill up his tables of data completely. There is no mention of these experiments later in his published writings. Nor did there need to be despite his view of mechanics as an experimental science. The most that can be said, and as a highly likely conjecture this much ought to be said, is that the instrument he constructed for measuring friction may very well have led him to set up problems in terms of torque and moments of rotation around an axis, and therefore towards the emphasis he placed upon the principle of moment–of–momentum in the *Essai sur les machines en général.*

However that may have been, the theoretical passages that were the heart of the memoir make clear how Carnot thought to approach the subject. He made no preliminary statements of basic principles of mechanics of any sort. Rather, he moved in imagination right into the interstices of hard body and pictured what the inflexible rods and inextensible wires were doing, mini–levers and mini–pulleys, micro–machines that transmit motion there where it starts, corpuscle to corpuscle. Once those interactions were clarified, they would be those ideally operating in the employment of any actual machines. That is what he thought about in the first instance: what one corpuscle was doing to another in a system of hard bodies. A motion was imparted to the system. The mutual interactions of the corpuscles transformed that motion into some other which it was the problem to determine. Designating by

- m the mass of corpuscle,
- V its virtual velocity (his expression was "[. . .] the velocity that it would have taken if it had been free, that is to say without the reaction it undergoes from the rest of the system [. . .]"[4]),
- u the velocity it really did assume, and
- y the angle between the directions of V and u,

then the following equation would hold:

$$\int mu\,(V\cos y - u) = 0,$$

which asserts that the sum of the products of the quantity of motion (mu) of each of the corpuscles multiplied by the velocity that it lost evaluated in the direction it took ($V\cos y - u$) is equal to zero (Gillispie 1971, Appendix B, § 29). For the formation of Carnot's idiom, the significant term was the one in parenthesis. It makes evident that from the very beginning he thought of interactions within a system in d'Alembert's terms of velocities lost or gained by its constituent parts, and thought of live force (which in later times would be seen as the energy involved in such interactions) as the product of momentum by a velocity or motion thus "lost" or "gained" (Gillispie 1971, Appendix B, § 27). The actual demonstration involved the physics of hard–body collision and equating the forces of action and reaction in pairs of contiguous molecules: in actual practice the equality of action and reaction was always the basic law of mechanics for Carnot.

In this memoir, Carnot claimed for the above equation the status of a fundamental theorem. It did duty for all the variants that he later elaborated in the *Essai sur les*

[4] Gillispie (1971), Appendix B, § 27.

machines en général–its determinate Equation (E) (see Eq. 2.1) for the general case, its transformation by way of geometric motions into the indeterminate Equation (F) (see Eq. 2.2), its deduction from Equation (F) of the principle of conservation of moment–of–the–quantity–of–motion with respect to any geometric motion, its derivation of the power Equation AA (see Eq. 2.3) concerning the forces applied to a system, its general principle of equilibrium and motion for machines in terms of the balance between moments–of–activity consumed and exerted by impelling and resisting forces. Carnot drew upon the general corpus of mechanics for all these refinements to his problem. In the early memoir only more or less definite foreshadowings of them occur scattered along a series of some 17 corollaries and remarks. Most were so elementary or trivial that one would never read them twice, and no emphasis was given to the ones that in the retrospect of Carnot's later writings seem significant, the kernels that later opened out into main positions. The most useful procedure will be to point out what elements he did then have of his later findings, and what the differences were.

The greatest difference in emphasis was that when writing of force in his early memoir Carnot began with the effects of accelerative or motive force in producing motions. In the *Essai sur les machines en général*, it will be recalled, accelerative force remained largely hidden from attention until near the end, when it appeared as one of the factors in moments–of–activity, i.e., as the force that does act across a distance or in a time. In the 1778 memoir, however, the first corollary concerns motive force. The (virtual) velocity V is now said to be the resultant of the motive force of the corpuscle m at a given instant and of its velocity u' in the previous instant. Evidently, then,

$$V \cos y = u' \cos z + pdt \cos x.$$

(Carnot was somewhat careless with his designations in this memoir. He identified z as the angle between u' and u, but not p as "*force motrice*" nor x as the angle between the directions of p and u. That is what he meant, however). The basic equation would then become

$$\int mu\left(u' \cos z - u\right) - \int mupdt \cos x = 0.$$

Carnot reserved to a later section (Gillispie 1971, Appendix B § 53) the proof that the resultant of several forces multiplied by the cosine of the angle that it makes with any given line equals the sum of the component forces each multiplied by the cosine of the angle that it makes with the same line. But it was in this discussion that he first employed this mode of representing and combining the directions of magnitudes of the type called vector quantities in later mechanics.

The second corollary (Gillispie 1971, Appendix B § 31) then stated that if there was impact, the term pdt became negligible compared to the value of u so that

$$\int mu\left(u' \cos z - u\right) = 0.$$

From this equation, Carnot observed, all the laws of impact could easily be deduced, although to do so would have been a digression from his purpose. For our purposes the interest lies mainly in the evidence that it was natural for him to begin with accelerative forces acting over time or distance. There was no hint of anything like a new concept to embrace that point of view. It is only what the problem involved from the outset.

Interestingly enough, a corollary about what were to become geometric motions was also couched rather in terms of force and activity than of imaginary displacements in a system. In language it resembled rather the dynamical treatment that Carnot finally gave the idea in his *Principes fondamentaux de l'équilibre et du mouvement* than the kinematical emphasis of the *Essai sur les machines en général.* It would appear, indeed, that the terms in which he thought about the problems developed from these conventional notions of force, through an intermediary stage where momentum was the subject, into the final analysis of power. The idea appeared in a way subsidiary to equilibrium. From the above fundamental equation he had derived the following consequence (Gillispie 1971, Appendix B, § 32, Corollary 3) for the case in which motion changes by insensible degrees (which he here described as that in which u differs infinitesimally from u'):

$$\int mpds \cos x - \int mudu = 0$$

(since $u' - u = -du$, and supposing $ds = udt$) or

$$\int mupdt \cos x - \int mudu = 0.$$

If equilibrium was considered to be the state of the system in which the forces p mutually "destroy" each other, this expression would reduce to the unhelpful identity, $0 = 0$. It was possible, however (Gillispie 1971, Appendix B, § 34, Corollary 5), to look at the situation in another way, and to conceive that there had been imparted to all parts of a system in equilibrium "motions such that no change resulted in the reciprocal action of the different parts of the system."

These would be motions that would be received "without alterations" (*Ibidem*), by the bodies on which they were impressed. The forces p would remain in mutual equilibrium. It would then be perfectly feasible to imagine these motions acquired in time, and the case of equilibrium would be assimilated to one in which the system changes by insensible degrees, where the equation for that case,

$$\int mpds \cos x = 0,$$

was not a mere identity. Though still unnamed, geometric motions were born of this passage and of Carnot's proclivity for thinking even of equilibrium as a state of opposing actions–most characteristically backing into a grasp on the concept of work by defining the situation in which forces did not work, or did merely putative work.

From that equation, Lazare Carnot next derived, in the case of systems in internal equilibrium, a series of corollaries that amounted to the conservation of linear momentum, the conservation of moment–of–momentum, the principle of Descartes, and the conservation of live force. To none did he give prior standing over the others. The first two he did not even identify in those terms. Two brief corollaries extended the discussion to elastic bodies, provided that the force of elasticity was included in the value of p (Gillispie 1971, Appendix B, § 40, Corollary 11). More interesting, they extended also to fluid bodies, which it was legitimate to consider as if composed of hard or elastic corpuscles (*Ivi*, § 41, Corollary 12). In a long digression (*Ivi*, § 42) Lazare Carnot derived the hydrostatical principle of equal pressure in fluids from his equation, and came finally to what his mind always naturally fixed on, the operation of machines handling weight, whether as load or source of power. After all, raising weights was the purpose of the greater number of all machines actually used (Gillispie 1971, Appendix B, § 43, Corollary 13).

It becomes immediately apparent that these were the considerations from which Carnot's approach took its distinctive features. Designating by

M the total mass of the system,
H the height from which its center of gravity falls,
t the time of fall of the center of gravity of the system,
V the velocity a body acquires in falling the height H,
h the distance which a molecule m falls in time t, and
K the initial velocity of the molecule m (although Carnot overlooked identifying this),

then since $ds \cos x = dh$, the basic expressions become

$$\int mpds \cos x = \int mpdh = MpdH$$

(the first summation is of individual molecules). Thus

$$2 \iint mpds \cos x = 2 \int MpdH = 2MpH$$

(the second summation is of elements of path), but

$$2MpH = MV^2,$$

and, therefore,

$$\int mu^2 = \int mK^2 + MV^2.$$ [5]

[5]Gillispie (1971), Appendix B, § 43, Corollary 13.

That elementary sequence illustrates very simply the arithmetical reasons that 50 years later led Coriolis to equate the newly named quantity work to one–half the live force, so that still later kinetic energy became defined as half the product of mass times velocity squared.

Thereupon, Carnot drew out as a passing consequence the principle that he was to propose for a fundamental axiom in the *Essai sur les machines en général*: in order to prove that the weights applied to a system are in equilibrium, it sufficed to show that the center of gravity did not descend. Now followed the remark that this principle could be applied to any machine by the device of exchanging a weight suspended over a pulley for any other force. Suppose that besides weights, other powers should be applied to the machine in order to move them, say manpower or horsepower.

The general equation would then become

$$\int mu^2 = \int mK^2 + 2 \iint mpds \cos x.$$

That last term, in turn, might be decomposed into two parts, of which one was MV^2, and the other was

> [...] double the sum of the products of each of the moving forces [i.e., the additional powers] multiplied by the element of path that it describes in the direction of that force.[6]

If F represents one such force, and u its velocity, then

$$\int mu^2 = \int mK^2 + MV^2 + 2 \iint Fudt.$$

Here was the germ of the work concept, therefore, and right alongside it that of a generalized process. Suppose that all bodies in the system were at rest at the beginning and end of the movement. In that case $u = 0$ and $K = 0$, and the expression became

$$2 \iint Fudt + MV^2 = 0,$$

or if H was the height to which the center of gravity was to be lifted, then the relation might be expressed

$$\iint Fudt = MpH,$$

[6]Gillispie (1971), Appendix B, § 46, Corollary 16.

which exhibited that in order to raise the center of gravity of a system of bodies to any height it was necessary that the quantity *Fudt* always be the same, whatever the route chosen, whatever the machine utilized, whether the parts of the system be raised separately or together, and whether a single force or several together be employed[7] (Gillispie 1971, Appendix B, § 47, Corollary 17). A further passage (*Ivi*, § 49) identified the factors, F, u, and t to be varied in achieving mechanical advantage and gave examples of processes in which it would be prudent to economize one factor or another. Carnot observed that it would result in a saving of force if each part of the system completed the process with zero velocity, or at least with as small a velocity as possible.

The second and third sections, comprising § 51 through 85, of the theoretical part of the memoir dealt respectively with equilibrium and with motion in the case of the seven classes of simple machines. Actually, Carnot remarked at the outset, "cords" (*Ivi*, § 51) and the lever were the only truly simple machines. It was even possible to reduce the former to the latter, but seven types of machines were customarily considered to be simple: cords, the lever, the pulley, the crank, the inclined plane, the screw, and the wedge. His treatment adapted his equation, $\int mpds \cos x = 0$, to the characteristics of each in turn in order to state its law of equilibrium while "[...] taking account of friction and the stiffness of cords as far as I could [...]" (Gillispie 1971, Appendix B, § 49). The expressions were artificial and the discussion forced, and perhaps that was its most important feature since the central purpose of the *Essai sur les machines en général*, a few years later, precisely was to achieve a general science of machines and to obviate the necessity for deriving laws of equilibrium and motion for each particular type of machine. The third section on motion was more summary, though no clearer, and ended with an appeal to the possibility of combining his principle with that of d'Alembert in order to reduce all questions about the theory of machines to problems of analysis.

Certain features of this discussion, however, did hold importance for Carnot's later work. He devoted the fullest treatment to the law of equilibrium for the "funicular machine" (*Ivi*, § 52), and it was there that he introduced his convention for projecting forces (he did not then speak of other magnitudes) upon other directions in terms of the cosine of the angle, balanced the moments of the resultants of clockwise against counter clockwise forces, and finally considered the projection of the machine onto a plane in which the projection of the applied powers became a two–dimensional force system. (These matters had so central a part in Carnot's later work that it seems wise to extend the reproduction of the theoretical portion of the 1778 memoir in order to include them (*Ivi*, §§ 52–60). It would seem that Carnot's thinking about vector quantity may have originated in his mental picturing of the distribution and transmission of forces by the cords of machines worked by ropes).

In summary, it may be said of this first memoir that Carnot had in it the germ of his central ideas, but not as yet differentiated from a central principle of

[7] It would be perfectly possible to see embedded in this Carnot's principle of continuity, but there is no evidence that he then saw there.

equilibrium into concepts themselves. He merely alluded in passing to the well–known principles of mechanics, such as conservation of momentum or conservation of live force, without emphasis on one as more relevant than others to his purposes. If any single concept may be called fundamental for him at this stage, it would be force in the sense of motive force or pressure, but only tacitly so–it was no more than what he meant when he alluded to force. Missing (except as fugitive hints or questionable clues) were two of the central features of the *Essai sur les machines en général*, the distinction between impelling and resisting forces and the notion of a generalized science of machines to be disengaged from the application of laws of mechanics to the classes of machines taken successively. The notion itself was not there, much less the recognition that the way to bring it to pass was to treat machines as themselves inertial bodies endowed with mass. (Seeing that is what impelled him in the *Essai sur les machines en général* to go over to terms of displacement and work.) It would be fairly farfetched, though perhaps just barely possible, to say that one discussion that took into account the weight of the lines in a funicular machine represented the first hint of what Carnot later developed into the main emphasis of his theory of machines (Gillispie 1971, Appendix B, § 60). It would be more reasonable to imagine that it was writing on theory of machines while repeatedly taking account of friction, instead of neglecting it, that led Carnot into the way of considering friction (like the weight of the machine) in the light of a force against which the motive forces had to work. In the concluding passage to Section I of the theoretical part Carnot said expressly that the force of friction and other "resistances" (*Ivi*, § 50) might be considered as "active" forces (*Ibidem*).

3.2 Argument of the 1780 Memoir

In reopening the subject for an award in 1781, the Academy reiterated its phrasing of the problem except that it now specified that the friction to be investigated should be that between the moving parts of simple machines:

"The theory of simple machines with regard to the friction of their parts and the stiffness of cords [. . .]" (Carnot 1780, Appendix C, in Gillispie 1971). A note in Carnot's own hand on a cover sheet indicates that he completed it at Béthune on 15 July 1780. In resubmitting his memoir, anonymously as was the Academy's practice, he identified it by the same motto from Titus Lucretius Carus (ca. 99 BC–ca. 55 BC) and observed in a note that having been accorded an Honorable Mention in the first competition, he had now "diminished its imperfections as far as I could", *Mémoire sur la théorie des machines* (Gillispie 1971, Appendix C, p 299).

In fact, he had enlarged and transformed it into a memoir that in important respects is the draft of the *Essai sur les machines en général* of 1783 even to the wording of many central passages. Indeed, comparison of the two prize memoirs makes it apparent that, though the genesis of Carnot's mechanical thought belonged to his education and to the years at Calais and Cherbourg, its formative period came between his first failure to win the prize (the announcement was made at Easter

1779) and the completion of the revised memoir in July 1780. During the interval of approximately a year he must have been reading, thinking, and living little other than the theory of machines.

True, he was still bound by the terms of the Academy's contest, and the organization followed that of the earlier memoir. The 1780 memoir also opened with experiments in Part I and went on to a Part II pretending to be their application to practice. The latter again consisted of two sections, one on "machines in general" and a conclusion on laws of equilibrium and motion in each class of simple machine. In deference (no doubt) to the Academy's predilections, Carnot did elaborate and complete his program of experiments. He introduced it with a broad discussion of friction in general, and this time classified the experiments according to whether the bodies touched point–to–point, line–to–line, or surface–to–surface. He devised new experiments, improved the account of the old ones, and dropped the most cumbersome of the formulas for variation of friction with velocity. Instead, he substituted a tabulation of coefficients of frictional interactions between certain actual materials: iron, copper, red chalk, beechwood, ash, yew, and elmwood. It remains difficult, nevertheless, to see that any of this material, or any of the succeeding section on cordage, had a life longer than that of this memoir, which there is no evidence that anyone ever read again between the time that the Academy awarded the prize instead to Coulomb and the writing of the present study. The attempt left Carnot himself without illusions. The concluding paragraph to the experimental part reads:

> I do not, therefore, conceal from myself that a great deal remains to be done before the knowledge both of friction and stiffness in cordage leaves nothing more to be desired. If someone has given a fully satisfactory account of the topics I have just considered, he will have rendered a great service to practical mechanics, and certainly so difficult a task could be worthily rewarded only by the recognition of the Academy.[8]

As was to be expected, therefore, the interest lies again in the first section of Part II, which consists of the second version of his theory of machines. The discussion was lengthier than in the first memoir, 60 paragraphs occupying some 40 folio pages, though still brief compared to the *Essai sur les machines en général*. In it he achieved notable modifications–in clarification of topics, in the enunciation of principles and definition of quantities, and in the articulation he had given to what the historian can now recognize to be a new mode in the science of mechanics and not merely the groping of a young engineer with a rather special set of interests and a point of view still latent. Indeed, without needing to speak of discovery in the sense of things or relations altogether strange and unknown, we may be justified in taking this text for the most reliable index to the aspects of Lazare Carnot's work that were most characteristically his own. The year since the failure of his first memoir was not sufficient for him to have read extensively in the literature or to have mounted elaborate experiments. It did suffice, however, for him to see through his problems, to seize their generality, and to elaborate his point of view in coherent propositions.

[8]Gillispie (1971), Appendix C, § 100.

Resubmitting his memoir to a jury of the Academy, he might have been expected to put into it that which would have reflected most credit on his capacity to contribute to the subject. On the other hand, he was not then writing, as he tried to do a few years later in the *Essai sur les machines en général*, for extrascientific readers for whom his own findings needed to be embedded in elementary exposition of the basic commonplaces of mechanics itself.

On the hypothesis that what was in the 1780 memoir, therefore, was largely Carnot's own, let us see what the arrangement and development of the main motifs were. First, he specified the subject to be the properties common to all machines and began the discussion, not in medias res as in the first memoir, but with the operational derivation of the center–of–gravity principle for equilibrium in a generalized machine, the same with which he later, in the *Essai sur les machines en général*, repaired the deficiency he there exhibited in the traditional statement.[9] As for the so–called Principle of Descartes, Carnot here stated it in passing as a consequence without identifying it by that or any name, and went on to assure the generality of the center–of–gravity principle by remarking that all forces could in principle be reduced to the action of weights.

A brief and clear diagnosis followed of the impediment that mechanics had created for the science of machines in abstracting from their mass. Let machines be considered instead as inertial bodies, and their science like that of all mechanics would be reduced to the problem: given the virtual motion of a system of bodies, what actual motion would ensue in consequence of their mutual interactions? The problem thus posed simply and directly, so were the conditions for a solution. The first assumption, though not so stated, was the comparative anatomy of hard body and elastic body. The second was the sole principle that Carnot always held to have been requisite for a resolution: the equality in opposition of action and reaction, a law simple, "incontestable" (Gillispie 1971, Appendix C, § 111) and of universal applicability.

On that basis alone Carnot proceeded to derive the two fundamental Equations (E) and (F) (see Eqs. 2.1 and 2.2) of the *Essai sur les machines en général.*[10] The wording was identical except that the latter he here called an arbitrary rather than an indeterminate. In the *Essai sur les machines en général* he corrected himself and held that it was the values to be attributed to the geometric motions u of Equation (F) that were arbitrary, not the relation itself. More striking are passages adumbrating the concept of geometric motion, identical in phrasing to those in the *Essai sur les machines en général*. Their role was that of auxiliaries allowing the derivation of the general indeterminate (F) from the determinate (E) restricted to elastic interactions. Only one feature of the later introduction of geometric motions was missing, the afterthought of the *Essai sur les machines en général* wherein Carnot included in

[9]See above Chapter 2, pp 20–22.

[10]See above Chapter 2, pp 25–28.

the class of geometric motions those that, while involving no virtual displacement, did no work.[11] At first reading, it is surprising that Carnot should not have seen the need for that provision. It seems closer to the dynamical emphasis that foreshadowed the concept of geometric motions in the 1778 memoir[12] than does the kinematical way in which he introduced the idea explicitly here and in the *Essai sur les machines en général*. Nevertheless, the final development he gave the notion in the *Principes fondamentaux de l'équilibre et du mouvement* of 1803 was a protoenergetic one. It may be, therefore, that the successive installments on geometric motions exemplify three phases through which Carnot's thinking about the whole subject passed. In the first phase, that of the 1778 memoir, it was motive force that was primary; in the second, that of the 1780 memoir and much of the first part on mechanics of the *Essai sur les machines en general* it was displacement; and in the third, that of the latter part of the *Essai sur les machines en général* concerned with machines and of the *Principes fondamentaux de l'équilibre et du mouvement*, it was power or its exercise in moment–of–activity or work.

Such must almost surely have been the order in which he evolved the science of machines, for comparison point by point of the 1780 memoir with the *Essai sur les machines en général* shows that in the latter published version he superimposed the more theoretical work passages and principles upon the arguments of the unpublished memoir without rethinking or reshaping them. The contrast comes out most strikingly in the absence from the memoir of the feature that Carnot added to and emphasized most strongly in the *Essai sur les machines en général* the conservation of the moment–of–the–quantity–of–motion, or moment–of–momentum. That principle was there introduced, it will be recalled, in a curiously repetitive manner. First, the conservation of angular momentum appeared unnamed and merely as a sample solution to the problem resolved by employing the indeterminate Equation (F) (see Eq. 2.2). It would be reasonable to suppose that in that form it represents the early phase of Carnot's thinking. Then there followed in the *Essai sur les machines en général* (but not in the 1780 memoir) the digression just recalled about the pseudogeometric status of motions that do no work, upon which Carnot came back to conservation of the moment–of–the–quantity–of–motion now magnified and. elevated to the status of the fundamental conservation law. Stating it in the form of three propositions,[13] it will be further recalled, he went on to derive from it well–known theorems of mechanics followed by his own equation designated (AA) (see Eq. 2.3):

$$\int Fu \cos Z = 0,$$

[11]See above Chapter 2, pp 19–23.

[12]See above Chapter 3, p 54.

[13]See above Chapter 2, p 30.

in which F, representing force in the sense of motive power, was substituted in effect for the product mU, the quantity of motion "lost" to internal interactions in the so-called fundamental Equation (F)

$$\int muU \cos z = 0.$$

Not so in the 1780 memoir, which said nothing of moment–of–the–quantity–of–motion. Having defined geometric motions in point of displacement only, Carnot simply obtained Equation (F) for the general solution and expressed it in various ways adapted to different needs. One proposition resolved the quantities into Cartesian coordinates. A second exhibited the analogy to conservation of live force in hard body interactions and included a term for its loss in impacts. A third concerned motion changing by insensible degrees over time and reverted to the expression Carnot had obtained in the 1777 memoir, i.e.,

$$\int mupdt \cos z = \int mud\,(V \cos y).$$

A fourth concerned the case in which the value of u is the real velocity V and found that the expression then reduced to conservation of live force. All these consequences came out rather parenthetically, however. Carnot's discussion in the 1780 memoir placed no emphasis on the principles of mechanics and made much less of the later claim that geometric motions permitted transforming live–force conservation in elastic bodies into an expression applicable to generalized hard–body interaction. These matters figured here rather as a sequence of problems than a deductive chain of principles, for it was a feature of the way in which Carnot's ideas developed that he tended in later work to dress up in the guise of principles what he had initiated as the solution to problems.

Also missing in the 1780 memoir was the analysis of power which in the *Essai sur les machines en général* made the transition between Part I on the principles of mechanics and Part II on Carnot's own subject, "machines strictly speaking". The omission is consistent, for it was in order to prepare the ground for those considerations that Carnot would then introduce the moment–of–the–quantity–of–motion into the *Essai sur les machines en général* in the interval between 1780 and its publication in 1783. All these differences, therefore, reflect the limitation on his thinking at the time of the 1780 memoir: there at the point where he was later to move on to the concept of power and work, he fell back on the traditional notion of dead force (Gillispie 1971, Appendix C, § 128). When the question was one of live forces applied to machines, he observed, the theory was related to that of impact, which sufficed to solve such problems. But the theory of machines conventionally understood envisaged only the forces employed to move them, i.e., dead weights, animal power, and wind or water. Accordingly he would limit himself to these forces for the remainder of the theoretical part of his memoir.

Nevertheless, limited though Carnot's vision may thus have been at the stage of the 1780 memoir, the evidence is interesting in showing that it was in connection both with such conventional problems and also with the later, more elaborate notions that he developed the concept of work. For if he did not have it fully disengaged in 1780, neither was it entirely missing. The quantity itself, moment–of–activity in 1783 in the *Essai sur les machines en général*, was called "quantity of action" in the 1780 memoir. There it was always coupled with impelling or resisting forces (for those terms also made their appearance) as the quantity that they produced or consumed. Carnot never referred to it without the qualifier and expressly warned the reader against confusing his quantity of action consumed or produced with that which mathematicians following Maupertuis meant by quantity of action of a system. Of action in that sense he, Carnot, would make no use in the present memoir (Gillispie 1971, Appendix C, § 132), and it must have been the obvious inconvenience that led Carnot to modify the terminology in the *Essai sur les machines en général*. It may, further, have been the emphasis there placed on moments in general, particularly that of the quantity of motion, that led him to redesignate his basic quantity "moment–of–activity" and to consider it in itself and not merely as a measure of the impelling or resisting forces it was consuming or producing.

Lacking a fully disengaged concept of power or work, together with the principle of conservation from which to deduce it, Carnot gave the sequence of propositions about machines themselves more episodically in the 1780 memoir than in the *Essai sur les machines en général* and failed to draw out of them his own principle of continuity, which appeared only in the form of a practical injunction in passages constituting the draft of the later scholium. In the *Essai sur les machines en général*, it will be re–called, he stated a single theorem for machines:

> Whatever the state of rest or motion of any system of forces applied to a machine, if a geometric motion be imparted to it without altering the forces, the sum of the product of each by the velocity of the point of application, evaluated at the first instant and in the direction of the force, will be zero.[14]

Or

$$\int Fu \cos z = 0,$$

from which he drew all the corollaries comprising the science of machines "strictly speaking", the fifth being what we have called the Work Corollary:

> *In a machine of which the motion is changing by insensible degrees*, the moment–of–activity consumed *in a given timeby the impelling forces* equals the moment–of–activity exerted *in the same time by the resisting forces*.[15]

---

[14] «THÉORÈME FONDAMENTAL. Principe général de l'équilibre & du mouvement dans les machines. XXXIV» (Carnot 1786, p 68).

[15] Carnot (1786), § XLI, pp 75–76.

The 1780 memoir had yet to achieve that unification of his point of view. In the absence of conservation of the moment of the quantity–of–motion, and therefore of work, he stated two parallel theorems for the science of machines, the first for equilibrium and the second for motion. Thus the former:

> When a machine is in equilibrium, if an arbitrary geometric motion be imparted to it without altering the applied forces in any way, the quantity of action produced in the first instant by the impelling forces will be equal to the quantity of action produced in the same infinitely short time by the resisting forces.[16]

That proposition he then stated in various ways (one of which is virtually identical to the phrasing just cited from the *Essai sur les machines en général* except that it did not embrace a machine in motion) and drew from it several corollaries. In those concerned with weight-driven machines, the equivalence of MgH to $\frac{1}{2}MV^2$ came out repeatedly.

The companion theorem concerning machines in motion was stated more awkwardly:

> If the actual motion of a machine is suddenly converted into any other geometric motion and the machine is left to itself, the conservation of live forces will obtain throughout the ensuing motion no matter what changes there may be in the motive forces.[17]

The corollaries with which Carnot followed this proposition were all correct, and the second of them essentially amounted to the Work Principle of the *Essai sur les machines en général* in the following form:

> When a machine is in uniform motion (i.e., when each point of the system has a constant velocity) the quantity of action produced in a given time by the impelling forces is equal to the quantity of action produced in the same time by the resisting forces.[18]

The statement was weaker than in the *Essai sur les machines en général,* but already Carnot identified it as the most useful of all the propositions in the theory of machines. The historical question that poses itself, therefore, is what difficulties still remained in his mind that prevented his having combined motion and equilibrium into a single principle as he seemed quite naturally able to do in the *Essai sur les machines en général*?

Several possible explanations suggest themselves. One is that lacking moment–of–momentum and thinking in terms of displacements, he had not seized on all the advantage that his geometric motions conferred in authorizing him to neglect internal forces of the system. That might be why he followed the theorem with this otherwise unnecessary justification–since live forces were conserved in a

---

[16] «*Théorème 1er*. Principe général de l'équilibre dans les machines» (Gillispie 1971, Appendix C, § 133).

[17] «*Théorème II*. Principe général du mouvement dans les machines» (Gillispie 1971, Appendix C, § 140).

[18] «*Corollaire II*. Des machines qui se meuvent uniformément» (Gillispie 1971, Appendix C, § 143).

system of which the motion was changing by insensible degrees, and since when a geometric motion was substituted for a real one, nothing was changed in the mutual interactions of the system, it followed that motion must still be changing by insensible degrees, and hence live forces must still be conserved "[...] at least for some time [...]" (Gillispie 1971, Appendix C, § 140) after such a transformation, no matter what motive forces should be substituted for those that were really influencing the parts of the system.

Another possibility is that Carnot had not quite succeeded in generalizing in his own mind relations that he had obtained from considering the behaviour of weights under gravity. That might be why the corollaries are clearer than the statement or discussion of the theorem itself, notably the first which concerns weight–driven machines:

> If the actual motion of a weight–driven machine be suddenly changed into any other geometric motion whatever, and the system be left to its own forces, the ensuing sum of the live forces at any instant is equal to the sum of the initial live forces (i.e., immediately after the change of motion) plus what the sum of the live forces would be if each point of the system had a velocity equal to that due to the height from which the center of gravity has fallen since the change of motion.[19]

Perhaps, finally, the obligation to conduct the argument toward a frictional application concealed the desirability of establishing the work principle in a unified treatment of motion and equilibrium. The reason he advanced in the 1780 memoir for considering it the most useful for the theory of machines was that in practice such ordinary devices as mill–wheels, pulleys, and capstans, once set in motion, soon reached a steady rate of operation. A final corollary brought all real machines within the scope of the discussion by specifying that friction was always to be classified as a resisting force, since its direction inevitably countered the real motion of the points whereon it acted.

It would be misleading, however, to emphasize shortcomings on the formal level of argumentation, and it is more illuminating historically to appreciate that what Carnot lacked as principle in the 1780 memoir, he already had as maxim. The final paragraphs of the theoretical section of the 1780 memoir were largely similar and in certain places identical with the *scholium* that concluded the *Essai sur les machines en général.* An interesting passage reveals the latency of his point of view. He had reverted to machines of which the state of motion changes by insensible degrees and pointed out that they had in common with periodic machines (i.e., pendulums) that live force was conserved and that in a given time, therefore, the quantity of action produced by the impelling forces equals that produced by the resisting forces. Even worthier of remark, a quantity one–half the increase in the sum of the live forces is identical with the quantity of action produced by the force of inertia of the bodies comprising the system. It followed that:

> [...] if we consider all forces, whether active or passive, applied to a machine in motion, including even the inertia of its matter taken itself as a real force that resists change of

[19]«*Corollaire Ier*. Du mouvement dans les machines a poids» (Gillispie 1971, Appendix C, § 142).

> state in the bodies, the quantity of action produced by the impelling forces in a given time will always equal the quantity of action produced in the same time by the resisting forces whatever be the motion of the machine [...].[20]

Here then, noted in passing in the 1780 memoir, was precisely the remark that he stated rather as a theorem in the *Essai sur les machines en général*, and that there appeared to be the principle in virtue of which he unified the discussion. Indeed, he even noted in the memoir that the entire theory of machines whether at rest or in motion could be reduced to that single statement. He simply did not yet do it, which shows that what he really needed to develop was less the content of theorem or propositions than the line of reasoning that connected them.

As for his own principle of continuity, it followed upon a discussion virtually identical with that in the *Essai sur les machines en général* about varying the factors of time, force, and velocity in achieving mechanical advantage. A footnote distinguishes between "[...] absolute work or labor of the agent [...]"[21] and "what I call here the work of the moving or impelling force [...]" the difference residing in the extra force consumed by the inertia of the machine itself. The things that he mainly had in view in all these remarks were the classical simple machines and their variants used on shipboard. It was for this reason, Carnot said further, that he had written only about those of which the motion changes by insensible degrees.

> For when the purpose is to produce the greatest effect possible from a machine, there is a decisive advantage to be gained in excluding percussion or sudden changes in the state of motion.[22]

Carnot went on to show how the intervention of impacts necessarily involved loss of a corresponding quantity of action. The argument was in much the same terms as that reproduced 3 years later in the *Essai sur les machines en général*, but the mode in the 1780 memoir was one of explanation rather than demonstration or proof (Gillispie 1971, Appendix C, § 157). It is, indeed, the most revealing feature of the document that this point, which later appeared to be the chief finding of Carnot's mechanics, so much so that it was called his law or principle, he first wrote down not as discovery of an unknown truth, but as a justification for limiting his subject to machines that in their operation embodied it.

For the rest, the 1780 memoir presents little interest. Like that of 1778 it concluded with a section purporting to apply the experimental results to determining expressions for each class of simple machine. Carnot had simplified the treatment over that of the earlier memoir by combining consideration of motion and equilibrium in each class of machine–all the more curious does it seem, therefore, that he did not do this in the preceding theoretical discussion–rather than by going through all seven classes once for the case of equilibrium and again for that of motion. He

---

[20] «*Scholie*» (Gillispie 1971, Appendix C, § 148).

[21] «[...] le travail absolu ou la peine [...]» (Gillispie 1971, Appendix C, § 153, footnote).

[22] Gillispie (1971), Appendix C.

recognized that his organization had the effect of separating the theoretical from the practical study of machines in general. No doubt the terms of the Academy's competition imposed on him the necessity thus to descend to particular cases for the practical part. But it does not appear that he had formed the project of such a science of machines as would dispense him from specifying particular conditions of motion and equilibrium characteristic of each class in practice. The separation of theory and practice was necessary, he observed, "[...] in order to avoid the confusion that would arise from mingling them [...]" (Gillispie 1971, Appendix C, § 191).

We can see that he had in mind with greater or less cogency all the relations or theorems that would comprise such a science. What he had still to accomplish was to graduate in his thinking from forces to powers and to seize on the conservation of the moment–of–momentum in order to make full use of his own idea of geometric motions in abstracting from the effects of internal forces in a system and hence overcoming the disjunctions between hard and elastic body, pressure and impact, dead force and live force. In a way that was conceptually subsidiary but probably more important in actuality, he needed to generalize from the manipulation of weights, whether as load or power, in order to embrace within his reasoning all processes involving work.

Such, at least, was the development that he gave his subject between the failure of his second memoir to win the academic prize in 1781 and the publication in 1783 of his *Essai sur les machines en général*, in which all the experimental and particular passages were eliminated, and the theory of machines alone was presented rather as an adaptation of the science of mechanics itself. It must, therefore, have been during this interval that he reviewed the whole science as it was known to an engineer in the late eighteenth–century, though how and through the agency of precisely which writings in that science he then seized on the unifying possibilities inherent in the principle of moment–of–momentum, the sources do not permit us to say: for the rest of it, all the development he gave was latent in his own approach in the two memoirs successively submitted to the Academy.

3.3 Argument of the *Principes fondamentaux de l'équilibre et du mouvement* (1803)

The remaining discussion of Carnot's science of machines will concern its immediate development and influence. Let us take first the revision that he himself gave the subject in his *Principes fondamentaux de l'équilibre et du mouvement* (Carnot 1803a) published just 20 years after the *Essai sur les machines en général* and in the same year with the *Géométrie de position* (Carnot 1803b). Most of the later writers who allude to Carnot's mechanics make reference to the *Principes fondamentaux de l'équilibre et du mouvement* rather than to the *Essai sur les machines en général*, which, if mentioned at all, is not always distinguished from the second, more renowned version.[23] Meanwhile, the development of the eighteenth–century French school of rational mechanics had culminated in the publication in 1788 of Lagrange's *Mécanique analytique*, a work that imposed itself immediately as the mathematical chef d'oeuvre of its subject. Lagrange's book was no isolated monument. On the contrary, the last two decades of the eighteenth century were a period of intensive activity in the formalization of mechanics, so much so that Carnot himself remarked in the preface to his *Principes fondamentaux de l'équilibre et du mouvement* that in the interval since his earlier writing, there had appeared so many others "[...] on all aspects of mechanics, so beautiful and so extensive that my own can scarcely be remembered" (Carnot 1803a, p v). (So far as can be determined, his modesty was accurate. The present writer has never seen a single mention of the *Essai sur les machines en général* in the contemporary literature of mechanics, nor any evidence that it was ever read).

Why, then, was a second edition wanted? Carnot justified it on the grounds that the *Essai sur les machines en général* had contained some ideas that were new when they first appeared, and that it was in any case always useful to envisage the fundamental truths of science from various points of view. Of the new version he said further, "[...] several scientists have strongly urged me to furnish it" (*Ibidem*). Unfortunately, we do not know who they were. Did they do so because Carnot was now a prominent statesman recently emerged from political eclipse and influential in the Institute? It would be unrealistic to suppose that such facts carried no weight, although neither is it necessary to impute base and much less baseless flattery to Carnot's advisers. Statesman though he now was, he was a statesman with a certain scientific reputation in virtue of his mathematical writings. He had published his book on the calculus in 1797 and his first book on geometry in 1801.

Are we to suppose, then, that the new engineering emphasis in technical education might have been thought to create a public for a book on mechanics from the point of view of a former engineer, since become famous? So it almost certainly was in Carnot's own mind, for the *Principes fondamentaux de l'équilibre et du mouvement*, unlike the *Essai sur les machines en général*, is at once textbook and treatise in the fashion created by the *École polytechnique*. The former quality flattens Carnot's already unassuming style and puts the book in the shadow when contrasted to the brilliant light in which the works of Lagrange and his school are illuminated by their formal elegance. At all events it must have been some such combination of personal and institutional motivations that encouraged Carnot to revise his science of machines. The explanation cannot be that his approach was

[23]See, for example, the following works: *L'entropie. Son rôle dans le développement historique de la thermodynamique* by Brunold (1930, pp 37–38), *Some factors in the early development of the concepts of power, work, and energy* by Cardwell (1967, pp 209–224), *Histoire de la mécanique* by Dugas (1955, pp 324–331) and *Lectures de mécanique. La mécanique enseignée par les auteurs originaux* by Jouguet (1924, II, p 72).

an answer to the problem structure in the science of mechanics itself, for (as will appear) if his work was no longer unknown, it was another 20 years before these concerns began to make any difference.

To turn to the content of the *Principes fondamentaux de l'équilibre et du mouvement*, it will be convenient to take it in its textbook aspect first, for Carnot had completely reorganized his presentation. The division between fundamental principles of mechanics and the theory of machines "properly speaking" had disappeared. The subject was still presented in two parts, but the distinction was now between its experiential and its rational aspects. The former reads like a glossary of mechanics. It comprised mainly definitions of the quantities and expositions of elementary laws of statics and dynamics for which Carnot claimed the sanction of experienced fact. Part II by contrast contained propositions that could be derived from those first principles merely by ratiocination–of them more in a moment, for they were those that were peculiarly Carnot's own either in conception or in function.

In observing that Part I now reads like a primer, the historian should intend nothing denigratory. It was a good primer. No franker source exists if he wishes to know what persons literate in mechanics at the end of the eighteenth century understood physically by its basic conceptions, such as force, motion, action, velocity, acceleration, etc.[24] Like many of the simplest matters generally taken for granted, these are not always easy to come by historically. In this respect, the *Principes fondamentaux de l'équilibre et du mouvement* exhibits a notable improvement in clarity over the *Essai sur les machines en général*, in which the modern reader often has to judge of Carnot's comprehension of terms and dimensions by the use he made of them rather than by an explanation. In the later work, for example, the discussion of projection and combination of directed quantities was explicit. It is extremely curious, however, that Carnot seems to have gone all the way back to his earliest, 1778, memoir for these proto–vectorial considerations[25] (Gillispie 1971, Appendix B, § 52–54). Was it his intervening work in geometry that encouraged him to see the merit of this representation, which he had merely taken for granted in the intervening work in mechanics? It may be so. At all events Carnot suggested the following notation for representing the projection Aa' of one velocity Aa upon an intersecting straight line AB:

$$\overline{Aa'} = \overline{Aa} \cos \overbrace{\overline{Aa} \quad \overline{AB}},$$

wherein the last term denoted the angle, a convention that would have been obvious although cumbersome if generally adopted. Here, for example, is his expression for

[24]Please see above Chapter 2, p 19, ft 7.

[25]The geometric figures missing from the 1778 Ms. may be reconstituted from Figs. 2, 3, and 4 of Pl. I of the *Principes fondamentaux de l'équilibre et du mouvement,* which fit the argument of the 1778 memoir.

the relation between initial velocity ($\overline{MW}$), altered velocity ($\overline{MV}$), and the velocity ($\overline{MU}$) "lost" when a body was constrained to change its motion[26]:

$$\overline{MW} = \overline{MV} \cos \overline{MW} \,\widehat{\quad\quad}\, \overline{MV} + \overline{MU} \cos \overline{MW} \,\widehat{\quad\quad}\, \overline{MU}$$

Writing of moments, Carnot now pointed out that the moment of a motive force, being the product of a force by a line, could always be reduced to a live force, and that the moment of a quantity–of–motion could be reduced to the product of a live force and a time. He now had clear that the latter was properly designated quantity–of–action of a mass. The concept of work he still called moment–of–activity, but his way of seeing these quantities seems to represent a certain development in the differentiation of the notion of energy (live force) from that of work (moment–of–activity). He distinguished between moments–of–activity "consumed" or "absorbed" (instead of consumed or produced). By the former he still meant the product of the force by the path described by its point of application, evaluated in the direction of the force; whereas the latter was now the product of the force by the velocity of the point of application, evaluated in the direction opposite to that of the force. It still comes out that the two quantities are identical numerically and of opposite sign since the angles are supplementary, but the distinction seems closer than it had done in the *Essai sur les machines en général* to that in later usage between the work done on or by a system (Carnot 1803a, §§ 60–61, pp 38–39). For there is the beginning of a distinction here between notions that in the next 50 years would evolve into the concepts of kinetic and potential energy. He laid it down that live force "properly speaking" had the dimensions MV^2 (as we would write it) and that when it took the form of the product of a motive force by a line (*PH* as he wrote it, or *mgh* as we would write it), it could be given the special name of "latent live force" (*Ivi*, § 64, p 41). In the latter case it was dimensionally identical with moment–of–activity (or work), and the significant thing is that Carnot saw it as conceptually distinct. For a system of bodies in which forces were operating over time, the following expression

$$mp \cdot udt \cos k$$

would give the moment–of–activity consumed during time *dt* by the force *mp* with respect to the velocity *u* (*k* being the angle between *u* and P). The moment–of–activity consumed by the whole system throughout the entire motion would then be given by

$$\sum \int mp \cdot udt \cos k$$

[26]For the Eq., see Carnot (1803a), § 43, p 26.

(of which expression one might note that Carnot had begun to distinguish between summation and integration, and also that he had lapsed into his old notation). That expression being a latent live force, it could be reduced to a quantity of the form MV^2:

> Whence it is easy to appreciate that the notion we have just given of moments–of–activity is encountered frequently in the theory of equilibrium and motion, either in the form of live force properly speaking, or in that of latent live force.[27]

From the definitions Carnot turned to a summary of what he now called "The hypotheses that can be admitted as general laws of equilibrium and motion" (Carnot 1803a, p 46), which he pretended to found on experience and reasoning, and which were in fact simply the classical laws of statics and dynamics. It would elucidate nothing not already evident about Carnot's reasoning to enumerate these propositions. Two remarks only may be curious to note. The first reflects on Carnot's own sense of historicity. He here attributed the principle of virtual velocities to Galileo, even calling it the "principe de Galilée", and went on to credit the lowest center of gravity condition for equilibrium to Torricelli as its consequence before rectifying it in his own operational statement. The attribution to Torricelli was in fact correct, though not that to Galileo, and one has the impression that even in the rather trivial respect of acknowledging the history of the concepts, Carnot's reworking of the book was shaped by the example of Lagrange. The second remark about Carnot's presentation of the elements is that he concluded it by deriving from the conservation of live force the relations that he needed in the various cases of change of motion. As usual, W was initial, V final velocity, and U the velocity lost in impacts. Then for hard bodies in the general case,

$$\int MW^2 = \int MV^2 + \int MU^2,$$

for hard bodies in which the motion changes by insensible degrees,

$$\int MW^2 = \int MV^2 \quad \text{(since } U \text{ is infinitesimal)}$$

For elastic bodies, he comes at it in reverse order. Since

$$\int MV^2 = \int MW^2 - \int MU^2,$$

and since the velocity U lost at impact is restored on rebound,

$$\int MV^2 = \int MW^2.$$

[27] Carnot (1803a), § 64, p 41.

These are statements remarkable only in that he could now thus simply set them down one after the other (Carnot 1803a, §§ 131–132, pp 104–105).

In Part II of the *Principes fondamentaux de l'équilibre et du mouvement* Carnot turned to a formal development of the so–called hypotheses considered as laws of nature. It is difficult, on the whole, to think the presentation felicitous. Ostensibly the goal was to achieve analytical expression for the laws, and the result was that the distinctiveness that Carnot brought to mechanics by directing attention to the study of machines was obscured by his attempt (or so it would seem) to imitate in some degree the style of Lagrange. The sequence of theorems swept so wide of the subject of machines that its denouement there would come as something of an anticlimax except that the climax itself, a generalization or vindication of the Principle of Least Action, was somewhat beside the point.

At the same time it would be wrong to underestimate all the theoretical aspect of the *Principes fondamentaux de l'équilibre et du mouvement* just because this part of the book may not have been an entire success or because he had already made all the significant points in the *Essai sur les machines en général*. He did achieve a greater clarity, most notably in the passages defining geometric motion:

> Any motion that, when imparted to a system of bodies, has no effect on the intensity of the actions that they exert or can exert on each other in the course of any other motions imparted to them, will be named geometric.[28]

Clearly, he had modified the definition, for he states it entirely in terms of the function of geometric motion, which it is now easier to recognize as what in later parlance would be called virtual displacements. Neither in the 1780 memoir nor in the *Essai sur les machines en général* had Carnot adapted his concept of geometric motions from the principle of virtual velocities. In the *Principes fondamentaux de l'équilibre et du mouvement*, however, he went on to recognize the analogy between such motions and that principle in the use Lagrange made of the latter. The difference was that virtual velocities being infinitesimal were inapplicable in problems involving the sudden changes of motion produced by impact, whereas geometric motions, being finite, generalized the principle so that it might apply to all cases, whether the change of motion be continuous or discontinuous, whether the bodies be hard, elastic, or fluid (Carnot 1803a, § 144, pp ix–x, p 115).

It is because of these passages that the historian will detect in the concept of geometric motion the forerunner if not the common ancestor of reversible processes and vector analysis, the former in that reversibility was the criterion by which the independence of such motions from the rules of dynamics might be recognized, and the latter in that these motions were, therefore, determined by the geometry of the system alone. Carnot nourished a prophetic if not always lucid vision of the prospect for a science that should be "[...] intermediary between ordinary geometry and mechanics" (Carnot 1803a, § 162, p 131). Both in the *Principes fondamentaux de l'équilibre et du mouvement* and in the *Géométrie de position* he alluded to hopes for

[28] «DÉFINITIONS» (Carnot 1803a, § 136, p 108).

thus uniting mechanics and geometry in a science yet to be created. It was mainly for lack of a theory of geometric motions that analytical difficulties frequently impeded the solution of mechanical problems. Carnot thought this to be specially true of hydrodynamics (Carnot 1803a, § 145, p 116), and the remark goes to strengthen the case for seeing in geometric motions the origin of reversible processes, since it was in connection with the fluid model of heat transfer that Sadi Carnot introduced the concept of reversibility.

As an earnest of such a theory, Carnot opened the "rational" part of his *Principes fondamentaux de l'équilibre et du mouvement* with theorems, of which the first nine about the properties of geometric motions in hard–body interactions. The sequence will prove somewhat puzzling to the reader not yet oriented to his purposes in mechanics itself. When he did finally come to that, he unfolded the argument in a manner different from that of the *Essai sur les machines en général.* Both editions prepared the ground for the application of mechanics to machines with the proposition that in the motion of a system of hard bodies, the net effect of mutual interactions of the parts is zero. In the *Essai sur les machines en général* (it will be recalled) Carnot stated that result in the guise of the general solution to the problem "given the virtual motion of any system of hard bodies, [...] find the real motion it will assume in the next instant"[29] and he symbolized the solution in the Equation (F) (see Eq. 2.2)

$$\int muU \cos z = 0.$$

which he had obtained through transforming an expression equivalent to conservation of live forces by means of the auxiliary idea of geometric motions. But in the *Principes fondamentaux de l'équilibre et du mouvement*, having elaborated the concept of geometric motions in a series of propositions, he stated the same thing in the form of a theorem:

> In the impact of hard bodies [...] the sum of the product of the quantities of motion lost by each of the bodies multiplied by its velocity after impact evaluated in the direction of the quantity of motion lost, is equal to zero.[30]

And he symbolized the relation in a mere corollary[31]

$$\sum MUV \cos \widehat{UV} = 0$$

The difference is indicative of that between the two books: in the *Principes fondamentaux de l'équilibre et du mouvement,* Lazare Carnot set out in a didactic order what in the *Essai sur les machines en général* he had established in a

[29] «*Problème* XX» (Carnot 1786, § XX, p 40; see also: *Ivi*, § X, p. 21).

[30] «THÉORÈME XI» (Carnot 1803a, § 168, p 139).

[31] For the Eq., see Carnot 1803a, § 169, p 143.

problematic one. A consequence was that he himself began the process of obscuring the distinctiveness of his own approach while improving its clarity. There was about the *Principes fondamentaux de l'équilibre et du mouvement*, moreover, not only an imitative quality in its relation to Lagrange but a regressive quality in its relation to theory. Carnot now tended to look back to the *Principle of Least Action* for the significance of his work concepts rather than forward to energy considerations. That he should have done so is natural enough for he never claimed to be an innovator, but to bring it out will require exemplification in a certain amount of detail.

Following the theorem just cited, a further proposition stated that in hard-body collision the sum of live forces prior to impact equals that following impact plus what the sum of the live forces would be if each of the bodies were moving freely with the velocity lost in impact:

$$\sum MW^2 = \sum MV^2 + \sum MU^2.$$ [32]

The point to notice here is that Carnot did not yet hold steadily in mind a concept of live force equivalent to kinetic energy. Sometimes what bodies were said to lose in collision was velocity or motion; sometimes it was live force itself, as in a corollary bringing out that the sum of live forces after collision was bound to be less than beforehand–but the case was restricted to hard body. He did have clearly in mind the convertibility of the quantities expressed in the dimensions of live force (kinetic energy) into those of moment–of–activity or latent live force (work or potential energy), but he was thinking of these matters dimensionally rather than conceptually. Only the concept of work did he have fully developed in all but name, and sometimes even in name. The role of energy he had to compound out of forces acting over distance (usually though not exclusively when it was a question of the measure of the force) or in time (usually though again not exclusively when it was a question of the operation of a machine process), and that necessity may explain why, the more he reflected, the more prominent the Principle of Least Action became in his thinking.

As he had done in the *Essai sur les machines en général*, Carnot turned to Least Action for the purpose of identifying the real motion among the infinity of geometric motions of a system of hard bodies. The motion that actually occurred would be that for which the sum of the products of the masses by the square of the velocity lost by each, was a minimum, or to put it more generally, such that

$$\delta \int MU^2 = 0.$$

Carnot considered that to be very beautiful and the fundamental law of hard–body collision. It might, he went on to show, be extended to cover elastic and even

[32]Carnot (1803a), §175, pp 145-146.

partially elastic bodies. When thus generalized it took the form,

$$\delta \int MUX = 0$$

(where X is the distance traversed by mass M with velocity U in time t), and this quantity, he pointed out, was that which Maupertuis had first denominated *Action*. Indeed, Carnot's view of his own accomplishment was that in the *Essai sur les machines en général* he had been the first to establish its applicability to the case of sudden or discontinuous change of motion in hard–body collision. Maupertuis had envisaged the matter so vaguely that he made no distinction between continuous and discontinuous change of motion. Euler had then distinguished the former case and made of it a rigorous, if still metaphysically justified, principle applicable to motion under central forces of attraction (Carnot 1803a, § 188, pp vi–ix, pp 163–164). Now in the *Principes fondamentaux de l'équilibre et du mouvement* Carnot thought himself to be exhibiting how superfluous metaphysical considerations were, and how generally the principle applied – only he did think preferable the form

$$\delta \int MU^2 = 0,$$

and that is the tantalizing aspect in later eyes since it does look very like the energy statement it could not yet be. He had not abstracted a notion of live force equivalent to energy from the properties of bodies: live force was lost completely in hard–body collision, partially lost in partially elastic collision; and Carnot was relying more fundamentally than ever on his workless geometric motions to bridge these differences. At the same time, he gave greater emphasis to the work principle at the expense of moment–of–momentum. In the *Essai sur les machines en général*, it will be re–called, conservation of the moment of the quantity–of–motion was the fundamental principle of mechanics sufficient for derivation of all laws of motion and equilibrium in all cases whatever.[33] Not so in the *Principes fondamentaux de l'équilibre et du mouvement*, where he invoked in its place a series of propositions about moments of rotation, moments of percussion, and d'Arcy's Principle of Areas,[34] and singled out instead as the fundamental conservative quantity the moment–of–activity of the general system with respect to any of its possible geometric motions, i.e., conservation of work (Carnot 1803a, § 197, p 176). But that appears to be something of an afterthought following the enthusiastic vindication of Least Action.

A final section headed "Considerations on the Application of Moving Forces to Machines" (Carnot 1803a, p 227) gathered together the remarks about machines

[33] See above Chapter 2, p 30.

[34] One wonders whether the papers of the chevalier Patrick d'Arcy had been an important source of inspiration to Lazare Carnot, but this is the only place he mentions them (Carnot 1803a, § 195, p 174). The relevant memoirs are: *Problème de dynamique* (d'Arcy 1752, II, pp 344–356) with an addendum (*Ivi*, pp 356–361), *Suite d'un mémoire de dynamique* (*Id.*, 1754, pp 107–108) and *Théorèmes de dynamique* (*Id.*, 1763, pp 1–8).

that Carnot had interspersed in the earlier edition to explain his reasonings, and combined them with the analysis of mechanical advantage and exhortations about the employment of power from its concluding scholium. The effect for the reader who knows only the *Principes fondamentaux de l'équilibre et du mouvement*, is to reduce to the level of an appendix what the reader of the earlier drafts can see to have been the motivation of the work as a whole. He had added nothing new to the earlier discussion about the factors of mechanical advantage or the way in which they might profitably be varied according to the economics of the process. He did here make one of his few references to the work of others, dissenting from Daniel Bernoulli's argument that a power source yielded much the same result whether the operator chose to augment the force or the speed. In the case of manpower and animal power, experiments by Coulomb had invalidated that hypothesis for the reason that it took no account of fatigue or boredom.[35]

A little further on Carnot also attributed to Daniel Bernoulli the maxim that in any proposed operation, the first thing to do is examine it in order to specify what effect intrinsically pertains to that operation, and avoid so far as possible producing any side effects. No work of Bernoulli was cited, but Carnot now said that it was pursuant to this principle that all shocks or sudden changes that are not essential to the construction of the machine are to be avoided, for shock always involves loss of live force and consumption to no purpose of a portion of the moment–of–activity developed. The only thing surprising here is that Carnot should thus have casually read back to Daniel Bernoulli the finding generally credited to himself, which remark illustrates further that what was involved in the development of engineering mechanics was less the knowledge of how things work than the articulation of the principles that all informed people knew more or less explicitly to be involved in processes.

In general, the argument about machine processes is briefer and easier to follow in the *Principes fondamentaux de l'équilibre et du mouvement* than in the *Essai sur les machines en général* or in the draft memoirs, and the concluding passage deserves quotation in order to exhibit further the gradual clarification in ideas that this edition reflects in what can be seen as the unwitting approach to energy considerations by way of the concept of work. After observing that the effect produced is always a live force, real or latent; comparable to the product PH of a weight P by a height H, or a force by a line, and that the moment–of–activity is, therefore, always the quantity to be economized for maximum effect whatever the process, whether it be a weight to be raised; a mill–wheel to be turned; a void to be created in the atmosphere, the sea, or some confined fluid; a machine to be started; or a system of bodies attracting each other in any proportion to the distances; and

[35]Carnot cited Coulomb's memoir on the employment of manpower, *Résultat de plusieurs expériences destinées a déterminer la quantité d'action que les hommes peuvent fournir par leur travail journalier suivant les différentes manières dont ils emploient leurs forces* (Coulomb 1799, II, pp 340–428). Coulomb first read this memoir to the Academy in 1778. See Gillmor (1971), chap. II. Carnot cited further a memoir of Euler, *de machinis in genere* (Euler 1751, III, p 254).

also whatever the motive agent, whether weights, wind, water, men, or animals in any combination at all, then:

> [...] whatever change may be occasioned in the system, the moment–of–activity consumed in a given time by the external powers always equals one–half the amount by which the sum of the live forces increases in the system to which they are applied during that same time, minus one–half. the amount by which that same sum of live forces would increase if each of the bodies moved freely on the curve it would describe, supposing it to be subjected at each point on the curve to the same force that actually does affect it: provided always that the motion changes gradually and that if spring–driven machines are involved, the springs at the end of the process are left in the initial state of tension.[36]

3.4 Comparing the Work of Sadi Carnot

Let us turn now to the memoir by Sadi Carnot, *Réflexions sur la puissance motrice du feu et sur les machines propres à développer cette puissance*, published in 1824, the year after his father's death and 21 years after the *Principes fondamentaux de l'équilibre et du mouvement*. The similarity has often been remarked between Lazare Carnot's observations there on hydraulic machines and his son's model attributing motive force or power to the passage of heat considered as a real fluid falling from a higher to a lower level of temperature (Brunold, pp 37–40). In fact, however, when Sadi Carnot's memoir is read in direct succession to Lazare's writings, the son's inheritance appears fuller than the mere adoption of an hydraulic model for the flow of heat. Indeed, the *Réflexions sur la puissance motrice du feu* may be taken both for the foundation stone it certainly became in the science of thermodynamics and for the final item in a series of Carnot memoirs on the science of machines beginning with the prize essay of 1778. It is the latter relation that concerns us here.

There is no reason to think that Sadi Carnot would have objected to such an attribution. It is one that may be supported on biographical as well as substantive grounds. Graduated from Polytechnique in October 1814 (after having fought with many of his classmates in the brief and vain defense of Vincennes in March), Sadi Carnot completed his training with two further years of study at the school of military engineering in Metz. Until 1819 he led the garrison life of a second lieutenant. He then arranged to go on inactive duty in Paris in order to devote himself to study and technical research. All the while his father and younger brother were in exile in Magdeburg, where Sadi Carnot was able to visit them for a few weeks in 1821.

It was evidently after this visit that Sadi Carnot began concentrating his attention on the principles of heat engines, and first the steam engine. In addition to the published *Réflexions sur la puissance motrice du feu*, there is extant the manuscript of a brief "recherché" entitled "Investigation of a formula for the Motive Power

[36]Carnot (1803a), § 293, pp 261–262.

of Steam".[37] Scholars hold differing opinions about the date of this memoir. To the present writer it seems probable that it was composed somewhat prior to the *Réflexions sur la puissance motrice du feu*. It was in a genre of which there were many examples in the years around 1820, and which will be discussed briefly in the next chapter. Sadi Carnot obtained a value for the motive power of 1 kg of steam expressed as a function of temperature by employing Clément's law for the pressure of saturated vapors and Dalton's table relating vapor pressure to temperature. The discussion was clear and the derivation clever. In the interests of generality, he identified three stages in the operation of a steam engine: as the steam passes into the cylinder, it may be considered as expanding isothermally in the first phase and adiabatically in the second while the third stage was isothermal compression in the condenser. (These terms were not yet coined). But though recognizing elements of Sadi Carnot's analysis, the reader of the *Réflexions sur la puissance motrice du feu* will see this schematization as a partial one. The cycle was complete only with respect to the return of the piston to the starting point and not with respect to the temperature of the steam.

What appears to be a residue of that incompleteness marred the reasoning in the opening part of the argument of the *Réflexions sur la puissance motrice du feu* itself. It seems probable that the closer attention Sadi Carnot might reasonably be expected to have paid to his father's work on receiving in that same year the news of Lazare's death was what showed him the way to overcome that incompleteness and to put his argument on a fully general basis. We know from Hippolyte's memoir of his brother that when he, Hippolyte, returned to Paris after Lazare died in 1823, he found Sadi Carnot at work on the manuscript and was made to read and criticize important passages in point of their comprehensibility to general readers.[38] The brothers could scarcely have failed to talk then of their father's science.

The very word "Réflexions" in Sadi Carnot's title recalls the ruminative vein of Lazare's book on the calculus, while in analysis and subject matter the genre was that of his science of machines. Like the *Essai sur les machines en général*, the *Réflexions sur la puissance motrice du feu* was a treatise nonetheless rigorous for being verbally expressed and nonetheless general for being an adaptation of science to the principles underlying the employment of machinery. Despite the gratifying state (Sadi Carnot began) to which the development of steam engines had attained in practice, there existed no theory in the light of which their further improvement might be guided. Was there any limit to the power that ingenuity might draw from heat? The question was rhetorical and implied that there must be. What, then, were the optimum conditions for the design and operation of heat engines? Might not

---

[37] *Recherche d'une formule propre à représenter la puissance motrice de la vapeur d'eau*– the manuscript, *Un manuscrit inédit de Sadi Carnot,* was published by Gabbey and Herivel (pp 151–166). For a discussion of the question surrounding the date, see footnote 42 below.

[38] Hippolyte's biography is contained in a reprinted edition of the *Réflexions sur la puissance motrice du feu* published in Paris in 1878. References in the notes that follow are to the pagination of the facsimile of the first, 1824, edition published by Blanchard in 1953 (Carnot 1953).

other materials prove preferable to steam for developing the expansive force of vapor? Atmospheric air, for example–and this possibility, to which Sadi Carnot recurred often enough to confirm its appeal to his imagination, indicates that he had at some point found suggestive the model air engines of Niépce and Cagniard, of which Lazare Carnot had written enthusiastic accounts many years earlier.[39]

Those questions could not previously have been answered (to paraphrase further the *Réflexions sur la puissance motrice du feu*) for the reason that the availability of motive power in heat had never been considered in generality but only with respect to particular types of machines. In order that a fundamental theory might be found, it would be necessary to abstract from all particular mechanisms and from the properties of all particular materials and to consider how heat might produce motion as a problem independent of all contingencies. The theory must confine itself neither to steam engines nor even to vapor engines: it must embrace every conceivable heat engine and must model itself upon the theory of the classic machines. For there was a science that did indeed have the character of a complete theory (so held his father's son), resting as it now did upon the principles of mechanics itself. And only when the laws of physics should be sufficiently extended to embrace all the mechanical effects of heat acting upon bodies of any sort would the theory of heat engines be in a comparably satisfactory state (Carnot 1824, pp 6–9).

Sadi Carnot's memoir has been so fully commented, studied, and paraphrased in recent years that no need exists to summarize the entire contents or to enlarge on its later importance in the history of thermodynamics. For the present purpose it will suffice to identify those elements of the argument that derived from the work of Lazare Carnot. Sadi Carnot began by calling attention to a circumstance that always accompanied the production of motion by a steam engine. He chose to see it as a restoration of equilibrium in the caloric, by which he meant that heat is always transferred from a hotter to a colder body, from boiler to condenser. The process, he emphasized, involved the movement and not the consumption of caloric, and Sadi Carnot's analysis depended upon adopting the point of view that availability of motive power from heat presupposed some prior disruption of equilibrium in the distribution of caloric, and that reciprocally wherever a difference of temperature occurred, there existed the potentiality of drawing motive power from the transportation of caloric that would restore the state of equilibrium. In principle the agent might be anything, a metallic bar, for example, the reason being that in any object a change in temperature always involved a change in volume that might be harnessed. The choice depended on efficiency.

The question of how the motive power of heat might vary with the nature of the agent chosen to realize it, whether steam, air, metallic bars, or whatever sort of body, could be discussed decisively only for a given amount of heat and a given drop in temperature. Suppose that a body A was maintained at a temperature of 100° and a body B at a temperature of 0°. What would be the motive power (or work–the term itself was only five years from adoption) that could be delivered by transferring a

[39]See above Chapter 1.

given quantity of heat from A to B? Was it more efficient to employ one substance than another? The obvious advantage of vapors was that the temperature of gases would rise on compression and fall on expansion so that it was possible to disturb the equilibrium of the caloric in the same substance as often as desired. Moreover–and here Sadi Carnot gave the first hint of the analytical use he was about to make of the idea of a reversible process–it would always be possible to consider that steam might theoretically be employed in a manner the inverse of that of a steam engine for the purpose of disturbing the equilibrium in the caloric, i.e., for transporting heat from a colder to a hotter body by the expenditure of motive power.

In the initial outline of such a process, Sadi Carnot made it occur in the three stages of his unpublished *Recherche d'une formule propre à représenter la puissance motrice de la vapeur d'eau*: (1) In the first, the body A discharged the function of a boiler generating steam at its own temperature, i.e., isothermally. (2) In the second, the steam expanded (adiabatically) in the cylinder until its temperature fell to that of body B. (3) In the third, the steam was condensed at constant pressure by contact with body B, which, therefore, was filling the office of the cold water injected into a condenser, except that it remained at constant temperature and did not mix.

Having thus to a degree idealized and schematized the functioning of a steam engine, even as Lazare had done for ordinary machines; Sadi Carnot observed that these operations "[...] could have been done in one direction and also in the inverse order." (Carnot 1824, p 19). Steam could, first, be formed by employing the caloric of body B at its temperature. Second, compressed until it rose in temperature to that of body A. Third, condensed by further compression at the temperature of body A. In both directions, therefore, the first and third stages were isothermal and the intervening step adiabatic–though it is to be emphasized that these terms had not then been coined. The former sequence of operations produced motive power (or work) and transferred the caloric (or heat) from the higher temperature of body A to the lower of body B. The inverse operation expended motive power (or work) and returned the caloric (or heat) from body B to body A. But if the same quantity of vapor were involved and if no motive power nor caloric had been lost, then:

> [...] the quantity of motive power produced in the first case will be equal to that expended in the second, and the quantity of caloric transported in the first case from body A to body B will be equal to the quantity restored in the second from body B to body A, so that one could perform an indefinite number of similar operations without there finally being either motive power produced or caloric transported from one body to another.[40]

Evidently, then, Sadi Carnot was introducing the idea of a reversible process in the same place in his argument that Lazare Carnot had done in his, at the outset, and for the same reason: as an auxiliary in the reasoning to permit comparison of the initial and final states of a system by eliminating from consideration internal changes of work or energy. In the *Essai sur les machines en général*, it will be recalled,

---

[40]Carnot (1824), p 20.

Lazare's initial definition of geometric motion had been in terms of reversibility.[41] If (as seems probable) Sadi Carnot had the idea of the three–stage cycle set out in his *Recherche d'une formule propre à représenter la puissance motrice de la vapeur d'eau* manuscript before his father's death,[42] it is specially significant that he incorporated the idea of a reversible process only in the text of the *Réflexions sur la puissance motrice du feu* discussed with Hippolyte upon the latter's return from Magdeburg.

The next comparison is equally telling. Discussing ideal hydraulic machines in the *Principes fondamentaux de l'équilibre et du mouvement*, Lazare Carnot had brought out that it was a condition for *maximum* efficiency that there be no motion in the millstream that was not transmitted to the wheel, since any residual velocity in the water could in principle be harnessed on egress to produce an additional effect (Carnot 1803a, pp 248–249). Sadi Carnot for his part went on from the formulation of reversibility in ideal steam machines (for perhaps the literal translation of "machine à vapeur" helps bring out the carry–over of ideas) to argue that if there existed any method for employing heat more advantageous than the pair

---

[41] See above Chapter 2, pp 25–28.

[42] It is agreed that Sadi Carnot must have composed the manuscript of the *Recherche d'une formule propre à représenter la puissance motrice de la vapeur d'eau* at some time between November 1819 and March 1827. In publishing the text (Gabbey and Herivel) Drs. Gabbey and Rerivel concluded that a date prior to 1824, when the *Réflexions sur la puissance motrice du feu* appeared, was more probable than a later one. Mr. James Challey suggests that 1823 is the most probable date of composition because of Sadi Carnot's use of the *dyname* as the unit of motive power, Dupin having coined the word in a report to the Academy in April of that year (Challey 1971, *Dictionary of Scientific Biography*, III, p 83, n. 3). For reasons indicated in the argument of this section, I agree. Dr. Robert Fox disagrees, however. He kindly provided for me the proofs in his article *Watt's expansive principle in the work of Sadi Carnot and Nicolas Clément* (Fox 1970, 233–253). In that paper he explores the relations between Nicolas Clément (1779–1842) and Sadi Carnot very carefully. The main point is to establish the importance to Sadi Carnot's work of contemporary power technology. This it does admirably. Dr. Fox argues further that the manuscript *Recherche* represents an attempt by Sadi Carnot to compute a formula for the motive power of steam that would be applicable in actual engines as the highly abstract and unrealistic reasoning of the *Réflexions sur la puissance motrice du feu* could not be, specifically because in the manuscript memoir Carnot computed the motive power that would be developed in the phase of adiabatic expansion between any two temperatures instead of restricting it to the unreal case of a span of 1°. Dr. Fox may well be right. I doubt it, or doubt at least that this is the whole explanation because it reverses the configuration both of his work and his father's, which moved from the analysis of machine processes to theory rather than from theory to specific engineering application. Further, it seems more likely that the inclusion of the three–stage cycle applicable only to steam in the early passages of the *Réflexions sur la puissance motrice du feu* reflects rather the carry–over of an unperfected stage of his analysis into the final product, than it does that he should afterwards have reverted to this imperfect analysis in order to base an applicable calculation on it. There is a possibility that we are both right, and that the manuscript memoir, which does indeed have all the appearance of having been finished, represents the final development that Sadi Carnot, after publishing the *Réflexions sur la puissance motrice du feu,* then gave to the idea of a three–stage cycle with which he had begun his thinking. If so, it remains a problem that he did not publish it. Since almost all his manuscripts were burned after his death it is unlikely that the question will ever be decisively resolved.

of processes just described, it would follow that some greater quantity of motive power could be drawn from the flow of caloric in the first or forward process. It would then be possible to divert the excess or a portion of it to the job of driving the reverse process, that is to restoring the caloric from body B back up the temperature scale to body A. If that could be done, it would amount to an indefinite creation of motive force without consumption of caloric or indeed of any agent at all. It would amount, in short, to perpetual motion. Such a conclusion being contrary to "the laws of mechanics and sound physics", it was inadmissible. Excluding that impossibility afforded Sadi Carnot the basis for a preliminary statement of what he called the fundamental theorem:

> [...] the maximum motive power resulting from the use of vapor is also the maximum motive power that can be realized [from heat] by any means at all.[43]

A moment's reflection, consequently, would exhibit the condition for realizing the maximum motive power in general: it was

> [...] that there should not occur in the bodies employed to realize the motive power of heat any change of temperature that is not due to a change of volume.[44]

The reader familiar with later thermodynamic reasoning will immediately notice that the argument was not complete: it omitted the step of an adiabatic compression on the return process. The vapor in this initial illustration being steam, Sadi Carnot could not have incorporated such a stage since to vaporize water by compression without input of heat was physically unimaginable. He recognized the difficulty and, in moving on to a full and rigorous demonstration employing an air engine, admitted that in this preliminary sketch the vapor had not been supposed to be restored to its initial state (Carnot 1824, pp 36–37). Commenting on the discrepancy, T. S. Kuhn in his *Sadi Carnot and the Cagnard Engine* (Kuhn 1961, p 571, ft 9) has conjectured that Sadi Carnot began with a steam engine because of its greater familiarity to his prospective readers.

It is plausible or to think that these reasonings, and the conclusions so far drawn, reflect the direct carry over from Lazare's theory of machines into Sadi's early thinking. His having resorted to the argument from gradual change is persuasive, since reversibility plays the part of geometric motion. So also is the ensuing paragraph, in which Sadi Carnot explicitly invoked the analogy between the flow of heat and the flow of water. It was probably at this juncture that Sadi Carnot started to go beyond his father. He here began citing the extensive experimental demonstrations of his generation on the thermal aspects of the physics and chemistry of gases, matters on which he enlarged in the main body of his memoir. It was also precisely here that he put forward the rigorous demonstration of the "fundamental proposition" – i.e. that the utilization of vapor was the means of realizing the maximum motive power from heat – and did so in a fully general form. It was

[43]Carnot (1824), pp 21–22.

[44]Carnot (1824), pp 22–23.

on the basis of that demonstration, the full Carnot cycle, that he stated the theorem later regarded as his fundamental contribution:

> The motive power of heat is independent of the agents put to work to realize it; its quantity is determined uniquely by the temperatures of the bodies between which the caloric in the final result passes.[45]

We shall give the demonstration in Sadi Carnot's own words:

> Let us imagine an elastic fluid, air for example, shut up in a cylindrical vessel, *abcd*, provided with a movable diaphragm or piston, *cd*. Let there also be two bodies *A* and *B*, kept at a constant temperature, that of *A* being higher than that of *B*. Let us picture to ourselves the series of operations which are to be described[46] (Fig. 3.3):

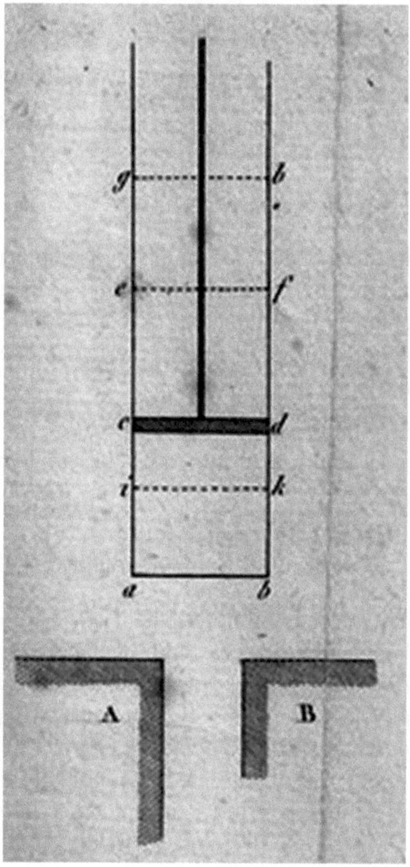

Fig. 3.3 One of Carnot's cylinders and operations in a cycle (Carnot 1824, p 118. With permission of the «Département de la reproduction, Bibliothèque Nationale de France, Paris»

[45]Carnot (1824), p 38.

[46]Carnot (1824), p 32.

1. Contact of the body *A* with the air enclosed in the space *abcd* or with the wall of this space – a wall that we will suppose to transmit the caloric (i.e. heat) readily. The air reaches, by virtue of such contact, the same temperature as the body *A*; *cd* is the actual position of the piston.
2. The piston gradually rises and takes the position *ef*. The body *A* is all the time in contact with the air, which is thus kept at a constant temperature during the rarefaction. The body *A* furnishes the caloric necessary to keep the temperature constant.
3. The body *A* is removed, and the air is no longer in contact with any body capable of furnishing it with caloric. The piston meanwhile continues to move, and passes from the position *ef* to the position *gh*. The air is rarefied without receiving caloric, and its temperature falls. Let us imagine that it falls thus till it becomes equal to that of the body *B*; at this instant the piston stops, remaining at the position *gh*.
4. The air is placed in contact with the body *B*; it is compressed by the return of the piston as it is moved from the position *gh* to the position *cd*. This air remains, however, at a constant temperature because of its contact with the body *B*, to which it yields its caloric.
5. The body *B* is removed, and the compression of the air is continued. Being isolated, its temperature rises. The compression is continued until the air acquires the temperature of the body *A*. The position of the piston passes during this time from the position *cd* to the position *ik*.
6. The air is again placed in contact with the body *A*. The piston returns from the position *ik* to the position *ef*; the temperature remains unchanged.
7. The step described under number (3) is repeated, then successively the steps (4), (5), (6), (3), (4), (5), (6), (3), (4), (5), and so on.

In these operations the piston is subject to an effort of greater or lesser magnitude exerted by the air in the cylinder. Its elastic force varies as much because of the changes in volume as of changes in temperature. But it must be noted that with equal volumes, the temperature is higher during movements of dilation than during those of compression.

> During the former, the elastic force of the air is found to be greater, and consequently the quantity of motive power produced by the movements of dilation is more considerable than that consumed to produce the movements of compression. Thus we should obtain an excess of motive power–an excess which we could employ for any purpose whatever. The air, then, has served as a heat engine; we have, in fact, employed it in the most advantageous manner possible, for no useless re–establishment of equilibrium has been effected in the caloric.[47]

The reader will notice that Sadi Carnot who like his father always uses ordinary language in his reasoning, nowhere employs the diagram that appears in many a physics text book to exhibit the Carnot cycle. His essay was largely ignored by physicists until the civil engineer, Clapeyron mathematicised it, *Mémoire sur la puissance motrice de la chaleur,* in 1834 (Clapeyron 1834, 153–190).

He there represented it graphically in a form that is too complicated to be worth explaining here and is included because it begins to look familiar (Fig. 3.4).

[47]Carnot 1824, pp 34–35. There is a translation by R. H. Thurston in Mendoza ed., *Reflections on the Motive Power of Fire and Other Papers on the Second Law of Thermodynamics by E. Clapeyron and R. Clausius* (Mendoza 1960).

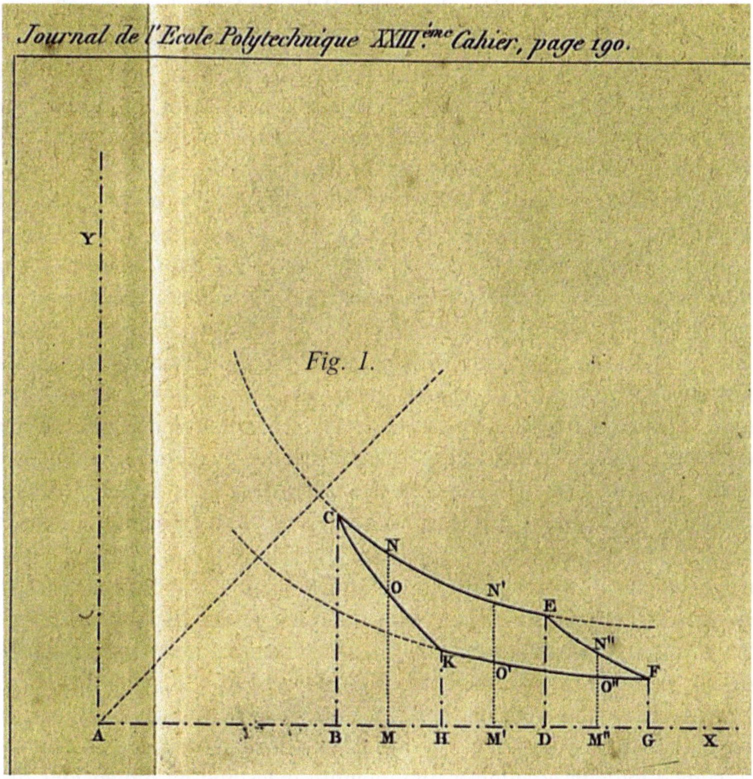

Fig. 3.4 Carnot's cycle in Clapeyron's diagram (Clapeyron 1834, Fig. 1, 190. «Collections École polytechnique Z 5 (1834)»)

That familiarity is complete in the graph given by Rudolf Clausius in his paper, *Über die bewegende Kraft der Wärme* (Clausius 1850, 368–397; 500–524) (Fig. 3.5).

Here the abscissa *oe* represents the volume and the ordinate *ea* the pressure on a unit weight of gas. The curve *ab* then represents the isothermal expansion, the

Fig. 3.5 Carnot's cycle in Clausius' diagram (Clausius 1850, 379)

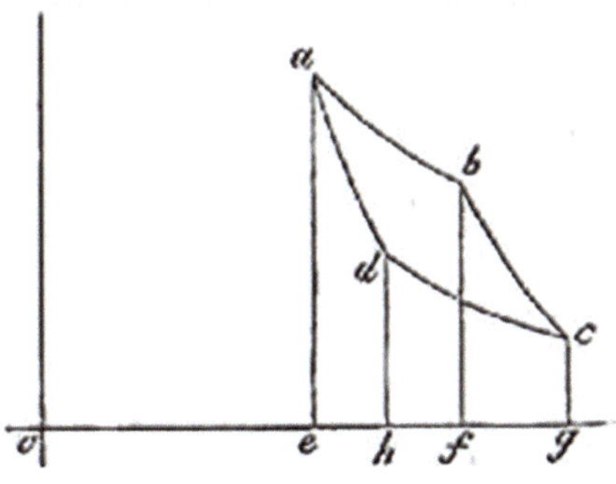

curve *bc* the adiabatic expansion, the curve *cd* the isothermal compression, and the curve *da* the adiabatic compression. The figure *abcd* is an equilateral hyperbola and represents the work done by the gas in the course of the process. (The terms adiabatic and isothermal were not yet coined). Sadi Carnot, of course, was not thinking in terms of this four cornered graph, but rather of his piston and cylinder. Its first isothermal expansion is only part of what will later be Step 6. But once his cycle is established, his Steps 3, 4, 5, 6 constitute the cycle as normally described.

The above proof of the motive power of heat is as far as we need to follow Sadi Carnot in order to exhibit the full inheritance of the son from the father. No doubt the most significant items were the development of Lazare's geometric motions into Sadi's reversible process and their application to fully cyclic processes. Besides that, there was the exclusion of perpetual motion, axiomatic in Sadi Carnot, demonstrative and tutelary in Lazare Carnot; the generalization from principles of operation in particular types of machine to the principles of machines in general, applied to heat engines by Sadi Carnot; the restrictive mode of reasoning in which the maximum possibilities inherent in the operations were determined and then the conditions for realizing them defined; the curious combination of quantitative mode and verbal expression, such that the reasoning moves from the ideal case of changes occurring continuously and infinitesimally to the physical reality of discontinuous and irreversible changes of state in a system; the discussion of force in terms of what it can do, taken usually over distance when it was a question of its measure and over time when it was a question of its realization in mechanical processes. It cannot be said that either Carnot brought this tacit distinction to a decisive differentiation between proto–concepts of work and energy. Yet both of them assumed the conservation of the quantity measured by dimensions of work and energy, Lazare balancing his accounts between moment–of–activity produced and moment–of–activity consumed or live force, Sadi Carnot between motive power produced and caloric transported.

There was even the long wait for recognition, and the question whether what developed was the state of knowledge and interest in the subject that permitted the cogency and applicability of their theories to be recognized. There is this difference, however. One doesn't want to say that Lazare Carnot's work was shallow. But it blended into the physics of work and energy, which would have happened anyway. Sadi Carnot's work was deep in a way that his father's was not.

It founded the science of thermodynamics.

Chapter 4
The Carnot Approach and the Mechanics of Work and Power, 1803–1829

The way in which the work of Lazare and Sadi Carnot made itself felt reveals a similarity between their writings as interesting as their substantive filiation and the analytic and stylistic parallels. The failure of Lazare's *Essai sur les machines en général* to attract contemporary attention has already been discussed. Although more often mentioned since, the *Principes fondamentaux de l'équilibre et du mouvement* fared little better when it appeared in 1803. It fell into the same obscurity, lasting another 15 years. The book was occasionally mentioned prior to 1818, but rather by way of noticing its existence than because its point of view affected the treatment of problems. When at about that time Lazare Carnot's mechanics of machines began to be comprehended in print, the recognition was of his distant paternity of principles or applications that might be taken for granted. Indeed, it does not appear that, apart from Sadi Carnot himself, others of the next generation of engineering mechanics actually read Lazare's writings, for what they credited to him was their own much simplified employment of the concepts and not his enunciation of them.

The explanation cannot well be that either father or son was an obscure or neglected personality (except in Lazare's Carnot early years). What seems likely, therefore, is that attention first to machines and then to heat and power developed in a largely verbal, pedagogical, and practical way. The subjects constituted a kind of engineering mechanics *avant la lettre* in which problems were posed, principles tacitly selected, and quantities employed because that was the way to get results. What further seems likely is that both Carnots participated in that development personally rather than through their books. Lazare Carnot in his later years was a kind of Nestor of engineering busying himself judging inventions for the Institute. Sadi Carnot kept up with his fellow polytechnicians and was a contemporary and in several cases classmate of those in whose writings recognition that his father's mechanics was of elementary importance now began almost spontaneously to appear: Petit and Navier, Coriolis and Poncelet, Clapeyron and Dupin, Chasles and Barré de Saint-Venant.

C.C. Gillispie and R. Pisano, *Lazare and Sadi Carnot: A Scientific and Filial Relationship*, History of Mechanism and Machine Science 19, DOI 10.1007/978-94-017-8011-7_4,

4.1 Early Engineering Mechanics

It supports this view to notice how largely the literature of early engineering mechanics was couched in the form of commentaries on earlier writings and new editions rather than in memoirs of research or systematic treatises. In this mode it was typical of engineering in general, wherein the sequence of specific researches, whether experimental investigations of novel phenomena or theoretical resolutions of particular problems, carried much less of the traffic than it did in science proper. There was such a literature, of course, and it grew in importance in the contribution of the first generation of polytechnicians. But even then engineering knowledge was growing, and perhaps mainly so, through the gradual codification of practices brought up to the surface for commentary and formalization. For the present purpose there is no need to go back beyond the works from which Lazare Carnot himself doubtless learned the elements of practical constructions, Bélidor's *Science des ingénieurs* and *Architecture Hydraulique.*[1]

The career of Bernard Forest de Bélidor was a colorful example of what, after all, had been open to a man of technical capacity in the Old Regime. Left a military orphan in childhood, he developed from an artillery instructor into an engineering expert and died a member of the Academy of sciences. The former of his books had a narrower and the latter a wider application than might be supposed nowadays from the titles. "Engineering" connoted mainly fortification and siegecraft, military works in short, while waterworks in the eighteenth century were important in transportation, the provision of power, and the embellishment and supply of cities, castles, and gardens–all problems that later fell into the domain of mechanical and civil engineering.

Bélidor began with an initial parade of founding practice on the principles of the science of mechanics. Most probably it was there that Carnot read of the principle of equilibrium that in much the same connection Bélidor and he both attributed to Descartes[2] (Bélidor 1737–1739, I, pp 27–31). The demand for Bélidor's works never slackened throughout the eighteenth century. When the copper plates wore out through constant reprinting, the publisher, Firmin Didot decided to have them re–engraved and to reissue both sets, and he engaged the services of C.–L.–M.–H. Navier, a rising light among the new breed of theoretical engineers, to bring the work up to date. The history of that edition is very revealing for the relation between practical construction and theory in mechanics. For despite the immense

[1]*Architecture hydraulique, ou l'Art de conduire, d'élever et de ménager les eaux pour les différents besoins de la vie* by Bernard Forest de Bélidor (1737–1739). There were further editions in different formats, and finally a new edition "avec des notes et additions" edited by Navier (Bélidor 1819); *La science des ingénieurs dans la conduite des travaux de fortification et d'architecture civile. Nouvelle éditions, avec des notes, par M. Navier, ingénieur ordinaire des Ponts et chaussées* (Bélidor 1729, [Navier edition] 1813).

[2]See Chapter 2, pp 20–22.

development of the science of rational mechanics in the 80 years intervening between Bélidor's first edition and Navier's reissue, it seemed wise not to make the slightest change in Bélidor's text lest the work lose its public.

That public consisted of contractors and builders who knew Bélidor for an engineering handbook in which they could look up specifications for such or such a foundation, revetment, doorway, cornice, or other standard construction, instructions covering not only architectural details but even the most efficient mode of organizing the work gang.

As for the theoretical part of the work, Bélidor's mathematical equipment had been faulty even for his own day, and the formal aspect of his presentation required complete revision. Navier undertook the task by composing for *Architecture hydraulique* a series of footnotes so extensive as to amount to a companion treatise that runs along the bottom of Bélidor's republished pages correcting, supplementing, and clarifying his discussion of principles and laws, yet somehow not supplanting Bélidor himself. Indeed, one is tempted to paraphrase the famous remark that L. J. Henderson lifted from J. T. Merz about the debt that science owed the steam engine. One feels that the science of mechanics owed more to Bélidor than Bélidor to the science of mechanics–it took from him the body that brought engineering mechanics out of it, and what it contributed in the way of refinement seemed insufficiently usable in actual constructions to find a public for a literature all its own.

Navier, a graduate of *Polytechnique* in 1804 and then of *Ponts et chaussées*, came like Carnot from a Burgundian background. His father, a lawyer and member of the Assembly of Notables in 1787, having died prematurely of "chagrin" in the Revolution,[3] Navier was brought up by an engineering uncle, one Gauthey whose *Traité de la construction des ponts* (Gauthey 1832) he then edited and published. His properly theoretical work belongs to the history of elasticity and the early adoption of Fourier analysis in mechanics. But he also engineered actual constructions: the Pont de Choisy, the Pont d'Argenteuil, the Pont d'Asnières, and the Pont des Invalides, a suspension bridge built in the style of the Tweed bridge and torn down in 1826 when a reservoir on the Butte de Chaillot burst, and the resulting flood undermined one of the piers. His writings, apart from particular memoirs on theoretical problems, practical constructions, and railway projects, developed out of courses he gave at *École polytechnique* and *Ponts et chaussées*. Just as he edited others, so in mechanics proper, others edited him. Thus, a *Résumé des leçons d'analyse*, consisting of lectures at *Polytechnique* for three successive semesters (1832, 1832–1833, and 1835–1836) was formally published under the editorship of J. Liouville who rearranged it and appended his notes.[4] Similarly the course on mechanics evolved from a resume for 1831, 1832, and 1833, and reached treatise form in *Résumé des leçons de mécanique*, 1841 (Navier 1841). Finally, a famous course on the science of machines given at *Ponts et chaussées* and first distributed

---

[3]On Navier see *Notice biographique sur M. Navier* by Gaspard Prony (1837).

[4](Paris, s.d.). A second edition "Revue par M. Ernest Liouville" (apparently the famous mathematician) was published in 2 vols in 1856 (Navier 1856).

in 1826 evolved in 1864 to a third edition of its first volume under the title *De la résistance des corps solides* (Navier 1826), and was furnished with extensive notes and appendices by Barré de Saint–Venant which themselves form an indispensable source for the history of mechanics, particularly the theory of elasticity and related topics.[5]

Together with Navier, Coriolis and Poncelet constituted the ruling triumvirate in the engineering mechanics of their generation, those who graduated from a *Polytechnique* still under the direct influence of its founders. Their writings fall into the same pattern. The most important single treatise was Coriolis' *Du calcul de l'effet des machines ou considérations sur l'emploi des moteurs et sur leur évaluation* in 1829 (Coriolis 1829), which gave Carnot's moment–of–activity its modern name of work. Of that, more in a moment. Coriolis then circulated in multigraphed form notes of his course on mechanics, at the *École centrale* for 1830–1831 and again for 1836–1837.[6] Evidently he intended a complete treatise of applied mechanics combining the science of machines with an up–to–date exposition of the fundamental principles of rational mechanics. He died before completing that synthesis, and the publishers had instead to reprint most of the 1829 memoir together with the later material on the mechanics of solid bodies.[7]

Poncelet for his part circulated his course at the *École d'application* at Metz in multigraphed form beginning in 1826. He observed à propos of Coriolis' 1829 treatise that his own course contained similar material, but that ill health had prevented, him from publishing it.[8] The course itself, or perhaps extensions of it given by Poncelet between 1827 and 1829, was then edited for publication by a captain of engineers, one Gosselin and published in 1828 under the title *Cours de mécanique industrielle*.[9] A second edition appeared in 1841, and a third (this was edited by X. Kretz) in 1870. Poncelet himself thought it useful to publish a review of his own scattered writings in a pamphlet undated but apparently of 1835 or thereabouts.[10] Finally, in 1874–1876 appeared a two–volume edition of

---

[5]3rd edition of *De la résistance des corps solides* "avec des notes et des appendices par M. Barré de Saint–Venant" (Navier 1864).

[6]*Leçons sur la mécanique. 1830–1831, 1836–1837* by Gaspard–Gustave de Coriolis (1830–1837), *Bibliothèque Nationale de France,* vols. 14276–14277.

[7]See *Traité de la mécanique des corps solides et du calcul de l'effet des machines* by Coriolis (1844).

[8]*Note sur quelques principes de mécanique relatifs à la science des machines.* It is an extract of *1er section du Bulletin universel des sciences*, *Bibliothèque Nationale de France*, Vp 5866 (Poncelet 1829).

[9]*Cours de mécanique industrielle, professé de 1828 à 1829.* (s.d). *Bibliothèque Nationale de France,* vol. 5562. See also *Note sur quelques principes de mécanique relatifs à la science des machines* (Poncelet 1829, p 323, ft 1).

[10]*Notice analytique sur les travaux de M Poncelet, Bibliothèque Nationale de France,* 4° Ln[27], vol. 16475 (Poncelet s.d.).

Cours de mécanique appliquée aux machines,[11] this too edited by Kretz, containing a useful introductory *résumé* of the evolution of the entire school. The original patron of the school, Monge himself, had difficulty in actually writing and finishing treatises of his own, and in fact never accomplished it. His immediate disciples put together his *Géométrie descriptive*[12] (Monge 1799). A work that he purportedly intended to publish on the mechanics of machines had instead to be undertaken by Hachette, a wheelhorse who replaced him at *Polytechnique* when he went off to swell Bonaparte's scientific train in Egypt (Hachette 1811). It was written, says the dedication to Monge, "[. . .] assisted by your advice [. . .]" (Hachette 1811, p vi).

In these early decades of the nineteenth century, two different modes may to a degree be distinguished in the conception of French treatises of practical mechanics. On the one hand, the literature that we have just been considering, despite its tendency for books to remain unfinished by the author, was generally theoretical in spirit and continued in increasingly mathematical form the intention of the two Carnots, i.e., to draw from the science of mechanics a theory of machines and of power at once general and applicable. That was the novel emphasis in content, and it was inspired by the actual development of machine–driven technology. Alongside this literature, there also continued to appear treatises and memoirs of a type that may be called manuals. To accounts of old machines they added those of new ones, and to the expression of classic principles they brought the newer formulations of rational and analytical mechanics. The distinction between the two traditions does not turn on the presence or absence of mathematics. Where a difference in the mathematics was relevant (it was not always so) it was on the kind of mathematics, the theoretical tradition being associated with the new geometry rather than analysis.

For the background of this second, more conventional approach, there is for present purposes no need to look further in the history of mechanics than to *Traité élémentaire de mécanique* of the abbé Bossut[13] (1775). Bossut had divided his subject into statics and dynamics. Statics was the science of equilibrium, which results from the mutual destruction of forces. The topics were general equilibrium principles; determination of centers of gravity; application of the foregoing to stating equilibrium conditions in each type of machine taken successively; and finally friction. The topics of Part II on motion were inertia, uniform motion, acceleration, fall, the motion of centers of gravity, and collision theory. The book was an excellent primer, and exhibits the subject matter at an elementary level as it was conceived when Lazare Carnot was an engineering student at Mézières.

[11]An earlier edition of this work was published in Liége in two volumes in 1845 (Poncelet 1845). It is not clear whether it had Poncelet's approbation. "Cet ouvrage n'a jamais été publié en France ni en Belgique" says a legend on the title page, and indeed *the Bibliothèque Nationale de France* has no copy.

[12]For Monge's interest in the science of machines, which derived from his early days at Mézières, see *L'œuvre scientifique de Monge* edited by René Taton (1951, pp 311–314).

[13]See also numerous later editions.

In the field of practical engineering and applied mechanics, the most pretentious work was Prony's *Nouvelle architecture hydraulique*[14] intended evidently to supplant Bélidor, which purpose it failed to accomplish. Prony was then (1790) inspector at the *École des ponts et chaussées* and one of the foremost engineers of the revolutionary period. It is not this work, however, that will make apparent to later readers the justice of the high reputation he undoubtedly enjoyed. It purported to be a general treatise of mechanics in a form applicable to the arts. In fact, the exposition of the standard topics of mechanics was intrinsically unclear and was continually interrupted by tables containing empirical formulas and descriptions of favorite designs for pumps, prime movers, roof vaultings, and incidental devices of all sorts. The second volume concerned steam engines–and here a further characteristic of this school of practical mechanics becomes evident. In it motors were thought of as what one employs "[...] to overcome resistance [...]" (Prony 1796, II, p 80) and successive writers normally harked back to the Coulomb memoir on friction, that which won the prize of 1781 over Carnot,[15] in estimating the magnitude of resistance. Prony never mentioned the work of Carnot in these volumes on practical mechanical engineering. The course he gave on mechanics at *Polytechnique* drew on Euler, Laplace, and Lagrange, and for hydrodynamics on Bossut.[16]

4.2 Applied Mechanics

As time went on, and Carnot became famous, other writers did allude to his *Principes fondamentaux de l'équilibre et du movement.* Those of the Coulomb–Prony succession did so, however, in the lip–service of prefaces or appendices. In 1810 a prominent engineer of the Lyonnais, André Guenyveau, a graduate of *Polytechnique* in 1800, published an *Essai sur la science des machines* (Guenyveau 1810). It is apparent that the subject was not yet synonymous with Carnot's lead. The body of the treatise passed from calculations about the state of equilibrium to practical application by drawing in Coulomb's results for friction (Coulomb 1785, pp 2–3). An interesting afterthought was printed in smaller type than the body of the text. The author there observed that motion might be communicated in two ways, by impact or by pressure, and from the difference he deduced Carnot's principle of the superior efficiency of continuous transmission. He did that without crediting it to Carnot, however, who was merely mentioned in the preface as one who had written a general treatise on machines. Hachette accorded him similar recognition in the preface to the 1811 *Traité élémentaire des machines* already mentioned (Hachette 1811). There Carnot's *Principes fondamentaux de l'équilibre et du mouvement*

---

[14]Two volumes (Prony 1790, 1796).

[15]See above, Chapter 3, ft 3.

[16]It was published as *Mécanique philosophique* in tome III of *Journal de l'École polytechnique de Paris* (Bossut 1800b).

came in for a brief encomium. Hachette observed that its closing pages contained the entire theory of machines and of motive forces applied to them, and called it the work of the "[...] deepest of thinkers and the most experienced of engineers. It is to the science of machines what the recent work of M. Poisson is to rational mechanics properly speaking." (Hachette 1811, p xix). But nowhere did the findings or analysis of Lazare Carnot, or indeed of Poisson, figure in the actual text or treatment of Hachette's course, which was no more than a compilation of designs for particular machines, a book of recipes for engines, pumps, gears, pulleys, and linkages of many sorts.

A more sophisticated contribution to this literature of applied mechanics was a memoir of 1818 by Petit *Sur l'emploi du principe des forces–vives dans le calcul de l'effet des machines* (Petit 1818, pp 287–305). No mere engineer, Petit was a rising young physicist–part author of the famous law of Dulong and Petit–who died young before giving his full measure. At first reading one is tempted to classify this memoir on the theoretical side of the division between the handbook approach and the tradition inaugurated by Lazare Carnot's mechanics of machines. It was exact, it was analytical, and it took the principle of live force to be that one among the general properties of motion that was most readily adaptable to the *calcul des machines*. Live force furnished in any particular case the most natural evaluation both of the motor and of the effect produced, so that the equation determining the relation between these two quantities would furnish the direct solution of the problem.

That application of the principle of live force had for a long time been well known to geometers, so observed Petit at the very outset, but then surprisingly went on to say: "The theory of machines, envisaged from this point of view, remains to be created almost in its entirety" (Petit 1818, p 287). Could he have said that because, unlike Hachette, he did not even know of Carnot's work? Or was it that, knowing it, he did not regard it as an adequate theory of machines from the point of view of live force? Or was it that by machines, he, like Sadi but unlike Lazare, was beginning to mean engines or motors? The first of these alternatives seems the most likely, although it is not impossible that the explanation lies in a combination with the latter two, for, as will appear, Petit was not in fact yet looking at the problem from the point of view of Lazare Carnot's theory; and if he did know it, he did not yet see, as Sadi was to do, its applicability to the study of heat engines.

Like Prony and Hachette, Petit considered that the fundamental function of a motor was to overcome resistance. He intended his essay to be an examination of particular cases of that general problem. For a machine in equilibrium all that need be taken into account was the intensity of the forces. For a machine in motion, however, regard had also to be paid to the distance traversed by the points of application. Petit illustrated the special applicability of the principle of live force to calculating the efficiency of machines through the equivalence dimensionally of MgH to $\frac{1}{2}Mv^2$. First, he adduced the classic example from the law of falling bodies. If the resistance was a weight of mass M raised to a height H, then the effect produced by the machine in raising it would be MgH. But since the velocity acquired by a body falling from height H was given by the relation $V^2 = 2gH$, the effect

MgH equalled $\frac{1}{2}MV^2$. No matter what the type of resistance, it would always be possible thus to express the effect produced by a machine in dimensions that reduce to a live force, the product of a mass by the square of a velocity. Hence a motor could always be regarded as containing a given quantity of live force. What was true of the resistance was also true of the motor: its expression also could always be reduced to a live force. Hence, the ratio of the live force employed by the motor to that communicated to the resistance (the terminology here is faithful to the text) would give an expression for the efficiency of the motor and permit determining the conditions under which it would be a maximum. The rule was:

> The live force communicated to the resistance is equal to that which the motor contained, less the live force lost in sudden change of motion, plus that retained by the motor after acting.[17]

Now then, although this reasoning and the consequent statement read very like Lazare Carnot's principle of continuity, which indeed would follow from them, Petit did not actually state such a principle. Probably he would not have thought to do so, and in order to see what differentiated their several sets of conceptions, we will need to follow the examples to which Petit applied his rule. For though the "moteurs" to which he alludes at the outset were the traditional waterfalls, springs, and animals, those in which he exemplified the rule were engines or prime movers, and the central thrust of the memoir was directed to devices powered by the expansion of vapor.

In such engines the live force developed by the fluid might not be computed in the direct manner of that developed by the flow of incompressible fluids such as water. Instead, we are to imagine that the expansible fluid was contained in a horizontal cylinder closed at one end and fitted at the other with a frictionless piston. Then let a be the length, b the cross-section of the cylinder, and h the height of a column of water of which the weight would be in equilibrium with the "elasticity" (he did not say pressure) of the gas. Since the thrust of the confined gas would give the piston an accelerated motion, it was easy to obtain equations for the live force generated by the expansion of the gas. The first case Petit considered was that of a movement in the absence of any external pressure on the piston.

1. Letting v be the velocity acquired by the piston after time t, and x the length of the gas-filled volume of the cylinder at that moment, the "elasticity" of the gas would then reduce to $\frac{ha}{x}$ and its motive force would be equal to $g\frac{\delta hab}{x}$, g representing the gravitational constant and δ the density of the water. But motive force was customarily represented by $m\frac{dv}{dt}$ or by $m\frac{vdv}{dx}$. Since m was the mass of the piston, it followed that

$$mvdv = g\frac{\delta abh}{x}dx.$$

[17] Petit (1818), p 290.

Integrating and choosing the constant so that v was zero when $x = a$, he found

$$mv^2 = 2g\,\delta hab \log \frac{x}{a}.$$

That is to say, when an elastic fluid occupying a volume a, and exerting a pressure equal to that of a column of water of which the height was h, expanded in the absence of external resistance, the live force that developed at the volume x would be capable of raising to the height h a mass of water of volume $a \log \frac{x}{a}$.

For the two other cases, perhaps it will suffice to state the laws without the algebra.

2. If instead of no external pressure, the piston had to overcome a constant external pressure equivalent to the weight of a column of water of height h' then the live force developed in expanding from volume a to volume x would be capable of raising to the height h, a mass of water of volume equal to $a \log \frac{x}{a} - \frac{h'}{h}(x - a)$.
3. If, finally, a piston was subjected to two uniform pressures exerted in opposite directions by gases of which the elasticities (following Petit's terminology) sustained in equilibrium columns of water of height h and h', the live force communicated to the piston would be capable of lifting to the height $(h - h')$, a mass of water equal in volume to that which the piston had traversed.

In an example of the application of these laws, Petit imagined the problem of comparing the live force that a given quantity of heat could produce, supposing it to be employed first to vaporize water and second to heat a mass of air. In the first case, a quantity of water weighing 1 g at 0° was vaporized at the temperature of 100°. It would occupy approximately 1,700 cm^3 as steam and exert a pressure equivalent to that of a column of water 10 m high. If it was then completely condensed, the live force developed would be capable of raising a weight of 17 kg 1 m. Turning now to the problem of heating a volume of air, the same heat required to vaporize 1 g of water would heat 666 g of water through 1°; or, taking the Bérard-Laroche value for specific heat of air of 0,267 it would heat 2,500 g of air at a pressure equal to that of a column of water of 10 m. The elasticity of the air would then increase by 0.0375 m, and the live force produced would lift to a height of 0.0375 m the weight of a volume of water equal to that occupied by the 2,500 g of air. That volume was 1,925 dm^3, so that reducing everything to the same units, the live force was sufficient to raise a weight of 72.2 kg to a height of 1 m–more than quadruple the live force in the first case.

It is very interesting historically that a paper containing these conclusions and written by one of the most intelligent physicists of his time should have appeared at just this date. What does it reveal?

First of all Petit's paper was evidently an important element in the immediate background of Sadi Carnot's thought–indeed, it was the application of the Carnot approach that enabled him to correct Petit's error by thinking of heat engines in terms of reversible process instead of overcoming resistance. One of the first points Sadi Carnot raised in the *Réflexions sur la puissance motrice du feu* was the possible superiority of air over steam engines, and further on he cited the Petit memoir after demonstrating that it makes no difference whether steam, air, or alcohol vapor be employed: what determines the motive power of a heat engine is temperature differential. Sadi Carnot there attributed Petit's erroneous conclusion to "[...] an altogether incomplete method of considering the action of heat" (Carnot 1953, p 86, ft 1). So, indeed, it was, but the passage in which Petit summarized his conclusion is nonetheless striking in its modest disclaimer of exactly the central deduction that it was the purpose of Sadi Carnot's memoir to draw. Thus Petit:

> Although I do not pretend to deduce from the comparison I have just made any consequences relative to the best method of employing the action of heat as motive force, it is nevertheless permissible to think that some advantage could well be drawn from perfecting those machines that, like the pyréolophore of MM. Niepce, employ air suddenly expanded by heat as a motor (1).[18]

That these confusions should have appeared in a paper of just this date, composed by one of the leading physical minds of his generation, is most instructive to the historian. For not only did Petit compare the energy consumed in vaporizing water to the work done by an expanding gas, but the paper exhibits to particular advantage the very necessity for a concept of work and its differentiation from energy. True, Petit had grasped the significance of live force for evaluating the operation of power machinery, but so had everyone interested – "[...] *la force vive est celle qui se paie*" in Montgolfier's words.[19] Nevertheless, he was still very far from a concept of energy, and was indeed confusing work with energy as did everyone else in what it is convenient to call the aggregative tradition in mechanics.

Since early in the eighteenth century it had been a standard practice to evaluate the capacity of a machine or an engine in terms of the height to which it could lift a given weight. The quantities of MgH and $\frac{1}{2}Mv^2$ being dimensionally reducible to each other in all such practical situations, Petit like most of his contemporaries tacitly came to regard them as alternative expressions for live force Their conception of what that term conveyed did not yet distinguish between (kinetic) energy and work, the one being the capacity to perform the other and the equivalence being a convertibility rather than a fundamental identity.

That distinction originated rather in Lazare Carnot's theory of machines in the difference between live force and moment–of–activity–not that he had it cleanly or consistently, but what distinguished the two modes in mechanics was that the one failed to perceive it and the other began to develop it. But Sadi Carnot need not have been so condescending about Petit, for in one respect he missed a point implicit

---

[18]Petit (1818), p 294. For the Niépce engine.

[19]The allusion is in Hachette (1811), pp xiv–xv. For the widespread recourse to the theoretical model of a water–pressure or "column of water" engine in the pertinent literature, and for its importance in the background of Sadi Carnot's analysis, see *Power technologies and the advance of science 1700–1825* by Cardwell (1965, pp 188–207).

in his father's convertibility of live force into moment–of–activity: the confusion between heat and energy in Sadi Carnot's theory was a further stage of that between work and energy in Petit and the others.

Navier was anything but condescending about Petit's paper. A few issues later he published in the same *Annales de chimie et de physique* a historical note reviewing the history of the live–force principle in the theory of machines.[20] It had, he said at the outset, the same purpose as Petit's: to publicize a set of notions that could be very useful in mechanical (i.e., engineering) practice by recalling what had already been done on that subject and attempting to revive certain procedures and results that had been almost forgotten. Navier had in mind here precisely Lazare Carnot's theory of machines, and it seems probable that it was this paper, followed more massively by his edition of Bélidor's *Architecture hydraulique*, that brought it centrally into view in a generation finally prepared by experience and environment, not just to allude to it as the writing of a famous man, but to develop it into real engineering physics and mechanics.

First, however, Navier gave a resume of the history of the principle from the beginning. He credited the earliest precise conceptions on the effects of machines to Galileo, who had established that a given force in a given time could achieve only a determinate effect and that its measure was the product of a weight and the height to which it was lifted.

That product of force into distance sufficed for estimating the effect of machines so long as they were supposed to have reached a state of uniform motion. When such was not the case, however, an auxiliary principle was needed in order to evaluate from the variation in motion the forces that had produced them and vice versa. The earliest principle to fulfill that requirement was the conservation of live force. It had been enunciated by Huygens although it amounted merely to an extension to a system of bodies of Galileo's proposition that a heavy body sliding freely down any curve always acquires the same velocity in descending from any given height.

The earliest paper that Navier cited as an attempt to evaluate the effect of machines in actual engineering was a contribution on water–wheels fitted with blades that Parent had published in 1704, *Sur la plus grande perfection possible des machines* (Parent 1745 [1704] 1706, pp 323–338). Parent had got his results, however, from the simple observation that effort expended on the blades varied with the relative speed of the wheel compared to that of the current. There ought to be some definite proportion between these velocities that would yield maximum power, and the problem was to determine it. Parent assumed quite arbitrarily that the effort or force exerted on the wheel was proportional to the square of the relative velocity, and found that the speed of the wheel ought to be one–third that of the current. As it chances, the assumption is not far from correct, and the result a good approximation. Nevertheless, Parent's absolute value for a maximum effect

[20]The work was *Détails historiques sur l'emploi du principe des forces vives dans la théorie des machines, et sur diverses roues hydrauliques* (Navier 1818).

was quite inaccurate, nor could such a value be determined precisely except by calculating the respective quantities of live force that the water communicated to the wheel and that it conserved in the runoff.

According to Navier, the first work actually to apply the principle of live force to the theory of machines was Daniel Bernoulli's (1700–1782) *Hydrodynamica* of 1738 (Bernoulli 1738). The association between the subject matter of hydrodynamics and the principle of conservation of live force began there. It took the form of the calculation that, abstracting from friction, the weight of water raised multiplied by the sum of the height to which it was lifted plus the height due to its final velocity was equal to the power employed to run the machine. Bernoulli's point of view was entirely neglected, however, in the famous writings on practical mechanics (or engineering) that appeared contemporaneously–Desaguliers' *Cours de physique expérimentale*[21] (Desaguliers 1751) and Bélidor's *Architecture hydraulique*. Scientists themselves ignored–it: Euler made no use of it in memoirs on the reaction wheel, the centrifugal force wheel, or the Archimedean screw.

Not until a brief memoir by Borda, some 30 years later did anyone carry on the application of the principle of live force to machines.[22] Adopting Bernoulli's approach, Borda improved on it in one respect. Bernoulli had given as the expression for the live force lost in percussion the quantity, $m\left(v^2 - v'^2\right)$ whereas Borda recognized that in accordance with the laws of collision it ought to be $m(v - v')^2$.[23] Some years later, in 1781, Coulomb published a memoir on windmills employing the same method for evaluating the output of machines and calculating the effect of percussions.[24]

According to Navier's resume, at the period of Daniel Bernoulli's *Hydrodynamica* the principle of live force was not considered to have been rigorously demonstrated, but was so considered by the time that Borda and Coulomb published their memoirs. What or who brought about that change he does not say, observing merely that Borda and Coulomb gave only particular applications in their memoirs and that no general theory then existed that could integrate the evaluation of machine processes with the principles of the science of mechanics.

In Navier's view it was Lazare Carnot's *Essai sur les machines en général* of 1783 followed by the *Principes fondamentaux de l'équilibre et du mouvement* of 1803 that made the junction. It will be germane to set down exactly what it was that Navier attributed to Lazare Carnot. (He made no distinction between the contributions of the two books, but saw them as successive versions of a

[21]It was translated from *A course of experimental philosophy* (Desaguliers 1734).

[22]*Mémoire sur les roues hydrauliques* (Borda 1770b, pp 270–287).

[23]Borda followed this memoir with a second in the following year, *Mémoire sur les pompes* (Borda 1770a, pp 418–431).

[24]On that see *Observations théoriques et expérimentales sur l'effet des moulins à vent, & sur la figure de leurs ailes* by Coulomb (1784, pp 65–81).

single work). First of all, Carnot demonstrated in a general manner the theorem on the evaluation of the loss of live force in the collision of inelastic bodies (in Navier that term had replaced hard body), a theorem that Borda and Coulomb had noticed only in the particular case that concerned them. Secondly, Carnot laid down in consequence the principles that must be observed in the design of hydraulic machines in order to achieve maximum efficiency–i.e., that the water should communicate all its motion to the machine, or at least retain no more than the minimum required to flow off; that there should be no turbulence or splashing in the fluid or percussion between the parts of the machine. And finally that the "[. . .] impelling forces [. . .]" should create no motion in the system that did not serve the purpose, the example being that of pumping water to a raised reservoir where it should arrive ideally at zero velocity.

No witness could now be better placed than Navier then was to judge of Lazare Carnot's work in mechanics. The items just cited were the specific results he singled out. Even more important in his view, however, was Lazare Carnot's way of looking at machines, and it is extremely interesting that Navier goes on to say that Lagrange himself adopted it in the *Théorie des fonctions analytiques*. There Lagrange explained in very precise and elegant fashion what may seem naïve to the modern reader when he encounters it in Carnot's *Essai sur les machines en général*–that the action of almost any force could be assimilated to that of gravity or of springs.[25] The evaluation of the live force produced by the former was easy–the product of a weight and a height. As to forces that act in the manner of springs, and here what was mainly in view were those produced by the alternate expansion and condensation of vapors, the rules were those established by Petit in the article just discussed.

For himself, Navier informed the reader, he would in a few months be submitting to the Academy the notes and additions he had prepared for the republication of Bélidor's *Architecture hydraulique*. Describing his purpose, he employed virtually the same phrase as that which he used in characterizing the importance of Carnot's *Essai sur les machines en général*: "I have tried to assimilate these considerations in the most elementary way to the first principles of mechanics [. . .]" (Navier 1818, p 152). It is significant that he had made no such comprehensive attempt in the notes and corrections he supplied for the republication in 1813 of Bélidor's *Science des ingénieurs*. Actually, such an enterprise would have been just as appropriate to that work. It must, therefore, have been developments in the interim that led him to undertake the ambitious and comprehensive commentary amounting to an analytical theory of machines accompanying the later edition.

Since his historical remarks were concerned mainly with reminding readers of what he said mechanics had largely forgotten, the work of Carnot on the theory

[25]On that see: *Théorie des fonctions analytiques contenant les principes du calcul différentiel, dégagés de toute considération d'infiniment petits, de limites et de fluxions, et réduits à l'analyse algébrique des quantités finies* (Lagrange 1881, pp 409–410). For the appreciation of Carnot's theorem that Lagrange included in the second edition, see Lagrange (1813), p 407.

of machines, it seems reasonable to suppose that the renewed appreciation in his own mind was what led Navier to edit *Architecture hydraulique* in the light of the theory of machines. In his notes to that work, he demonstrated the principle of conservation of live force for a single mass point and extended the result to a system of points by means of a conventional argument from d'Alembert's Principle. The discussion was in terms of virtual velocities. From the conservation of live force, he then derived Carnot's theorem, which he now described as completing the principle of conservation of live force in the case of discontinuous change of motion, and of which he gave the statement

> [...] that the sum of the live forces in the system after a sudden change is less than what it was beforehand, and that *the system has lost a quantity of live force equal to what it would have been if all the bodies were moving with the velocities that they lost in this change of motion.*[26]

Earlier on Navier credited another device to Carnot, one which except for this remark the modem reader might not think to notice as distinctively his, the trigonometric practice of evaluating the projection of any directed quantity on another as the product of its multiplication by the cosine of the angle between them (Bélidor 1819, pp 17–18), which may be taken to have been a small step in the direction of vector analysis. It is a possibility that gains plausibility from the probable importance of Carnot's geometry in the background of that subject.[27]

4.3 The Concept of Work

Since our purpose is not to recount the history of mechanics, but only to identify the points of entry of Lazare Carnot's engineering science into mechanics, the remaining episodes may be briefly summarized. They culminated in 1829 in the composition of a book generally recognized to have been the foundation of nineteenth-century work in engineering mechanics, Coriolis' *Du calcul de l'effet des machines* (Coriolis 1829). In that work Coriolis undertook to gather all the general considerations that might illuminate problems of economizing force or mechanical power. To the historian of science the most interesting passage is that in which Coriolis gave the quantity "work" the name by which it has since been known in physics and distinguished it from live force. His discussion is worth paraphrasing for it exhibits how usage had come into existence and changed.

By the word "work" (travail) Coriolis meant the quantity called variously mechanical power (puissance mécanique), quantity–of–action, or dynamical effect (Coriolis 1829, p III), all of them variations on Lazare Carnot's original moment–of–activity. (For its unit Coriolis further suggested the designation "dynamode"

[26] Navier notes in Bélidor's *Architecture hydraulique* (Bélidor 1819, pp 111–113).

[27] See below, Chapter 5.

(Coriolis 1829, p III, p V, pp 33–34), which, however, did not take.) A further innovation he described as "slight" although it has proved equally significant. He would call by the name live force the "[...] product of a weight by the height due to its velocity [...]". Under that definition, live force would be only half of the product that had previously been meant by that term, the mass by the square of the velocity, and both Coriolis' reason for making the change and his discussion of how the confusion had arisen are revealing.

Charged with a course at *Polytechnique*, Coriolis proposed the change out of his experience of the burden that a badly defined dimensionality imposed on students. In their problems in mechanical engineering they were forever coming upon the quantity $\frac{1}{2}mv^2$ in its equivalence to MgH and they inevitably found it confusing that the name live force should refer to double that most commonly recurrent of quantities. That it had come to do so was owing to the difference between the traditions of rational mechanics and mechanical practice. In times past, rational mechanics never envisaged work and did not mean the product of weight by velocity in employing the term live force. More recently, however, all practical engineers had come to mean by the term live force precisely the work that could be produced by the velocity a body had acquired. Coriolis admitted that rational mechanics could equally well have got on with its accustomed usage if mathematicians preferred, and insisted only that his redesignation would simplify discussion of the theory of machines. Nevertheless, the outcome makes evident what it was that by then had become the dominant set of concerns, for surely it was there in the redesignation and not 30–odd years later in its renaming as kinetic energy that live force became differentiated from work.

As for the term "work" Coriolis observed that it came so naturally to mind that it had already been employed for the same purpose in passing by Navier in his notes on Bélidor and also by Prony in a memoir on a new machine. He might have added Lazare Carnot himself (Gillispie 1971, Appendix B, § 27, ft c, pp 272–273).[28] A little while previously the question of designating units for the study of power and machines had been raised in the Academy. A commission had been appointed. Laplace had been a member, and his opinion was that the Academy would be unwise to take the initiative in proposing names. He thought–was it with distant memories of the commissions on the metric system?–that it could properly do no more than sanction usage once it was established, and Coriolis invoked this weighty support.

The treatise itself was impressive testimony to the development its subject had undergone. It was a comprehensive manual of engineering mechanics and treated the problems arising in the design and employment of power machinery in theoretical terms. Coriolis wrote at a level that presupposed in the reader familiarity with the elements of mechanics and of the differential and integral calculus. He invited readers who did not have these qualifications to persevere, and he had taken pains to insure that the verbal passages alone should suffice to convey the principal notions of the theory of machines. It is difficult to imagine readers who could in fact have

[28]See also above, Chapter 2, § 2.5.

profited from that invitation, however, although it is true that the mathematical treatment served to express the propositions and not to generate theorems or even insights. The work was not intrinsically mathematical like eighteenth–century rational mechanics or the generalized mechanics of the later nineteenth century.

Nevertheless, comparison of Coriolis to Carnot's *Principes fondamentaux de l'équilibre et du mouvement*, not to mention the *Essai sur les machines en général*, exhibits an enormous development in comprehensiveness, sophistication, and idiom; its canon had become established in the intervening 25 years. Since, except for Navier's notes to Bélidor, there were no treatises that had contributed to that development, it is obvious that it must have come about mainly in an educational milieu, and in the form of courses of instruction in engineering schools: Coriolis taught at *Polytechnique*, Navier at *Ponts et chaussées*, and Poncelet at the *École d'artillerie et du génie* in Metz; and multigraphed lectures formed the literature of the subject.

Inevitably when a number of alert people were working along similar lines under such pressures, questions of priority arose. They were handled with restraint in this instance, and interest us less for themselves than for the evidence the record affords that Navier, Coriolis, and the other protagonists now recognized the pioneering role of Lazare Carnot. Coriolis testified in the preface that when he first became interested, he knew only the works of Lazare Carnot and Guenyveau, and conceived what were then several original ideas. Simultaneously, however, Petit published his note on live forces followed hard after by Navier's edition of Bélidor, which between them contained everything he had thought to be novel in his own studies, the only differences being in mode of treatment. He already had in draft certain parts of what he was publishing in 1829 and had communicated them in 1819 to Mallet, Bélanger, and Drappier all graduate engineers, and also to several students. In 1820 he was in communication with Ampère on the same subjects (as, indeed, Petit had already been), and in 1824 with Poncelet.[29]

The latter exchange created the conditions for a priority dispute that was skirted if not quite averted by the good sense and self–restraint of all concerned. In 1825 Poncelet began offering a course in applied mechanics at the *École d'artillerie et du génie* in Metz, the text of which he circulated in multigraphed form in 1826. Besides supplying it to his students, he sent copies to a few interested persons in Paris, and in 1827 an account of its contents was given to the Academy of sciences, *Rapport suivant sur un mémoire de M Poncelet intitulé Cours de mécanique appliquée aux machines* by Arago and Dupin (1827, VIII, pp 527–531). The topics and treatment were in significant instances similar to the contents of Coriolis' treatise.

Coriolis learned of Poncelet's chagrin, and on 1 October 1829 wrote a note to the editor of the *Bulletin des sciences mathématiques* apologizing that in consequence of a misunderstanding between Poncelet and himself over dates of

[29]A note in Petit's paper (Petit 1818, pp 297–298, ft 1) remarks that Ampère had been studying the application of the principle of live force to the theory of machines and had told Petit that he, Ampère, had already worked out certain of the findings of Petit's paper.

earlier correspondence, certain passages in the original version of the preface to his treatise might be construed as implying that Poncelet had taken advantage of private communications, and advising that persons who already had bought his treatise should substitute for the first two pages of the preface as printed a revised version that he was making available at the publisher's (Poncelet 1829, XII, pp 322–323). In Poncelet's view, the shoe was on the other foot. Not that he accused Coriolis directly of having appropriated his, Poncelet's, findings, but in a lengthy note of 8 October he did take issue with the report on Coriolis' treatise composed by Navier, *Rapport suivant sur le Mémoire de M. Coriolis, intitulé Du calcul de l'effet des machines*, who was secretary of the commission to which the Academy referred it[30] (Prony et al. 1829, IX, 266–272). Navier gave a full and enthusiastic summary. Poncelet neither disputed the merit of the work itself nor enlarged on all the similarities it exhibited with his own. He took exception mainly to Navier's ascription of originality to Coriolis on two specific topics, both of which had already been worked out in his own course, and one of which had then formed the subject of a further note by Cauchy.[31]

The point concerned a general expression for what Poncelet called the "[...] virtual moment of friction in gears [...]" (Poncelet 1829, p 324) and Coriolis called the work consumed by friction in gears (Poncelet 1829, pp 324–325). Poncelet's discussion is interesting in that it constituted a criticism both of d'Alembert's principle and of Carnot's theorem on the grounds that neither took account of the tangential resistances in a system. He described d'Alembert and Carnot, together with Poisson, Navier, and Petit, as the best authors on the subject of impact. He had criticized the theorems, not from the point of view of the inner structure of bodies with its distinction between elastic and hard or inelastic, but from that of the interactions of the moving parts of machinery, from which he then deduced the interactions of the molecules of bodies.

Poncelet's own analysis had shown how to evaluate in general the alterations of velocity and losses of live force caused by impact among the bodies constituting a machine; these occurred in the first instance in consequence of the direct and reciprocal reactions of the bodies themselves, and in the second in consequence of the passive resistance deriving from reaction wherever friction arises between moving parts or bearings. For effects of the first class, Poncelet set up differential equations of motion for each part of a machine, and his criticism consisted in showing that, neglecting tangential resistances and frictions, these equations expressing the normal forces of compression and their reactions reduced to a single linear equation from which, given simple assumptions, first the principle of d'Alembert, and then the theorem of Carnot fell out. The latter he held to be of no practical use, since to apply it presupposed the prior evaluation of the loss or gain

[30]The commission consisted of Navier, Prony, and Girard. Their report was also published in *Bulletin des sciences mathématiques* (Navier, Prony and Girard 1829, XII, 103–116).

[31]*Note sur quelques principes de mécanique relatifs à la science des machines* (Poncelet 1829, XII, pp 323–325).

of velocity in each part of the system according to d'Alembert's principle. In other words, given Poncelet's own more operational and more general point of view, the two principles were tautological nor did he appear to see them as historical stages in the approach to his own point of view.

The second criticism was rather of their adequacy. In the collisions wherein external resistances were in some degree a function of the forces of compression within the system, a further term had to be introduced into the expressions. If, as often happens, that function was not rational and the resulting equations were non-linear, then the results were different from those obtained in applying d'Alembert's principle on the equilibrium between the total variation of the quantities of motion taken to be simple forces of pressure or percussion. In practice, to be sure, d'Alembert's principle was usually a good approximation, and the remainder of Poncelet's note consisted of an analytic method for handling the radicals that then appeared in the equations to be integrated, and a criticism of Coriolis' looseness in failing to provide for the effects of friction and tangential or passive resistances.

Cauchy for his part approached the same difficulty from the point of view of theory[32] He attacked Carnot directly:

> In the various treatises of mechanics it is taught that live forces are lost every time bodies undergo a sudden change in velocity, and that this loss of live force is the sum of the live forces due to the velocities that are lost. But this proposition, which has been named Carnot's theorem, is evidently inexact as is the demonstration on which it purportedly rests.[33]

Essentially, Cauchy's criticism came down to a mathematician's refutation of the physicist's or engineer's distinctions between sudden and gradual change of motion and between elastic and hard body. All that sudden change of relative velocity really meant was that it occurred in a time too short to measure; similarly, the only difference between elastic and inelastic bodies was that in the former the interactions between the molecules depended only on the distances by which they were separated, whereas in the latter the interactions depended both on time and distance. Actually, one might equally well describe as a sudden or instantaneous change of motion any change that occurred in a very short time. Were that to be allowed, however, there would be instantaneous change of velocity in elastic collision–to which Carnot's theorem did not apply. In order to evade that difficulty, writers on mechanics had been forced to say that velocity varied continuously in elastic and discontinuously in inelastic collision. But this was nonsense, and so consequently was the statement that the loss of live force was the sum of the live forces due to losses of velocity in sudden changes. Pursuing the argument towards the goal of a properly general statement, Cauchy arrived at what he considered a preferable proposition:

---

[32]This memoir, *Mémoire sur un nouveau principe de mécanique* (Cauchy 1829, XII, pp 116–112), antedates the Coriolis–Navier–Poncelet interchange. Cauchy read it to the Academy on 21 July 1828 and reported in *Procès–verbaux* (Cauchy 1828, IX, p 94).

[33]Cauchy (1829), p 116.

> When in a system of material points, the velocities change suddenly in consequence of molecular actions produced by impacts between certain parts of the system, the sum of the virtual moments of the quantities of motion gained or lost during impact is zero whenever a virtual motion is being considered in which the velocities of the molecules that act on each other are equal among themselves.[34]

Which principle Cauchy then went on to formulate in analytic terms. But although Cauchy regarded this as refuting, and Poncelet referred to it as superseding, Carnot's theorem, the historian of Carnot's scientific work might prefer to point out that both of them in imagining the interactions of the molecules in a system employed the concept–and Poncelet even used the phrase–of geometric motions–in Poncelet's words:

> [. . .] a geometric or virtual common motion such that they cease from then on to react on each other and their system assumes a form that is stable for each body, one that at that moment allows the ordinary conditions of liaison.[35]

Cauchy observed of his principle that if the impact terminated at the exact instant when any material point that had acted upon another coincided with that other, then the principle yielded all the equations needed to determine the motion of the molecules of the body at that moment, and that one of those equations, specifically the one obtained in supposing the virtual velocities to coincide with the actual velocities after impact, would contain the consequence that the loss of live force was the sum of the live forces due to the velocities lost (Cauchy 1829, p 119). But this consequence, Poncelet was pointing out, justified the principles of d'Alembert and Carnot for what was improperly called the impact of hard bodies, since what was envisaged there was that neighboring molecules should either be joined or else retake invariable distances and that the virtual velocities should be equal and parallel. This came to the same thing as supposing that at the moment of impact the "[. . .] different molecules of the bodies acquired a geometric or virtual common motion such that they cease from then on to react on each other [. . .]" (Poncelet 1829, p 333). In a word, it supposed that after impact, the bodies did not separate or else that they travelled together without any velocity relative to each other along the normals common to their points of contact – and this assimilated Cauchy's analysis to the same terms that he had written off in his course of 1826.

It would appear probable, therefore, that neither Cauchy nor Poncelet had actually read Carnot, for these statements, far from refuting him, in effect repeated in terms of virtual displacements the analysis he had started in terms of geometric motions.

[34] Cauchy (1829), p 119.

[35] Poncelet (1829), p 333.

Chapter 5
An Engineering Justification of Algebra and the Calculus

The geometer Michel Chasles, one of the nineteenth–century personages who was both a master and a masterly historian of his favorite subject, discussed the contributions of the "illustrious Carnot" after the descriptive geometry of Monge and before the projective geometry of Poncelet in the sequence to which he attributed revival of the methods of Desargues and Pascal and a consequent renaissance of geometry following its eclipse by eighteenth–century analysis (Chasles 1875, p 210). It was a central feature of the careers of those Chasles singled out that each of them had his place in the tradition of engineering mechanics and occupied himself in the science of machines concurrently with the new geometry.

In theme and discourse Carnot's mathematical writings are as characteristic as his mechanics of the scientific personality of one interested in things for the purpose of operating with them correctly and most effectively. Carnot came at the mathematical problems that concerned him from even farther behind the front lines than he did in mechanics. The questions he posed had more to do with the foundations and procedures than the content of mathematics, and were largely about the meaning of infinitesimal and negative quantities and their significance in the distinction between analysis and synthesis. He put them with the bold simplicity of an engineer for whom any mathematical expression had merely auxiliary standing unless it represented an operation that could be carried out metrically or physically. In the *Géométrie de position* of 1803 (Carnot 1803b) he distinguished the notion of a true quantity from that of a "value," which pertained merely to algebraic functions: "Every quantity," he asserted, "is a real object such that the mind can be seized of it, or at least its representation in calculation" (Carnot 1803b, p 7).

It is evident from the scientific manuscripts surviving at Nolay that the problems occupying Carnot's mind in the years following his return from exile in 1800 were mainly those of geometry and technology. The family archives there contain notes on the inventions referred to his judgment by the Institute, including the Niépce and the Cagniard engines, together with the manuscripts of reports submitted for the *Procès–Verbaux*. They also contain numerous drafts on what actually motivated most of his writings in geometry, the anomaly of negative quantity, together with

C.C. Gillispie and R. Pisano, *Lazare and Sadi Carnot: A Scientific and Filial Relationship*, History of Mechanism and Machine Science 19, DOI 10.1007/978-94-017-8011-7_5,

fragmentary correspondence with his German translators that gives the historian welcome insight both into his purposes and the state of his mathematical reputation. As remarked in the preface, it was through his exchanges with J.–K.–F. Hauff, who translated the first, 1797, edition (Carnot 1797) of *Réflexions sur la métaphysique du calcul infinitésimal* into German (Carnot 1800b) in 1800 that we learned of the origin of that work in the prize essay submitted to the Berlin Academy in 1785. In a reminiscence Hauff acknowledged that when he had initially seen the title listed by the bookseller, he had paid small attention. The author, though widely known to be one of the most distinguished of practitioners –"eines der vorzüglichsten Pratiker"– had no reputation in theoretical mathematics. Hauff would, indeed, have expected little from a French metaphysicist on the subject of geometry, as would the Hofräth Kästner to whom he happened to mention it. Of this initial prejudice he had been disabused by Lacroix, who informed him that, on the contrary, Carnot's was a mathematical intelligence to be taken seriously. His misapprehension thus corrected by the most widely read of contemporary mathematicians, Hauff set to work to master the argument of the *Réflexions sur la métaphysique du calcul infinitésimal* and, having done so, to translate it, since the work was one that provided all the foundation the subject might require.[1]

There is more distinguished testimony to his contemporary reputation in geometry. In 1810 Professor H. G. Schumacher of the University of Copenhagen published a German translation of the *Géométrie de position*. He had consulted Gauss on the importance of the work and in the preface mentions with gratitude Gauss's good opinion of it–a good opinion not (it may be remarked) easily given.[2] Among admirers qualified to judge later in the century, Chasles was not alone in the appreciation mentioned at the beginning of this chapter. In 1854 Giusto Bellavitis referred to the same work as the origin of his own method of equipollencies, and Felix Klein in the well–known lectures published in 1926 referred to it as a "noteworthy book".[3] It is odd, therefore, that Carnot's mathematical works should have fallen into such obscurity since the earlier twentieth century that few mathematicians can say what they contained.

His writings fall into two groups, one consisting in the geometric works and the other in the successive versions of *Réflexions sur la métaphysique du calcul infinitésimal*. It will be convenient to consider the geometric writings first since their relation to his work in mechanics is more explicit, though not necessarily deeper, and

---

[1] Hauff's translation appeared under the title, *Betrachtungen über die Theorie der Infinitesimalrechnung von dem Burger Carnot* (Carnot 1800b). The relevant papers are in carton 28 in the Carnot family archives at Nolay.

[2] *Geometrie der Stellung oder über die Anwendung der Analysis auf Geometrie,* 2 vols (Carnot 1810, II).

[3] Ettore Carruccio, "Bellavitis", *Dictionary of Scientific Biography* (Gillispie 1970–1980, I, pp 590–592). See also *Vorlesungen über die Entwicklung der Mathematik im 19* (Klein 1926, I, p 79). Lazare Carnot introduced the concept and word "equipollence" (Carnot 1803b, § 82, pp 83–84) to signify the equivalence, not just of values, but of any mathematical objects such as points or curves that could be substituted for one another.

then to go on to an account of his justification of the calculus. That was the subject of his earliest work in mathematics, the Berlin dissertation of 1785, and also of his last: he published the final revision in 1813.

5.1 Geometric Analysis and the Problem of Negative Quantity

Lazare Carnot published *De la corrélation des figures de géométrie* in 1801 (Carnot 1801), a year after leaving ministerial office under Napoleon. It contains the initial working out of the views with which he reentered scientific work after nearly a decade in public life. He opened the subject with the observation that between the parts of a geometric figure two sorts of relations obtain, relations of magnitude and of position. His topic was limited to the latter. Its problems were to determine whether a point was situated above or below a given line, to the right or left of a given plane, inside or outside a given circumference or curved surface, etc. The method would be to compare every figure he was investigating to another of known properties, to exhibit the modifications point to point by means of a systematic notation, and thus to "establish the correlation of the figures" (Carnot 1801, § 2, p 1).

The approach will be familiar to the student of Carnot's mechanics. The reader was to imagine any system whatever of variable quantities, whether or not geometric, and to consider it in two different states. The one taken for the basis of the comparison he called the primitive and the other the transformed system. In the analysis of the latter the difference between any two quantities was "direct" when the greater and the lesser corresponded to their analogues in the primitive system. In the contrary case, the difference was "inverse." (In these comparisons, Carnot always meant real quantities and not fictitious values of merely algebraic significance.)

As was his wont, Lazare Carnot began developing his point of view with an elementary example.

Fig. 5.1

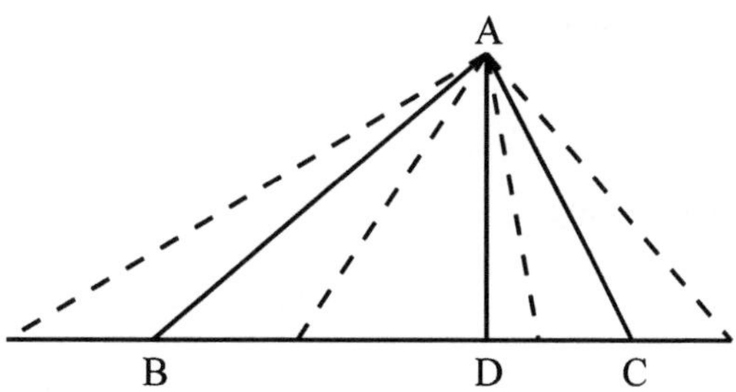

In the triangle *ABC* (see Fig. 5.1) a perpendicular $\overline{AD}$ is lowered from the apex to the base $\overline{BC}$. In the initial construction, the point *D* falls between *B* and *C*. This is the primitive system. It may be transformed by moving the point C to the left. In effect the figure constitutes a system in which the segment $\overline{CD}$ is a variable and $\overline{BD}$ is a constant. So long as the point C approaches but does not reach the point D, it remains true of both primitive and transformed systems that $\overline{BC} > \overline{BD}$. Their difference $\overline{CD}$ is a direct quantity, and $\overline{BC}$ and $\overline{BD}$ are in direct order in the two systems. By definition, therefore, the correlation is direct. Let the point C pass to the left of D, however, and then in the transformed system $\overline{BD} > \overline{BC}$. Their difference $\overline{CD}$ is inverse, and the quantities $\overline{BC}$ and $\overline{BD}$ are in inverse order compared to the primitive system. The correlation, therefore, becomes indirect.

By correlative systems Lazare Carnot meant those that could be considered as different states of a single variable system undergoing transformation by insensible degrees. But it is important to notice that the question was not simply one of stretching or rearranging the parts. It was not required that all correlative systems should actually have been evolved out of the primitive system. It sufficed that they might be assimilated to the primitive system through mutations that could be imagined to occur by insensible degrees. We are back with Carnot's favorite reasoning device: a comparison of systems between which the nexus of change is a continuum. Such were correlative systems, and the quantities that corresponded were correlative quantities.

When the correlation was direct, any train of reasoning that was valid for the primitive system would hold for the correlative. For example, in the above system *ABCD* (see Fig. 5.1) we know that in the right triangle *ABD*,

$$\overline{AB^2} = \overline{BD^2} + \overline{AD^2}$$

and that in the right triangle *ACD*,

$$\overline{AC^2} = \overline{CD^2} + \overline{AD^2}.$$

Subtracting,

$$\overline{AB^2} - \overline{AC^2} = \overline{BD^2} - \overline{CD^2}.$$

Furthermore,

$$\overline{BD} = \overline{BC} - \overline{CD}.$$

Substituting,

$$\overline{AB^2} - \overline{AC^2} = \overline{BC^2} - 2 \cdot \overline{BC} \cdot \overline{CD}.$$

This same reasoning would be true of all the transformed systems as C moved towards D so long as CD was a direct quantity.

Once C had passed the point D, however, CD would become an inverse quantity and the correlation indirect. The final equation would then be

$$\overline{AB^2} - \overline{AC^2} = \overline{BC^2} + 2 \cdot \overline{BC} \cdot \overline{CD},$$

in which the sign of the last term was inverted in the correlative system. It was–to generalize the example–the characteristic of indirectly correlative systems that the formulas of the primitive system might be applied to the transformed system by virtue of changing the signs of all the variables that had become inverted.

Reciprocally, these procedures of correlation might be used for solving problems, and that is why they held the appeal for Lazare Carnot that led him to develop their application to analysis in general. Reasoning on the same example, suppose it was given that the three sides were in the proportion,

$$\overline{AC} = \overline{BC} = \frac{2}{3}\overline{AB},$$

and the problem was to find the segment $\overline{CD}$. It is unknown whether C falls to the right or to the left of D. Trying the former hypothesis first, would give

$$\overline{AB^2} - \overline{AC^2} = \overline{BD^2} - \overline{CD^2},$$

which works out

$$\overline{CD} = -\frac{1}{12}\overline{AB},$$

a negative value, and hence anomalous as a solution. What the minus sign actually signified was that the initial formulation had been wrong, and that in fact C fell to the left of D. When the right assumption is made, the problem yields a positive value in the result, and if the calculation has been valid, this confirms the correctness of the choice among the possible conditions of the problem (Carnot 1801, § 33, pp 18–19).

So much by way of sample from Carnot's *De la corrélation des figures de géométrie* will show how the anomaly of negative quantity led him into the problem–solving reaches of geometry, the branch of mathematics to which he was in any case best suited by temperament and training. Apparently Carnot's contemporaries did not find it easy to form a clear idea of precisely what his theory of negative quantity was amid all the geometry with which he argued it, for to the last of his original publications, the essay of 1806 developing the theory of transversals, he appended a brief *Digression sur la nature des quantités dites negatives* (Carnot 1806a, pp 96–111) in *Mémoire sur la relation qui existe entre les distances respectives de cinq points quelconques pris dans l'espace, suivi d'un essai sur la théorie des transversals* (Carnot 1806b). He there explained that he had written it at the behest of several "savans du premier ordre"–one may guess that Lacroix may well have been among them–who had urged him to disengage his theory of negative quantity from its extensive detail "[. . .] eliminating from the discussion everything that was not strictly elementary [. . .]" (Carnot 1806a, p 96). In 1810 Schumacher attached a German version to the first volume of his translation of the *Géométrie de position*, and quoted a letter from Carnot to the effect that many duties and weakening health no longer permitted him to return to these subjects, and that in any case this epitome was the best he could do (Carnot 1810, I). We may conveniently follow Schumacher's example and guide ourselves on this abstract of the argument before passing on to an account of other aspects of the larger of the works that it informed, the *Géométrie de position*.

Doing algebra involves manipulating negative values at every turn. What might they signify? Clearly the signs $+$ and $-$ merely indicate operations to perform, and addition presents no difficulty since $+a$ can always be regarded as $0+a$. As for negative quantity, however, two schools had developed in the course of the many discussions the problem had evoked in the eighteenth century. One interpretation accepted the notion of quantities less than zero. The other considered that a minus sign meant that a quantity was to be evaluated in the direction opposite to that of a positive quantity. Lazare Carnot found both explanations unconvincing. As for the former, he asserted roundly that the notion of something being less than nothing was absurd. It was here that he introduced his distinction between a quantity properly speaking and an algebraic value, the latter being a merely fictitious entity introduced for purposes of calculating. If it was permissible in a calculation to neglect quantities of no magnitude (as it was), then it ought surely to be justifiable to ignore those less than zero if such there were. But everyone knows that negative terms cannot be neglected. Thus, whatever they signify, it cannot be quantity (Carnot 1806a, p 98).

Advocates of the second view, on the other hand, never defined what they meant by a direction opposite to the positive. Sometimes they adduced the amount of a debt opposed to that of a credit. Sometimes they referred to a left ordinate opposed to a right one. But these were examples, not definitions, and they left difficulties. Why, for example, is it impossible to take the square root of a negative quantity? There is no difficulty in extracting the square root of a debt or a left ordinate. Why, to consider a related point, did negative quantities dominate in the multiplication of unlikes, giving their sign to the product? Moreover, there were exceptions. The secant to an arc of a circle in the third quadrant cannot be distinguished from the secant to the opposite arc in the first quadrant in either magnitude or direction, and it should accordingly be positive in sign. In fact it is negative.

In the place of such interpretations, none of them satisfactory, Carnot proposed two senses in which negatives could be properly understood. In the more obvious, a negative quantity was the magnitude of a value governed by a minus sign. Strictly speaking this usage was correct only when the value was preceded by a positive one of greater magnitude, and was, therefore, of relatively trivial importance. The deeper and more revealing sense was that a negative quantity is a magnitude governed by a sign that was wrong. What Carnot meant by wrong may be illustrated in the problem just cited in which a negative result indicates that the initial assumption about the correlation of the figures was mistaken and that in the conditions of the problem *BC* was smaller than *BD*, whereas it had been initially taken to be larger (see Fig. 5.1).

In the *Digression sur la nature des quantités dites négatives* Lazare Carnot distinguished between the two senses of negative quantity by means of a trigonometric example. Consider the formula (Carnot 1806a, p 99)

$$\cos(a+b) = \cos a \cdot \cos b - \sin a \cdot \sin b \; \ldots (A) \tag{5.1}$$

Here the last term was negative in the first or literal sense. In the second sense, however, the same term was positive so long as *a*,*b* and $(a+b)$ were each less than 90° because the sign had to be correct if the equation were to be exact in the first quadrant. But if $(a+b)$ were greater than 90°, the equation would be incorrect. For if the problem were investigated directly and synthetically, it would turn out that (Carnot 1806a, p 99)

$$\cos(a+b) = \sin a \cdot \sin b - \cos a \cdot \cos b \ldots (B) \tag{5.2}$$

If Equation (A) (Eq. 5.1) were to apply to this case, it would be necessary to regard the first term, $\cos(a+b)$ as governed by a sign contrary to what it ought to have been, for changing it would have given (Carnot 1806a, p 103)

$$-\cos(a+b) = \cos a \cdot \cos b - \sin a \cdot \sin b \ldots$$

which would have reduced to Equation (B) (Eq. 5.2). In effect, therefore, extending Equation (A) (Eq. 5.1) to the case in which $(a+b)$ was greater than 90° subjected the term, $\cos(a+b)$, to the wrong sign and made of it a negative quantity in the second sense. Obviously these two meanings of negative were quite different, and for the second sort of quantity, those bearing a wrong sign, Carnot preferred the term "inverse quantity"[4] in contradistinction to "direct quantities"[5] governed by their right and proper sign. The usage is that of the correlation of geometric figures, which would frequently be the means of establishing whether a quantity was plus or minus, direct or inverted.

How, then, did these inversions get into the process of calculation? Carnot said by error, the kind of mistake or misapprehension about the basic conditions of the problem illustrated in the example of correlation cited above. If in setting up a calculation we mistake a credit for a debt, then the algebraic expression of what is to be paid or received is opposite to what it should be, and we recognize the error practically through knowledge of whether money is owed or owing. Similarly, if we mistake a right for a left ordinate in formulating a problem, we get the wrong sign in the result and know it to be so since the solution makes no sense physically. In general,

---

[4]On "inverse quantity" see: Carnot (1806a, p 101, p 102, p 103, p 104, p 111).

[5]On "direct quantity" see: Carnot (1806a, p 100, 102, p 111).

> Any inverse quantity can be considered as the difference of two direct quantities of which the greater has been taken for the smaller, and reciprocally.[6]

A skeptic might be forgiven for asking the use of introducing into calculation quantities governed by a false sign. The answer should be that it is often necessary to risk doing so in order to formulate a problem at all. The ordinate of a curve might be required without the geometer's knowing in which quadrant it was located. He would then simply make an assumption. That would permit him to give an absolute value. If then he had been wrong about the sign, the error would show up as an absurdity in the result, and he could change signs without redoing the calculation. The procedure turns out to involve no real risk of final inaccuracy, therefore. Since only a positive value ever represents an intelligible solution to a genuine problem, a negative solution merely reveals an inconsistency between the actual initial conditions and the hypothesis that had been adopted in framing the equations.

Trigonometry was to Carnot in mathematics what hydrodynamics was in mechanics–the part of the science in which his mind ran on the problems exemplary of the general issues.[7] Discussing the transition from an inverse to a direct quantity and vice versa, he invoked the algebraic principle that a variable in changing sign passes through zero or infinity. The point of view was one that naturally embraced, and in his own mind probably grew out of, the change of sign in trigonometric functions. Carnot thought that phenomenon to be the decisive one in destroying the position of those who held negative quantity to signify nothing more than ordinary magnitudes taken in reverse direction. Why, for example, should the cosine of an obtuse angle necessarily be an inverse quantity? It was so only because the primitive system, i.e., that on which the reasoning in trigonometry had originally been founded, was one in which all angles were acute. The instrument best adapted to analyzing relations of this sort was the geometry of position, and that was why it was the approach to geometry most suited to clarifying its application to analysis and algebra in general.

The *Géométrie de position* was published in 1803, the same year as *Principes fondamentaux de l'équilibre et du mouvement*. It is an imposing book, a book more important for its approach to geometry than for the theoretical point it purported to

[6] *Digression sur la nature des quantités dites négatives* (Carnot 1806a, p 102).

[7] It is interesting that the first published hint of the correlation of figures, and also the first original scientific work to which Carnot seems to have turned after the Revolution, appeared in the form of a "Lettre du citoyen Carnot au citoyen Bossut, contenant quelques vues nouvelles sur la trigonométrie". It is appended to Bossut's *Géométrie et application de l'algèbre à la géométrie* (Bossut 1800a, II, pp 401–421). Bossut recounts that just as his book was about to go to press, Carnot, then serving as Minister of War, mentioned to him several ingenious and stimulating ideas on trigonometry. At Bossut's persuasion, he agreed to write them up for publication in a way suited to an elementary text. He had leisure to give only the results, in any case, leaving the proofs to the students as an exercise. The subject is that which he developed fully in section 4 of *Géométrie de position,* the relation between the linear quantities of a triangle eliminating all angular quantities from the expressions.

be establishing. It cannot be said that Carnot's theory of negative quantity entered into the texture of later mathematical thinking, but writing the *Geometrie de position* (Carnot 1803b) carried him far beyond his original intention of giving a second, somewhat fuller edition of *De la corrélation des figures de géométrie* (Carnot 1801). He enlarged the examples and treated the rules of correlation in an altogether more comprehensive manner. Two years seem a short time for the development he gave the subject, and it would appear both from internal evidence and from the number of drafts of particular problems in the family archives that he must have had much of this material long in mind and that he worked very hard to set it out in order. No longer limiting the scope to correlations of particular geometric systems, he now proposed to compare and unify the two main types of geometric relations and thus to associate in a single treatment relations of magnitude with relations of position[8] (Carnot 1803b, p 1).

Once past the first section on negative and imaginary quantities, the work contains much substantial geometry and trigonometry, which, like any actual mathematics, does not lend itself to summary as do questions of principle, method, and lineage. The second section proposes a notation and mechanism for exhibiting the modification that figures undergo through changing the relative positions of their significant elements–points, lines, angles, arcs, areas, etc. Consider, for example (see Fig. 5.2) the correlative triangles $ABCD$ and $A'B'C'D'$ (in principle the same figures as in the earlier work)[9]:

Fig. 5.2

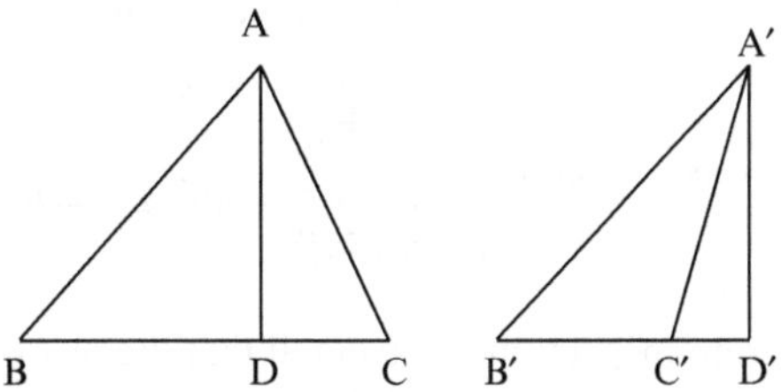

For the points, two categories of correlation are relevant, correlations of construction and of position. Points that correspond under the correlation of construction are

First system	$ABCD$
Second system	$A'B'C'D'$.

[8]In a letter to Lacroix of "4 Nivôse an 10 [25 December 1801]" Lazare Carnot observes that he is spending his leisure developing the ideas in the *De la corrélation des figures de géométrie*, which had suffered from the haste with which it was composed, for a new edition containing a chapter on the applications of algebra to geometry and another on the theory of curves, neither of them a treatise in itself, but applications of his theory of positive and negative quantity and of transversals (*Bibliothèque de l'Institut de France,* Mss. 2397).

[9]This example is worked in great detail on Carnot (1803b, pp 81–87).

The correlations of position, however, are

First system	$\overline{BDC}$
Second system	$\overline{B'C'D'}$.

(The bars signify that the points are on a straight line.)

As for lines, angles, arcs, and areas in the two systems, further correlations are needed to complete the comparison, specifically correlations of absolute value and of signs. Given the correlation of construction just cited, correlations of absolute–value may be written down directly:

$$\text{First systems}\ \begin{cases} \overline{AB},\ \overline{AC},\ \overline{BC},\ \overline{BD},\ \overline{CD},\ \overline{AD},\ B\hat{A}C, \\ \overline{A'B'},\ \overline{A'C'},\ \overline{B'C'},\ \overline{B'D'},\ \overline{C'D'},\ \overline{A'D'},\ B'\hat{A}'C', \end{cases}$$

$$\text{Second system}\ \begin{cases} A\hat{B}C,\ \ A\hat{C}B,\ \ B\hat{A}D,\ \ C\hat{A}D,\ \ \overline{\overline{ABC}},\ \overline{\overline{ABD}},\ \overline{\overline{ACD}}. \\ A'\hat{B}'C',\ A'\hat{C}'B',\ B'\hat{A}'D',\ C'\hat{A}'D',\ \overline{\overline{A'B'C'}},\ \overline{\overline{A'B'D'}},\ \overline{\overline{A'C'D'}}. \end{cases}$$

(The double bars signify an area or surface, and such expressions as *BAC* the angle *BAC*). From the correlation of positions, it is evident that there is an inversion in the order of the points *D*, *C* and D', C'. This inversion establishes the correlation of signs, and the general tabulation of correlations of the two systems is thus:

$$\text{First system}\ \begin{cases} +\overline{AB},\ +\overline{AC},\ +\overline{BC},\ +\overline{BD},\ +\overline{CD},\ +\overline{AD},\ +B\hat{A}C, \\ +\overline{A'B'},\ +\overline{A'C'},\ +\overline{B'C'},\ +\overline{B'D'},\ -\overline{C'D'},\ +\overline{A'D'},\ +B'\hat{A}'C', \end{cases}$$

$$\text{Second system}\ \begin{cases} +A\hat{B}C,\ \ +A\hat{C}B,\ \ +B\hat{A}D,\ \ +C\hat{A}D,\ \ +\overline{\overline{ABC}},\ \ +\overline{\overline{ABD}}, \\ +\overline{\overline{ACD}} \\ +A'\hat{B}'C',\ +A'\hat{C}'B',\ +B'\hat{A}'D',\ -C'\hat{A}'D',\ +\overline{\overline{A'B'C'}},\ +\overline{\overline{A'B'D'}}, \\ -\overline{\overline{A'C'D'}} \end{cases}$$

From inspection, we can tell that the primitive system *ABCD* has been transformed into the correlative $A'B'C'D'$ in consequence of a movement of the point *C* on $\overline{CB}$ so that it is now situated between *B* and *D*. The table also tells us that none of the quantities $\overline{AB}$, $\overline{AC}$, $\overline{BC}$, $\overline{BD}$ or $\overline{AD}$ passed through zero or infinity, and consequently that these are direct quantities in the correlative system and there carry a positive sign. Not so the quantity $\overline{CD}$, however, which did pass through zero in becoming $\overline{C'D'}$, and therefore becomes inverse and bears (as the table shows) a minus sign. So it is, not to insist further on the obvious, for all the other quantities in the two systems–their relations and signs may be found by inspection of the table, which gives at a glance the transformation undergone by any quantity of the primitive system in becoming its correlative.

The tables of correlation that Carnot established for certain more complex systems cover several pages of text. Among the more obviously useful was a complete tabulation of the correlations in all four quadrants of what he called lineo–angular quantities, i.e., sines, cosines, tangents, etc., which exhausts the entire theory of the signs of trigonometric functions. He appears to have thought of these complications as the nucleus of a sort of engineering handbook of geometric systems that would permit resolving problems by considering unknown systems as correlatives of the set of primitive systems of which the properties were known. The formulas expressing the relations were to contain only real and intelligible expressions–no imaginary and no inverse quantities. Then the procedure was to ascertain for a given correlative system what mutations the primitive system underwent in the transformations that produced it, a change that he always considered to occur continuously and one that never modified the basis of the original construction but only the relative positions of the elements.

The body of the *Géométrie de position* consists of various sorts of problems resolved by the technique of forming tables of positional properties. We shall have to limit ourselves to a few among a great many examples. Thus,

> Three right lines being drawn in the same plane, and given the angles formed by the directions of any two of them with the direction of the third, find the angle formed between the first two directions.[10]

That constitutes the first problem in a lengthy section devoted to the techniques of determining relations among systems of straight lines without having recourse to trigonometric functions. A further section applies the method to various problems of elementary geometry–e.g., given a circumference and two tangents, construct between them a third tangent equal to a given line. A final section adapts the correlation of figures to the solution of problems by the transformation of coordinates.

The contents consist in good, hard geometry approached algebraically. As time went on, it was the train of concrete problems and theorems, rather than the theory of negative quantity inspiring it, that won esteem for Carnot as one of the founders of nineteenth–century geometry. The most original propositions concern the theory of transversals, on which Carnot published a further small treatise, his last, 4 years later (Carnot 1806b). Felix Klein singled out for appreciation the theorem sometimes called Carnot's in the nineteenth century. It asserts the equality of the products of the segments created by a transversal intersecting the three sides (prolonged as need be) of a triangle: the product of any three segments always equals the product of the remaining three provided no two segments with the same angle of the triangle or point of the transversal for extremity enter as factors into either product.[11] Otherwise, Klein complained of the elementary nature amounting almost to triviality of many of the propositions, while acknowledging that the book was full of good ideas. It was, he said acutely enough, a reflection of Carnot's personality: honesty without genius. Chasles' final judgment was that Carnot's work formed only a series of powerful inductions and that he never did achieve a rigorous, absolute and fundamental demonstration of the principle of correlation of figures (Chasles 1852, p v).

---

[10] Carnot (1803b, Problem XXIV, § 189, p 255).

[11] *Vorlesungen über die Entwicklung der Mathematik im 19. Jahrhundert* (Klein 1926, I, pp 79–80). This proposition appears as "Theorem V" (Carnot 1803b, § 219, pp 276–278).

Lazare Carnot himself recognized the affinity of his method with that of the porisms of the ancients (for he was well informed about the history of geometry).[12] He acknowledged that the greater part of the problems in this, his fullest work, belonged to elementary geometry but considered the characterization no matter for apology. That having been the only geometry known to a Napier, a Viète, a Fermat, a Galileo, a Huygens, that having been the mathematics preferred by a Halley, a Maclaurin, and a Newton its fundament might even be allowed to offer certain advantages. Elementary though it was in subject matter, it was certainly not so in point of difficulty or intellectual value. It was important, moreover, to take a wider and more generous view of its matter. It was absurd, for example, that the science of projections, the descriptive geometry renewed by Monge, should form a science separate from the elements of geometry. To plead that its problems were applied geometry was a quite inadequate excuse, for in actual fact to separate principles categorically from applications was an impossibility. Euclid's *Elements* nowhere contain instructions for the techniques of projection that, applications though they might well be, embodied principles as theoretical as any in the elements themselves. A similar judgment could be made of the study of polygons and polyhedrons, which might be thought comprised in trigonometry, but so implicitly that to disengage it amounted to starting a new science.

Indeed, Carnot's handling of geometry exhibits throughout what we have come to recognize as his hallmark, the gathering of elementary subject matter under a point of view at once general and operational and in this combination of qualities novel. The statement of the problem of mechanics that opened the *Essai sur les machines en général* (Carnot 1786, § X, p 21) is echoed in the definition of the problem of which the study constituted geometry:

In any system of right lines whether lying in the same plane or not, and some of them, or else of the angles formed by the assembly either of the lines themselves or of the planes

Fig. 5.3 Simplified performance model adapted by Klein (1926)

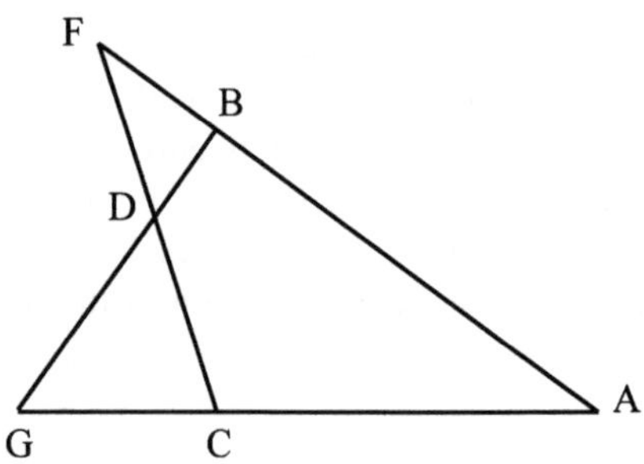

The triangle *GBA* being intersected by the transversal *CF*,

$$\overline{AF} \cdot \overline{BD} \cdot \overline{GC} \cdot \overline{AC} \cdot \overline{BF} \cdot \overline{GD}$$

[12]Carnot (1803b, p 483).

containing them, being given in sufficient number so that the entire figure is determined; then find all the rest of it.[13]

Putting the problem of his book thus, Carnot permitted himself one of the few self–assertive asides that his scientific writings contain, remarking, "It is a long way from what is ordinarily called the elements of geometry to this general problem" (Carnot 1803b, p xxxiij).

More than the style carries over from mechanics into Carnot's vision of a new, a more vigorous geometry. At the outset he distinguished the "géométrie de position" from "géométrie de situation" (Carnot 1803b, pp xxxv–xxxij) the subject (since called topology) started by Leibniz although no comprehensive treatise yet existed. Carnot thought that the term "géométrie de transposition" would have been more appropriate since movement among the parts of a system was an essential aspect of all its problems. Although it was distinct from his own subject, he imagined a potential relation between them. A geometry of transposition would be to the geometry of position what motion was to rest. At that it would be but the least part of a very extensive, highly important subject, one that never had been treated for itself, the general theory of motion abstracted from consideration of the forces that produce or transmit it. The new science, in a word, would be that of geometric motion (Carnot 1803b, p xxxvij).

To the reader of the *Essai sur les machines en général* it will be obvious that Carnot had more in mind than mere kinematics here, and indeed that he was coming at his favorite notion from a point of view complementary to that of his mechanics. He gave a further hint of what he had in mind at the conclusion of a series of theorems about a point called the center of mean distances, better known to mechanics as the center of gravity. It seemed to Lazare Carnot that it would contribute to the vitality of geometry to recapture that topic from mechanics. For geometry was impoverished in being limited to its conventional elements and to determinations of equilibrium, that is to say, to statics. Strictly speaking, it ought to be geometry that considered motions per se, mechanics being rather the science of the communication of motion. The idea of motion was, after all, as simple as that of dimension. It might even be genetically inseparable. Pedagogically, a line is said to be the path of a point and is in fact traced out by a pencil in motion. In turn the motion of a line is said to produce a plane and that of a plane, a solid.

Why should not the method be extended in order that geometry consider abstractly what is produced by the motion of a solid in space? For mechanics cares nothing about that as motion. It limits itself to what happens in encounters with other bodies along the way. More comprehensively, three categories of spatial relations between bodies might be imagined. In the first, bodies are entirely separate: conventional geometry pertains to these relations. Second, they may interact by collision, pressure, or traction: this is the proper domain of mechanics. Third, they may be in contact at several points without in any way interacting: and it was determination of these relations that would form the subject of a new science of

[13]Carnot (1803b, p xxxiij).

geometric motions that would be transitional between the other two. To determine those conditions presented a problem independent of mechanics since, by definition, no motion was communicated or destroyed within the system, and Lazare Carnot here restated the criterion in terms of reversibility that he had given in the *Essai sur les machines en général* in 1783:

> [...] when such a motion exists in a system of bodies, the contrary motion is always possible; which is not the case when the motion is not geometric.[14]

Once the theory of such motions was fully penetrated, it could be counted on to rid mechanics and hydrodynamics of their analytical difficulties, because it would then be possible to base both sciences entirely on the most basic general principle of communication of motion, the equality in opposition of action and reaction, always the starting point of Carnot's analysis. From these aspirations it becomes clear that the appearance in the same year of both *Géométrie de position* and *Principes fondamentaux de l'équilibre et du mouvement* was the result of no mere coincidence or fit of industry. They were parts of the same program. The introduction to *Géométrie de position* promises the reader another work on the idea of geometric motion, "the theory of which is the transition from geometry to mechanics" (Carnot 1803b, pp xxxv–xxxvi) and the *Principes fondamentaux de l'équilibre et du mouvement* redeems the promise, recasting the entire theoretical part of the science of machines in terms of geometric motions. Not that he flattered himself actually to have created the science to which all this was a summons–his research extended only to the principal properties of these hypothetical motions (Carnot 1803a, § 145, p 116).

5.2 The Compensation of Error in Infinitesimal Analysis

From the Berlin dissertation of 1785, it is obvious that Lazare Carnot's thinking about infinitesimal analysis long antedated any recorded interest in the problem raised by the role of negative and imaginary quantity in finite analysis. In the essay on that document that Professor A. P. Youschkevitch (1906–1993) has contributed to the first edition of the present work (Gillispie 1971), which completes its discussion of Carnot's mathematical work, he compared this earliest version to the first edition of the *Réflexions sur la métaphysique du calcul infinitésimal* of 1797 and to the further revision that Carnot gave the book in its second edition of 1813 (Gillispie 1971, pp 149–168). These views, too, he was developing concurrently with his preoccupation with the mechanics of machine processes, and although his treatment has never commanded great interest on the part of professional mathematicians, its immediate translation into many languages and republication from time to time for over a century, is evidence of its appeal to readers professionally concerned with

[14] Carnot (1803b, § 298, p 338).

mathematics. The text of Lazare Carnot's final version, that of the 1813 edition, was unchanged in later printings and forms the basis of the account that follows.[15]

Such a reader of mathematics, one whose own education in the calculus recapitulated like that of many an engineering student its progress from l'Hôpital to Cauchy or thereabouts, may peruse the opening passages of Lazare Carnot's *Réflexions sur la métaphysique du calcul infinitésimal* with the gratifying feeling that here is an author declaring what one thought one had noticed oneself but never dared to come out and say. To Lazare Carnot, it was obvious that the infinitesimal calculus had produced the speediest and most fortunate of revolutions in the mathematical sciences. Decomposing bodies right down to their elements, it seemed to give entree into the internal structure and very organization of things. (Notice that Carnot began his analysis in mathematics as he did in his earliest mechanics memoir in the instinct that the texture of reality is corpuscular).[16] Yet those who handled this keen and powerful tool left unclear what those elements actually consisted in, those

> [...] singular beings that at one time play the role of true quantities and at another are to be treated as absolutely null and that seem by their equivocal properties to occupy some middle ground between magnitude and zero, between existence and nullity.[17]

Not that the difficulty had impeded discovery, for mathematicians had robustly gone on to employ infinitesimals without penetrating the question of their nature or dissipating its obscurity. Philosophers for their part remained divided not only in their views, but even in their way of seeing the problem. In this situation, what he was proposing was to bring the existing points of view together, show the relation between them, and justify the procedures of the calculus. In good eighteenth–century style, he began genetically.

The very notion of the calculus, in Carnot's view, had arisen pragmatically out of the necessity to make approximations in calculating. Given the normal difficulty and frequent impossibility of formulating the conditions of a problem precisely in solvable equations, it would have been natural to neglect embarrassing quantities when they were so insignificant that eliminating them might be supposed to introduce only trivial error into the result. Such was the ancient geometric practice of treating curves inaccessible to theory as if they were polygons composed of a very large number of very short sides. To take the most familiar example, that of a regular polygon inscribed in a circle, when the number of sides is increased sufficiently, it becomes possible without appreciable error to attribute to the circle the properties of the polygon. A particular calculation might then be simplified by neglecting the length of one of these now very small sides by comparison with that of some given

---

[15] Page references are to the most recent edition, that published in Paris in 1921 in the series *Les maitres de la pensée scientifique* (Carnot 1803b). Paragraph numbers apply equally to other, earlier French editions.

[16] See above, Chapter 3, pp 66–67.

[17] Carnot (1921, I, § 1, p 2).

line, say the radius. It might be a great convenience to be able to substitute the length of the latter for the sum of the radius and the side. If the latter was small enough, the error would not be worth calculating.

But how legitimate was such a shortcut?

In arguing that the metaphysics of the calculus rested upon its rigorous legitimacy, Lazare Carnot continued the illustration. Suppose the problem is to construct a tangent at the point M to the circumference MBD (see Fig. 5.4) of a circle of center C and radius r (Carnot denotes the radius by a but it will denature nothing to alter that).

Fig. 5.4

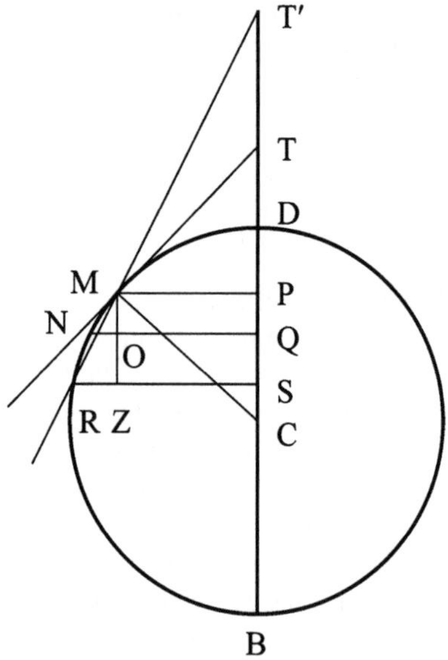

The axis DCB serves as abscissa and D as origin, so that $DP = x$, the corresponding ordinate $MP = y$, and the point M has the coordinates (x, y). TP then is the required subtangent.

To find it, consider the circle to be a polygon with a very large number of sides. Let MN be one of the sides D prolonged to intersect the axis at T'. Evidently MN will be the tangent in question since by definition it does not penetrate the interior of the polygon. If, then, the perpendicular MO be erected intersecting at point O the line NQ parallel to MP, we should have

$$MO : NO :: TP : MP$$

or

$$\frac{MO}{NO} = \frac{TP}{y}.$$

Further, since the equation of the curve for the point M was

$$y^2 = 2x\,(r - x),$$

for the point N, the equation would be

$$(y + NO)^2 = 2r\,(x + MO) - (x + MO)^2.$$

Subtracting the latter equation from the former and reducing,

$$\frac{MO}{NO} = \frac{2y + NO}{2r - 2x - MO},$$

so that, setting the two values for $\frac{MO}{NO}$ equal to each other and multiplying through by y,

$$TP = \frac{y\,(2y + NO)}{2r - 2x - MO}.$$

If MO and NO were known, we should have the value of TP and the construction of the tangent. In actuality, however, these quantities MO and NO were by hypothesis smaller than the side MN, itself very small. They were, therefore, negligible by comparison to the lengths, $2y$ and $2\,(x - r)$, to which they were added. Hence, the equation would be reduced to

$$TP = \frac{y^2}{r - x}.$$

The result would thus be acceptable in practice even though not absolutely exact. But–and this is the central tactic of Carnot's argument–the relation was not only close to being true: it was entirely true, and could be demonstrated geometrically. There was no need to appeal to the doctrine of infinitesimals. The point could be proved even to someone who had never heard of that kind of analysis. The radius of a circle being perpendicular to the tangent at its extremity, the triangles CPM and MPT were similar, whence

$$CP : MP :: MP : TP$$

so that again

$$TP = \frac{MP^2}{CP} = \frac{y^2}{r - x}.$$

In short, the synthetic geometric demonstration confirmed the validity of the analytic method of infinitesimals. The advantage of the latter was that once allowed, any curve could be considered as divided into a very large number of very small sides, any surface into strips, any body into corpuscles, and any quantity into parts of itself. But it was not to be thought that the immense facility thus won for calculation rested on any inexact procedure. Infinitesimal analysis was to be viewed not as a method for merely reducing error to tolerable insignificance, but as a rigorous process that when rightly applied compensated perfectly for errors it admitted into the reasoning in order to simplify the solution of problems.

Returning to the example and to the equation

$$TP = \frac{y\,(2y + NO)}{2r - 2x - MO}$$

it was necessary to neglect *NO* and *MO* in order to get the solution. To do so was deliberate. The way in which the problem was formulated must necessarily have introduced some error, since strictly speaking a circle could never equal a real polygon. And since we know from the geometric demonstration that actually

$$TP = \frac{y^2}{r - x}$$

what we have done in neglecting *NO* and *MO* in the infinitesimal analysis has exactly compensated the error introduced into the statement of the problem in order to facilitate the reasoning.

A complementary way of looking at the problem confirmed what Lazare Carnot meant by the compensation of errors. Consider the circle now as a true curve rather than a polygon, let the line *RS* be drawn parallel to *MP*, and construct the secant RT' through *R* and *M*. Obviously, then

$$T'P : MP :: MZ : RZ,$$

and

$$T'P = TP + T'T = MP\frac{MZ}{RZ}.$$

That being clear, and if then *RS* be translated parallel to itself toward *MP*, the point T' would simultaneously approach *T*, and one could diminish the line $T'T$ as much as one pleased without distorting the proportionality. If, therefore, in solving the problem the quantity $T'T$ were neglected, an error would certainly be introduced into the equation

$$TP = MP\frac{MZ}{RZ},$$

but it was an error that could be made as small as might be wished, and the equation therefore could be approached as closely as might be wished to equality.

Similarly, the equation

$$\frac{MZ}{RZ} = \frac{2y + RZ}{2r - 2x - MZ}$$

was perfectly exact whatever the location of *R*; and as *RS* is brought closer to *MP*, the error introduced by setting

$$\frac{MZ}{RZ} = \frac{y}{a - x}$$

might similarly be made as small as desired. From these equations, then, Lazare Carnot by simple substitution obtained the solution

$$TP = \frac{y^2}{r - x},$$

which we know from the geometry is rigorously correct–even though the two equations from which he has here drawn it, i.e.

$$TP = y\frac{MZ}{RZ} \quad \text{and} \quad \frac{MZ}{RZ} = \frac{y}{r - x},$$

are in fact false, since the perpendicular distance between *RS* and *MP* was at no point assumed to be zero, nor even very small, but only equal to some arbitrary line that could be made as small as one willed (Carnot 1921, I, §§ 2–10, pp 4–12).

Such, then, was an example of the process of compensating for error that made the infinitesimal analysis rigorous. Later in the work Lazare Carnot extended it to the integral calculus. The procedure was not at all to be confused with a method of approximation (unless in its hypothetical genesis). It involved rather the use of infinitesimal quantities in order to find relationships between given quantities, and its results contained no errors at all, not even infinitesimal errors. Lazare Carnot went on to generalize the elements of the system in a series of definitions that distinguish between the types of quantity that figured in calculation. To the conventional division into determinate and indeterminate quantities, Lazare Carnot preferred a tripartite classification, distinguishing between quantities:

1. that were invariant and determined by the conditions of the problem; to this class belong parameters such as the radius, *MC*, in the problem of the tangent just worked out;
2. that, though variable by nature, acquired determinate values by reason of conventions or hypotheses introduced into the problem; to this class belonged what were ordinarily meant by variables, the coordinates, subtangents, or normals to a curve like the lines *DP*, *MP*, *TP*, and *MT* of the example;
3. that were always variable and indeterminate and to which definite dimensions were never assigned; to this class belonged all infinitely small and infinitely large quantities and also those involving the addition of a finite and an infinite or infinitesimal quantity. Instances in the sample problem were the lines DQ, NQ, TQ, $T'P$, $T'Q$, MZ, RZ–infinitesimals (*MZ* and *RZ*) or functions of infinitesimals.

Quantities of the first two classes Lazare Carnot defined as "designated" and those of the third as "non–designated." What characterized the last was not that they were of minute size, but that they could be varied at will. Yet at the same time non–designated quantities were not purely arbitrary. On the contrary, they were related to quantities of the second class by one system of equations just as the latter were to quantities of the first class by another, not unrelated system of equations. These systems containing only designated quantities Lazare Carnot called complete or perfect. Equations containing terms in non–designated quantity by contrast he called incomplete or imperfect. What characterized them was the nature of those quantities: not that the error in the equation was insignificant or the quantity approximately zero, but that at the will of the calculator either could be made as small as he pleased. Another way of looking at the infinitesimal calculus, therefore, was to say that it consisted in transforming insoluble or difficult complete equations into manageable imperfect equations and then managing the calculation so as to eliminate all non–designated quantities from the result. Their very absence from the result was the evidence that it was rigorously correct (Carnot 1921, I, §§ 12–18, pp 14–21).

In the way that he developed these comparisons the congruence between Carnot's points of view in the calculus and in mechanics appears to particularly good advantage. Consider, he asked the reader, any general system of quantities, some constant and some variable, and suppose the problem was to find the relations between them. Let any specified state of this general system be regarded as a datum. Its quantities and any variables depending exclusively on them would then be the "designated" ones. If now the system be considered in some different state invoked for the purpose of determining more readily what relations did obtain between the designated quantities of the fixed system, this latter state would serve an auxiliary role and its quantities would be auxiliary quantities. If further the auxiliary state be approached to the fixed state so that all the auxiliary quantities approach more and more closely to their designated analogues, and if it is in our power to reduce these remaining differences as far as we please, then the differences would be what is meant by infinitesimal quantities. Since they were merely auxiliary anyway, these arbitrary quantities must not appear in the solution to the problem, which was to determine the relations between the designated quantities. Reciprocally, it was a proof that the result was correct if no arbitrary quantities did occur in it. Thus, the error willfully introduced in order to solve the problem had been eliminated. If it did persist it could only be infinitesimal. But that was impossible since the result contained no infinitesimals. Hence the procedures of the calculus had somehow eliminated it, and that was their peculiar genius.

Such, then, was an engineer's view of the calculus. Perhaps it will do no violence to paraphrase a later mathematician, Lewis Carroll on words to the effect that the question is not what quantities mean, but who is master, we or they? In Carnot's sensibility, mathematics was a tool to the mind. Yet the attitude implied no disdain. Like a good workman he was respectful of his tools. He admired those who, like Lagrange, handled them with greater elegance and virtuosity than he did, but his admiration did not inhibit him from saying how he saw matters himself. A few months after the appearance of Carnot's *Réflexions sur la métaphysique du calcul infinitésimal*, Lagrange published his own account of the calculus, which he acknowledged having composed in treatise form in consequence of the obligation to expound the subject before the students of the *École polytechnique*[18] A story–almost surely apocryphal–in Carnot's entourage had it that when Lagrange was preparing the text for the press, Prieur de la Côte-d'Or, encountering Lagrange, urged him to look up Carnot's book, and that Lagrange declined, saying enigmatically, "We can meet without seeing each other." When he later did read the *Réflexions sur la métaphysique du calcul infinitésimal* (so the tale goes), Lagrange wrote Carnot that if he had known the book, he would not have needed to undertake his own account (Carnot H, I, pp 13–14). Perhaps, though probably not–in any event, the parallelism of their interests is as notable as the difference in their styles both in mathematics and mechanics, a case of the high road and the low road.

Lazare Carnot never expressed anything like jealousy in the matter. To the later edition of the *Réflexions sur la métaphysique du calcul infinitésimal* as to his *Principes fondamentaux de l'équilibre et du mouvement* he added remarks congratulating himself that his more homespun arguments had found confirmation in the analytical formulations of the foremost mathematician of their times. In that edition, indeed, he quoted in conclusion a paragraph from Lagrange of which he said, "There is my entire theory summarized with greater clarity and precision than I could have given it myself" (Carnot 1813, I, § 37, pp 47–48):

> It seems to me that, since in the differential calculus as it is used, we consider and in effect calculate with quantities that are or are supposed to be infinitesimal, the correct metaphysics of this calculus consists in the circumstance that the error arising from this false supposition is retrieved or compensated by that which is created by the procedures of the calculus itself, in accordance with which only infinitesimal quantities of the same order are retained in differentiating. For example, in regarding a curve as a polygon with an infinite number of sides, each one infinitesimal and the prolongation of which is tangent to the curve, it is clear that an erroneous supposition has been made. But the error is corrected in calculation by the omission we there make of the infinitesimal quantities. The point is easy enough to see in particular examples although it is one of which it would perhaps be difficult to give a general demonstration.[19]

So much constituted the central thrust of the argument of Carnot's *Réflexions sur la métaphysique du calcul infinitésimal*. In the definitive, 1813 version what had been a brief summary of the rules of differentiation included in the 1797 edition has been expanded into an entire second chapter expounding the notations and practical operations of the calculus in textbook fashion. The juxtaposition with the doctrine of compensation of errors invests the book with that same combination of novelty in point of view with elementariness in scientific content that characterizes the later writing in the mechanics of machines and the *Géométrie de position*. The third

[18]On that see *Théorie des fonctions analytiques contenant les principes du calcul différentiel, dégagés de toute considération d'infiniment petits, de limites et de fluxions, et réduits à l'analyse algébrique des quantités finies* (Lagrange 1797).

[19]Carnot (1921, I, § 37, pp 39–40).

and final chapter is historical and comparative, and purports to show how Carnot's position reconciles the two main traditions of the calculus, the Leibnizian method of infinitesimals and the Newtonian doctrine of the limiting ratio of nascent or vanishing quantities. It is clear from this discussion that he was widely read in the literature of the subject. He held the classical method of exhaustion to be the common ancestor of both traditions, the lineage coming down through Cavalieri's analysis of indivisibles and Descartes' method of indeterminates.

For himself he now professed to stand in the line of Leibniz and the justification of infinitesimals by the principle of continuity (an emphasis consistent with his habitual recourse to continuity of motion in mechanics). It might very well be, he admitted, that Leibniz had failed to define an infinitely small quantity before employing it, and that he even left it ambiguous whether he himself had taken the calculus to be absolutely rigorous or merely a method of approximation. Never mind whether Leibniz had had his doubts on these points, however. The prodigious success of the algorithm in his own hands and later in those of l'Hôpital, the Bernoulli brothers, and their successors reduced these objections to the dimensions of quibbles. Indeed, all that Lazare Carnot now claimed to be explaining, with the credentials of a century behind the calculus, was how Leibniz might properly have replied. He might have insisted that the method was perfectly rigorous once the infinitesimals were eliminated from the solution so that they contained only finite algebraic quantities. His critics had put themselves in a worse quandary by insisting that infinitely small quantities should logically be treated as having disappeared. For if so, all the terms in their equations would vanish leaving only a string of zeros with which to calculate relations (Carnot 1921, I, § 30–32, pp 29–33, and II, § 174, pp 73–74).

Not that Lazare Carnot was insensitive to the advantages of the Newtonian method of limiting ratios expounded in Sect. 5.1 of the *Philosophiæ Naturalis Principia Mathematica* (Book I, Section I). He was, indeed, more respectful of its algorithm, the fluxional calculus, and the eighteenth–century British mathematics in general than have been most historians of science, who have tended to repeat as history the contumely heaped upon it by Babbage, Peacock and other British mathematical reformers of the early nineteenth century. Both in the *Réflexions sur la métaphysique du calcul infinitésimal* and in the *Géométrie de position* Carnot (and Chasles after him) made appreciative reference to the work of Maclaurin and of several later geometers, notably Robert Simpson, John Landen, and Matthew Stewart, appealing for the translation of their writings into French.[20] Lazare Carnot himself would have appreciated the point made recently by Youschkevitch that if the European mathematicians constructed the house of analysis in the eighteenth century, the British synthetic and critical tradition made greater progress in the

[20]*Réflexions sur la métaphysique du calcul infinitésimal* (Carnot 1921, II, § 158, pp 59–60); *Géométrie de position* (Carnot 1803b, § 435, pp 481–483); *Aperçu historique sur l'origine et le développement des méthodes en géométrie* (Chasles 1875, pp 170–186).

study of the foundations[21]–an odd reversal (it may be remarked) of what is usually taken to be the characteristically British role of pragmatism over against continental preoccupation with deep principle.

In Lazare Carnot's view, Newton's method of first and last ratios of nascent and vanishing quantities had modified the ancient method of exhaustion by obviating the necessity for reducing every problem to a proof from the absurd (Carnot 1921, II, § 112, pp 10–11). In this classic technique the properties of a problematic system would be determined through their correspondence to the known quantities of an auxiliary system that was closely enough related to permit approaching the one to the other. Newton's method had substituted the principle of continuity for reduction to the absurd, and made an enormous gain in economy. In view of Carnot's own approach both to mechanics and to mathematics, it seems probable that it was through reflecting on the Newtonian doctrine of first and last ratios, its geometrical basis and its strong–minded voluntarism, that he first developed the justification he thought to bring to the complementary Leibnizian doctrine of infinitesimals. He professed to see no difference between them in point of rigor, although in fact he admitted the justice of the criticisms addressed to Leibniz on that score whereas of Newton's method he reported no reproach except a certain clumsiness, the consequence of its not being directly expressible in an algorithm. And it may well be that his account, perfectly correct what is more, of Newton's reasoning will serve as a clue to where he found his own:

> When any two quantities approach each other more and more closely in value so that their proportion or quotient differs less and less and as little as one wills from unity, these two quantities are said to have for last ratio a ratio of equality. In general, when it is supposed that diverse quantities approach respectively and simultaneously to other quantities considered as fixed, until they all simultaneously differ as little as one wills, the relations between the fixed quantities are the last ratios of those supposed to approach them in value, and those fixed quantities are themselves called the limits or last values of those which thus approach them.[22]

Lazare Carnot argued that the Newtonian difference between a quantity and its limit or final ratio was what Leibniz meant by an infinitesimal quantity, and that, therefore, it will appear to the reader that it was he who wished to bring them to the same thing by his Newtonian definition of an infinitesimal as that which we can render as small as we will. He invoked the authority of d'Alembert, and indeed of Newton himself, to deny any distinction between the method of first and last ratios and that of limits (Carnot 1921, II, § 134, p 37). It is surely significant, moreover (as Professor Youschkevitch pointed out in the essay on the Berlin dissertation), that in the initial, 1785 version Carnot's reasoning draws heavily on the theory of limits. Why he should thereafter have progressively deemphasized it, no text permits us to say. His doing so was consistent, however, with a desire evident elsewhere in his work,

---

[21]In his introductory essay to the Russian translation of the *Réflexions sur la métaphysique du calcul infinitésimal* translated by Soloviev N.M., with a critical introduction by A. P. Youschkevitch and a biographical sketch of Lazare Carnot by M. E. Podgorny (Carnot 1933, pp 16–18).

[22]Carnot (1921, II, § 129, pp 33–34).

particularly in the *Principes fondamentaux de l'équilibre et du mouvement* of 1803, to accommodate his scientific style more closely to the prevailing analytic taste.

5.3 Analysis and Synthesis

A concluding note appended to the final 1813 version of the *Réflexions sur la métaphysique du calcul infinitésimal* compares the compensation of error in the infinitesimal calculus to the theory of negative quantity that Carnot had developed in his geometric writings to account for algebraic operations. In both branches, finite and infinitesimal, though in somewhat different ways, analysis consisted essentially in introducing auxiliary quantities that realistically speaking were erroneous in order to facilitate the process of calculation. The difference was that the procedures of infinitesimal analysis themselves eliminated the errors in the course of computation, whereas in finite analysis they remain in the solution where we recognize them by comparison with a rational or physical reality and correct them accordingly.

The true character of analysis, its distinctive genius and the secret of the advantage it enjoyed over synthesis, lay in its capacity to employ negative, imaginary, and infinitesimal forms. In Carnot's view any other distinction between the two main modes of reasoning in mathematics was illusory. Conventionally it was said that synthesis moved from the known to the unknown whereas analysis supposed the solution known, but in terms that are unknown, the finding of which constitutes the problem. That widely adopted distinction Carnot considered to be false. If any, truly subtle work of synthesis were perused, the reader would find that the essential procedure was the same as in analysis. The author would be reasoning on the unknown as if it had been found. The only real difference between synthesis and analysis was that, once the equation was formulated, and that was always a synthetic job, analysis permitted itself to employ inverse and imaginary quantities and physically unreal or metrically impossible operations, and synthesis did not.

In the synthetic method the object of the reasoning could never be lost to view. It had to be kept constantly before the mind's eye, real and clear, together with whatever combinations it might enter into. If synthesis sometimes employed signs and symbols, it did so only as abbreviations or shorthand. Not so analysis, which disposed of all the resources of synthesis and also of others. It made combinations with objects that do not exist. It represented them by symbols and mingled real beings with imaginary ones. Then in the course of solving problems, it transformed equations systematically and in the course of the calculation eliminated the imaginary entities after they had served the purpose of auxiliaries. In the end, therefore, whatever was unintelligible was made to disappear, and there remained only what could equally have been discovered synthetically. But it had been obtained more easily and directly, almost one might say mechanically.[23]

[23]Carnot (1921), Note Relative au n° 162 de L'Ouvrage précédent, § 20, pp 103–104. Lazare Carnot incorporated a more extensive comparison of analysis to synthesis in the *Digression sur la nature des quantités dites négatives* (Carnot 1806a).

Recognizable now are the elements of the Carnot style, like father, like son, in mathematical as in mechanical reasoning. Lazare Carnot adverted to something of the sort in a set of passages in the *De la corrélation des figures de géométrie*, where it comes down to the point that we make progress in knowledge by breaking complex problems into simple ones and analyzing the relations of systems that are unknown by comparison to those that are known (Carnot 1801, §§ 45–51, pp 26–30). Reduced to those terms the point might seem like banality itself, and only in the execution is it evident that the mode of handling problems inaugurated a combination of the engineering with the rational in a genre that was novel in its quality if not in all or perhaps any of its elements except reversibility in cyclic processes. The manner and the matter seem incompatible in a way–verbal in expression and yet exact in substance, redundant in argumentation and yet general in bearing. But the signet is the location of the reasoning in the ideal system and the control in the concrete system, the one relevant to determining the other by its approachability through a continuous change of state. It is from mechanics that we know the laws of the abstract machine, from hydrodynamics of the ideal heat engine, and from geometry of extended quantity. All are sciences of experience in a genetic sense, to be sure, and therefore operable in accordance with the more actual experience through which one knows that something is lost in impact and that a heat exchange must go one way. Perhaps it was in the mathematical reasoning, however, that Lazare Carnot most resolutely converted the imperfections of the actual into analytical assets. In the discussion of the calculus he treated inequalities as if they were equalities. He treated quantities that were approaching a goal as if they had reached it. What mattered was not the unavoidability of error, but that we should be masters of eliminating it once incurred. Thus could imprecise or "imperfect" equations furnish precise and rigorous results, and inverse quantities real ones, not by some fortuitous cancellation of errors, but by their compensation or inversion, which really meant the calculator's power of compensating for them[24].

[24]It is very interesting that his first discussion of the method should have occurred in the concluding passages of the theoretical portion of the 1780 prize memoir on the theory of machines (Carnot 1780 in Gillispie 1971, Appendix C, §§101–160, pp 299–343). His hope clearly was to qualify the very abstract reasoning of his memoir for the contest by relating it to the effects of friction and binding. This, the "useful part" (*Ivi*, § 160, pp 337–340) of the problem could be stated, "The forces applied to a given machine being known and such that neglecting friction and stiffness in cords there would be equilibrium, determine what must be added to the impelling forces to put the machine on the point of moving so that if these forces were increased, motion would begin" (*Ibidem*). A rigorous solution being extremely difficult, the method that Carnot was proposing would deliberately introduce an erroneous supposition in order to achieve a solution exact enough to satisfy "any reasonable man" (*Ibidem*). The supposition was that friction on a point is proportional to pressure, and the resulting error, a small one at worst, was then to be reduced "as far as one wishes" (*Ibidem*) by successive applications of Carnot's empirically determined formulas for the relation of friction to pressure. See *Lazare Carnot Savant* (Gillispie 1971, Appendix C, §§101–160, pp 299–343).

Chapter 6
History and Historiography of Sadi Carnot's Thermodynamics

La logique explore de nouvelles voies pour tenter d'analyzer la créativité scientifique qui se manifeste dans l'invention et la decouverte.

(Hallyn, p 7, line 3; s. transl. below)

By considering the vast biographical literature accumulated in the past century and today, here we prefer to focus solely on some particular aspects strictly related to Sadi Carnot's science.

In the first part of this section we will deal with the history and historiography around the birth of Sadi Carnot's theory taking into account the scientific context of the nineteenth century.

The second part is dedicated to the current methodology used to historically investigate some crucial aspects of Sadi Carnot's reasoning regarding the foundation of his thermodynamic theory. We will propose an *excursus* on significant examples in the history of science in which categories of investigations are used to distinguish historical methodology, interpretation and scientific theory to better pre-present historical epistemology dealt with in the following chapters (see Chapter 9).

Some biographical comments on Sadi Carnot will conclude this section. In particular, we will focus on Sadi Carnot's cultural background listing his homework as a student–engineer at *École polytechnique de Paris*.

6.1 A Historical *Excursus*

6.1.1 On the Science of Sadi Carnot's Time: The Mechanics

In order to understand what appears to be a logical difficulty in Sadi Carnot's book, we should understand why it was *mainly* and *naturally* understandable by scientists at that time: Newtonian principles in physics and mathematics (1687).

C.C. Gillispie and R. Pisano, *Lazare and Sadi Carnot: A Scientific and Filial Relationship*, History of Mechanism and Machine Science 19, DOI 10.1007/978-94-017-8011-7_6,

The alternatives to Newtonian physical theory were mainly d'Alembert's dynamics (d'Alembert 1743), Lazare Carnot's mechanics (Carnot [1783] 1786, 1803a, b) and some aspects of Lagrange's mechanics (Lagrange 1788). It has often been assumed that Newtonian mechanics was capable of encompassing – without any consequences – every theoretical and deductive mechanics formulation of the eighteenth and nineteenth centuries (e.g., above cited Lazare Carnot, Lagrange et al. . . .). This kind of assumption led to the conclusion that the various formulations of mechanics are equivalent; and "new ideas" based on "*actual facts*" and alternatives to exclusive Newtonian ideal mechanics were widely obscured.

> Purely mechanical phenomena do not exist [. . .]. On the other hand, thermal, magnetic, electrical, and chemical conditions also can produce motions. Purely mechanical phenomena, accordingly, are abstractions, made, either intentionally or from necessity, for facilitating our comprehension of things. The same thing is true of the other classes of physical phenomena. [. . .] The view that makes mechanics the basis of the remaining branches of physics, and explains all physical phenomena by mechanical ideas, is in our judgment a prejudice. Knowledge which is historically first, is not necessarily the foundation of all that is subsequently gained. As more and more facts are discovered and classified, entirely new ideas formed. We have no means of knowing, as yet, which of the physical phenomena *go deepest*, whether the mechanical phenomena are perhaps not the most superficial of all, or whether all do not go *equally deep*. Even in mechanics we no longer regard the oldest law, the law, of the lever, as the foundation of all the other principles. The mechanical theory of nature, is, undoubtedly, in an historical view, both intelligible and pardonable; and it may also, for a time, have been of much value. But, upon the whole, it is an artificial conception. Faithful adherence to the method that led the greatest investigators of nature, Galileo, Newton, Sadi Carnot, Faraday, and J.R. Mayer, to their great results, restricts physics to the expression of *actual facts*, and forbids the construction of hypotheses behind the facts, where nothing tangible and verifiable is found. If this is done, only the simple connection of the motions of masses, of changes of temperature, of changes in the values of the potential function, of chemical changes, and so forth is to be ascertained, and nothing is to be imagined along with these elements except the physical attributes or characteristics directly or indirectly given by observation.[1]

Let us consider, for instance, that since Newtonian principles refer to a single–material–point–body, they cannot deal with extended–body–systems, which include bonds rather than energetic matters without adopting the collision theory. Table 6.1, presents the novelty of Sadi and Lazare Carnot's theories in contrast to Newtonian theory in the history which prevailed at that time[2]:

In Table 6.1, let us note Carnot's columns. The key concepts of Newtonian mechanical paradigm, absolute space and time as temporal (and mathematical) variations of physical quantities, are absent. Moreover, two Carnots use the impossibility of perpetual motion as the basis of their mechanical and thermodynamics theories. One might ask: *what is the theoretical organisation in the two Carnots? On what principles is it based?*[3]

[1]Mach ([1883] 1996), p 495, line 5 (*Italic* style from the author).

[2]Pisano (2010).

[3]Please see below Chapter 7.

Table 6.1 Summary of the fundamental concepts in Mechanical Newtonian theory, Lazare and Sadi Carnot's scientific theories

Main concepts	Newtonian theory Mechanics	Lazare Carnot (1753–1823) Mechanics	Sadi Carnot (1796–1832) Thermodynamics
Space,	Infinite and absolute	Limited and relational (volume);	Limited and relational (volume);
Time	Infinite and absolute	*idem*	*idem*
Bodies	Mathematical points	Global, machines	Global, machines
Inertia	Perpetual	Impossibility of perpetual motion	Impossibility of perpetual motion
Basic–concept	Acceleration	Transformations	Transformations
Interaction	Force–Cause	Work	Work [moment–of–activity]
Setting of theory	The law in the dynamics theory, $F = ma$	The laws in *Collision theory*[b] $\sum m_i \vec{W}_i^2 = \sum m_i \vec{V}_i^2$	Integration of *dq/t*
Techniques	Differential equations; infinitesimal	Geometric motion[c]; vector calculus $\sum m\vec{U}_i \cdot \vec{u}_i = 0$	Cycle
Solutions	Any possible solution Trajectories from *minus* to *plus* infinitum	Invariants; geometric motions for mechanical machines	Maximum efficiency of heat machines

Lazare Carnot assumed the hypothesis that $V_i =$ velocity after interaction is always the same. Next he provides a generalization for all bodies by an *n*–elasticity index (see below Chapter 11)

[a]Mainly it refers to Newtonian paradigm in the history.

[b]*m* mass of the body, *W* velocity before interaction, *V* velocity after interaction, $U = \mathrm{W} - \mathrm{V}$, *u* arbitrary geometric motion.

[c]m_i mass of the ith body, U_i velocity (lost by that body) during the collision, u_i velocity called "mouvement géométrique". In Lazare Carnot's original words: "[...] j'appellerai ces mouvements *mouvements géométriques* : c'est–à–dire que *si un système de corps part d'une position donnée, avec un mouvement arbitraire, mais tel qu'il eût été possible aussi de lui en faire prendre un autre tout à fait égal & directement opposé ; chacun de ces mouvements fera nommé mouvement* (1) *géométrique* ; [...] je dis qu'en vertu de ce mouvement géométrique, les corpuscules voisins qui peuvent être censés se pousser par une verge, ou se tirer par un fil [...]" (Carnot 1786, pp 28–30). Lazare Carnot after calculation by a mean *geometric motion* concept, assumed the hypothesis that $\vec{u} =$ constant, that is, the same translation–geometrical uniform motion for all bodies. Please, see details in Chapter 11

We should also note that since there is an absence of collision theory (e.g., friction et al.), Newton's theory is different from Lazare Carnot's mechanics (Gillispie 1971; Drago and Perno 2004). In fact, the latter included collision theory in operative mathematics. In this way, e.g., a quantity of motion is interpreted by non–infinitesimal mathematics and using a *potential infinity*. One possible approach to understanding the history of the foundations of science[4] combines historical and epistemological aspects by means of a logical and mathematical inquiry called *historical epistemology*. Previous epistemological research reveals that in the history of science we encounter both logical *axiomatically organised* theories (AO–theories) and theories whose organization involves the use of non–axiomatic principles as a method for solving a given problem. Thus, these theories can be defined as logical *problematic organization* (PO–theory). In brief, an AO–theory is developed by "self–evident" principles and is generally followed by the use of *sophisticated mathematics* (e.g., in Newtonian mechanical theory).

Below (see Table 6.2) we present an example–hypothesis of the scientific traditions (based on their foundations and) classified by two choices, AO and PO (Pisano 2008).

Table 6.2 Different choices for mechanical traditions

Engineering mechanical traditions (PO)	1° *classical* mechanical traditions (AO)	2° *classical* mechanical traditions (PO)
Niccolò Tartaglia (1499?–1557)	Galileo Galilei (1564–1642)	Galileo Galilei (1564–1642)
Galileo Galilei (1564–1642)	René Descartes (1596–1650)	Christiaan Huygens (1629–1695)
Johann Bernoulli (1667–1748)	Isaac Newton (1642–1727)[a]	Gottfried Wilhelm von Leibniz (1646–1716)
Jean Charles de Borda (1733–1799)	Leonhard Euler (1707–1783)	Jean Baptiste Le Rond D'Alembert (1717–1783)
Lazare Carnot (1753–1823)	Pierre Simon de Laplace (1749–1827)	Lazare Carnot (1753–1823)
	Joseph Louis Lagrange (1736–1813)	

[a] 4 January 1643 according to the Gregorian calendar; 25 December 1642 according to the Julian calendar. Newton's country switched from the Julian Calendar to the Gregorian Calendar in September 1752.

Lagrange's formulation is interesting (Panza 2003; Pisano and Casolaro 2011b). It can be placed as an intermediate stage, due to its author's tendency to engage *Actual Infinite* (AI) in mathematics, which indeed is intended to contribute to the search for a new mathematical technique suitable for any mechanical problems. In particular, Lagrange used AI in mathematics but at the same time he organized his theory around the main problem of searching for a new mathematical technique in order

[4]In regard to methodology and logics one can consult: Sarton (1952), Nickles (2001), Bevir (1999), Batens and Meheus (2006), Agazzi (1980), Beth (1961), Dalla Chiara (1985), Destouches (1948, 1951, 1966), Gauthier (1991), Giannetto (1989), Giles (1970, 1979), Przelecki (1969), Rothstein (1956) and Weizsacker (1973).

to solve any kind of problem in mechanics (Capecchi and Drago 2005). We should remember that the metaphysical nature in Newton's theory (e.g., absolute space and time and the law of *inertia*, the use of infinitesimal objects to mathematically interpret masses as material–points, and the concept of force) was emphasized (since its birth) by important scholars (e.g., Leibniz, Berkeley ... up to D'Alembert, Lazare Carnot, Lagrange, Mach, and Poincaré). For example, it is interesting to mention that Lazare Carnot utilized (Carnot 1813, pp 12–15) the *théorie de la compensation des erreurs* à la Berkeley[5] (1685–1753) for the infinitesimal calculus: "deux erreurs [two errors]" are made, which nevertheless are algebraically annulled in the end, and "[...] by a compensation of errors: which compensation, however, is a necessary and certain consequence of the operations of the calculus."[6] (Carnot 1832, p 105, line 20). On the other hand

> The mechanics of Newton are purely *geometrical*. He deduces his theorems from his initial assumptions [AO principles] entirely by means of geometrical constructions.[7]

In the history of mechanics, Lazare Carnot emerges as the principle scholar linking the two mechanical traditions (see Table 6.2): artisanal–engineering (e.g., Bernoulli 1738; Borda 1770a, b) and classical. Lazare's mechanics is very general and never touches upon infinitesimal forces or other such concepts. In this sense, Lazare's mechanics (Gillispie 1971; Drago 1991; Koetsier 2007) produced a fundamental alternative to Newtonian physics. Moreover, Lazare Carnot's dialectic between mechanics–mathematics changes because it is anti-metaphysical and devoted to experimental data. That establishes an approach to science which is differently from the common paradigm at his time (Dhombres and Dhombres 1997). Finally, the generalization of Lazare Carnot's *Principle of virtual work* is historically very important because it precedes Lagrange's approach in his *Mécanique analytique* (Lagrange 1788). Later, Lazare's son Sadi Carnot, constructed classical thermodynamic theory (Erlichson 1999) exploiting the same approach and idea of in–out work,[8] et al. with the heat machine and French technology of the nineteenth century (Redondi 1976; Kuhn 1960). "What his father made for ordinary the son made for the mechanics of heat".[9]

---

[5]Berkeley's lemma claims: "If with a View to demonstrate any Proposition, a certain Point is supposed, by virtue of which certain other Points are attained; and such supposed Point be itself afterwards destroyed or rejected by a contrary Supposition; in that case, all the other Points, attained thereby and consequent thereupon, must also be destroyed and rejected, so as from thence forward to be no more supposed or applied in the Demonstration" (Berkeley 1734, Sect. XII–XIV).

[6]"[...] par la compensation de leurs erreurs, compensation cependant qui est une suite nécessaire et infaillible des opérations du calcul [...]" (Carnot 1813, p 189, line 10). The quotation in the running text concerns an official translation on 1832 (Carnot 1832). See also Chapter 9.

[7]Mach ([1883] 1996), 465, line 29; *Italic* style from the author.

[8]On the concept of *work*, Sadi Carnot is essentially led by Lazare's mechanics (Drago and Pisano 2007) even if definitions appeared in the first years of 1800s (Grattan–Guinness 1990b, pp 1046–1121).

[9]Saint–Robert (1870c), pp 447–448, line 29.

6.1.2 On the Science of Sadi Carnot's Time: The Theory of Heat

A thermometry developed first. Scientific studies concentrated on the construction and improvement of thermometers that measured the so–called *degrees of caloric* (Morveau et al. 1787; Lavoisier 1789, I, Chap. 1, III, Chap. 3). The two major accredited theories were: the mechanical theory, which conceived heat as an *accident of the matter*, taken from Roger Bacon (1214–1294) and Johannes Kepler (1571–1630), and the *fluidist* theory which conceived heat as a *substance* (a fluid), with chemists associating heat with element *feu* (Pisano and Bussotti 2012). The mechanical theory of heat (Callendar 1910) coexisted for quite some time with the *fluidist* theory of heat based on the previous theory of the *phlogiston* studied by Georg Ernst Stahl (1660–1734). The *fluidist* theory was welcomed by scientists and philosophers because they found it coherent. Leonhard Euler (1707–1783), Henry Cavendish (1731–1810), Joseph Priestley (1733–1804), Adair Crawford (1749–1795), Johann Mayer (1752–1830), Immanuel Kant (1724–1804), and Alessandro Volta (1745–1827), considered *phlogiston* neither heat nor temperature; but as something which produced heat when it freed itself into the bodies. We note that research on heat phenomena mainly implicated magnitudes such as heat Q and temperature t. It is logical that the interest focused on the ratio dQ/dt, which seemed to be the basic theoretical physical magnitude. Joseph Black (1728–1799) was crucial for the development of the theory (Black 1803; Guerlac 1970–1990, pp 173–183). He started with Georg Wilhem Richmann's[10] (1711–1753) formula,[11] extended it to fluids different from water and conducted initial studies on what today we call thermometry and calorimetry, latent heat and specific heat. In 1789, chemistry produced a real revolution and Lavoisier, like other chemists of his time, searched for the basic principles of this new theory in a revolutionary fashion.

Antoine Laurent Lavoisier (1743–1794) assumed (Lavoisier 1789, 1862–1893) two new elements such as *chaleur, calorique*[12]; the research on *Mémoires sur la chaleur*[13] carried out by Lavoisier and Pierre Simon Laplace (1749–1827) is also of remarkable importance (Lavoisier and Laplace 1784). In 1802 Joseph–Louis Gay–Lussac (1778–1850) formulated a law of gases (Gay–Lussac 1802); then Pierre Louis Dulong (1785–1838) and Alexis Thérèse Petit (1791–1820) (Dulong and Petit 1816; Dulong 1829) in 1816–1819, showed that specific heat depended on temperature (in some cases). However, we have to wait for the second half of the 1800s to give validity, not only experimental, to the laws of gases known at the

[10]He wrote "caloris" that correspond to our modern magnitude temperature (Richmann 1750, 171–172).

[11]Cfr.: "De quantitate caloris, quae post miscelam fluidorum, certo gradu calidorum, oriri debet, cogitationes" (Richmann 1750).

[12]Lavoisier attempted an early distinction between heat and temperature. To be precise, he also he also discussed the concept of "*lumière*" (Lavoisier 1789, pp 12–17; on history of optics, one can see a recent book by Darrigol (2012)).

[13]First reading on 18th June 1783.

time studied by theoretical physicists (and chemists). In particular, the adiabatic law (Poisson 1823b, pp 5–16; Laplace 1822) caused some confusion and various laws were formulated; even when, e.g., Siméon Denis Poisson (1781–1840) formulated the correct equation, most scientists did not consider it. At this point, scientific knowledge of the matter took two paths, one based on the properties of gases (kinetic model of gases) and another based on the efficiency of heat machines (Zeuner 1869), which naturally included the gas theory which would later become thermodynamics. In this context Sadi Nicolas Léonard Carnot (Payen 1968, pp 18–30; Fox 1971a, pp 67–72; Taton 1976; Mendoza 1959, pp 377–396; Carnot 1980) relying on his father Lazare's (Carnot 1786; Gillispie 1971; Gillispie and Youschkevitch 1979) theory of mechanical machines, chose the latter. He was certainly one of the most important and ingenious scientists of the nineteenth century.

> Nicolas–Leonard–Sadi Carnot was, perhaps, the greatest genius, in the department of physical science at least, that this century has produced. By this I mean that he possessed in highest degree that combination of the imaginative faculty with intellectual acuteness, great logical power and capacity for learning, classifying and organizing in their proper relations, all the facts, phenomena, and laws of natural science which distinguishes the real genius from other men and even from the simply talented man. Only now and then, in the centuries, does such a genius come into view. Euclid was such in mathematics; Newton was such in mechanics; Bacon and Compte were such in logic and philosophy; Lavoisier and Davy were such in chemistry; and Fourier, Thomson, Maxwell, and Clausius were such in mathematical physics. Among engineers, we have the examples of Watt as inventor and philosopher, Rankine as his mathematical complement, developing the theory of that art of which Watt illustrated the practical side; we have Hirn as engineer–experimentalist, and philosopher, as well; Corliss as inventor and constructor; and a dozen creators of the machinery of the textile manufactures, in which, in the adjustment of cam–work, the highest genius of the mechanic appear.[14]

His *Réflexions sur la puissance motrice du feu* (Carnot 1978) and last consideration with regard to the theory on gases (Carnot 1978, pp 223–225; Fox 1971b; Reech 1853, pp 357–378) appeared in quite a short period of time, approximately 3 years. It was a total overturning of the physics–mathematics relationship (Drago and Pisano 2007, pp 497–525; Pisano 2011a), which was a new theory which had its origins in engineering practices. However, Jean Baptiste Joseph Fourier (1768–1830) had already adopted a strong analytical approach when he formulated his partial differential equation for heat propagation in solids (1807); this was before Sadi Carnot's contribution and when the difference between heat and temperature was still a subject of debate. Fourier introduced many innovations (Pisano and Capecchi 2009) because the science of heat and the theory of differential equations were all in an early stage of development in his time.

Following Fourier's theory some 50 years later, Gabriel Lamé (1795–1870) wrote *Leçons sur la théorie de la chaleur* (Lamé 1861a, b, 1836; Pisano and Capecchi 2009). At this point, the theory of heat is a well–defined science and several important conclusions are reached on thermodynamics, whose field of

---

[14]Carnot (1943), p 1, line 1.

applicability in physics is even studied by mathematicians.[15] Let us note that Lamé was not interested in classical thermodynamics and devoted himself to heating transmission, a topic that now has its own framing in a new branch of physics. It was essentially carried out by mathematicians for whom the physical problems provided a hint to addressing the complex problems of mathematical analysis. Among them, in addition to the partial differential equation there is the potential theory. Lamé is surely a transitional figure towards what will become the physical mathematics of the nineteenth century. Finally, we remark that from a historical epistemological point of view, the first research on heat seemed to echo the scientific programs in *Traité de mécanique céleste* and *Exposition du système du monde* (Laplace 1805, 1836 [1984]) by Laplace, concentrating on central forces and offering the possibility of achieving differential equation solutions.

Below, we provide an *incomplete* list of scientists who possibly could have studied heat and machine topics and could be considered as the intellectual background of Sadi Carnot's scientific thought on thermodynamics theory (See next page Table 6.3).

6.1.3 Toward Physical Mathematical Theories

A certain mechanical formulation, aspiring to construct a new physical–mathematical relationship (Pisano 2011a; 2012a, b) in thermodynamics, brought about a dispute regarding a crucial theoretical point: *can the thermodynamics of energy re–establish physical theory entirely, including mechanics*? The classical energists attempted to demonstrate this by using mathematics; that is to say by deriving the entire concept of mechanics from the conservation of energy, using the first principle of thermodynamics as the basis of physical theory. However, in 1880 Helm and Vogt (Helm, pp 38–42) clarified that the mathematical problem suggested by the first principle of thermodynamics (the exact differential of work) was solvable in only a few cases in mechanics; this precluded establishing the generality sought in order to re-establish every physical theory regarding energy. In light of this premise, the comparison to mechanics was poorly formulated; it was based more on the great innovation of energy (only outsiders of little consequence had acknowledged it before 1847) than on the new theory that truly distinguished thermodynamics from every other physical theory, entropy. This debate minimized the relevance of thermodynamics to that of a theory whose importance was certainly secondary to mechanical theory. Following this intellectual confusion, the new theory appeared essentially as the thermodynamics of heat devices which included heat engines. Or rather, it appeared as a tool for the development of *empirical physics* (thermometry and calorimetry) which, while very

[15]Some of them: Liouville (1836), Dirichlet (1837), Reech (1853), Riemann ([1861] 1868, pp 391–404; Spivak 1979, pp 179–182). See also: Chasles (1839).

Table 6.3 An incomplete list of Sadi Carnot's precursors

Year	Scientist	Opera
1740	Mariotte	*Œuvres de M. Mariotte, de l'Académie royale des sciences, comprenant tous les traités de cet auteur ... La Haye, Neaulme. T. 1, pp. 148–194. De la nature de l'air. Du chaud et du froid.*
1797	Fourier	*Mémoire sur la statique, contenant la démonstration du principe des vitesses virtuelles et théorie des moments par le citoyen Fourier. Journal de l'École polytechnique, An VI, cinquième cahier, T. II, pp 20–60.*
1813	Delaroche and Berard	*Mémoire sur la détermination de la chaleur spécifique des différèrents gaz. Annales de chimie ... T. 85, pp 72–110, pp 112–182, pl. p 224.*
1818	Petit	*Sur l'emploi du principe des forces vives dans le calcul de l'effet des machines. Annales de chimie et de physique, c. VIII, pp 287–305.*
1818	Edwards	*Description de la machine à vapeur à haute pression importée en France, par M. Hemphray Edwards. Bulletin de la Société d'Encouragement pour l'Industrie Nationale, vol. CLXXIV, pp 365–386.*
1819	Heron de Villefosse	*De la richesse minérale ... Paris, Imprimerie Royale, T. 3, 3ème partie, pp 50–114. Machines à vapeur, machines à double effet ou de rotation.*
1819	Despretz	*Note extraite d'un mémoire lu à l'Institut le 29 Novembre 1819. Loi de Dalton, ébullition d'un liquide. Annales de chimie et de physique, 1821, T. XVI, pp 10 5–109.*
1822	Gay–Lussac	*Extrait d'un mémoire sur le froid produit par l'évaporation des liquides. Annales de chimie et de physique, T. XXI, pp 82–92.*
1822	Fourier	*Théorie analytique de la chaleur par M. Fourier. Paris, Firmin–Didot, XXII–639, p 2 pl.*
1823	Ørsted	*Sur la compressibilité de l'eau ... Annales de chimie et de physique, T. XXII, pp 192–198.*
1823	Fourier and Ørsted	*Sur les quelques nouvelles expériences thermoélectriques faites par le Baron Fourier et M. Oersted. Annales de chimie et de physique, T. XXII, pp 375–389.*
1823	Cagniard de Latour	*Nouvelle note sur les effets qu'on obtient par l'application simultanée de la chaleur et de la compression à certains liquides. – Annales de chimie et de physique, T. XXII, pp 410–415.*
1823	Perkins	*Sur les nouvelles machines à vapeur. Annales de chimie et de physique, T. XXII, pp 429–432.*
1823	Despretz	*Extrait d'un mémoire sur les chaleurs latentes de diverses vapeurs. Annales de chimie et de physique, T. XXIV, pp 323–331.*
1824	Clément and Desormes	*Chimie appliquée aux arts. Journal du cours de 1825 (1826–1827) au Conservatoire des arts et métiers par J.–M. Baudot (suivi de) Cours de 1826 à 1827, suite de cours 1824 (sic) à 1826, vol. 1, pp 160–280, vol. 2, pp 161–350, pp 62–114.*

Cfr.: "Catalogue des documents exposés à l'occasion du 150ème anniversaire de l'édition "1824" des *Réflexions sur la puissance motrice du feu, June 1974, Archives de l'Académie des sciences de Paris*" important, could not compete (for example) with the sophisticated analytical theory of heat, to which the luminary physical mathematicians at the time such as Poisson, Cauchy and Lamé contributed decisively. The birth of chemical physics in 1878, however, hailed thermodynamics as a theory capable of very important theoretical developments, even more so than the analytical theory of heat, which would instead lose significance.

Therefore, thermodynamics could have played a foundational role. Consequently, in order to affirm its autonomy, the young theory had to construct a foundational apparatus *ex novo* with the new concepts of energy (from Mayer and Grove on) and of entropy (from Clausius, who, however, was not included until Planck cited him in 1897). In 1882, in *The Science of Mechanics* (1883), Ernst Mach (1838–1916) outlined a personal reflection which had begun in 1871 and became influential in the 1880s, mostly through his books on mechanics and thermodynamics. Although Mach philosophically considered thermodynamics as the most important theory for the foundations of physical theory, in reality he proposed the independence of the two theories; therefore, he argued for the overturning of the traditional secondary role of thermodynamics compared to mechanics. However, he opposed Boltzmann, who proposed elimination of the innovation of thermodynamics, including in mechanics.

6.2 A Historiographical *Excursus*

6.2.1 On Sadi Carnot's Historiography

According to Challey (Gillispie 1970–1980, vol. III, pp 79–84) some historians thought that Sadi Carnot's thermodynamics theory was too abstract to be considered an essay on engineering[16]: especially because the function for heat machines' efficiency, presented at the end of the work, did not easily and clearly (Pisano 2001; Drago and Pisano 2005) define how it depended on temperature. Furthermore, Sadi Carnot's theoretical efficiency was disproportionately higher than that which was achievable at the time.

Other reviews suggested the opposite, considering Sadi Carnot's thermodynamics too practical and therefore less valuable for theoretical physics in general; that was also the opinion of Sadi Carnot's contemporaries, who were accustomed to infinitesimal analysis. They concluded that Carnot's book was naïve because it lacked the basic concepts for a nineteenth century scientific theory: absolute space and time, cause–force, differential equations, material–point, etc. It was revealed that

[16]Robert Fox suggested that a technological study of *Réflexions* only, cannot explain the originality of either theory or reasoning (Carnot 1978, 1986; see also Fox 1988, 2012).

in the *unknown manuscript* (Gabbey and Herivel 1966; Carnot 1978, pp 223–225) Carnot had been in favour of the mechanical theory of heat[17] (Callen 1974, pp 423–443; Callendar 1910) and that he had even obtained the value of the heat equivalent – 20 years before Mayer (1814–1878) (Mayer 1842a, b). In this sense, he discarded the theory of conservation of caloric and presented the conservation of ("puissance mécanique") energy (Carnot 1978, p 248). In spite of all this, Carnot's theory seemed of little interest for over a century, essentially, because it was denounced as being based on caloric theory, which was considered erroneous in modern times. Other studies however, have considered the young French scientist's thermodynamics fascinating, both for the simplicity of the language used and for the air of *mystery* produced by the lack of mathematical calculations in this work.[18]

> But the fact remains that the central problem of the *Réflexions sur la puissance motrice du feu* is an engineering problem, and that the most of the solutions that are offered on the book have their roots firmly in the engineering tradition. This helps to explain why the argument has none of the elaborate mathematics that Carnot's contemporaries were accustomed to finding in the work of such men as Laplace, Poisson and Fourier, all of whom wrote on the theory of heat, as mathematical physicists, between 1800 and the 1830s. [...] For although the *Réflexions sur la puissance motrice du feu* lacks the mathematical sophistication of much of the physics of the period, its unassuming presentation conceals an argument of extraordinary depth and, in parts, of baffling complexity.[19]

Friedrich Wilhelm Ostwald[20] (1835–1932) was among those historians sympathetic to Carnot. He observed that Carnot used two different words to indicate heat: "chaleur" and "calorique". Particularly, he (Ostwald) affirmed that in Sadi Carnot's theory, "calorique" indicates "heat changed without variation in temperature". Thus, Carnot's "calorique" should be equivalent (due to the constant value of temperature) to entropy. Carnot's simple language had left the appearance of an unsound historical hypothesis. Ostwald proposed replacing *calorique* with *entropy*, thereby giving Carnot's theory validity within modern thermodynamics.

In the 1940s, through the carefully considered re–construction of Sadi Carnot's biography (Aries 1921; Picard 1927; Kerker 1957, pp 143–149, Costabel 1976) and within the context of his time, Léon Rosenfeld (1904–1974) presented (Rosenfeld 1941, pp 197–212) an important interpretation of Carnot's and his own principles. Rosenfeld also tentatively proposed a correlation between Carnot's theory and the

---

[17]We remark that in *Notes sur les mathématiques, la physique et autre sujets* Sadi Carnot clearly introduced his "thèse générale" on energy: "[...] la puissance motrice est en quantité invariable dans la nature, qu'elle n'est jamais à proprement parler ni produite, ni détruite" (Carnot 1878a, folio 7r; Picard 1927, p 81, line 14; see also Fox 1986, p 191). On that see Chapter 11.

[18]Sadi Carnot only wrote mathematical notes in a footnote (Carnot 1978, pp 73–79, ft 1). On the footnote, see Pisano (2001), pp 205–230; Drago and Pisano (2005), pp 37–58.

[19]Robert Fox in Carnot (1986), pp 2–3, line 7.

[20]He translated Carnot's book into German, aiding its dissemination. We also note that Lavoisier, in his famous *Traité élémentaire de chimie* (Lavoiser [1789] 1937), had already dedicated some pages to the definition *calorique* and *chaleur* according to their different contexts (Lavoisier [1789] 1937, pp 12–17, line 37).

technological developments of the industrial revolution[21] (Fox 1971a, b; Singer [1954–1958] 1993). One can certainly notice and acknowledge engineering in Carnot's theory, which focused on the fundamental problem of finding the upper limit to machines' efficiencies. We know that Sadi Carnot based his work on the analogy between the gravitational fall of water in the hydraulic wheel and the caloric fall during the cycle of heat driven machines.[22] Additionally, it is significant that Sadi Carnot also sought to provide a general theory, which could be applied to any machine and could work with any fluid. The theoretical physical content of such a theory is actually very significant, going far beyond the engineering study of machines. Carnot constructed a theoretical model, based on equations of state and functions of state, which helped him reach some important theoretical results on gas laws, known at the time only through empirical observations. During the 1970s and 1980s, Truesdell (1970) and Hoyer[23] (1976, pp 221–228) tried to interpret the foundationsof thermodynamics using this new historical–foundational approach. The different historical interpretations of Sadi Carnot's theory appear similar to the two main existing interpretations of the birth of modern science: optics and mechanics in the 1600s.

Some historians, such as Alexandre Koyré (1892–1964) argued about the relevance of the choice (of the kind) of infinity (Koyré 1957) in mathematics (Capecchi and Pisano 2010b; Pisano 2007b). Carnot's work however, is far from considering the infinite in mathematics (as we understand it today). Even though he used infinitesimal analysis in a mathematical footnote in *Réflexions sur la puissance motrice du feu* (Pisano 2001), we see that in most of his work it is only discursive. In this sense, the young French scientist seems to use only finite (and positive) quantities. Nevertheless, if we accept that Sadi Carnot's essay was developed by means of intellectual factors, these cannot be the factors put forward by Koyré (1965, 1961a, b, 1966). It is well known that Koyré brilliantly examined the birth of modern science and its historical disciplines, which he summarized with two phrases: "destruction of the cosmos" and "geometrization of space".[24] Other historians who wrote on the same topic considered artisans' work and inventions to be the sole catalyst for the birth of seventeenth century science. Alexandre Koyré however proposed the opposite antipodal thesis: even building upon the work of the most brilliant artisan, the result would inevitably be undermined by the inaccuracy of measurement:

---

[21] See also: Gillispie (1960), Chaps. 9 and 10, Nagel (1961) and Kragh (1987).

[22] This is only valid if the heat, as well as the caloric, remains constant. For Carnot's cycle and related arguments, one may refer to: Kuhn 1955, pp 91–94, 1961, pp 567–574; Mendoza 1963, pp 262–263; Klein 1976, pp 213–219; Montbrial 1976, pp 333–335.

[23] He demonstrated, by using the cycles, that Sadi Carnot's caloric theory and modern theory, coincide for an infinitesimal processes: that is $\Delta t \to 0$. This is correct in the first and second approximation in thermodynamics theory. (See below Chapter 8).

[24] Of course, the reference to Mach, Koyre et al. is only to show how it is possible to investigate the history of science by categories and not to include them in the historiographies on Carnot.

> The new science, we are told sometimes, is the science of craftsman and engineer, of the working, enterprising and calculating tradesman, in fact, the science of rising bourgeois classes of modern society. There is certainly some truth in these descriptions and explanations [. . .]. I do not see what the *scientia activa* has ever had to do with the development of the calculus, nor the rise of the bourgeoisie with that of the Copernican, or Keplerian, astronomy theories. [. . .] I am convinced that the rise and the growth of experimental science is not the source but, on the contrary, the result of the new *theoretical*, that is, the new *metaphysical* approach to nature that forms the content of the scientific revolution of the seventeenth century, a content which we have to understand before we can attempt an explanation (whatever this may be) of its historical occurrence.[25]

Hence, Koyré observed that the birth of modern science cannot be explained through human–works–categories alone but that a conceptual and theoretical element is also essential:

> I shall therefore characterize this revolution [the birth of modern science] by two closely connected and even complementary features: (*a*) the destruction of the cosmos and therefore the disappearance from science – at least in principle, if not always in fact – of all considerations based on this concept, and (*b*) the geometrization of space, that is, the substitution of the homogeneous and abstract – however now considered as real – dimension space of the Euclidean geometry for the concrete and differentiated place–continuum of pre–Galilean Physics and Astronomy.[26]

Alistair Crombie (1915–1996), who studied (Crombie 1959, 1994) the birth of modern science as an evolution of artisans' skills, conducted further interesting research.[27] That is what Charles Singer (1876–1960) (Singer [1954–1958] 1993), Trevor Williams and Thomas Derry (Derry and Williams 1960, I), sought to do during the twentieth century. An interpretation of Sadi Carnot*à la* Crombie would however require a specific synthesis, demonstrated through all the pertinent technological facts of the 1800s.

It is also well known that Thomas Samuel Kuhn (1922–1996) in *The Structure of Scientific Revolutions* (Kuhn 1962) describes the historical evolution of science as punctuated by sporadic revolutions, that is by alterations or changing of a previous pattern. Unfortunately, despite the title of his book, Kuhn did not identify any "revolutions" other than the birth of modern (Newtonian) science[28] and the attempts on the crisis[29] at the beginning of the twentieth century.

Another historical interpretation, one that appears to be more extensive, is based on the scientific relationship between Sadi Carnot and his father Lazare (Gillispie 1971, 1976; Drago and Pisano works). The most significant works which were

---

[25]Koyré (1965), pp 5–6, line 25.

[26]Koyré (1965), p 6, line 17.

[27]E.g., one can see: Duhem (1905–1906, 1977), Gille (1964), Pisano (2008), Grant (1971), Drake and Drabkin (1969), Renn (2000), pp 25–36; Clagett and Moody (1960), Clagett and Murdoch (1958–1959), Brown (1967–1968), Pisano and Capecchi (2008), pp 120–129; Pisano and di Pietrantonio (2007), pp 197–219; Pisano (2013).

[28]Cfr.: Kuhn (1963), pp 347–369, (1955, 1970), pp 231–278, (1974), pp 459–482.

[29]This shortcoming has already been expounded by historians (Kvasz 1999).

derived from this path of research include those by Koenig, Scott and Gillispie–their theories all being based on analysis of the same themes within Carnot's writings: the discussion of gas theory, the impossibility of perpetual motion and the efficiency of a machine. For instance,

> Sadi's work can be seen as an extension [of the father's mechanical theories] to the heat machine inspired by Lazare's thought.[30]

Ten years later, in 1971, Scott (1971) noted that the concept of *cycle*had already been implied by Lazare Carnot's theory of mechanical machines, when they include springs.[31] In fact, according to Lazare, efficiency can be calculated only after the springs have returned to their original state. That is, when the entire machine has completed a full cycle of operation. In *Réflexions sur la puissance motrice du feu* Carnot attributes the concept of *cycle* with no significant application in theoretical physics until the nineteenth century, while it played an important role in his own work. It is noteworthy that, in later work, Carnot's father Lazare referred to a state in a mechanical system in which the resting position was restored when the springs had returned to their initial positions (Scott 1971, pp 82–103, pp 212–241). Then, *where did the cycle idea come from*? Previous researchers have hypothesized that the typical work of the electric current in Volta's battery (see Chapter 8) could have inspired the concept of the *cycle* in Sadi Carnot's theory (1745–1827) (Pisano 2003). Moreover, since Lazare Carnot had a clear idea of kinetic momentum and energy, he probably understood that work can be carried out independently from the specific job performed: especially when the resting position at the beginning and that at the end are the same in regards to gravity.[32]

Charles Coulston Gillispie published an excellent and crucial overview with regard to science at the time of the post French revolution (Gillispie 1980, 2004) and a special and suggestive critical edition of Lazare Carnot's work on geometry, analysis and mechanics, devoting a chapter of his book to a comparison between Lazare and Sadi Carnot's theories (Gillispie 1971, pp 90–100; 1976, pp 23–34). In his work, he also proposed (Gillispie 1980, 2004) an overview with regard to science after the French revolution (see also Drago 1989a, b) along with a comprehensive and insightful critical edition of Lazare Carnot's works on geometry, analysis and mechanics. Gillispie devoted a chapter of his book to discussing Lazare and Sadi Carnot's filiation (Gillispie 1971; Gillispie and Youschkevitch 1979).

---

[30]Koenig (1959), pp 57–111.

[31]Carnot (1782, 1803a, b, 1813); Please, see also "Note" in Carnot (1813), 213–257. Poisson (1781–1840) also wrote about a cycle in his *Traité de mécanique* (Poisson 1833, pp 5–16; 1823 II, pp 552–554). One can also see critical discussions in Truesdell's works (Truesdell 1970, pp 36–38, 1980, pp 208–235). However, these works do not seem to be solely inspired by Sadi Carnot's studies.

[32]Naturally, as long as there are no collisions and the shifting is done "par degrees insensible", that is, in Sadi Carnot's reasoning, "rétablissement [...] d'équilibre dans le calorique" (Carnot 1978, p 33, line 9), reversibility, in modern times.

COLLOQUE SADI CARNOT

PIECES MISES A LA DISPOSITION DE L'ECOLE POLYTECHNIQUE PAR Mr P-S CARNOT.

1) Portrait de Nicolas Léonard Sadi CARNOT par BOILLY, peint en 1813. Sadi Carnot est en uniforme de polytechnicien (22 cm 8 x 17 cm,2.- 39cm x 24 cm, cadre compris).

2) Portrait non encadré de Sadi CARNOT par DESPOIX (29 cm x 22,6cm), peint en 1830.

3) Un exemplaire de la première édition des Réflexions sur la puissance motrice du feu (1824; 22cm x 13cm,5)

4) Facture (2 Juin 1824) des imprimeurs (Guiraudet et Gallay, 315 rue St Honoré) des "Réflexions sur la puissance motrice du feu". Cette première édition (la facture en fait foi) a été tirée à 600 exemplaires.

5) Manuscrit de Sadi Carnot: "Notes sur l'économie politique et autres sujets de morale politique". (39 feuillets, non datés - titre inclus - de 23cm x 18cm).

7) Photocopie de l'acte de naissance de Sadi Carnot (14 Prairial, An 4e de la République) 27cm x 21 cm.

8) Photocopie d'une lettre adressée, de St Omer à Nolay (Côte d'Or), le 13 Pluviose an 10, par Lazare Carnot, père de Sadi Carnot, à son frère Jean-François-Reine Carnot, maire de Nolay (21cm x 17 cm)

9) copie à la machine d'un extrait des "Mémoires domestiques" de Joseph Carnot, frère de Lazare Carnot: "mémoires rédigés entre le 18 Fructidor An V et le 18 Fructidor An IX." (21cm x 5cm)

10) Lettre de Lazare Carnot (11 Ventose an 8 de la République) à son frère Jean-François-Reine Carnot. Il lui parle de la solitude de son fils Sadi "qui est tout seul", et le prie de lui confier un de ses enfants, qui deviendrait le camarade de Sadi Carnot. (33 cm x 25 cm.)

11) Lettre du Gouverneur de l'Ecole impériale polytechnique, faisant savoir à Sadi Carnot qu'il est reçu à l'examen d'entrée à Polytechnique (24 cm x 18 cm.) Lettre du 29 7bre 1812.

12) Lettre de l'oficier de semaine de l'Ecole royale polytechnique, annonçant à Sadi Carnot qu'il a été admis, à sa sortie de Polytechnique, à être admis dans le service du Génie militaire. Lettre manuscrite du 7 Octobre 1814 (18,6 cm x 12,1 cm)

13) De la main de Sadi Carnot: copie d'une adresse à Napoléon, envoyée en 1814 à l'Empereur

14) ~~Photocopie d'~~ Reproduction d'une lettre adressée à son fils Sadi le 10 Avril 1814 par Lazare Carnot, alors gouverneur d'Anvers. Sadi Carnot s'était battu à Vincennes en Mars 1814

15) Passeport daté du 23 Avril 1814, permettant à Sadi Carnot de se rendre à Anvers (35,5 cm x 25,5 cm)

16) (écriture d'Hippolyte Carnot) :a) en un feuillet; petite biographie de Sadi Carnot
b) " " " :anciens camarades de Sadi Carnot
c) en deux feuillets: "Sadi Carnot, sur son livre d'adresses)
d) en deux feuillets: Notices biographiques sur Sadi Carnot.

17) Photocopie de l'article nécrologique paru dans la Revue encyclopédique, T.56, p. 528 et suivantes, sous la signature de Robelin. Ce dernier, d'après Hippolyte Carnot, était le meilleur ami de Sadi Carnot.

18) Photocopie d'un article sur la musique (article signé S.C.) paru dans la Revue encyclopédique, T.56, p.167) D'après Hippolyte Carnot, le texte est de Sadi Carnot.

19) Agrandissement photographique du feuillet 7, recto, des Notes manuscrites de Sadi Carnot. Le texte, forcément antérieur à la date de la mort de Sadi Carnot (24 Août 1832) prouve jusqu'à l'évidence que c'est bien Sadi Carnot qui a découvert le premier principe de la thérmodynamique (loi de conservation de l'énergie.)

20) Photographie (17 cm,5 x 9 cm) de la tombe de Sadi Carnot, à Ivry-sur-Seine

Fig. 6.1 The page of the documents presented by *École polytechnique* at 150 anniversary from publication of the *Réflexions su la puissance motrice du feu* (© *Collections archives de la bibliothèque de l'École polytechnique de Paris*)

Let us turn to the memoir by Sadi Carnot, *Réflexions sur la puissance motrice du feu* sur [...] published in 1824, the year after his father's death and twenty–one years after the *Principes fondametaux de l'èquilibre et du mouvment* [by Lazare Carnot]. The similarity has often been remarked upon between Lazare Carnot's observations there on hydraulic machines and his son's model attributing motive force or motive power to the passage of heat considered as a real fluid falling from a higher to a lower level of temperature.[33]

At the *Parisienne Sadi Carnot Colloquium* (Taton 1976) Gillispie proposed the following fundamental remarks:

> It is an honour to be giving the opening at this colloquium. [...] 1) Is it correct that both Lazare and Sadi Carnot produced bodies of work which deriving from their training as engineers, went largely unnoticed by physicists for a generation, even though the work introduced concepts that proved fundamental for physics when later taken up by other hands? The question has a subsidiary aspect, which the elements make for a contrast rather than a comparison between father and son: is it further correct that Lazare Carnot offers the unique example in the history of science of an innovator who inadvertently owed the scientific notice he eventual won to political prominence? [...] 2) The second question is more substantive, and concerns the analysis itself. Is it correct to see the origin of Sadi's idea of a reversible process in the use that his father made of what he called geometric motion in his analysis of machines? [...] 3) My third question grows out of these considerations and concerns the method, and shall be brief about it. Is it appropriate to read the *Réflexions sur la puissance motrice du feu* both as the foundation it later became of the science of thermodynamics, and also as the final item in a series of Carnot memoirs on the science of machines, in which the method of analysis derives from engineering instead of from the rational mechanics of the 18th century? [...] My [4] final question remains, however: was the method of analysis itself novel with the two Carnots?[34]

That was essentially the state of the art expressed in our book (see previous page Fig. 6.1).

6.3 On the Methodology used to Investigate Sadi Carnot's Science

In this section, we outline our main ideas of historical epistemology based on the use of logical–historical categories, which we will adopt in the following paragraphs to investigate Sadi Carnot's scientific thought.

One of the main characteristics of this kind of investigation is an attempt to contextualize key inferences presented by a given author with respect to crucial aspects of the theory, e.g., in the *principles*, *problems* and *methodological procedures*. Of course, the epistemological enquiry is applied *a posteriori* and focuses on the *creativity* and the *genesis* of scientific discourse such as original theorems–reasonings in respect to predecessors or to *normal science* (Kuhn 1962) in that period.

> The logics explores new ways to try to analyze scientific creativity that manifests itself in the invention and discovery.[35]

---

[33]Gillispie (1971), p 90, line 35. Gillispie claimed 15 points establishing a hypothesis of intellectual and scientific link between Sadi and Lazare. (Gillispie 1971, pp 90–100).

[34]Gillispie (1976), p 24, line 1, line 13; 26, line 40; p 31, line 27; p 32, line 45.

[35]Hallyn p 27, line 3 (Transl. is ours: RP). See also *Une analyse rhétorique de Sadi Carnot* (*Ivi*, pp 227–257).

The scientific sense of *creativity* is often placed in the *problematical systems* (Meyer 1979, 1989) *reformulation* and *theoretical advancements*. The discourse is historical and mainly focused on the history of foundations of the science. Thus, we avoid reducing the historical investigation to logical propositions or complex symbolisms. It is not our aim. On the contrary it concerns historical epistemology of science. An overview on the *specificity*, *scientificity* and *categories* is proposed.

6.3.1 On **Specificity** *and* **Scientificity** *in Historical Discourse*

Historians (both scientific and humanist) produced a vast history of science through particular methods and instruments: history of foundation historical–social consequences in the history of technologies (Singer [1954–1958] 1993) et al. It originates a learning process that allows critical feedback[36] of scientific *facts* and *events*. *However, how can we decide on one method rather than another?* The modern historiography of science usually considers the history of science by means of its disciplinary topics (physics, mathematics, biology et al..), practices (methods, theories et al..) and its own historical development (the history of the history of science, academic disciplines et al..). In historical studies, these aspects generate another important matter, which we attempt to address in this section: a certain kind of rigor in the process of control and scientific knowledge is required. The same is required by theoretical and experimental science. In this sense, considering historical and epistemological inquiry, *does a historical judgment, able to monitor historical propositions and their theses by means of internal resources, exist?* Of course, historical facts and opinions are not reproducible in a laboratory; and a historian should build his remarks on facts he has never seen or directly lived. Therefore, another problem arises: *how could the history of science and its historiography clash with historical scepticism?* Several styles of thinking[37] re–dimensioned historical truths to produce narrative discourses and in the best cases, re–dimensioned them to be interpretations only. Between the nineteenth and twentieth centuries, Ernst Mach generated a radically new vision of the foundations of (Mach [1883] 1996) of physics:

> 1. That branch of physics which is at once the oldest and simplest and which is therefore treated as introductory to other departments of this science, is concerned with the motion and equilibrium of masses. It bears the name mechanics. 2. The history of the development of mechanics, is quite indispensable to a full comprehension of the science in its present condition. It also affords a simple and instructive example of the process by which natural science generally is developed.[38] [...]. They that know the entire course of the development of science, will, as a matter of course, judge more freely and more correctly of the

---

[36]E.g: (Thackray 1970; Koyré 1965; Clagett and Moody 1960; Clagett (ed) 1959; Gille 1964; Grant 1971; Gillispie 1960, 1971, 1980; Carnot 1986).

[37]E.g., one can consider the successors of structuralism.

[38]Mach ([1883] 1996), p 1, line 1.

> significance of any present scientific motion than they, who limited in their views to the age in which their own lives have been spent, contemplate merely the momentary trend that the course of intellectual events takes at the present moment.[39]

Alexandre Koyré emphasized the history of science and the role played by mathematics (Koyré 1957) in the history of scientific thought. According to the Russian historian, we can consider that[40]: (1) the history of scientific thought has never been entirely separate from philosophical thought. (2) the most important scientific revolutions are always determined by replacing philosophical speculations. (3) The history of scientific thought (for physical sciences) has not developed in a *vacuum*, but moves within a set of ideas, foundational principles and axiomatic evidence, which have usually been specifically considered to pertain to philosophy. Koyré's intellectual matrix has been made clear through the intuition that the fundamentals of scientific theories contain two basic choices. Thus, ultimately, we are interested in trying to understand *the importance of making inquiries in regard to the history of science by using historical categories applied to a scientific method of investigation.*

Generally speaking, a point of view can be concluded through a historical hypothesis. It emerges from an investigation developed by means of techniques of inquiring *ad hoc*. In this sense, a discussion regarding *specificity* by means of some model of *scientificity* could be plausible; that is to say, historical inquiry proceeds by methodological inquiry thanks to the choice of interpretative and historical categories. The latter also control the intermediate steps of the historian's job. They are also not an additional component; they are part of the entire inquiry. A history of concepts (Jammer 1957, 1961) and their conflicts, without categories, could not emerge. For example, the written testimonies, architectonical and artistic works, coins, photos and verbal narrations, just to name a few, are examples. In order to establish a certain *specificity* in historical discourse, the testimonies must necessarily undergo scientific analyses and criticism. Let us see an example. Some research, such as that of Pierre–Maurice Duhem (1861–1916) (Duhem 1905–1906), Roberto Marcolongo, (1862–1943), Clifford Truesdell (1919–2000) and Bertrand Gille (1920–1980) suggested a review of Leonardo da Vinci's (1452–1519) role as a genius (Pisano 2009, pp 165–182; Pisano 2013; Capecchi and Pisano 2007, 2008, 2010b). They stood in favour of a more human figure of a *learned man,* endowed with a quick intelligence: e.g., not all his designs about machinesemerged straight from his fantasy. Modern historiography reached the conviction that Leonardo obtained his results partially from other sources, rather than writing them together with other authors (Gille 1972, p 128). We can reasonably make the hypothesis that the abundance of documents about his manuscripts – and the lack of them in other cases – could also be due to great care when searching for the brilliant scholar's documents. This new point of view developed from scientific inquiry based on da Vinci's foundations and those of his contemporaries. Following this

[39]Mach ([1883] 1996), p 7, line 27.

[40]A conference (1954, Boston) of *American Association for the Advancement of Science*. Cfr.: *The scientific Monthly*, 1955 or Koyré (1971).

point of view, one can also consider a different dichotomy to interpret the passage between Renaissance science (Capecchi and Pisano 2010b; Pisano 2009a, b, c) and modern science. The first was developed by Duhem who considered a continuity between two times. The second was developed by Alexander Koyré (1965) who assumed a discontinuity and consequently a revolutionary birth of modern science few scientists believed (the scientificity). Both of them used continuity/discontinuity as meta-categories to investigate the crucial problems in theories. For the sake of brevity, we only note Koyré's choices:

> The new science, we are told sometimes, is the science of craftsman and engineer, of the working, enterprising and calculating tradesman, in fact, the science of rising bourgeois classes of modern society. There is certainly some truth in these descriptions and explanations [...]. I do not see what the *scientia activa* has ever had to do with the development of the calculus, nor the rise of the bourgeoisie with that of the Copernican, or Keplerian, astronomy theories. [...] I am convinced that the rise and the growth of experimental science is not the source but, on the contrary, the result of the new *theoretical*, that is, the new *metaphysical* approach to nature that forms the content of the scientific revolution of the seventeenth century, a content which we have to understand before we can attempt an explanation (whatever this may be) of its historical occurrence.[41] [...] I shall therefore characterize this revolution [the birth of the modern science] by two closely connected and even complementary features: *(a)* the destruction of the cosmos and therefore the disappearance from science – at least in principle, if not always in fact – of all considerations based on this concept, and *(b)* the geometrization of space, that is, the substitution of the homogeneous and abstract – however now considered as real – dimension space of the Euclidean geometry for the concrete and differentiated place–continuum of pre–Galilean Physics and Astronomy.[42]

Therefore, during an inquiry into history of science, *are we understanding the objective or the completeness?* In this regard, in 1961, for the *centenaire de la naissance Émile Meyerson*, Koyré gave a great contribution:

> Indeed we do not know, or at least we know just as we think: the phenomenological description is a difficult thing, and in current thinking the form and the content are inextricably intertwined. So not only it is difficult to give an account of its unconscious assumptions, of the underlying axiomatic that carries and informs it. But it is almost inevitable to confuse the present form, which can be and will be short, with its essential form and structure. And it is much easier to identify it, analyzing and studying a thought that is foreign to us, theories that are no more ours, above all if we continue this analysis throughout long periods of time, so that the variety of contents would let emerge the unity of operations.[43]

Paul Karl Feyerabend (1924–1994) was another important historian who also specifically explained his categories using his idea of *anarchism* in science as meta–category. He claimed that in science:

---

[41] Koyré (1965), pp 5–6, line 25.

[42] *Ivi*, p 6, line 17.

[43] Koyré (1986), p 138. One can also see: "Message d'Alexandre Koyré à l'occasion du centenaire de la naissance d'Emile Meyerson" (Koyré 1961b).

> Nothing is ever settled, no view can ever be omitted from a comprehensive account. Plutarch, or Diogenes Laertius and not Dirac, or Von Neumann are the models for presenting a knowledge of this kind in which the history of science becomes an inseparable part of the science itself – it is essential for its further development as well as for giving content to the theories it contains at any particular moment. Expert and laymen, professional and dilettanti, truth–freaks and liars – they all are invited to participate in the contest and to make their contribution to the enrichment of our culture. [...]. After Aristotle and Ptolemy the idea that the earths moves – that strange, ancient, and "entirely ridiculous"[44] Pythagorean view – was thrown on the rubbish heap of history, only to be revived by Copernicus and to be forged by him into a weapon for the defeat of its defeaters. The Hermetic writings played an important part in this revival, which is still not sufficiently understood[45] and they were studied with care by the great Newton himself.[46] Such developments are not surprising. No idea is ever examined in all its ramifications and no view is ever given all the chances it deserves. Theories are abandoned and superseded by more fashionable accounts long before they have an opportunity to show their virtues.[47]

Of course, the process of knowledge also implies a definition of the object of interest; the definition of method, the choice of instruments; an approximation of an objective in the inquiry. Without beginning a discussion on the quantitative critical and statistical methods etc. which are not necessarily addressed in this section, we shall briefly focus on only the critical method. We consider it very important in complex historical investigation.

In order to investigate the history of science, one could assume two preliminary operations: (1) examine how this kind of inquiry unites other kinds of investigations in other fields. (2) try to understand the differences between historical inquiry and its peculiarity. However, a question arises: *can history assume a certain degree of scientificity?*

Generally speaking historians can improve their technique, but may also not assume an epistemic and impartial method. Nevertheless, one can agree with the *objectivity* of a particular fact born from a historical inquiry and based on categories which others can control. Then, following this point of view, a thesis adopted by a historian should be like the scientific ones born out of a constantly controllable inquiry. In this sense, a critical method with categories seems to be a valid apparatus ora crucial component of investigation. When one considers the historiographical function, one would assume a separation between the method and its content. In practice, this attempt of separation does not work and perhaps it is not permissible or acceptable.

Science is usually a process of knowledge based on comparing and evaluating models and observables and above all, the validity of bibliographies. Likewise, historical hypotheses based on models need to be internally consistent and ought to be falsifiable, too. In the Table 6.4, we list some notes on crucial aspects of scientific and historical discourse:

---

[44]See also Ptolemy quoted by Feyerabend in footnote 4, p 35.

[45]See also footnote no. 5 by the author.

[46]See also footnote no. 6 by the author.

[47]Feyerabend (1989), p 35, line 6.

Table 6.4 On methods

Method in mathematics	Method in physics	Scientific (?) method in history
Understanding and choice of infinite	Experience and observation	Primary sources.
Analysis and implementation	Hypothesis, Interpretation	Use of categories
Synthesis	Deductive and inductive, (formal) prediction from the hypothesis, modelling	Formal interpretation by mathematical–logical approach
Review and Extend	Experiments and feedback (to hypothesis)	(Eventual) Comparison with same result obtained by not–use of categories

In historical (and historical epistemology, see Chapter 9) discourse, bibliographies are emphasized. The criticisms of manuscripts (primary and secondary) can produce a methodological reflection contributing to the establishment of a degree of rigour for inquiry: the historian is placed in an evaluative position with respect to eventual uncorrected falsehoods, forced interpretations, conceptual mistakes etc. Nevertheless, it is important not only to eliminate mistakes, but to evaluate them in terms of the cultural context and other scientific theories; even if mistakes could often suggest parallels and unexplored roads of history worthy of attention. Therefore, in order to assume *scientificity* in historical discourse, adequate inquiry techniques adaptable to scientific content must be adopted.

6.3.2 On **Interpretation** *in Historical Discourse*

We have now arrived at the center of the question: the use of interpretative categories in history and in the history of science. In order to answer the previous questions, we further circumscribe the problem, moving from previous considerations regarding methods to precise methodological choices: the categories and their epistemological capacity to interpret. When one decides to investigate the past, in order to understand temporal segments, it is generally possible to assume some concepts of interdisciplinary ideas and discoveries. In the first analysis, we can call them categories. It is possible that they are hidden or are visibly declared since the beginning of an inquiry. On the contrary, the absence of declaring could prejudice the *specificity* and *scientificity* of an historical investigation especially with regard to scientific facts. Obviously, history is not only referred to in terms of scientific concepts. Sometimes, the categories emerge from other crucial aspects such as social, economical and cultural ones. Let us see some examples.

Generally speaking, the placement (of facts or events) is the first thing that a historian examines. If one refers to modern history, three major categories are possible: *century, decades,* and *generations*. The next step refers to attributing a sense and significance to a series of events distributed at the time. In fact, a classification of facts and events by means of *centuries* and *decades* can help to localize some crucial aspects of certain problems. For example, one can consider the crisis in the 1930s and the *totalitarianism* of the past century; *idem* for the Second World War in the

1940s; *idem* for the Cold War and the process of colonization in the 1950s, etc. The first two aforementioned categories, *century* and *decades,* are well suited to historical processes, but they cannot provide details regarding men, revolutionary groups and countries. Thus, one can also include *generations*, which with respect to the previous two categories, seem to be more attuned to detail. For example, historians often consider the concept of the *generation*, e.g., young people in 1917 who went to serve in the army at just 18 years old and those who protested during the student revolts of 1968. The first example is a *generation* composed of young people of the same age; the second example, independently from age, is based on a mass of people who protested for common ideals. Therefore, the *generation* of 1917 is characterized by age, while that of 1968 is characterized by a cultural event. Let us see another example. If we consider the very long history of the Roman Empire, a historian usually does not recount all of the events and details, but focuses on the most important events and facts. Therefore, one makes some historical choices; however, history could also be produced by the effects of cultural choices. Therefore, if we read chronologies it is not a surprise that, for the same events, different classifications or interpretations are applied. Of course, this plurality is welcome, and in other cases, the *objectivity* (Gillispie 1990) is preferred. Let us see another example. The English and American cultures easily extend a special category from modern history to the Second World War or up to the beginning of the 1st years of the Cold War; provided that the term *late modern* is added to refer to the period from 1789 to 1989. Therefore, the definition of *contemporary history* is not sufficient to explain a diffuse acceptable scanning. Moreover, the chronological items do not usually circumscribe to 1900. In fact, the years following the Great War offer other interesting situations. In the end, a historian who chooses his categories (or follows another historian's interpretation) is performing a *translation of conscience of the past* by means of crucial options, which could also be merely assumed at the time of investigation.

6.3.3 On **Categories** *in History of Science*

Is it possible to choose which categoriesare adequate with respect to inquiry in the history of science? The use of logical categories has ineludible choices, both implicit (or explicit) and independent of the grade of awareness and accuracy defined. Let us see some examples in the history of science. Space and time could also be respectively translated into geography and chronology. A science usually moves from mythical, magical and subjective assumptions of phenomena, to objective interpretations of scientific processes of knowledge. This passage is very evident in the history of science investigated by means of space and time concepts (Jammer1957, 1961) from a mysterious and divine (space), to be linked with symbolic ritual (time), to further measurable dimensions.[48] Therefore, space and time are privileged wit-

[48]Let us keep in mind that it is not a small step from the novelty of the measurement to the mathematical interpretation by space and time (abstract concepts).

nesses of the birth of modern science and its development: space and time during the birth of modern science are also considered to be mathematical objects. Let us think of the interpretation of a physical phenomenon. The (first and second mathematical) variation of a magnitude in time or space is necessary to obtain (e.g.) velocity and acceleration. Through previous arguments, it is obvious that the idea to let the birth of science coincide with the passage from subjective to objective produces a slightly naive image of the facts and events. At the same time, to object to the impossibility of solving the problem of the subjectivity and relativity of historical inquiry, produces an epistemological discourse in which a naive line of demarcation between science and history or science and epistemology can be traced. Therefore, one is continuing to work within scientific studies when using categories, particularly within the history of the foundations of science. We will now advance some details that will be addressed in the following chapters. Based on recent inquiry[49] one can see that in the history of science we can encounter both logical *axiomatically organised* theories[50] (AO) as well as those whose organisation requires non-axiomatic principles suggesting a method for solving a given problem for a theory which is thus logically *problematically organized* (PO theory). In brief, an AO theory is developed by "self–evident" principles and is generally followed by advanced mathematics (e.g., in Newton's theory). A PO theory is based on logic and methodological (architectural) principles (Pisano 2010) which indicate a direction for the development of the theory (Drago and Perno 2004, 2007). Theoretical choices concerning the two options (AO or PO) allow a proper detection for the foundations of the scientific setting of a theory and choices of its kind of infinity in mathematics. Such theoretical choices also contribute to the emergence of foundational differences among different theories. One can consider above all that, among the intellectual factors, a basic role is played by the choice of what kind of infinity to use in mathematics, potential or infinite. For example, one can consider the historical development of both mathematics and physics, which never regularly appeared continuous in their scientific organizations. For example, let us think of the role played by geometry and algebrain the *mensura* from French Nicolas d'Oresme (1323?–1382) (Oresme 1966) to German Nikolaus Krebs[51] (also called Cusanus (means born in Kues): 1401–1464), up to modern science (seventeenth century). Furthermore, a relevant concept should be highlighted: the *incommensurability among different theories*. As previously mentioned, the latter concept was intuitively introduced by a historian and a philosopher of science, namely Kuhn (1962) and Feyerabend (1989). Their definition of this concept is rather approximative and is based on few historical concepts: e.g., Newtonian mechanics in relation to special relativity or quantum mechanics.

[49]E.g., Drago (1991), Drago and Pisano (2000), *Id.*, (2004), Capecchi and Pisano (2008), pp 1–29. Pisano and Gaudiello (2009a), Pisano (2010); et al. by idem authors.

[50]AO: *axiomatical organization* in a theory. PO: *problematical organization* in a theory. (Cfr.: Drago 1991 and Drago and Pisano works). Generally speaking one can assume a non-classical logic within intuitionistic logic.

[51]E.g.: *De mathematicis complementis* (1453–1454); *de mathematica perfectione* (1458).

In the light of our previous considerations, an investigation by category allows the kind of foundations for research to stand out (e.g.) an alternative to the Newtonian and idealistic (mathematical) paradigm: e.g., Leibniz (1646–1716), d'Alembert (1717–1783), Lazare Carnot (1753–1823), Sadi Carnot (1796–1832), Faraday (1791–1867). Logical and mathematical foundations could inquire about a scientific theory. Two or more scientific theories are different for continuity–discontinuity or commensurability when they are:

1. Methodically (structurally and semantically) organized
2. Mathematized
3. Different (at least) for one of the two choices

Traditional historiography of science identifies two types of history of science, resulting from an investigation of: (a) internal[52] historiographies, which tend to provide an explanation of the history of science in terms of the variables belonging to the science itself or to the scientist's mind, and (b) external historiographies, which tend to underline the decisive role played by social components through the development of science. This classification is clearly suggested by the specific nature of the subject investigated. Throughout the nineteenth century, theories with a different approach with respect to the Newtonian paradigm, suggested that *subjective history* (e.g. history thought and experienced by scientists) could be different from *effective history* (e.g. history relying on the fundamental choices made by scientists who influenced the interpretation of history by means of crucial choices). Similarly, *subjective history* was also different from *objective history* presented in textbooks as a list of data concerning mathematical laws, argumentative techniques and objective concepts (Table 6.5):

Table 6.5 Some aspects of Objective, Subjective and Effective history

Objective history	Subjective history	Effective history
Historical facts, experiments, dates of discoveries	Foundations of scientific theories	Investigation by means of two logical–mathematical categories
Primary sources and early theories	Biography and correspondence	Logical organization of a scientific theory
Birth of new experimental apparatus	Concepts (e.g. intuitive and surrogate concepts)	The choice of kind of mathematical infinite used in a scientific theory
Academic context, societies, academies et al.	Mathematical content of a theory	Changing infinite → change physical concepts

See also: Pisano and Gaudiello (2009b)

From a methodological point of view, the investigation on Sadi Carnot's foundations will be conducted in the manner previously mentioned and using the notion of

[52]Koyré wrote notes regarding these two types of historiography (Koyré 1963, 1973) in a response to Henry Guerlac's (1910–1985) talk (Guerlac 1963). The papers were published in the proceedings of *Symposium on the History of Science* in 1961 and edited by Crombie (1963).

effective history. For example, an incomplete list of historians and philosophers who used explicit categories can be presented[53] (Table 6.6):

Table 6.6 Some historians and their choices

Historian	Main keywords	Main categories	Main subjects
Mach (1838–1916)	Foundations	Economy of thought.	Mechanics and heat
Duhem (1861–1916)	Continuity	Geometry, infinite, cosmos, active	Statics
Koyré (1892–1964)	Discontinuity	The destruction of the cosmos and geometrization of space	Mechanics
Crombie[a] (1916–1996)	Regularity	Individual regularity and regularity of population	Mechanics
Khun (1922–1996)	Paradigm	Normal, anomaly	Classical physics
Drago	Foundations	Organization of a scientific theory and choice of mathematical infinite in theory	Classical and modern Physics–mathematics
		Surrogates concepts: *the evanescence of force–cause and discretization of matter*	Classical chemistry and thermodynamics.
Thackray	Foundations	Inertial homogeneity of matter and short–range forces	Classical chemistry–physics

[a]*Six major categories–headings*: (1) Arguing by means of analyses and synthesis (postulation). (2) Exploration by means of controlled experiments, observation and measure. (3) The construction of hypothetical modeling. (4) The taxonomy. (5) The method of historical derivation (genetic method) was applied first to languages and human cultures, then to geological history (evolution). (6) Probabilistic and statistical analysis (Crombie 1994)

Let us look at an example. Arnold Thackray in his *Atoms and Powers* (Thackray 1970) clearly expressed his historical categories (Pisano 2007a):

> The theory [Newtonian chemistry] has two essential components–belief in the inertial homogeneity of all matter and its possession of an "internal structure", and "acceptance of attractive and repulsive forces as proper categories of [interpretation] explanation [historical]". [...] A third and more ambiguous Newtonian category, the ether, though often referred to or hinted at, did not feature prominently before 1740s.[54]

The "inertial homogeneity of matter", quoted by Thackray, refers to the Newtonian conviction of a hierarchically ordered and strictly structured matter whereas the second category, the admission of "short–rayed forces", refers to the fact that (according to such a view of science), for chemistry as well as for celestial and earthly mechanics, a quantifying method is necessary.

[53]That example is discussed in deep details in: Pisano and Gaudiello (2009a, b).

[54]Thackray (1970), p 122, line 9. Author's quotations.

6.4 Biographical Sketch of Sadi Carnot

6.4.1 *Notes on the Philological Aspects of Réflexions sur la puissance motrice du feu*

Given the number of extensive studies on Sadi Carnot's biography, here we do not include standard information regarding Sadi Carnot's life and refer the reader to the widely recognized secondary literature (Carnot 1986, pp 160–166, 1978, pp 273–312; Challey 1971, pp 79–84; Costabel 1976; see also Grinevald 1976, pp 393–395). The substantial collections belonged to Monsieur Pierre–Sadi Carnot and now belong to Carnot's family–Monsieur Gaetan Carnot. In this book, we will use and study all the documents related to his scholarly–life at *École polytechnique* and two main dossiers conserved at *École polytechnique* and *Académie des sciences*.[55] Some unpublished images and homework are presented.

Nicolas Léonard Sadi Carnot[56] (Paris, 1 June 1796 – Paris, 24 August 1832) was certainly one of the most important and ingenious scientists of the nineteenth century. The birth of thermodynamics, which encompasses the concept of reversibility, occurred along with the exposition Sadi Carnot included in his book. An edition in four exemplars of *Réflexions sur la puissance motrice du feu* 1824 edition is conserved at the library of the *École nationale des ponts et chaussées de Paris*.[57] Carnot's handwritten inscription in the book reads: "Offert à M de Prony" (See Fig. 6.2).

Carnot had 600 copies of *Réflexions sur la puissance motrice du feu* printed at his own expense (June 12th, 1834). It was edited by Bachelier – the leading French publisher – and sold for three francs. The book was presented to the *Académie des sciences* (July 26th), and in the same year it was reviewed very positively by Pierre Simon Girard (1765–1836) in *Revue encyclopédique*, an important pro-republican journal devoted to several branches of literature.

[55] They also include Lazare Carnot's dossier. Others are also archived at *Service Historique de l'Armée de Terre* at the *Château de Vincennes*. In this regard, below we list his main works and current library localization.

[56] Mendoza (1959), pp 377–396; Payen (1968), pp 18–30, Fox (1971a), pp 67–72, Carnot (1980), Taton (1976).

[57] 1 pl. - 1824 - 1 vol.in-8°, 4 ex. - 2 ex. relies, 1 ex. broche fonds prony/corancez – (8.5447). See Fig. 6.2.

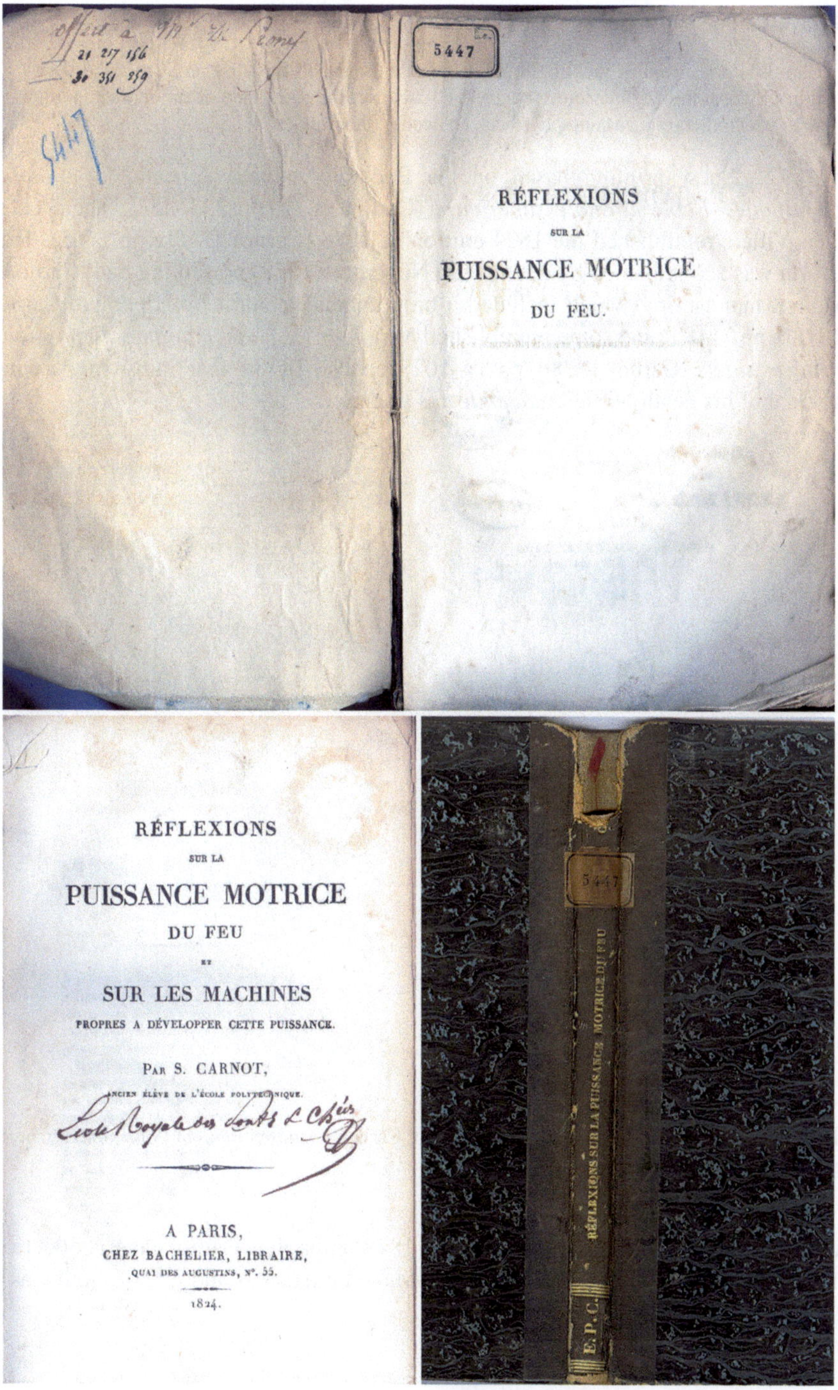

Fig. 6.2 Plates from the original exemplar of 1824 edition (*Collections de Fonds ancien et archives, école nationale des ponts et chaussées, École des ponts ParisTech*, Champs sur Marne–la Vallee. On the *left-hand side*, Carnot's handwriting on Prony)

Girard states:

> Carnot is not afraid of tackling difficult questions; and in this first production he shows himself capable of going into matter which has become today one of the most important with which theoreticians and physicist can occupy themselves.[58]

In 1872, a new printing based on the Bachelier edition appeared in *Annales scientifiques–École normale supérieure* (Carnot 1872, pp 393–457). Then, Gauthier–Villars republished the 1824 edition in 1878 (Carnot 1878b, pp 1–65). This edition was accompanied by a letter (30 November 1878) presenting Sadi Carnot's original manuscript (which included a biographical memoir by Hippolyte Carnot (1878b, pp 71–87) and some unpublished *Notes sur les mathématiques, la physique et autres sujets* (Carnot 1878b, pp 89–102; see also 1878a) that Sadi wrote around the time of his death) to the *Académie des sciences*.

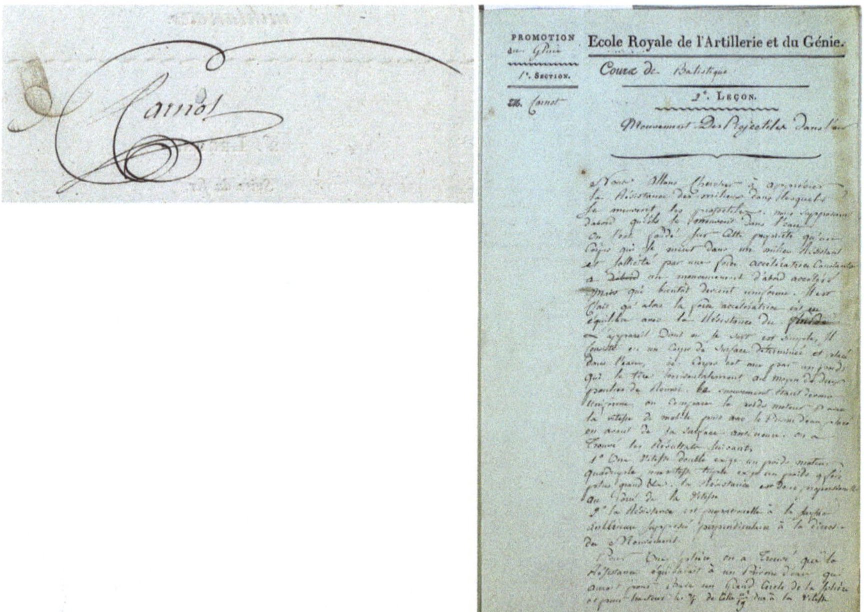

Fig. 6.3 Plates from the two original exemplars of Carnot's handwritings (© *Collections Archives de la Bibliothèque de l'École polytechnique de Paris*)

At the end of the same year (16 December 1878) Hippolyte Carnot gave a collection of Sadi Carnot's manuscripts to the *Académie des sciences* and one of them was a

[58]Girard (1824), p 414, line 11. (See also: Mendoza 1960, xii). For Sadi Carnot's commemoration see Robelin (1832), pp 528–530.

proper version[59] of the *Réflexions sur la puissance motrice du feu* (Carnot 1878a). Several editions were subsequently published and all of them were essentially based on the 1824 edition. Below, we provide Figs. 6.3 and 6.4 in which the first pages of the 1824 edition and 1878 manuscript (Carnot 1878a) are presented:

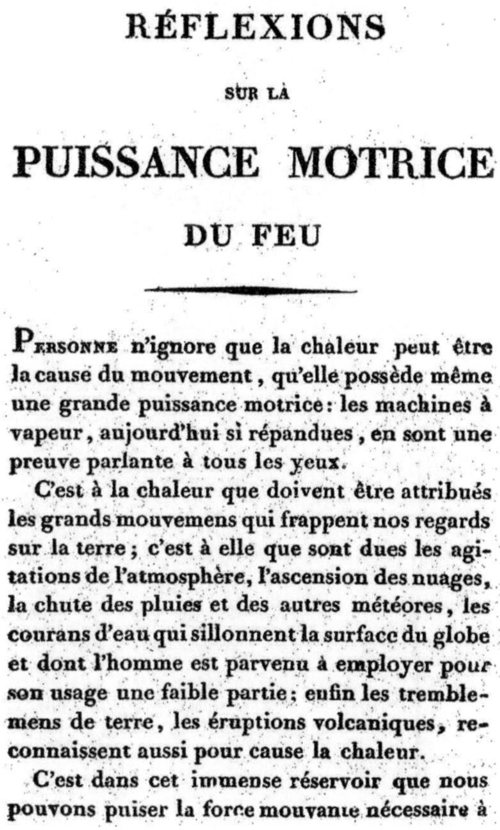

RÉFLEXIONS

SUR LA

PUISSANCE MOTRICE

DU FEU

PERSONNE n'ignore que la chaleur peut être la cause du mouvement, qu'elle possède même une grande puissance motrice: les machines à vapeur, aujourd'hui si répandues, en sont une preuve parlante à tous les yeux.

C'est à la chaleur que doivent être attribués les grands mouvemens qui frappent nos regards sur la terre; c'est à elle que sont dues les agitations de l'atmosphère, l'ascension des nuages, la chute des pluies et des autres météores, les courans d'eau qui sillonnent la surface du globe et dont l'homme est parvenu à employer pour son usage une faible partie; enfin les tremblemens de terre, les éruptions volcaniques, reconnaissent aussi pour cause la chaleur.

C'est dans cet immense réservoir que nous pouvons puiser la force mouvante nécessaire à

Fig. 6.4 The first page of Sadi Carnot's Edition (1824) (Carnot 1824, p 1)

We can remark that the first page of the 1824 edition (Fig. 6.4) is slightly different from the 1878a manuscript which Hyppolite Carnot gave to the *Académie des sciences* (Fig. 6.5). In the 1878a manuscripts, one can read several sentences and passages that were removed; and this seems obvious if we consider that it is a manuscript and the other one is an edition. E.g., "Personne n'ignore que la chaleur peut être le cause du mouvement, qu'elle [...]" (Carnot 1824, p 1) and "Tout le monde sait que la chaleur peut être la cause du mouvement, qu'elle [...]" (Carnot 1878a, folio 1 [3v], see below Fig. 6.5). Of course, scientific contents

[59]"dons de la famille Sadi Carnot le 16–12–1878". It is conserved at *Académie des sciences, Institut de France,* Paris. Ms. in 4 *cahiers* for a total of 92 *folia r/v*.

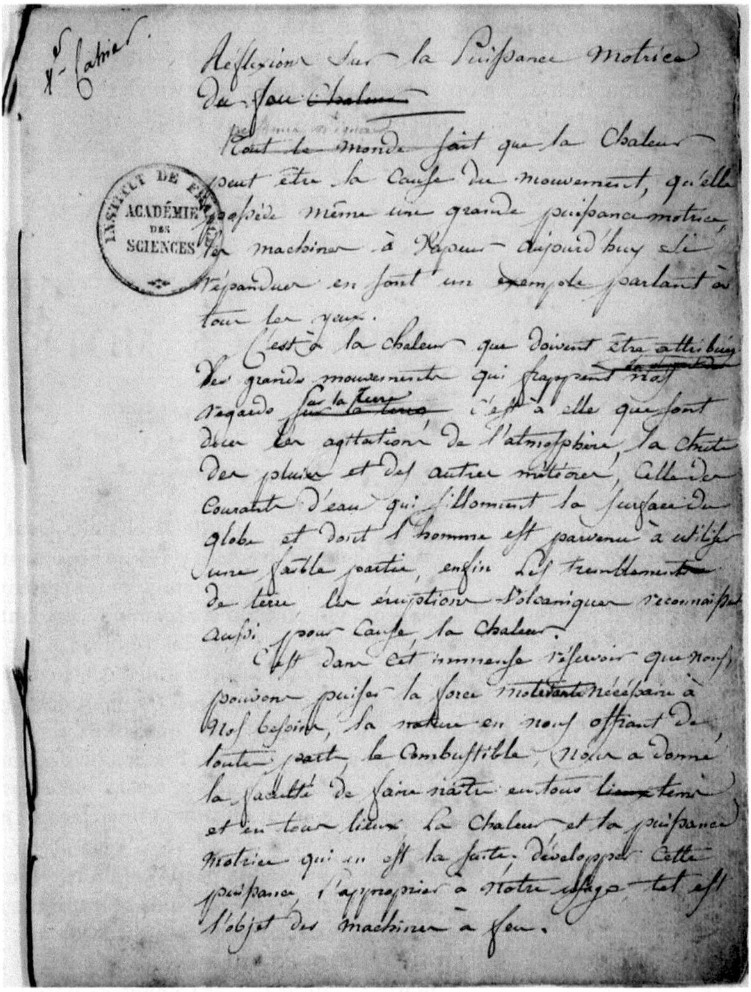

Fig. 6.5 The first page of Sadi Carnot's manuscript (1878a) (Carnot 1878a, folio 1 [3v]. With permission of the *Académie des sciences*)

expressed by Sadi Carnot's do not generally[60] change.[61] To conclude these short philological reflections, we have a final observation. In the 1824 edition and most of

[60]However, we should remark that, sometimes, the contents are different. Given the importance of its content, we like to provide an example into the original language: "La loi fondamentale que nous avions en vue de confirmer nous semblerait exiger cependant, pour être mise hors de doute, des vérifications nouvelles [...]". (Carnot 1978, p 89). In the 1878–version we read: "La loi fondamentale que nous avions en vue de vérifier nous semble d'ailleurs mise hors de doute, tant par les raisonnements qui ont servi à l'établir que par les rapprochements qui [...]". (Carnot 1878a, 4° Cahier, folio 5)

[61]Nevertheless, studying original manuscripts without corrections in order to see if new epistemological results emerge could be very interesting.

the following editions based on it, a set of figures of cylinders[62] appears at the end of the book (e.g., Carnot 1953, p 118, 1978, p 118). There are five figures (placed on one page) which express Carnot's cycle by means of cylinders. In particular, the latter, in the version published in 1878, are distributed in the running text of the book (Carnot 1878b, Fig. 1, p 17; Figs. 2–3, p 21; Fig. 4, p 46, Fig. 5, p 55); *idem* situation in the running text for the edition published at *École normale supérieure* (Carnot 1872, Fig. 1, p 409; Figs. 2–3, p 413; Fig. 3, p 438; Fig. 4, p 447). However, we note that the 1878a manuscript (Fig. 6.4) does not include any cylinders. We can see figures and only one (small but evident) cylinder in *Notes sur les mathématiques, la physique et autres sujets* (Carnot 1878b, folio 21) (Fig. 6.6).

In 1927, Emile Picard (1856–1941) published a facsimile edition of the complete *Notes sur les mathématiques, la physique et autres sujets, Sadi Carnot, biographie et manuscrit* (Picard).

Fig. 6.6 The first page of Sadi Carnot's *Notes sur les mathématiques, la physique et autres sujets* (1878b). With permission of the *Académie des sciences*

In particular, we also remark that Gauthier–Villars' 1878 edition (Carnot 1878b) does not include all the parts of the *Notes sur les mathématiques, la physique et autres sujets* (Carnot 1878a) whereas they are fully included in Blanchard's edition (Carnot 1953, pp 125–151) and in another work published by Émile

[62]Please, see also the discussion below, Chapter 7.

Picard (1856–1941) (Picard 1927, pp 33–94; see also Carnot 1978, pp 237–271). In the end, we might claim that the following editions of *Réflexions sur la puissance motrice du feu* are slightly different (not in physical appearance) from Hyppolite's manuscript (Carnot S 1878a) and we should also note the role played by the absence of the figures. *Maybe an editorial deal took place? Or some change in the research program?* Let us see.

Lazare Carnot's scientific program was quite vast: mechanics, geometry, mathematics, fortifications, military strategies, et al… theory of mechanical machines, and above all, trying to organize a general theory of (all kinds of) machines. Thus, one might reasonably presume that (potentially) a project on heat machines belonged to Lazare who later proposed this ambitious project to Sadi, asking him to complete the theory of mechanical machines including heat machines (see Chapter 11). We also remark that Sadi Carnot's book, that is to say the whole theory of the efficiency of heat machines, seems to have been fully compiled by S. Carnot in about 2–3 years. In others words, one might claim that since Lazare Carnot was very much interested in heat phenomena related to machines, he had already drafted a copy of *Réflexions sur la puissance motrice du feu* which he showed to his son around 1821 when they met in Magdeburg (Fig. 6.7).

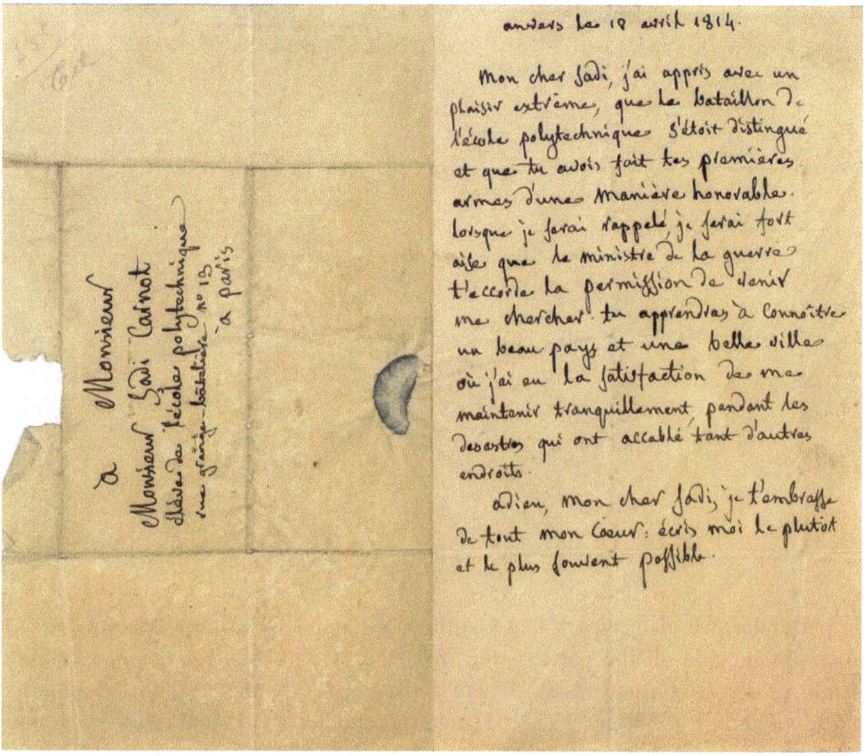
a Monsieur
Monsieur Sadi Carnot
élève de l'école polytechnique
rue grange-batelière nº 13
à paris

anvers le 19 avril 1814.

Mon cher Sadi, j'ai appris avec un plaisir extrême, que le bataillon de l'école polytechnique s'étoit distingué et que tu avois fait tes premières armes d'une manière honorable. lorsque je serai rappelé je serai fort aise que le ministre de la guerre t'accorde la permission de venir me chercher. tu apprendras à connoître un beau pays et une belle ville où j'ai eu la satisfaction de me maintenir tranquillement pendant les desastres qui ont accablé tant d'autres endroits.

adieu, mon cher Sadi, je t'embrasse de tout mon cœur: écris moi le plutot et le plus souvent possible.

Fig. 6.7 Plate from the original addressed to Sadi Carnot (© *Collections archives de la bibliothèque de l'École polytechnique de Paris*)

We also remark that Lazare often wrote about the role played by the new physical magnitude, heat. Moreover, at the end of the eighteenth century he was very busy reviewing the new projects on heat machines which were presented to the *Institute de France.* At that time, he also wrote on aerostats, *Lettre sur les aérostats* (Carnot 1784), a memoir addressed to the *Académie royale des sciences* on 17 January 1784, and preserved in its archives. However it appears to have gone missing (Figs. 6.8 and 6.9).

Fig. 6.8 Sadi Carnot's last permanent residence in Paris In agreement with Robert Fox, it is very probable that it was here that Sadi Carnot wrote his famous book. He lived in the *Marais* arrondissement (1, Rue du Parc Royal) for more than 10 years. (Credits photos: RP, 2011)

On the other hand, we have no historical indications that Sadi Carnot published anything regarding his interest in heat phenomena prior to 1821. Instead, we know that he was (or he was because his father was) interested in familiarizing himself with the state of the art of heat machines presented by other scholars, as we can see by studying several translations from English to French included in his folder (Fig. 6.10; Table 6.7).

Fig. 6.9 Source of the first experiment–human fly in Paris (The photos are of the back and front of the monument at Place Paul Verlaine, 13eme. (Credits photos: Ilaria Gaudiello))

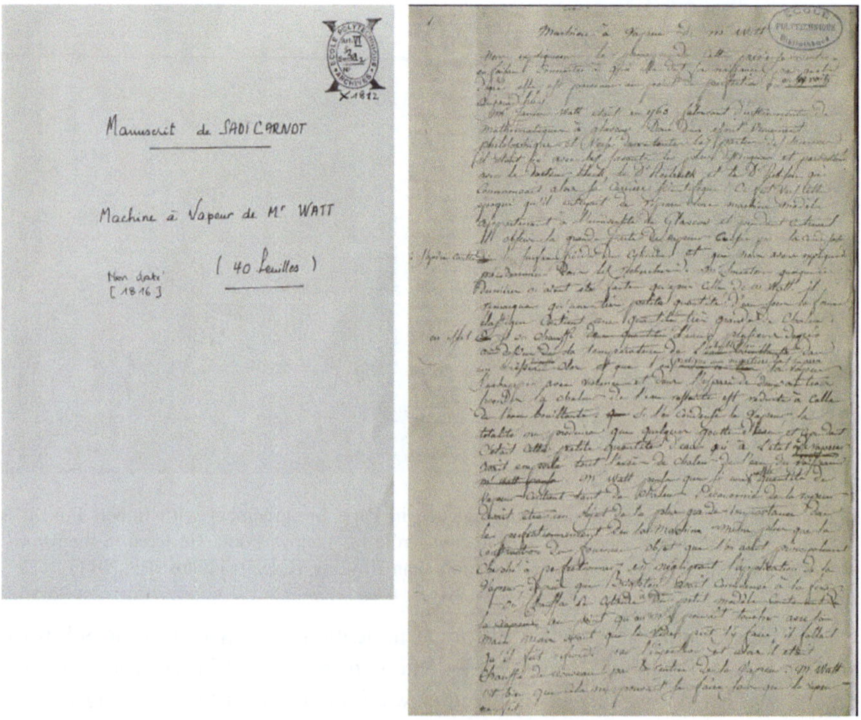

Fig. 6.10 The first and second page of Sadi Carnot's manuscript on Watt's machines "Pièces mises à la disposition de l'École polytechnique par Mr P–S Carnot". *Archives de l'École polytechnique route de Saclay de Paris*. Some manuscripts on Leibniz, correspondence and ancient letters in Sadi Carnot's folder are currently the object of our (RP) research

Table 6.7 A list of translated documents on heat machines

Year	Title	Our comments
X 1812	*Machine a vapeur de M Watt*	On heat machines
1816[?]	[17 folia: 1–17r/v no r/v from folio 17]	[translated from English Watt's work to French]
		No mathematics
		No figures
		No cycles
X 1812	*Machine d'expansion de M Watt*	On heat machines
1816[?]	[8 folia: 17–25r/v no r/v from folio 17]	[translated from English Watt's work to French]
		No mathematics
		No figures
		No cycles
		2 tables [folio 19; folio 21]
X 1812	[4 folia: 26–30]	On heat machines
1816[?]		[translated from English Watt's work to French]
		No mathematics
		No figures
		No cycles
X 1812	*Description de la machine*	On heat machines
1816[?]	*simple . . . complète de MM Watt* [1 folio; 25–26]	[translated from English Watt's work to French]
		No mathematics
		No figures
		No cycles
X 1812	*Machines a double cylindre de M Hornblower*	On heat machines
1816[?]	[30–34 + 35*–41*] *Without number of page	[translated from English Watt's work to French]
		No mathematics
		No figures
		No cycles
		1 table [folio 36]
X 1812	*Machines a double cylindre de M Woolf*	On heat machines
1816[?]	[41*–48*] *Without number of page	[translated from English Watt's work to French]
		No mathematics
		No figures and no cycles

We can also suppose that just after his father's death (18 November 1822), during his applied courses at *Conservatoire des arts et métiers*, he precisely studied the applications of heat theory to machines: e.g., efficiency, gases such as a working substance et al. In fact, when he returned to Paris, he attended the *class on heat machines* at the *Conservatoire des arts et métiers.* Here, Sadi Carnot met Nicolas Clément (1779–1842) who later may have deposited the text at the library of the *Conservatoire des arts et métiers,* where it was rediscovered in 1966. He learned the importance of using adiabatic transformation and also the simple and general way of developing the theory of heat machines; a way that, in fact, refers to Lazare Carnot's general machines theory. With regard to the time spent at the *Conservatoire des arts et métiers,* as previously stated, Clément's lectures were very important for the development of the scientific culture of a very young Sadi Carnot. The latter studied the theories and applications of heat machines using the most popular and commonly accepted laws of the time. In particular, Sadi Carnot used Clément's *uncertain* law on the pressure and temperature of suture vapour to produce very original reasoning and calculations; but his results would prove *uncertain* themselves (Fox 1988, p 294) (Fig. 6.11).

a

(98)

grande chute du calorique. **La vapeur prenant naissance sous une pression plus forte se trouve aussi à une température plus élevée, et comme d'ailleurs la température de la condensation reste toujours à peu près la même; la chute du calorique est évidemment plus considérable.**

Mais pour tirer des machines à haute pression des résultats vraiment avantageux, il faut que la chute du calorique y soit mise à profit le mieux possible. Il ne suffit pas que la vapeur prenne naissance à une température élevée: il faut encore que par l'extension de son volume elle arrive à une température assez basse. Le caractère d'une bonne machine à vapeur doit donc être non seulement d'employer la vapeur sous une forte pression, mais *de l'employer sous des pressions successives très-variables, très-différentes les unes des autres, et progressivement décroissantes* (1).

(1) Ce principe, véritable fondement de la théorie des machines à vapeur, a été développé avec beaucoup de clarté, par M. Clément, dans un mémoire présenté à l'Académie des sciences, il y a quelques années. Ce mémoire n'a jamais été imprimé, mais j'en ai dû la connaissance à la complaisance de l'auteur. Non seulement le principe y est établi, mais il y est appliqué aux divers systèmes de machines à vapeur actuellement en usage; la puissance

b

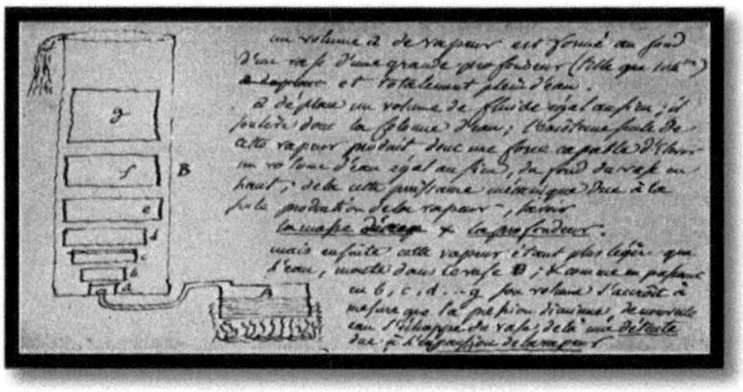

Fig. 6.11 (**a**) Sadi Carnot quotes Clément in his *Réflexions sur la puissance motrice du feu*. (Carnot 1978, ft 1, pp 98–102[*bis*]; see also: *Ivi*, pp 65–66. In 1813, Clément married with Desormes' daughter, but at the beginning of the 1820s he could have changed his surname to "Clément–Desormes". That is the reason because Sadi Carnot (maybe) only cited "Clément" manuscripted in few years and published in 1824. (Cfr.: Carnot 1986, p 48, ft 50)). (**b**) Clément's lectures' at *Conservatoire des arts et métiers*. (Desormes and Clément 1819, pp II5–II8)

Nevertheless, this new systematical knowledge potentially aided him in continuing his father's project. Thus, a theoretical advancement of *Réflexions sur la puissance motrice du feu* from Lazare's outline to a composition by both Carnots might seem historically plausible.[63] This could also explain both the lack of references

[63]It is still an object of research by one of us (RP).

to Lazare's works in *Réflexions sur la puissance motrice du feu* and his sudden abandonment of heat machines after 1824.

Sadi Carnot would not become a scientist by profession. He dealt with this topic for a few years. Moreover, even if the work was indeed reviewed by the *Académie des sciences*, we see that it was also ignored for almost 25 years. Sadi Carnot's friends (although they were students at *École polytechnique, Paris*) affirmed that it was difficult to understand the book's scientific content. In fact, it contained astonishing innovations: the idea of the cycle, a limit to the efficiency of heat machines, a theorem demonstrated by an absurdum proof, and new gas laws. The work seemed even more difficult to interpret because of Sadi Carnot's surprising obtainment of results, which were almost entirely accurate, despite his alternative dependence either on caloric theory or heat theory. Moreover,

> The book is particularly rich in what have been generally supposed to be the discoveries of later writers, and in enunciations of principles now recognized as those forming the base and the supporting framework of that latest of the sciences. As stated by Tait,[64] in his history of Thermodynamics, the "two grand things" which Carnot originated and introduced were his idea of a "cycle" and the notion of its "reversibility" when perfect. "Without this work of Carnot, the modern theory of energy, and especially that branch of it which is at present by far the most important in practice, the dynamical theory of heat, could not have attained its now enormous development". These conceptions, original with our author, have been, in the hands of his successors, Clausius and other Continental writers, particularly, most fruitful of interesting and important results; and Clapeyron's) happy thought of so employing the Watt diagram of energy as to render them easy of comprehension has proved a valuable aid in this direction.[65]

In the following list, we summarize some of the editions[66] of *Réflexions sur la puissance motrice du feu*:

Carnot S (1824) Réflexions sur la puissance motrice du feu et sur les machines propres à développer cette puissance. Bachelier, Paris.

Carnot S (1878a, b) Réflexions sur la puissance motrice du feu et sur les machines propres à développer cette puissance. Reprint edition including Hippolyte Carnot's letter of 1878 to the French Academy of Sciences and some of the previously unpublished Notes written by Sadi Carnot after publication of the Réflexions. Gauthier–Villars, Paris.

Carnot S (1890) Reflections on the Motive Power of Heat. Thurston RH (ed), London Macmillan and Co. This translation was also republished in 1960 by the American Society of Mechanical Engineers.

Carnot S (1943) Reflections on the motive power of heat and on machines fitted to develop this power. Thurston RH (ed), Waverly Press, Inc., Baltimore–Maryland, USA

Carnot S (1953) Réflexions sur la puissance motrice du feu et sur les machines propres à développer cette puissance. Blanchard, Paris. Photographic reprint of

---

[64]Tait (1877).

[65]Carnot 1943 (Quotations from the author).

[66]Indispensable are: Taton (1976) and Fox's editions (Carnot 1978, 1986). See also Redondi (1980) for a summary on French technology and Sadi Carnot's book in the nineteenth century.

the original 1824 Réflexions, including Hippolyte Carnot's letter of 1878 to the French Academy of Sciences and an abridged version of the Notes.
Carnot S (1960) Reflections on the Motive Power of Fire and Other Papers on the Second Law of Thermodynamics by E Clapeyron and R Clausius. Mendoza E (ed). Dover, New York.
Carnot S (1978) Réflexions sur la puissance motrice du feu sur les machinés propre à développer cette puissance, édition critique par Fox Robert. Vrin J., Paris.
Carnot S (1986) Reflexions on the Motive Power of Fire: A Critical Edition with the Surviving Scientific Manuscripts. Fox R (ed). The Manchester University Press.

The only other known works by Sadi Carnot are *Recherche d'une formule propre à représenter la puissance motrice de la vapeur d'eau* (Carnot 1978, pp 223–234; see also Carnot S–EP[67]) and *Un manuscrit inédit de Sadi Carnot* published by Gabbey and Herivel (1966, pp 151–166) (Fig. 6.12).

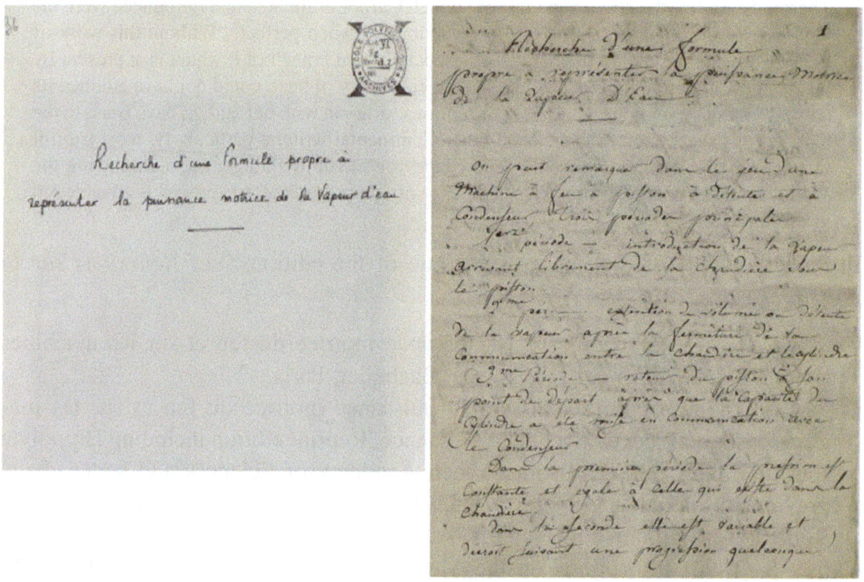

Fig. 6.12 The first and second page of Sadi Carnot's manuscript on *Recherche d'une formule propre à représenter la puissance motrice de la vapeur d'eau* (© *Collections archives de la bibliothèque de l'École polytechnique de Paris*. (The manuscript was also rewritten in the same original language in Carnot 1978)

[67]We mainly worked with a primary copy of the original manuscript. Thus, to be brief, we refer to it as Carnot S–EP because we have a plate from the original which was very kindly offered by *École polytechnique* (Carnot S–EP).

There is a considerable secondary literature dealing with Sadi Carnot's works and their context and importance for the development of the science of thermodynamics. For the sake of brevity, we refer the reader to the references listed at the end of this work.

6.4.2 On Sadi Carnot's Studies at École polytechnique

Sadi Carnot had good knowledge in mathematics, physics, hydraulics, military strategies and applied engineering sciences. E.g., during his young studies he produced several unpublished lecture notes and study projects on fortifications, topography, hydraulics at *École polytechnique* (Belhoste 2003; Kuhn [1961] 1980). In the following we report some of the projects and studies on fortifications, hydraulics, applied chemistry, tables for attach and defend strategies, studies on chemistry applied to weapons, descriptive geometry, various militaries' arguments, utensils, mechanical machines,[68] trigonometry, ballistic theory, pure chemistry. In the following, we present some plates from the original documents and arrange a list of them that report these abilities (Fig. 6.13; Table 6.8):

[68]In particular a manuscript on machines (6 r/v folia in ca. A3–format) that included an interesting mathematical approach.

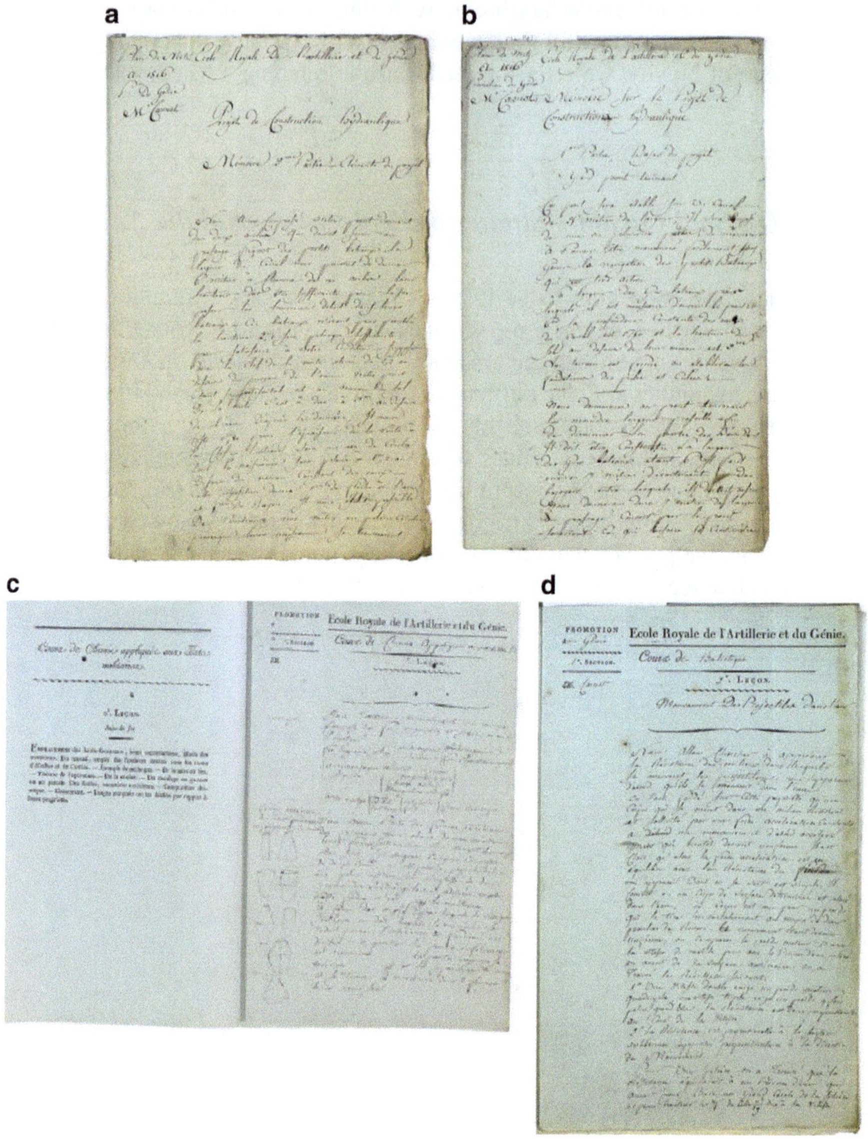

Fig. 6.13 (**a**) on *hydraulique,* (**b**) on *hydraulique,* (**c**) on *chimie appliquée,* (**d**) on *balistique*

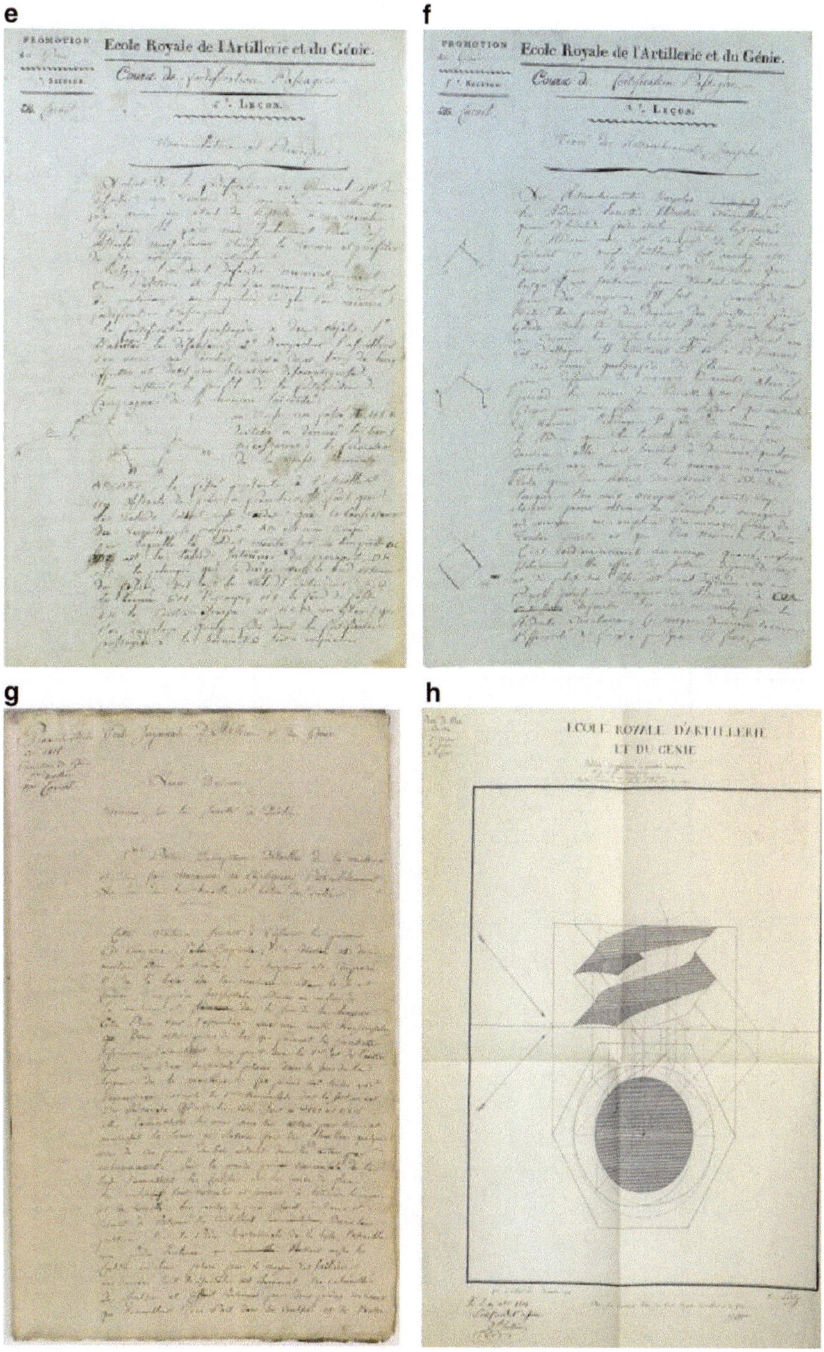

Fig. 6.13 (**e**) on *fortifications,* (**f**) on *fortifications,* (**g**) on *lunette,* (**h**) on *écrous de vis*

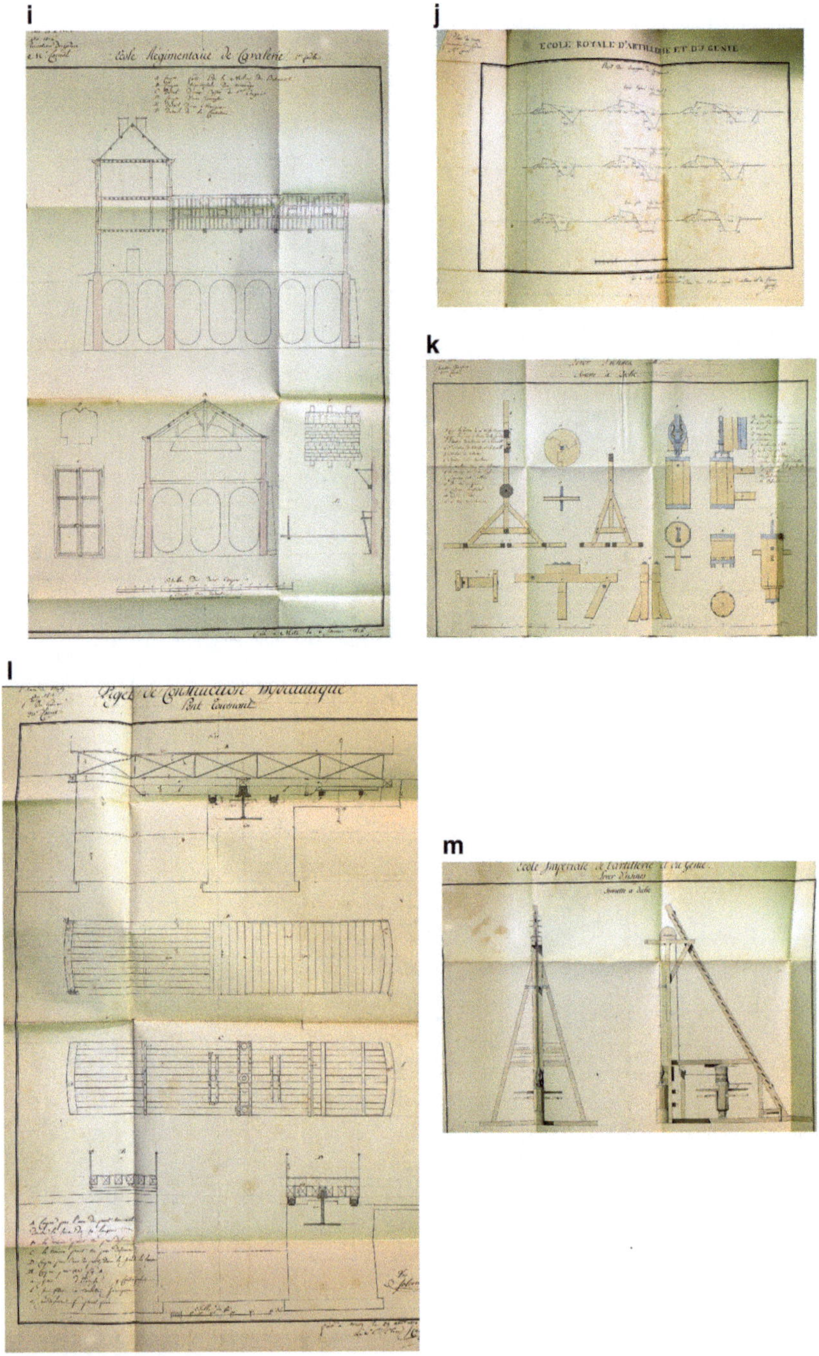

Fig. 6.13 (**i**) on *école de cavalerie,* (**j**) on *machines,* (**k**) on *machines,* (**l**) on *pont tournant,* (**m**) on *machines*

n

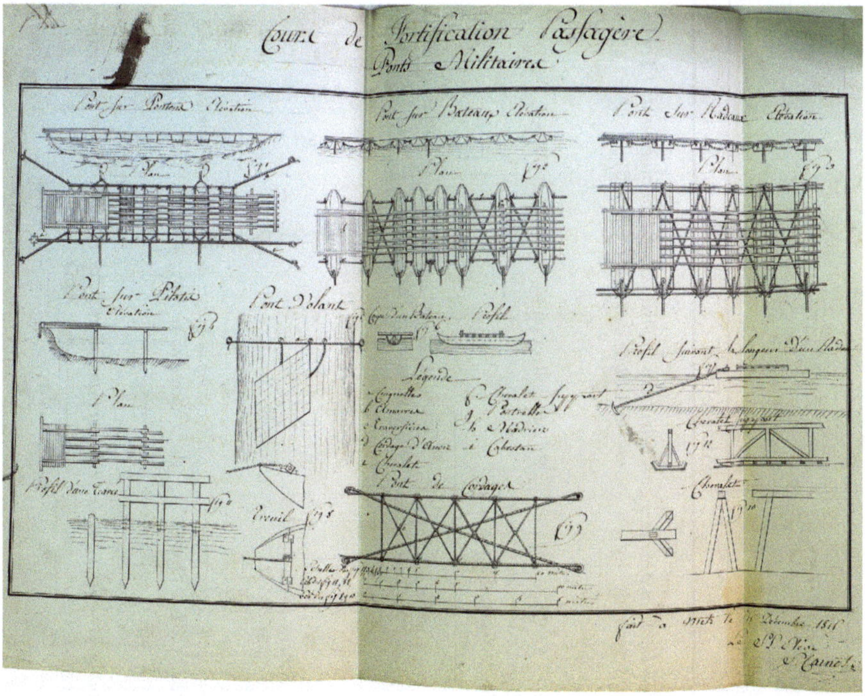

Fig. 6.13 (**n**) on *ponts militaires*

Below, we provide a list of Sadi Carnot's scientific works (Table 6.9) and a list of some documents announcing–commemorating his death (Table 6.10).

The causes of Sadi Carnot's death are uncertain. The prevailing opinion is that he died from a disease, which appears to be "choléra" (Arago and Gay–Lussac 1833, LII, p 211; Henrion, p. 167). Some of his friends submitted an obituary (Girard 1824, pp 411–414; Robelin 1832, pp 528–530; Gondinet 1833, 46; Redondi 1980 pp 195–196).

Nevertheless, we should keep in mind that he probably developed mental illness when the cholera epidemic of 1832 hit Paris. However, we have no historical proof of this.

He was 36.

He had no heirs.

France lost a Special Engineer, and Science lost a Great Scientist.

He was the first "inventeur de la Thermodynamique" (see Fig. 6.14).

Table 6.8 A list of Sadi Carnot's documents related to his military and scientific aptitude

Year	Title of the document (Sadi Carnot X 1812)	Archive
Sadi Carnot, *à l'École polytechnique*, Paris		
1811	*École impériale polytechnique. Programmes de l'enseignement de l'École impériale polytechnique arrêtés par le Conseil de Perfectionnement, dans sa session de 1811–1812. Paris, Imprimerie Impériale, 1811.* (In 4°, 18,5/24,5 cm, 63 p 2 dépl.)	*Collections archives de la bibliothèque de l'École polytechnique de Paris* (X 2 b/10)
1812	*École impériale polytechnique. Lettre signée par le Gouverneur de l'École impériale polytechnique, LACUEE, Comte de CESSAC, annonçant à Sadi Carnot son admission à l'École, et le convoquant avant le 2 Novembre prochain, Paris,le 29 Septembre 1812.* (1 Feuillet, 18/24 cm)	Private collection by Pierre Sadi Carnot
1812	*École polytechnique. Registre matricule.* (Feuillet 58)	*Collections archives de la bibliothèque de l'École polytechnique de Paris* (X 2 c/2)
1812–1814	*École impériale polytechnique. Registre des notes.* *Promotions 1812–1813* *1813–1814* *Sont portées en regard du nom, les notes de Sadi Carnot pour les années 1812–1813, 2ème division, 1813–1814, 1ère division.* (In 4°, 24/37,5 cm, 2 vols.)	*Collections archives de la bibliothèque de l'École polytechnique de Paris* (X 2 c/8)
1812–1814	*Note* [by Lazare Carnot] *sur le séjour de Sadi Carnot à l'École polytechnique, ses frais de scolarité, etc. ... rédigée par son père Lazare Carnot (1812–1814). Don du Professeur Paul Carnot.* (1 feuillet, ms., 14/22 cm).	*Collections archives de la bibliothèque de l'École polytechnique de Paris*
1813–1814	*École royale polytechnique. Tableau présentant l'ordre alphabétique des élèves des deux divisions et leurs notes, sur chaque partie d'instruction, d'après les examens intérieurs. Année scolaire 1813–1814.* (12 feuillets, 24/36 cm).	*Collections archives de la bibliothèque de l'École polytechnique de Paris*

Sadi Carnot, *à la bataille de Paris*, 1814		
1814	*École polytechnique. Liste de 113 élèves absents de l'École à l'époque du 23 Avril 1814 (où figure le nom de Carnot).* (4 feuillets, ms., 20,5/31 cm)	*Collections archives de la bibliothèque de l'École polytechnique de Paris*
1814	*Lettre de Lazare Carnot félicitant son fils d'avoir fait ses premières armes d'une manière honorable à la bataille de Paris, le 30 Mars 1814. – Anvers, 18 Avril 1814.* *Don du Professeur Paul Carnot.* (1 feuillet, 20,5/24,5 cm).	*Collections archives de la bibliothèque de l'École polytechnique de Paris*
Sadi Carnot, *Officier du Génie Élève à l'École d'artillerie et du génie à Metz*		
1814	*École royale polytechnique. Lettre annonçant à Sadi Carnot son admission "dans le service du Génie militaire". Paris, le 7 Octobre 1814.* (1 feuillet, ms., 12/18,5 cm).	Private collection by Pierre Sadi Carnot
1814	*École d'artillerie et du génie à Metz. Élèves sous–lieutenants du Génie. Promotion du ler Octobre 1814. (Liste de 10 noms parmi lesquels est mentionné celai de Carnot).*	Dépôt de l'Artillerie L.C.A., Arcueil
1814–1815	*Mémoire sur le problème de géométrie descriptive, trouver les ombres portées par un écrou de vis à filets triangulaires, cet écrou étant supposé coupé par le milieu.* (5 feuillets, 1 plan)	*Collections archives de la bibliothèque de l'École polytechnique de Paris*
1814–1816	*École royale de l'artillerie et du génie.* *Promotion du Génie l'ère Section. Cours de l'École de Metz, Manuscrits, 21/33 cm. (Notes prises au cours signées "Sadi Carnot").* *Cours de fortification passagère,* *1814–1815, 34 feuillets* *Cours de balistique, 1815, 23 feuillets* *Cours de Chimie appliquée aux Arts et militaires, 1816, 68 feuillets* *Cours de pyrotechnie militaire,* *1816, 9 feuillets* *Cours d'attaque et de défense des places,* *1816, 42 feuillets*	*Collections archives de la bibliothèque de l'École polytechnique de Paris*

(continued)

Table 6.8 (continued)

Year	Title of the document (Sadi Carnot X 1812)	Archive
	Cours d'artifice de guerre, *1816, 6 feuillets* *Cours de constructions,* *S.d. (1816), 2 feuillets* *Cours de géodésie,* *1816, 4 feuillets* *Don du Professeur Paul Carnot*	
1814–1817	*Travaux exécutés par Sadi Carnot lors de sa scolarité, 1814 à 1817. École impériale de l'artillerie et du génie – Place de Metz – Promotion du génie. manuscrits 21/34 cm, datés et signés.* *Don du Professeur Paul Carnot.*	*Collections archives de la bibliothèque de l'École polytechnique de Paris*
1815	*Mémoire sur le projet de fortification passagère no 9– Redoute avec traverse défensives.* (7 feuillets, 2 plans)	*Collections archives de la bibliothèque de l'École polytechnique de Paris*
1815–1816	*Mémoire sur la machine à tirer des tuyaux de lunettes* (14 feuillets, 3 plans.)	*Collections archives de la bibliothèque de l'École polytechnique de Paris*
1815–1816	*Lever d'usines: Mémoire sur la sonnette à déclic* (6 feuillets, 3 plans)	*Collections archives de la bibliothèque de l'École polytechnique de Paris*
1815–1816	*Projet de bâtiment militaire: École régimentaire de cavalerie*	*Collections archives de la bibliothèque de l'École polytechnique de Paris*
1816	*Projet de fortification permanente* (10 feuillets)	*Collections archives de la bibliothèque de l'École polytechnique de Paris*
1816	*Lever trigonométrique* (4 feuillets, 1 tableau)	*Collections archives de la bibliothèque de l'École polytechnique de Paris*
1816	*Projet de construction hydraulique: Grand pont tournant* (17 feuillets, 4 plans)	*Collections archives de la bibliothèque de l'École polytechnique de Paris*
1817	*Lever d'usines: Mémoire sur la buanderie militaire* (8 feuillets, 3 plans)	*Collections archives de la bibliothèque de l'École polytechnique de Paris*

Sadi Carnot, *Officier du Génie Carrier Militaire*

1821	*État de service de M. Carnot (Sadi), né le ler Juin 1796 à Paris. Ministère de la Guerre – 3ème bureau –Bureau du Génie.* *(dernière situation portée sur cet état: Admis dans l'arme du Génie par Ordre du 25 Mai 1821)* (1 feuillet, ms., 20/31 cm, s..d)	Archive du Service Historique des Armées
1825	*Travaux de Topographie exécutés par Sadi Carnot, Lieutenant d'État–major, sur l'ordre du Ministre de la Guerre.*	*Collections archives de la bibliothèque de l'École polytechnique de Paris*
1828	*Démission sous l'autorisation du Roi : Le Ministre secrétaire d'État de la Guerre accepte la démission offerte par M.. Carnot (Sadi), Capitaine au Corps Royal du Génie* [. . .] *Paris, le 19 Mai 1828.* (1 feuillet, 20/31 cm)	Archive du Service Historique des Armées

Cfr.: "Catalogue des documents exposés à l'occasion du $150^{\text{ème}}$ anniversaire de l'édition "1824" des *Réflexions sur la puissance motrice du feu, June 1974, Archives de l'académie des sciences de Paris*"

Table 6.9 A list of Sadi Carnot's scientific works

Year	Title	Archive
s.d. [1878]	*Réflexions sur la puissance motrice du feu. Manuscrit non daté, 16,5/21,5 cm constitué de quatre cahiers* *le 1er formé de 22 feuillets* *le 2ème formé de 24 feuillets* *le 3ème formé de 20 feuillets* *le 4ème formé de 26 feuillets* *(Dons de la famille Sadi Carnot le 16–12–1878).*	Archive Académie des sciences, Institut de France, Paris
1824	*Réflexions sur la puissance motrice du feu et sur les machines propres à développer cette puissance, Paris, Bachelier, 1824. – In 8°, 12,5/19,5 cm, 118 p., 1 pl.* *Edition similé à la librairie scientifique Hermann, Paris, 1912.*	Private collection by Pierre Sadi Carnot
1872	*Réflexions sur la puissance motrice du feu et sur les machines propres à développer cette puissance, Publié en 1872 Annales scientifiques École normale supérieure. Deuxième série, Tome premier, pp 393–457.*	*Collections archives de la bibliothèque de l'École polytechnique de Paris* (Z 2 a/11)
1878	*Réflexions sur la puissance motrice du feu et sur les machines propres à développer cette puissance*. Reprint edition including Hippolyte Carnot's letter of 1878 to the French Academy of *Sciences and some of the previously unpublished Notes written by Sadi Carnot after publication of the Réflexions.* Gauthier–Villars, Paris. *La l'ère partie est la réimpression de l'édition originale parue en 1824 chez Bachelier, Paris.* In 4°, 21/27 cm 103.	*Collections archives de la bibliothèque de l'École polytechnique de Paris* (C 2 b/31)
1824	*Imprimerie de Guiraudet et Gallay. Mémoire des impressions faites au compte de Monsieur Carnot pendant le mois de Mai .. Paris, le 2 Juin 1824* *1 feuillet, 18/24 cm.* *(Facture de l'impression de Réflexions sur la puissance motrice du feu, où il est précisé que l'ouvrage a été tiré á 600 exemplaires).*	Private collection by Pierre Sadi Carnot
s.d.	*Nicolas Léonard Sadi Carnot. Notes sur l'économie politique et autres sujets de morale, politique, etc. ...* *Manuscrit non daté,* *37 feuillets, 17,5/22,5 cm.*	Private collection by Pierre Sadi Carnot

s.d. [1878]	*Sadi Carnot. Notes sur les mathématiques, la physique et autres sujets.* *Manuscrit non daté.* *16 feuillets, 16,5/22 cm* *7 feuillets, 15,5/20 cm* *(Notes de Sadi Carnot relatives à la nature de la chaleur et à sa transformation en puissance motrice.* *Don de la famille Carnot, 16–12–1878*	Archive Académie des sciences, Institut de France, Paris
s.d. [1816]	*Sadi Carnot. Machine à vapeur de M. Watt* *suivi de Machine d'expansion à double cylindre de M. Woolf. Manuscrit non daté,* *40 feuillets, 21,5/33 cm.* *(Traduction par Sadi Carnot de deux articles rédigés en anglais)*	*Collections archives de la bibliothèque de l'École polytechnique de Paris*
s.d.	*Sadi Carnot. Recherche d'une formule propre à représenter la puissance motrice de la vapeur d'eau.* *Manuscrit non daté, 17/21,5 cm, 21 p.* *Don du Professeur Paul Carnot*	*Collections archives de la bibliothèque de l'École polytechnique de Paris*
s.d. [1966]	*Gabbey W.A. and Herivel J.W.* *Un manuscrit inédit de Sadi Carnot. Paris, P.U.F., 1966* *In 8°, 15,5/24 cm, 16 p.* *(Revue d'histoire des sciences et de leurs applications, T. XIX, no 2, Avril–Juin 1966, pp. 151–166. Les auteurs présentent et commentent le texte du manuscrit indiqué ci–dessus : Recherche d'une formule propre à représenter la puissance motrice de la vapeur d'eau.*	*Collections archives de la bibliothèque de l'École polytechnique de Paris* (Z 2 c/250)
s.d. [1927]	*Académie des sciences/Dir. Sadi Carnot, biographie et manuscrit publiés sous le haut patronage de l'Académie des sciences, avec une introduction de Paris, Gauthier–Villars, 1927* *In 4°, 21/27 cm, 95* *(Cet ouvrage contient un facsimilé du manuscrit: Notes sur les mathématiques, la physique et autres sujets, suivi du texte imprimé)*	*Collections archives de la bibliothèque de l'École polytechnique de Paris* (X 3 b/377)
1868	*Paul de Saint–Robert, Sadi Carnot. Notice biographique par le Comte Paul de Saint–Robert.* *Turin, Imprimerie Royale* *In 8°, 15,5/22 cm, 22 p.*	Archive Académie des sciences, Institut de France, Paris

Cfr.: "Catalogue des documents exposés à l'occasion du 150ème anniversaire de l'édition "1824" des *Réflexions sur la puissance motrice du feu, June 1974, Archives de l'académie des sciences de Paris*"

Table 6.10 A list of Sadi Carnot's obituary/documents commemorations

Year	Document	Archive
Friday 24 August 1832	*Faire–part du décès de Sadi Carnot, le 24 Août 1832. 2 feuillets, 20/26,5 cm*	*Collections archives de la bibliothèque de l'École polytechnique de Paris*
Monday, 27 August 1832	*Annonce du décès de Sadi Carnot, lors de l'épidémie de choléra. Le Moniteur Universel, n° 240, lundi 27 Août 1832*	*Collections archives de la bibliothèque de l'École polytechnique de Paris* (Z4/33)
	Tombe de Sadi Carnot au cimetière municipal d'Ivry–sur–Seine. Photographie, 12/17 cm.	Private collection by Pierre Sadi Carnot
1894	*École polytechnique. Livre du centenaire, 1794–1894. Paris, Gauthier–Villars, 1894. In 4°, 3 vol.*	*Collections archives de la bibliothèque de l'École polytechnique de Paris* (X 2 b/36)
	Notice sur la vie et l'œuvre de Sadi Carnot par M. Lévy, T. 1, pp. 181–189.	
1926	*Henry Le Chatelier. Sadi Carnot et la mécanique chimique. Discours prononcé à la Société des Ingénieurs Civils, en présence du Président de la République, le mercredi 20 Janvier 1926, (pour fêter le centenaire de la publication du mémoire de Sadi Carnot sur la puissance motrice du feu).*	*Collections archives de la bibliothèque de l'École polytechnique de Paris* (Z4/99)
	Bulletin mensuel polytechnicien, 6ème année, n°8, 25 janvier 1926, pp. 139–142.	
1932	*Inauguration d'une plaque commémorative de la naissance de Sadi Carnot au Palais du Petit Luxembourg à Paris, sa maison natale, le vendredi 8 Juillet 1932.*	*Collections archives de la bibliothèque de l'École polytechnique de Paris* (Z 1 a/9)
	Notices et discours. Institut de France, Académie des sciences, Tome I, 1924–1936, pp 402–414.	
	Discours de Robert Bourgeois, Émile Picard, Jean Perrin.	
	Une plaquette existe aux Archives de l'académie des sciences de Paris, Institut de France.	
1946	*Casimir Monteil. Sadi Carnot fondateur de la thermodynamique, séance du 7 Mai 1946. Paris, Société d'encouragement pour l'industrie nationale, 1946.*	*Collections archives de la bibliothèque de l'École polytechnique de Paris* (Z 1 a/25)
	Brochure spéciale, 84 p.	

Adapted from a larger list in "Catalogue des documents exposés à l'occasion du 150ème anniversaire de l'édition '1824' des *Réflexions sur la puissance motrice du feu, June 1974, Archives de l'Académie des sciences de Paris*"

NOMS	PRENOMS	Promotions	Dates: naissance	Dates: décès	Classement: entrée	Classement: sortie	Corps de sortie	Situation (en principe fin de carrière), titres - Observations
CARNOT	N. L. Sadi	1812	1796	1832	24	10	G.	Inventeur de la Thermodynamique
CARNOT	M. F. Sadi	1857	1837	1894	5	7	P.C.	Député de la Côte d'Or - Ministre - Président de la République - Assassiné
CARNOT	M. Adolphe	1858	1839		12	5	M.	Insp. gén. M. - Dir. ENSM - m. Institut (S)
CARNOY	Joseph M. B.	1911	1890	1916			A.	Lt - MPF
CARNUS	Jean-Michel B.	1975	1955				GREF	
CARNY (ALLOUARD-)								voir ALLOUARD
CARO	Emmanuel	1855	1835		117	67	A.	G.B.
CARON	Pierre François	1813	1796		43	44	A.	Col.
CARON	Louis Félix Joseph	1815	1795		28		D.	
CARON	Honoré L. M.	1841	1823		176	127	A.	Lt Col.
CARON	Jules	1847	1829		83	25	M.E.	Admin. à la Dir. Gle des M.E.
CARON	Jean Charles Aug. Alb.	1863	1843		56	90	G.	Lt Col.
CARON	Alphonse Edouard	1881	1862	1951	38	49	A.	G.D.
CARON	Jacques Lucien	1919N	1900	1983			D.	Dir. Chf Maroc
CARON	René L. E.	1936	1914				A.	Dir. Imprimerie Banque de France
CARON	Pierre F. L.	1942B	1923				F.C.	
CARON	Hervé I. M. P. Y.	1956	1937				D.	
CARON	Daniel Joël	1962	1942				D.	

Fig. 6.14 Plate from the original source of the History of *École polytechnique de Paris* (Callot p 390)

6.5 An *Ad Hoc* Index of *Réflexions sur la puissance motrice du feu*

We conclude this introduction to Sadi with a brief overview of his book. Below we list the sequence of presentation adopted by Sadi Carnot in his *Réflexion sur la puissance motrice du feu*, where we provide an *ad hoc* index,[69] adding hypothetical chapters, paragraphs and some terms (in *italics*) to complete a panorama of the original arguments the young French scientist addressed (Table 6.11).

[69] Adapted from an old non-published work presented by one of us in his *lauream* dissertation in physics, *Sadi Carnot's thermodynamics: a new interpretation based on logics and mathematics* at Faculty of Physics (and Mathematics and Natural Sciences), University of Naples "Federico II", Italy, Chap. I. The dissertation is archived both Faculty of Physics, University of Naples "Federico II" and The National Florence Library.

Table 6.11 An *ad hoc* index of *Réflexions sur la puissance motrice du feu*

Chapter	Section	Paragraph	Page
Chapter I	1.1	Cosmological conception of heat[a].	1
On the Heat	1.2	Social progress due to development of motive power of heat machines: mines, industry and transport.	1
	1.3	*Historical* development of the motive power of heatmachines.	5
Chapter II	2.1	*An incomplete theory. Examples of unsolved problems.*	6
Premises to the theory of heat machines	2.2	The absence of theory comparable to the theory of mechanical machines.	7
	2.3	Presentation of *running* heat machine.	9
	2.4	The motive power is generated by there–establishment of equilibrium of caloric when it transfers to two–different–temperature bodies.	10
	2.5	*Any body as* possible *workingsubstance* for heat engines.	12
	2.6	*The heat such as motion through changes of volume.*	12
	2.7	The motive power depends on the working substance used?	14
Chapter III	3.1	*How to produce heat.*	15
Sadi Carnot's theorem	3.2	*The thermal work was born from a difference in temperatures which is caused by re–establishment of equilibrium of caloric.*	16
	3.3	*First proof of Sadi Carnot's theorem.* Presentation of a cycle (incomplete *in 3 phases*) having vapor as a working substance.	17
	3.4	Presentation of the reverse cycle.	19
	3.5	Ad absurdum reasonings upon the best motive power generated by other working substances.	20
	3.6	The maximum of motive power resulting from the employment of steam is also the maximum of motive power realizable by any means whatsoever.	21
	3.7	The conditions to obtain the maximum efficiency *of motive power; Δt is necessary*	22
	3.8	*Non-ideal conditions.*	25
	3.9	*Introduction of a cycle (incomplete, in 3 phases between two sources) within a very small–temperature–range and* closing *of it by* infinitesimal isochors.	27
	3.10	Closing *the cycle (incomplete, in 3 phases between two sources) by means of* finite isochors.	27
	3.11	Analogy between *motive power of heat and* falling water.	28
	3.12	*The problem of the function of efficiency of the heat machine.*	28

(continued)

Table 6.11 (continued)

Chapter	Section	Paragraph	Page
Chapter VI	6.1	*Applications of Carnot's theorem*; The motive power produced is independent by different temperatures.	68
Temperature difference and the maximum motive power	6.2	The efficiency of a heat machine: *the fall of caloric produces more power at lower temperatures than higher temperatures.*	72
	6.3	*Examples* of the calculation of the motive power of *heat*	74
	6.4	Mathematical speeches. *Research for a function of the work; $C(t)$ –dependent only on temperature.*	76
Chapter VII	7.1	Case–studies on the production of motive power having heat machines that use different working substances to carry out heat.	88
On the working substance	7.2	Comparison of different working substances used for production of motive power.	89
	7.3	Study of the action of heat on solids for production of motive power.	91
	7.4	Study of the action of heat on liquids for production of motive power.	92
	7.5	Study of the action of heat on gases and vapors for production of motive power.	93
	7.6	*Advantages of* heat high–pressure–machines on low–pressure–machines.	97
	7.7	Overview of the working principle of a high–pressure–machine.	100
	7.8	*Watt's steam machine.*	102
	7.9	Reflections on the use in heat machines of various permanent gases and vapors from water vapor; advantages and disadvantages of air machines with respect to vapor machines.	103
	7.10	Closing remarks (*operatives*) on the motive power of heat machines *and comparison between the running motive power produced and theoretical power produced*	112

[a]Sadi Carnot praised three main factors: *chaleur dans la nature*, *bateaux à vapeur*, *cosmolgique, economique et communicationelle de la chaleur*. (Cfr.: Cardwell 1971, pp 89–120, 186–192). These are important preliminary premises of his theory. In fact, we will see in the next chapters the crucial roles played by *communication* and *production* in Sadi Carnot's work (see Chapters. 7, 10) and in the two Carnots' work (see Chapter 11)

Below we provide three images of the students of the *École polytechnique* at Sadi Carnot's time.

Fig. 6.15 *X Élèves de l'École polytechnique*, 1812 (*Collections Archives de la Bibliothèque de l'École polytechnique Paris* (Ref.: Élève de l'École polytechnique. 1812/dess. Joseph Louis Hippolyte Bellangé et sculp. Rouget, Paris : s.n., 1812; 1 gravure : coul. ; 26 × 17 cm. Cote: A3 P11 IA))

Fig. 6.16 X Élèves at *Siege de Paris*, 1814 (*Collections archives de la bibliothèque de l'École polytechnique de Paris* (Ref: A1 P5 ID))

Fig. 6.17 *École royale polytechnique et une visite (la seule) de Napoléon le 28 avril 1815* (*Collections Archives de la Bibliothèque de l'École polytechnique Paris* (Ref.: S.l.: s.n., 1815–1820/dessiné et gravé par Peronard. S.l.: s.n., 1815–1820. 1 grav.; 47,5 × 55,5 cm. Cote: A1 P7 IA))

Chapter 7
On Principles in Sadi Carnot's Thermodynamics

Une semblable théorie manque évidemment pour les machines à feu

(Sadi Carnot, *Réflexions sur la puissance motrice du feu*, p 8).

In 1824 Sadi Carnot published *Réflexions sur la puissance motrice du feu* in which he almost entirely founded thermodynamic theory. Two years after his death, his friend Clapeyron' introduced the famous *PV* diagram to analytically represent the famous Carnot cycle, one of the main and crucial ideas presented by Carnot in his booklet. Twenty–five years later, in order to achieve the modern version of the theory, Kelvin and Clausius had to reject the caloric hypothesis, which had influenced a few of Carnot's arguments. Relying on the possibility of studying the history of science by means of logical investigation, in this section we shall propose historical–epistemological research on Sadi Carnot's original thermodynamic theory. In this theory, the French scientist presents more than two principles, all of which are expressed by double negative sentences (generally speaking) within non–classical logic.

7.1 An Outline of the Problem of History and Logic

Strictly speaking, a principle can be considered the first element in the development of a scientific theory and it may not be mathematically confuted or experimentally demonstrated. In some previous papers (Drago and Pisano 2000) on the history and foundations of physics, it has been shown that the discursive part of Sadi Carnot's *Réflexions sur la puissance motrice du feu* of 1824 (Carnot 1978) does not include any principles (e.g., like in Newton's theory) but it presents more than 60 Double Negative Sentences (DNSs). Since DNSs are not equivalent to the corresponding

C.C. Gillispie and R. Pisano, *Lazare and Sadi Carnot: A Scientific and Filial Relationship*, History of Mechanism and Machine Science 19, DOI 10.1007/978-94-017-8011-7_7,

affirmative sentences, they belong to non–classical logic; typically the law of double negation, $\neg\neg A = A$.[1]

In this section, in order to investigate Carnot's logical reasoning by means of historical categories and foundations, we will follow a rigorous method dealing with this kind of analysis and show that Sadi Carnot's reasonings DNSs are based on non–classical logic. We also remark that DNSs used by scientists cannot be dealt with using classic first–order logic (Hodges 1983). In this sense, we only use the logical elements as special categories and not in the sense of a logical theory of categories or simply of a new theory. Let us see some examples.

7.1.1 On the Mathematics

One could discuss that if an object is shown through an *absurdum* proof, its existence is not proven mathematically. While, in classical logic:

$$A \text{ is true when } \neg\neg A \text{ is true},$$

because, within classical logic, one should only verify that a contradiction does not emerge. In *multi–valued logic*, however, more than two truth–values can exist. Let us see an example. Let a be a number in decimal form:

$$0, a_1 a_2 a_3 \ldots a_n$$

$$a_n = \begin{cases} 0 \Leftrightarrow 2n = p_1 + p_2 \\ 9 \end{cases} \qquad \text{where } \{p_1, p_2\} \in \textit{prime numbers}$$

The property "$2n = p_1 + p_2$" is valid but we do not know if it is valid for every integer. In fact, in the eighteenth century the mathematician Christian Goldbach (1690–1764) conjectured[2] that every even integer greater than 2 (*Goldbach number*) can be expressed as the sum of two primes numbers.[3] Therefore we can write, e.g., $10 = 7 + 3$, $14 = 11 + 3$, $18 = 13 + 5$. As often occurs with conjectures in mathematics, one can read a great number of supposed proofs of the *Goldbach conjecture*, but they are not currently accepted by the mathematical community. To be brief, it is not possible to prove truth content in the *Goldbach conjecture*. Moreover, a counter–example is also impossible. Thus, generally speaking, one can write that

> $a = 0$ *(for all of its figures) cannot be claimed because scientific proofs are lacking.*

---

[1] $\neg\neg A$ in logic: *non non A*. This chapter's arguments are mainly based on Drago and Pisano (2000), Pisano (2010).

[2] In 1742, the *Goldbach conjecture* was proposed in a letter addressed to the Swiss mathematician Leonhard Euler (1707–1783).

[3] Let us remark that it is a curious property because prime numbers cannot be deduced by division, while odd numbers and the sum of two odd numbers concern another operation.

Let us note that its negative ($\neg(a = 0)$) sentence cannot be claimed. In fact, if the latter could be proven then that would indicate that one was able to claim it when $a \neq 0$; however, this means that we should also be able to present a counter–example (e.g., negative of a negative) of the *Goldbach conjecture*. Thus, one should conclude that

$$\neg\neg\ (a = 0)\,\textit{fails}.$$

7.1.2 On the Physics

Generally speaking, physical laws establish a mathematical–physical relationship among numerical data and physical magnitudes. Based on German physicist George Simon Ohm's (1787–1854) laws one can claim:

$$V = Ri. \tag{7.1}$$

That is, V is equal to the resistance R multiplied by the current i. In general, in physics, we know that the scientific validity of an affirmation depends on the physical system, the adopted theoretical–experimental model and the theory's field of applicability. Thus, one could claim that:

> *a given proposition $B(s_1, \ldots s_n)$ where B expresses a mathematical relationship among the values of the quantities $(s_1, \ldots s_n)$ is scientifically valid with respect to a given model*[4] *M, when its results are scientifically valid with respect to all the physical systems belonging to M where the quantities represented by $(s_1, \ldots s_n)$ are measured.* (Adapted by Dalla Chiara and Toraldo di Francia 1999, p 87).

We return to (7.1). This law – belonging to physics – expresses a particular logical action according to which (a) the truth of a conjunction (compared to model M) would imply the truth of both members in (7.1); while the opposite would not be logically applicable (*Ibidem*). Moreover, generally speaking, the (b) *non–truth* of a given proposition does not imply the truth of its negation. Therefore, given two quantities A and B, one can write:

(a) $(A \wedge B)$ true$\Rightarrow$ A true, B true
A true, B true $\neq(A \wedge B)$ true
(b) $\neg(A) \neq \neg\neg(A)$

---

[4]In order to present a strictly physical example, I only consider propositions $B(s_1, \ldots s_n)$ and models where the magnitudes $(s_1, \ldots s_n)$ are physically measurable. (The example given is adapted from: Dalla Chiara and Toraldo di Francia 1999, pp 88–89). On the foundations of science, model and reasoning, one can also see: Da Costa and French (2003).

In other words, this result would weakly violate the validity of the principle of *Tertium non datur.*[5] Thus, *what does it mean to measure by* (7.1)? In fact, by considering the previous discussion, in order to measure (7.1) a simultaneous measurement of the three quantities *V*, *R*, and *i*, should be required. In this sense, in order for (7.1) to be experimentally *true*, it is necessary for three real corresponding numbers *a*, *b*, and *c*, to exist respectively for the measurements of *V*, *R*, and *i*. Thus, one can write:

$$a = bc. \tag{7.2}$$

The measurement is obviously never *perfect* (or to be more precise, the experimental data should coincide with the theoretical data only within the limits of the experimental errors). Therefore, due to the *uncertainties* of the devices and error of measurement, in the same range of measurement of *a, b and c,* another real triad a', b', c', with $(a' \neq a)$, $(b' \neq b)$, $(c' \neq c)$, should exist, so that:

$$a' \neq b'c'. \tag{7.3}$$

In brief, both (7.2) and its negation (7.3) would be true with respect to a given experimental situation. Nevertheless, the conjunction of (7.2) and (7.3) would not be true, because a real triad $a,\ b,\ c$, should exist so that one should simultaneously obtain:

$$a = bc \ \wedge \ a \neq bc$$

which is false in classical logic. These two examples could be expressed by double negation, demonstrating that their scientific content should follow, generally speaking, non–classical logic.

[5]In classical mathematical logic, three main laws are claimed *a priori*: *The principle of non–contradiction:* "*x* and non *x*". One thing cannot both be and not be at the same time and in the same respect. The same proposition cannot be both true and false. *The principle of Tertium non datur*: Either "*x*" is "*y*" or "*x*" is not "*y*". Either a thing is or it is not, there is no third possibility. *The principle of identity:* Every being is that which it is. Each being is separated in its existence from other beings. On *The principle of Tertium non datur* and related topics see: Kolmogorov (1932, 1991, 1925), Drago 2005, pp 57–81).

7.1.3 On the Physics Mathematics

LAWS. *of Natural Philoſophy.* 19

Axioms or Laws of Motion.

LAW I.

Every body perſeveres in its ſtate of reſt, or of uniform motion in a right line, unleſs it is compelled to change that ſtate by forces impreſs'd thereon.

PRojectiles perſevere in their motions, ſo far as they are not retarded by the reſiſtance of the air, or impell'd downwards by the force of gravity. A top, whoſe parts by their coheſion are perpetually drawn aſide from rectilinear motions, does not ceaſe its rotation, otherwiſe than as it is retarded by the air. The greater bodies of the Planets and Comets, meeting with leſs reſiſtance in more free ſpaces, preſerve their motions both progreſſive and circular for a much longer time.

LAW II.

The alteration of motion is ever proportional to the motive force impreſs'd; and is made in the direction of the right line in which that force is impreſs'd.

If any force generates a motion, a double force will generate double the motion, a triple force triple the motion, whether that force be impreſs'd altogether and

C 2 at

Let us see the Newtonian principle of inertia (NPI):

DEFINITION III. *The* vis insita, *or innate force of matter is a power of resisting, by which every body, as much as in it lies, endeavours to preserve in its present state, whether it be of rest, or of moving uniformly forward in a right line.*[6]

Axioms; or Laws of Motion. Law I. *Every body perseveres in its state of rest, or of uniform motion in a right line, unless it is compelled to change that state by forces impressed thereon.*[7]

Nowadays, one can read:

Every body will persist in its state of rest or of uniform motion (constant velocity) in a straight line unless it is compelled to change that state by forces impressed upon it.

[6] Newton (1803), I, p 2; *Italic style* and capital letters belong to the author.

[7] *Ivi*, I, p 19 (Author's *italic*).

It has been remarked (Nagel 1961) that all physical laws can be expressed by means of a proposition preceded by two *universal* and *existential* quantifications[8]: (∀) ("for all") and (∃) ("there exists" or "for some"). A possible logical expression of the Newtonian principle of inertia can be:

$$A = \forall x \exists y : P\left(x, y\right)$$

A	= the proposition of the principle of inertia.
X	= a body.
y	= a complex system constituted by an inertial system, a closed system, and a clock.
P(x, y)	= a predicate concerning *x* and *y*: "if *x* is in *y*, than it is of its state of rest or of uniform motion".

Nevertheless, this discussion pertains to classical mathematical logic which – by nature – is not interested in the *effective calculability* of its functions with respect to operative–experimental–data. In this sense, the content of the first principle is lacking in experimental and *calculable evidence*. In detail:

> The Newtonian principle of inertia claims that *y exists*, but it does not claim how one can find it.
>
> The "whether" (or commonly "unless") contained in the proposition is not an *operative situation*. It only explains the changes of the state of motion which occurred in the body *a posteriori*.
>
> A precise distinction when $\vec{v} = 0$, and when $\vec{v} \neq 0$, is required by the NPI.
>
> A precise knowledge when for $\vec{v} =$ *constant* in orientation (direction and versus) and in scalar–magnitude for the entire path is required by the NPI.
>
> A precise knowledge of absence–forces or of a non–zero net force is required by the NPI.
>
> The Newtonian principle of inertia is valid subordinately to validation of $\sum_i \vec{F}_i = 0$ (for material–point and on the entire path).

Every physical variable should be subjected to its measurement. If the measurement cannot be applied, the scientific content generates *uncertainties* in scientific knowledge. For this reason, NPI's content, as mentioned above, can be expressed by a DNS, ¬¬*A*: *It is not true that* $\vec{v} = 0$ *is not equal to* $\vec{v} \neq 0$. Thus, all of the examined experimental–logical–*ambiguity* listed can be found in the Newtonian principle of inertia within a non–classical logic investigation.

[8]At the beginning of the past century, Thoralf Albert Skolem (1887–1963) suggested a technique to formalize the existential quantification on *y*–variable of a given predicate into a constructive mathematical function (Skolem [1920] 1967; Hodges 1997).

7.1.4 *On the Logic and Structure*

Another crucial feature in scientific language regards the paradox of the formalization of logic (Carnap 1943). In order to express the axioms and to construct a *meta–discourse* about them, we should use natural language, which is not formalized; we cannot formalize it in advance because we risk producing a regression *ad infinitum*. Moreover, it is not natural to state the axioms of logic and to consequently deduce all the rest from them (*Ibidem*). In brief, in the following we present a possible historical-epistemological method to investigate the foundations of science (Pisano 2011a, 2012b):

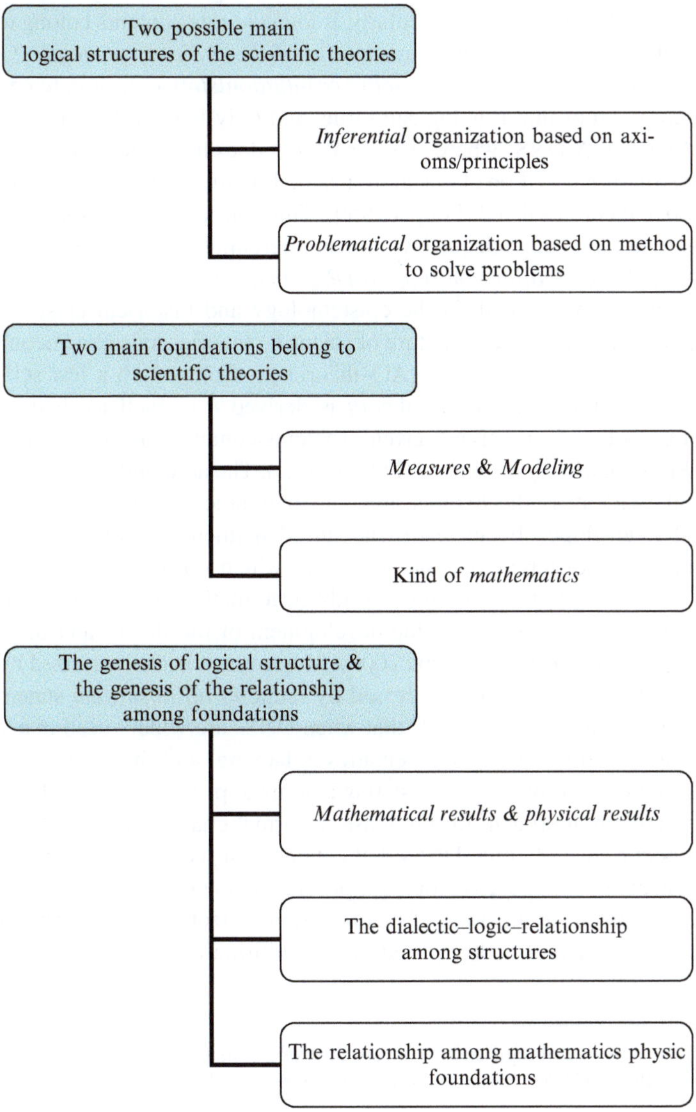

In fact, in everyday life this is never done. When we think, we usually proceed from temporary premises, and then introduce or remove logical elements of natural language. This is a natural inference, which can express both classical and non–classical logic (Prawitz and Melmnaas 1968; Haack 1974). In mathematical–classical logic, so–called *well–formed–statements* are assumed to be either true or false, even if we do not have proof of either. In fact, if we consider an inferential system in terms of classical logic (e.g. a list of inferential propositions) we can obtain a sequence *free from self-contradiction*. In this kind of logical system, a theory must be testable if it is *scientific*, e.g., it should be subject to falsification (Popper 1963). Let us note that in such a system of reasoning, it is not possible to obtain *undecidable contents,* e.g. (apparently) those generated by scientific DNSs belong to non-classical logic. Particularly, if *undecidable* contents belong to a given principle of the theory, then we have an *out of the ordinary principle*. Generally speaking, in non-classical (or *constructive* or *intuitionistic*) logic, a statement is only true if there is proof that it is logically true, and only false if there is proof that it is logically false. A DNS (where $\neg\neg A = A$ *fails*) does not depend on an inferential scientific structure based on classical logical dichotomy of theses, e.g., obtained by the deductive theorems listed (Popper 1959). Thus, one could think that *undecidable* contents are not logically adequate within an inferential system. Nevertheless, *what kind of logical organization can support DNSs in a scientific theory?*

As previously mentioned, in the epistemology and historical epistemology of science related to the logical structure of scientific theories, one can encounter both *Axiomatically Organised* theories (AO–theories) – from which a few self–evident principles (or axioms) the whole theory is derived – as well as theories whose organization is based on solving given problems contained in the theories, which are thus *Problematically organized* (PO–theories). The assumed principles are often not as self–evident as the axioms are non–axiomatic principles. An AO theory is generally developed by means of advanced mathematics (e.g., mathematics–physics theory by differential equations, et al.) which starts its derivations directly from the axioms. A PO theory uses less advanced mathematics with the principles that only indicate a direction for the development of the theory and they may be *methodological* in nature (Kieseppä 2000). Moreover, it is characterized by the use of DNSs and most results are expressed by *reductio ad absurdum* statements. In previous studies, it has been noted[9] that when DNSs are widely used in a scientific theory, a use of sophisticated mathematics is lacking, and the theory is based on declared problems to solve without stating axioms or principles typically pertaining to an Aristotelian approach. In this sense, a formal characteristic of POs appears to be the occurrence of some DNSs that cannot be turned into equivalent positive sentences because the operative tools for proving them do not exist. In other words, as previously stated, in this kind of theoretical organization, the scientific contents of DNS "¬¬A" cannot be converted into an affirmative sentence corresponding

[9]More details on the use of logic and the organization of the theory in historical investigations are in: Drago and Pisano (2000), Pisano and Gaudiello (2009a, b), Capecchi and Pisano (2010a, b).

to "A" because the latter lacks scientific proof. Therefore, DNSs, within scientific theory, characterize a particular approach to science. Following this point of view, a borderline between classical logic and most non–classical logic is represented, not by the law of the excluded middle, but by the law of double negation. Generally speaking, when the law of double negation fails, we are arguing outside classical logic and, *in–first–approximation*, we are arguing, within intuitionist logic (Prawitz 1977).

Finally, two general ways to organize a scientific theory can be claimed: (1) a former one, e.g., based on Aristotle's arguments, is organized through axioms (AO) and its logic is classical, (2) a *Problematic Organization* (PO), belonging to non–classical where a result could also be "fuzzy". In the next section, we will deal with these aspects of the historical epistemology investigation in Sadi Carnot's *Réflexions sur la puissance motrice du feu*.

7.2 Sadi Carnot's Theory and its Logical Organization

With regard to the categories of investigation announced in the previous chapter, here we will deal with Sadi Carnot's principles by means of logic referring to previous studies (Pisano's works on the history of thermodynamics; Lervig 1972, 1982). In particular, we will see that Lervig proposed an interesting analogy between modern thermodynamic principles and those of Sadi Carnot, focusing on entropy and caloric. Here we will explain and use the following categories: (a) logic (DNSs) to understand the logical organization of the theory and (b) the choice of infinite in mathematics used by the author to verify the relationship between mathematics and physics and the related interpretation–formalization of certain physical phenomena expressed by crucial statements. In this way, starting with Lervig's arguments, we will try to show that it is possible to discuss a potential development based on non-classical logic with respect to Lervig's interpretation based on (non–declared) classical logic.

7.2.1 On the DNSs and Logical Investigation

Around 1824 the caloric theory of heat remained, in France at least, the rising theory of heat.[10] The *Académie des sciences* reviewed *Réflexions sur la puissance motrice du feu* only as the result of a commemorative essay, written in 1834 by the late

[10]Mendoza (1960), Buchdahl (1966), pp 23–26, 113–116, 121–124, 174, Cropper (1987), pp 120–122.

Sadi Carnot's friend Émile Clapeyron[11] (1799–1864). After this review (Clapeyron 1834), *Reflections* was almost universally ignored for 25 years (Pisano 2007a); but Rudolf Clausius (1822–1888) and William Thomson (1824–1907) wrote more important works (Clausius 1850, pp 368–397, 500–524; Thomson 1848, pp 66–71, 1851a, pp 100–106, 1851b, pp 175–183) for a theoretical advancement of the theory, based on advanced mathematics.

Here below, Clapeyron and Clausius interpret Carnot's cycle to show how a young Sadi's original reasoning was reformulated in some crucial aspects of his thermodynamic theory (Figs. 7.1, 7.2 and 7.3).

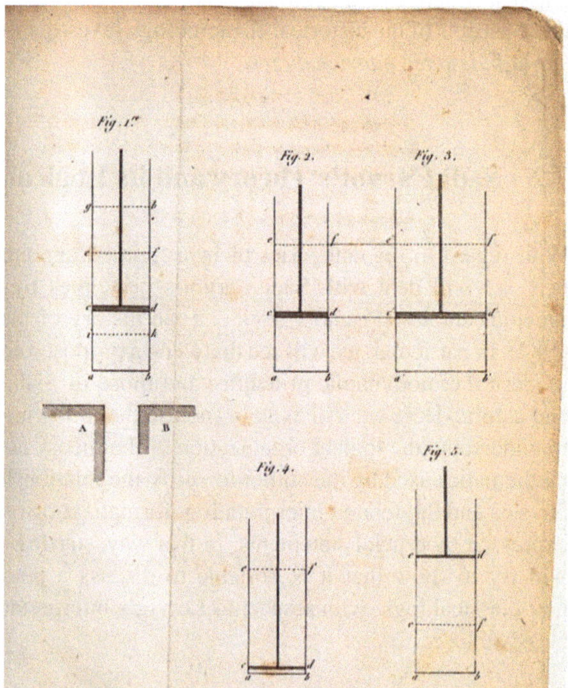

Fig. 7.1 S. Carnot's cylinders–cycle (1824) (Carnot 1824, p 118. "V–33929, gravure face, p 118. Département de la reproduction, Bibliothèque nationale de France, Paris. Numéro de BL: LIV–1006–006278 Date: 23/06/2010 Client: RP CLI–048007 Devis n° : DEV–1005–008353")

[11]"L'idée qui sert de base à ses recherches me paraît féconde et incontestable [...]" (Clapeyron 1834, p 154, line 19). The work was, however, very different from Sadi Carnot's original work. In fact, Clapeyron's work was an analytical essay and introduced the famous *PV* diagram to represent Carnot's cycle. See also: Girard (1824), p 414, Robelin (1832), pp 528–530.

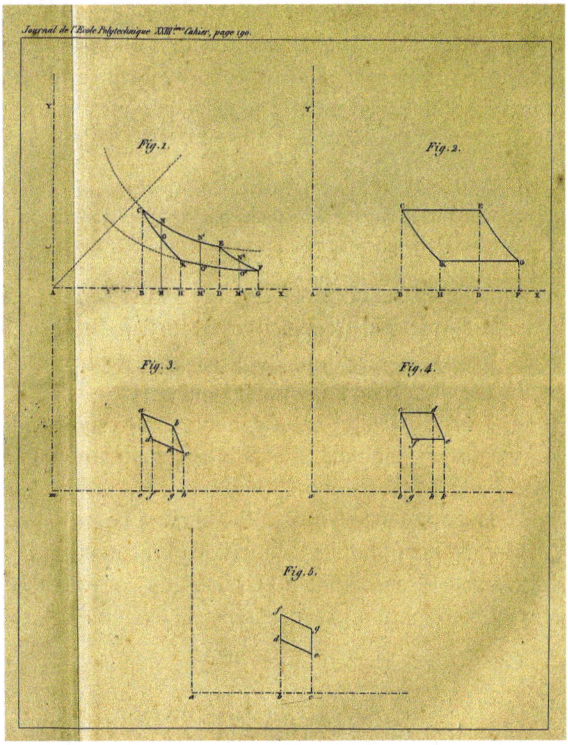

Fig. 7.2 Clapeyron's diagram (1834) (Clapeyron 1834, Fig. 1, p 190. (Collections École polytechnique Z 5 (1834))

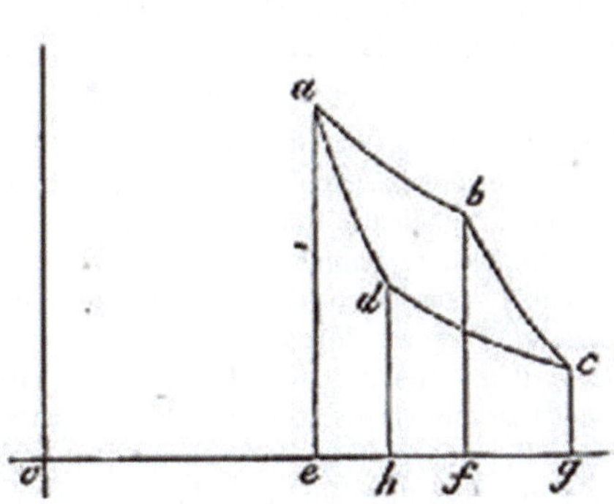

Fig. 7.3 Clausius' diagram (1850) (Clausius 1850, p 379. See also Clausius 1865a, b)

Sadi Carnot's friends (students at the *École Polytechnique de Paris*) considered it a *difficult book*.[12] The work does indeed contain some surprising innovations: the

[12]Let us note that a kind of "[. . .] indifférence de la plupart des contemporains [. . .]" circulated around Sadi Carnot's publication (Fox 1988, p 288).

idea of the cycle, an upper limit to the efficiency[13] of heat machines, a *reductio ad absurdum* theorem proof, and new law of gases. Its interpretation is also difficult because, surprisingly, Sadi Carnot obtained results that were almost (all) exact in spite of being based on erroneous heat conservation (caloric) theory. On the other hand:

> However, the novelty of the *Réflexions sur la puissance motrice du feu* was of a kind that it was difficult to appreciate, so much so that the book probably bred incomprehension rather than excitement among the few contemporaries who read it.[14]

As previously stated, a PO–theory is essentially based on methodological (also called architectural[15]) principles, which indicate a direction for the development of the theory. In addition, the absence of sophisticated[16] mathematics is abrogated by the use of double negative sentences (DNS) in which the law of double negation $\neg\neg A \Rightarrow A$ fails. Therefore, *A* cannot be directly stated: proposing that problem with an (idealistic) sentence, A would abrogate problem *A* itself. Recent works (Drago and Pisano 2007; Pisano 2005, 2006a, b, 2007b), demonstrated that Sadi Carnot's thermodynamic theory was a PO–theory, based on problems regarding the validity of caloric theory and the calculation of the maximum efficiency[17] of a heat machine (Drago and Pisano 2000, 2002). We can consider the principle of the impossibility of perpetual motion in mechanics, an example of crucial importance in the history

---

[13]With regard to the efficiency, second law and its epistemological studies, one can also see: Barnett (1958), Bailyn (1985), Armstrong (1960), p 564, Lemons and Penner (2008), Brønsted (1939), Buchdahl (1987), Pisano (2001), Cheng (1998), Dartnal and Reizes (2005), Kostic (2008).

[14]Robert Fox in Carnot (1986), p 22, line 11.

[15]We also remark that the distinction AO–and–PO–theories and principles in AO–and–PO–organization are epistemologically different from distinguishing between *a priori* and a *posteriori* principles, e.g. in the Kantian approach or before in Aristotelian science on the role played by *definitions*, *principles* et al. and their peculiar syllogisms (Hodges 1998). E.g., Aristotle, particularly, considered *postulates* as well as *first propositions* in science as containing truth and necessity. They are different from *proper principles* as *propositions* belonging to a particular theory. (Cfr.: Aristotle 1853, Book I, chap VIII, p 263, chap. IX, p 265, chap X, p 266). See also: "That this Treatise is useful for three purposes" in *The Topic* (Aristotle 1853, Vol. II, Book I, chap II, p 36; Aristotle 1996, Book I, chap I, p 9). Thus, the term "methodological" used in this paper, expresses that Sadi Carnot builds his theory by means of principles, like the impossibility of a perpetual motion, which are different from those of Aristotle, such as (e.g.) Newton's mechanical principles. In our work, from here forward we will use the *Principle of theory*. The theme is also developed in recent works (Capecchi and Pisano 2008; Pisano 2007a, pp 107–112, 2007b, pp 1–29, 2007c).

[16]Of course, as one of the referees accurately notes, other scientists used non–sophisticated mathematics in their theories. E.g., Poncelet, in the end of 1820 and 1830 developed innovations in mechanics using a *discursive* approach based on the *absence de calcul infinitésimal*. *Idem* situation for: Charles Dupin (1784–1873) for his course of mechanics (Chatzis 2009, pp 99–113).

[17]"Kuhn (1959), pp 530–535, Fox (1970), Drago (1995), Décombe (1919), Kelly (1964), p 643.

of physics (previously used by Simon Stevin (1548–1620). See also Koetsier 2010), in the following form:

> [...] ipsique globi ex sese continuum et aeternum motum efficient, quod est falsum.[18]

If we refer to the previous sentence as "*A*", we can logically write:

1. A = motion has an end
2. $\neg A$ = motion has no end
3. $\neg\neg A$ = Impossibility of perpetual motion

In particular, the corresponding affirmative sentence "[...] motion has an end" is empirically unsound. Its verification would require either infinite observation time or a proof that friction will not eventually and definitively vanish. Consequently, the principle of the impossibility of perpetual motion is better expressed by employing a DNS (case–3). We should note that this principle cannot play the role of an *a priori* sentence, e.g., evident to reason but instead comes from common experience. In this way, a particular case study can be said to belong to non–classical logic[19] (Drago and Pisano 2002, pp 130–131). Therefore, the formal characteristic of PO–theories seems to be the use of DNSs to express important *problems* of the main arguments of the theory. DNSs entail an organization of the entire theory which is distinct from the deductive–axiomatic organization theorized by Aristotle (AO); PO–theories aim to discover the methodological solution of a problem rather than to generate theorems from abstract principles. In conclusion, we can see two ways of organizing a scientific theory; the first, the PO type, uses non–classical logic; the second, the AO–type, is organized by axioms and uses classical logic. Using the first type, one can produce a cycle of logical arguments around *A*, as Sadi Carnot argued in his theory. Conversely, classical logic can never yield cycles, since the chain of theorems is essentially infinite and each *A* sentence can be cyclical, being equal to the result of a cycle of two negations, i.e., $\neg\neg A$.

7.2.2 *The Distribution of DNSs in Sadi Carnot's Book*

Let us consider the crucial problem of thermodynamic theory (whether heat is connected to work and how). At Sadi Carnot's time, most industrialists and technicians believed the following sentence to be true: "heat is work". However, they did not consider that the sentence is partially true since at the time heat engines gave an unprecedented quantity of work with respect to the work obtained from both animals and man. Instead, from a theoretical point of view, it was already

---

[18] Stevin [1605] 1608 p 35. "It is not true [*falsum*] that the globe moves by itself and has not end [*aeternum*]". See also: Dijksterhuis (1955), pp 174–179, 507, 509, Clagett and Moody (1960). Lazare Carnot (1786, p 94, line 18).

[19] Generally speaking one can consider the *intuitionism* in logic (Dummett 1975; Prawitz and Melmnaas 1968, pp 215–229).

known that heat is not equal to work. At the time, the prevailing theory of heat was caloric, according to which Q is a function of state (or, in physical terms, it is an imponderable fluid), while work is not. According to this theory, the production of work – or its absorption – does not change the state of heat. That corresponds to a new sentence: "Heat is not work". This second sentence negates the previous one. When the second sentence is also negated, a DNS results; it expresses the modern theory of the equivalence between heat and work: "It is not true that heat is not work". Which sentence is true among the three? Let us notice that a problem formulated in this way constitutes a highly theoretical theme (i.e., what is the nature of heat?) and at the same time a highly technical one (i.e., how much work is possible to extract from a given quantity of caloric by means of a heat machine?). By recognizing the double nature of the problem we remove the aura of a mere engineer's theory from Sadi Carnot's theory; instead, his theory of machines – similar to his father's theory with respect to mechanical machines (Carnot 1778, 1780, 1786) – constitutes the best background in order to ponder these highly theoretical problems.

We conclude that in those times, it was very difficult to construct a theory on thermal phenomena because its fundamental sentence was principally a theme of logic. If the theory had been founded on caloric theory, then the caloric notion would have suggested the first principle of an AO–theory and this principle would express the nature of heat itself (an imponderable fluid). Therefore, it would constitute a kind of metaphysical principle that is typical of an AO. In this case, Sadi Carnot would have been able to deductively build the theory in classical logic (Clapéyron tried to do this in 1834). It is evident that in S. Carnot's text he did not do so. Thus, Sadi Carnot's theoretical work was very difficult because he had to base his theory on caloric theory without believing it due to some discrepancies with experimental evidence (mathematical footnote in Carnot 1978, pp 73–79); therefore he had to hypothesize the theory of equivalence as well – i.e. "It is not true that heat is not work" – is a DNS belonging to non–classical logic. Therefore, Carnot implicitly proposes the problem (unknown in those years) of which kind of logic to follow: either classical logic through caloric theory, or non–classical logic through equivalence theory. Perhaps in light of this logical difficulty – at that time it was not possible to overcome – a famous *Memoire* by Lavoisier and Laplace ([1780] 1784) developed the theory by laying out the two previous hypotheses in parallel. This logical problem influenced the whole development of Thermodynamics, which even today preserves double negations in all the statements of its second principle. In fact, Sadi Carnot organized his theory in a problematic way, following a non–classical logic. It is interesting to notice that the writing in which Sadi Carnot used infinitesimal analysis (in the unpublished *Manuscript*[20] (Carnot 1978, pp 223–234, for example)) lacks double negative sentences. Previous studies (Drago 1991) show that mathematics, which is entirely expressed by equalities, is not compatible with a DNS, which instead does not admit equality between $\neg\neg A$ *and* A. This fact

[20]In the unpublished *Manuscript* (Carnot 1978) we can also find irrelevant DNSs. E.g.: "Jamais la différence n'excédera [$\neq$ difference is inside]" 1/40 (Carnot 1978, p 230, line 25).

establishes a clear choice by Sadi Carnot in *Réflexions sur la puissance motrice du feu*, i.e. to present the whole theory in a discursive way, by relegating the mathematics to a footnote. This choice is confirmed by the fact that he founded his theory by arguing through an *ad absurdum* theorem; that is, through impossible experimental situations. It is surprising that an impossible situation constitutes the foundation of an entire physical theory, where one wants rather to affirm the truth of hard facts. Nevertheless, we know that the consequences of S. Carnot's reasoning through his *ad absurdum* theorem hold true in both *Réflexions sur la puissance motrice du feu* and theoretical physics in general. The analysis (Drago and Pisano's works) of *Réflexions sur la puissance motrice du feu* shows 63 occurrences of double negative sentences. They are listed in the Appendix. Here we are interested in DNSs that are scientific in nature. By excluding the remaining ones (whose serial numbers, in the Appendix, are in parentheses; only two kinds of his sentences occur and they belong to an early discursive part). Moreover, let us note that of the 63 scientific DNSs, most (51) occur in the theoretical part of Carnot's book. Thus, such a great number of DNSs – representing an argument in non–classical logic and playing a crucial role in theoretical elaborations – constitutes decisive evidence for the originality of S. Carnot's way of elaborating a theory, in comparison to the traditional organization, likely as Euclidean geometry; it relies upon non–classical logical reasoning. The following Table shows the distribution of DNSs in Sadi Carnot's *Réflexions sur la puissance motrice du feu* (Carnot 1978), including footnotes; the numbers of the DNSs' footnotes are in brackets (Table 7.1).

Table 7.1 Distribution of DNS's in *Réflexions sur la puissance motrice du feu*

Réflexions sur la puissance motrice du feu	DNS's total number
pp 1 ÷ 10: study of an ideal heat machine	32 (+18)
pp 40 ÷ 48: study of gas theory	9 (+6)

7.2.3 *The Possible Theoretical Roles Played by a DNS*

If we read DNSs one after another (see Appendix below), we notice that their sequence may synthetically express the entire development of Sadi Carnot's scientific thought (see, e.g., previous *ad hoc* index, Chapter 6). In order to verify this hypothesis, it is necessary to list how many roles a DNS may play; for example, it is necessary to distinguish between a problem and a general presupposition of the theory etc. Based on previous investigations within the historical epistemology of science (Drago and Pisano 2001; Pisano and Gaudiello 2009a, b; Capecchi and Pisano 2010a, b; Pisano 2010a; see also Hallyn), we propose the following six historical roles played by a DNS within a scientific PO–theory (see Table 7.2).

Table 7.2 List of possible roles played by a DNS within a scientific PO–theory

Role	Meaning of the role	Abbreviation
Problem	Fundamental problem of the theory to be solved	[PR]
General principles	General principles of theoretical Physics or mathematics	[GP]
Theoretical premise i.e. principles upon the relationship with reality	Author's introduction on the theoretical or experimental arguments	[TP]
Methodological principle of the theory	Reasoning and proposition which indicate a direction to the solution	[MP]
Ad absurdum proof	Technique of argumentation	[*]
Operative principle i.e. principle of resolution of a problem to which theory applies	Reasoning or proposition on actions and techniques to be performed	[PO]

In fact these six roles might cover the possible roles played by DNSs in Sadi Carnot's theory. However it appears dubious as to which category to attribute to some DNS's (Appendix); for instance, DNSs 50 and 51 (Appendix) are operative principles, although such sentences are mentioned in a reasoning concerning Carnot's famous theorem (*ad absurdum proof*) on the independence of motive power from the kind of gas; moreover, DNSs 10 and 11 are expressed by means of *ad absurdum* arguments; even though they are theoretical presuppositions. However, these ambiguities are few in number with respect to the total number of DNSs.

7.2.4 Sadi Carnot's Arguments Through his DNSs

Now we can gather all of the DNSs according to their theoretical roles and then connect the resulting six groups according to the following Fig. 7.5.[21] We put a DNS whose physical meaning is similar to that of a previous DNS in parentheses. The table offers a global vision of the development of *Réflexions sur la puissance motrice du feu* while providing a particular summary of the theoretical content of the book (Fig. 7.4).

Let us remark that the set of DNSs is well distributed in all cases; the DNSs play all the possible theoretical roles within a scientific theory. Therefore, there is no remarkable aspect (not including the mathematical part) of Sadi Carnot's theory that is not represented by DNS's. After DNS 49 (independence of W_{max} from the kind of gas) there are 13 DNS's that express operative principles outlined by the entire theory; except for DNSs 61 and 62 which constitute Sadi Carnot's afterthought regarding the validity of caloric theory.

[21]The DNSs serial numbers in the flow–chart cases are those listed in the Appendix.

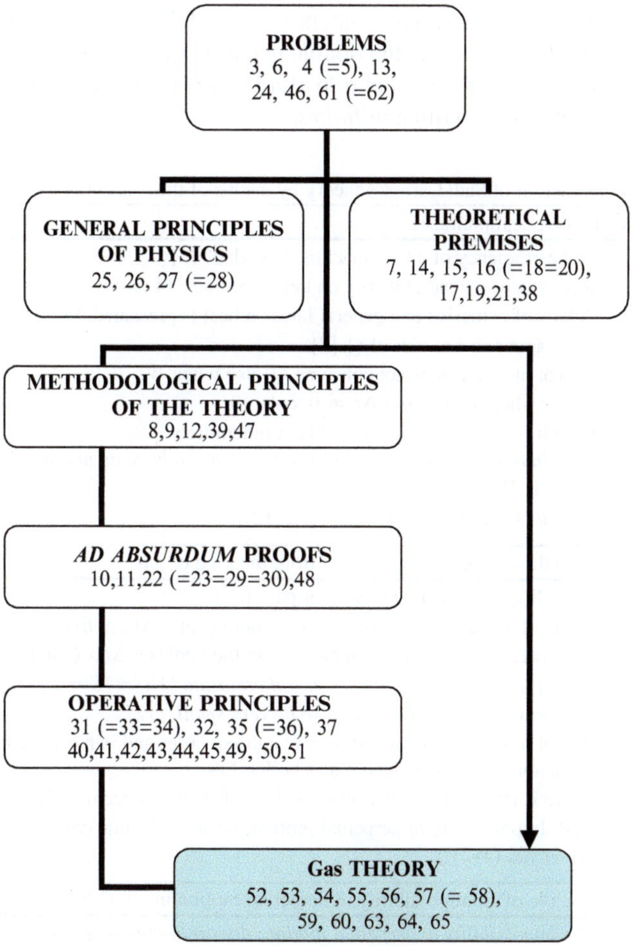

Fig. 7.4 Carnot's theory–flow–chart through all DNSs

7.3 Reconstruction of Arguments and Some Cycles of Reasoning Expressed by DNSs

Now we will verify the sequential consistency of DNS's by listing them according in the order of their occurrence. We sever the sequence in some groups of DNS's, anytime we meet a DNS in the role of a problem. The resulting groups are marked by *B1, B2 ... B6* symbols; we obtain the following Table *Bs* which illustrate the synthetic theoretical content of *Réflexions sur la puissance motrice du feu*. Some of the *Bs* constitute real cycles of reasoning by beginning with problems and then ending in operative principles; we may consider the remaining (*B.4* and *B.6*) as sub–cycles of arguments regarding reversibility and operative principles, respectively.

The titles of the *Bs* are ours. The content of each DNS is reiterated by a short sentence, (which often preserves the double negation of the original sentence). As previously mentioned, the theoretical role of a DNS is expressed with an acronym. The DNSs in footnotes are written in *Italics.*

Table 7.3 Reconstruction of Sadi Carnot's theory by means of the sequential list of DNS's

B.1 Problems and theoretical premises	
Problems	Is the efficiency of a heat machine bounded? [3]
	Research of a general theory on heat machines. [4 (=5)]
	A lack of extensive and general laws on heat is presented. [6]
PT	The vapour as a carrier of Q. [7]
MP	A heat machine does not consume Q, also when $W>0$. [8]
MP	$W>0$ when there exists $\Delta t \neq 0$, only. [9]
Ad absurdum proof	(PT) It is impossible to give off heat indefinitely [10]
	(PT) In order to produce work it is necessary to have an unlimited source of water [11]
(MP)	*Q causes both motion and $\Delta V>0$.* [12]
B.2 First reasoning on the independence of W_{max} from the kind of gas	
Problem	Does W_{max} depend on the kind of gas? [13]
PT	Study of a heat transformation where both Q and Δt are fixed [14]
PT	Both percussion and friction produce in the bodies a $\Delta t>0$ in the bodies. [15]
PT	*The water capacitor can work as a thermostat.* [16 (=*18*=*20*)]
PT	*Δt between two bodies may also be very small. [17]*
PT	A heat machine can operate both in a direct order and in a reverse order; a water capacitor may be considered as a boiler. [19]
PT	There is reversibility only when no loss of W and Q occur. [21]
Ad absurdum proofs	(GP) Impossibility of perpetual motion, i.e. an indefinite creation of motive power [22 (=23)]
B.3 Proving the principle of an impossibility of perpetual motion in all of theoretical Physics	
Problem	*The impossibility of perpetual motion: does it hold true outside of theoretical mechanics? [24]*
GP	*Both electric phenomena and heat production are caused by motion. [25]*
GP	*Electric phenomena are also governed by mechanical laws. [26]*
GP	*Impossibility of perpetual motionin every kind of phenomenon. [27 (=28)]*
Ad absurdum proofs	*Impossibility of perpetual motion, i.e. an indefinite creation of motive power [29 (=30)]*
B.4 Arguing on reversibility	
OP	To efface a Δt without a $\Delta V>0$ is a waste [31(=33=34); 32, 35 (=36); 37)]
PT	*dt as a theoretical possibility.* [38]
MP	Validity of all reasoning even when $\Delta t \rightarrow 0$. [39]
OP	To reach the initial state through an isochoric transformation implies a direct contact between the bodies of different t, and hence irreversibility. [40]
OP	*It is possible to use a given quantity of heat again. [41]*
OP	*$\Delta t=0$ in Gay–Lussac's experience. [42]*

(continued)

Table 7.3 (continued)

OP	During the compression of a gas, we try to avoid low temperature (in the gas). [43]
OP	*Maybe there is an anomaly in gas law when ΔV causes Δt. [44]*
OP	It is profitable to avoid useless *Q(>0)* passage *between* $\Delta t > 0$. [45]
B.5 *A posteriori* argument the validity of caloric theory	
Problems	Is *Q = 0* in a cycle? *[46]*
	(MP) In a cycle Q = 0.[47]
Ad absurdum proof	*(MP) We cannot overthrow caloric theory. [48]*
B.6 Conclusions from B.2 (S. Carnot's theorem) and B.4; some operative principles regarding gas theory and uncertainties of caloric theory	
OP	Independence of W_{max} from the kind of gas. [49]
	DNS 49 holds true in the case $\Delta V \neq 0$ when $\Delta t \neq 0$ only. [50]
OP	(OP) Never allow contact between bodies having a $\Delta t \neq 0$. [51]
OP	In any transformation of a gas for $\Delta t = 0$, in which ΔP and ΔV are fixed, *W* does not depend on the kind of gas. [52][a]
OP	It is difficult in a gas to measure *Q* when $\Delta V \neq 0$. [53]
OP	Air transformation in a non–variable *V* , when $\Delta t \neq 0$. [54]
OP	Heat capacity of gases changes according to $\Delta V \neq 0$ and perhaps $\Delta t \neq 0$. [55]
OP	The difference $c_P - c_V$ does not change [(56)]
OP	*Hypothesis regarding the constancy of specific heatof a gas when* $\Delta t \neq 0$. [57, *(=58)*]
OP	*Hypothesis regarding the constancy of specific heatof a gas when* $\Delta t \neq 0$. [57, *(=58)*]
OP	When $\Delta t \neq 0 c_P$, c_V and c_V/c_P all are not constant. *[59]*
OP	*Different working substances do not give more advantages. [60]*
Problem	Doubts about the validity of caloric theory. [61, (=62)]
OP	*In Perkins's machine, no effects without space* ($\Delta V \neq 0$). *[63]*
OP	*Watt's mistake: when t decreases, no proof that vapour cannot condense. [64]*
OP	*Possibility of new technological requirements of the furnace. [65]*

[a]From here Sadi Carnot starts his reasoning on gas theory

The Table 7.3 present the discursive exposition of Sadi Carnot's entire theory divided into six steps, of which three steps represent cyclical arguments ending in *ad absurdum proofs*.

- In *B.1* the starting point is represented by problems regarding the efficiency of heat machines in general; then, he states some methodological principles, from which he derives, *via* an *ad absurdum proof*, a crucial sentence about heat and water: i.e., caloric causes motion and variations of volume.
- In *B.2* S. Carnot brings forth a more specific problem: does W_{max} depend on the kind of gas? After a series of theoretical principles concerning the roles played by both the thermostat and condenser in a heat machine, he presents his celebrated *ad absurdum proof*, based on a cycle of operations; the absurd statement is the impossibility of perpetual motion (DNS's: 16 = 18 = 20, 19, 21). The list of DNSs in *B.2* substantiates a cycle of reasoning.

- In *B*.3, all DNSs 24–30, are included in a long footnote. Sadi Carnot examines a problem suggested by *ad absurdum proofs* in *B.2* on the impossibility of perpetual motion; is this kind of impossible motion present in different theories other than theoretical mechanics? He argues by means of general principles of theoretical physics; then, he concludes – again through an *ad absurdum proof* – the impossibility of perpetual motion in general. The sequence until DNS 30 constitutes a cycle of reasoning. The set of following DNSs in the text (31–37) expresses operative principles, which are not direct consequences of this cycle; they rather follow from the content of *B.2*.
- In *B.4,* by considering the extracted work of a cyclical heat machine, Sadi Carnot introduces the notion of reversibility: *to efface a* Δt *without a* $\Delta V > 0$ *is a waste*. Then, he introduces an important principle of the theory; he states that a theoretical argument holds true even when $\Delta t \rightarrow 0$; hence an infinitesimal *dt* represents a theoretical possibility in his reasoning (in a cyclical heat machine). He then closes a heat cycle through infinitesimal isochors; that is correct (also in modern terms) because an infinitesimal change preserves the reversibility of a heat machine. The set of four DNSs located in three footnotes represents some additions to the main reasoning. Even by excluding the footnotes, we see that the set of DNSs in *B.4*, represents a unity of reasoning, but is not a cycle yet, as *B.2* is; it is rather a completion of previous cycle *B.2*.
- In the brief *B.5,* Sadi Carnot argues on the plausibility of caloric theory, by returning to the problem already examined in the beginning of *Réflexions sur la puissance motrice du feu*; through an *ad absurdum proof* he concludes that he accepts this theory for lack of more adequate suggestions. The set of these DNSs constitutes a cycle of reasoning.
- In *B.6,* S. Carnot offers some applications of his theorem (resumed by DNS 49 – the independence of W_{max} from the kind of gas). Being sure of the content of his theorem, he approaches gas theory by aiming to discover (theoretically and practically) the mathematical function for representing the efficiency of a heat machine. Then, Sadi Carnot outlines the operative principles regarding the variation of the heat capacity of a gas and deduces the hypothesis of constant specific heat; at the end, he examines the use, in a heat machine, of a gas different from (water) vapor; he ends *B.6* with considerations on the detailed technological requirements of the furnace in a heat machine (here, Sadi Carnot considers Perkins' and Watt's machines). In *B.6,* the logical thread represented by the sequence of DNSs appears broken; in others words, the results presented by the DNSs in *B.6* do not form a consistent system. However, let us remark that in this part of the theory (*B.4* and *B.5*), Sadi Carnot argues for mathematically determining function *C(t),* representing the extracted work and depending on temperature only; therefore, for the presented reason about the relation between DNSs and mathematics, DNSs are no longer capable of expressing the thread of the reasoning (which now, is a mathematical reasoning, as well) followed by Sadi Carnot. In other words, the set of DNSs points out a temporal sequence of problems, but they do not manifest a gradual logical development.

In the following chart, we represent a global view in order to show the relationships between *Bs* in Carnot's book (Fig. 7.5).

Fig. 7.5 The relationships among previous Bs–tables

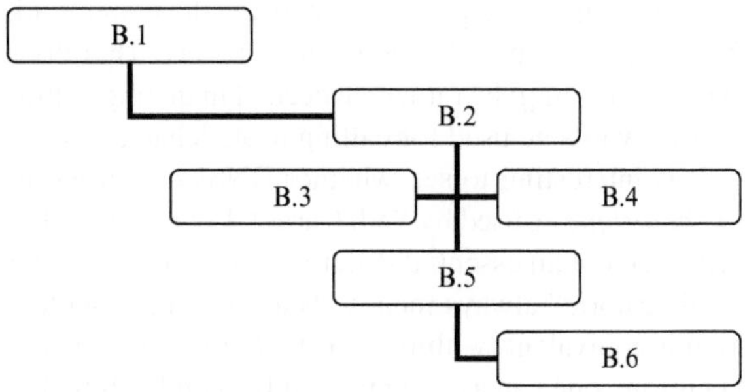

In conclusion, the list of DNSs in *Réflexions sur la puissance motrice du feu* allow us to illustrate for the first time a very detailed structure of Sadi Carnot's arguments, adequately representing Carnot's original scientific thought; no essential content of *Réflexions sur la puissance motrice du feu* is missing. Moreover, the list of *B* tables together with the list of DNSs, follow the line of thought according to which Sadi Carnot constructed his thermodynamics. We conclude that his theory has to be qualified (for the first part, surely) as a logical theory.

7.4 Sadi Carnot's Thermodynamics as PO Theory. Its Non–classical Logic and its Central Problem

The large number of DNSs in Sadi Carnot's book, his celebrated *ad absurdum theorem*, and the cycles of arguing, all provide evidence that overall, his theoretical system represents a kind of logical work; more precisely, an attempt to clarify a theoretical system using non–classical logic (perceived by him in an intuitive way, through DNSs). He tried to introduce his readers to this kind of argument; but scientists judged *Réflexions sur la puissance motrice du feu* to be a very difficult book, and even his friends (by reasoning in classical logic) considered it incomprehensible (Robelin).

The earlier features of his theory show that Sadi Carnot's theory is not AO, but it is a PO–theory. Then, in the following, we try to reconstruct his whole theoretical thought by specifying two main characteristic features of a PO theory; which have been drawn by a comparative examination of several theories in History of Science; i.e., the crucial problem each one express as a DNS and the synthetic method of reasoning. Since a PO–theory relies upon a crucial problem, we inspect S. Carnot's theory using this point of view. We easily see the first problem in his theory: Is caloric a valid theory? (DNSs 48, 61 (=62)). While DNSs 3, 4 (=5), 6, 13, 24, 46 illustrate that Sadi Carnot's subordinate problem was the relationship between heat and work, (we saw it in Sect. 7.3). If Sadi Carnot had abandoned caloric theory that follows classical logic, for assuming equivalence theory (which originates DNSs), he had to reason in non–classical logic through DNSs. This move, in fact, was performed by him by writing 63 DNSs; but he did not link the use of these DNSs to the principle of equivalence between heat and work (which is itself a DNS) and therefore he did not abandon caloric theory. Therefore, he was unable to mould his logical reasoning; he rather succeeded in distinguishing (in a sharp way) in his mind the two words he used for calling heat, "chaleur" and "calorique".

It is interesting to see whether DNSs can properly shed some light on *chaleur* and *calorique* as used by Sadi Carnot. Let us remember that Ostwald advanced[22] the hypothesis of an essential difference. He suggested that in Sadi Carnot's theory, the word "caloric" always means "heat exchanged without variations in temperature"; so, it is equivalent (within a constant, given by the value of temperature) to entropy. Therefore, one can ask whether and which DNS Sadi Carnot, in the development of his theory, expressed such a difference. Moreover, one can even ask whether DNSs mark the passages where Carnot committed errors, due to incorrect calculations or to the introduction of some uncorrected scientific hypotheses at the time, or to an incorrect formulation of the theory. Yet all of these passages contain DNSs. Therefore, no evidence emerges from DNSs regarding either of these historical questions or Carnot's theoretical mistakes. However, neither can one conclude that Carnot committed errors without any doubt, nor that the difference between heat and caloric caused him to reflect in a particular way, that is without DNSs. We should also note that more than 30 years earlier, Lavoisier attempted an early distinction between "chaleur" and "calorique" et al. magnitudes in his *Traité élémentaire de chimie* (Lavoisier 1789, pp 12–17), thus one can also think that it was not necessary to include new definitions for a book so closely related to chemical arguments; arguments so far from mechanical (Newtonian) foundations. Nevertheless, the latter explanation could justify the use of the two words without particular indications, but it cannot explain the difference Ostwald suggested (Ostwald 1892). Finally, on the ambiguity caused by the two words, we present some examples in which Sadi Carnot (1978) uses the term "chaleur" but (from a physical point of view) means "calorique" (see also Carnot 1996):

Page	*Errata*	*Corrige*
17, line 13	quantité de chaleur	quantité de calorique
70, line 10	quantités de chaleur	quantités de calorique
70, line 11	quantité de chaleur	quantité de calorique
70, line 14	quantité de chaleur	quantité de calorique
71, line 12	quantités de chaleur	quantités de calorique
71, line 14	quantités de chaleur	quantités de calorique
71, line 18	quantité de chaleur	quantité de calorique
71, line 20	quantité de chaleur	quantité de calorique

[22]Let us also note that the same analogy between caloric and entropy was provided by Mendoza in one of the ten added footnotes in his edited book on Sadi Carnot (Mendoza 1960, ft 4, p 17; see also la Mer).

We have already seen that in the end, Sadi Carnot lays out an even more specific problem in a concrete manner: which efficiency W_{max}/Q (see below Chapters 9 and 11) should be used for a heat machine (DNS 3)? In a PO, a DNS starts a theory which proceeds through arguments which are DNSs and concludes the problem either by means of an *ad absurdum proof* (which is based on a material cycle) or states its kind of mathematics, (likely his father Lazare's mechanics (Carnot 1803a); which is also a PO–theory; in fact, through DNSs, a new mathematical technique is introduced: symmetries that give invariants). Sadi Carnot's theory does not introduce the subject of mathematics as all scholars of thermodynamics did at the time. Instead, he develops a great part of his general theory through verbal reasoning on one cycle only; for this reason at the end of his discursive arguments, it was difficult for him to determine the mathematical function of efficiency for a heat machine; nevertheless he then attempted to exploit a mathematical sub–theory about gases for this purpose.

7.5 On Sadi Carnot's Theory: a Case Study

Philip Lervig is a scientist at the *Institute of Physics, Aarhus University* (Århus, Denmark) interested in the history of science. In the past century, he proposed a possible analogy between Sadi Carnot's thermodynamics and the mechanical theory of heat (Pisano and Capecchi 2009, pp 83–90). Particularly, in two articles published during the 1970s and 1980s Lervig (1972, pp 222–239, 1976, pp 199–212, 1982, pp 85–122) gave an interesting contribution to this debate and an interpretation of Sadi Carnot's work by means of a very perceptive historical–comparative[23] analysis of caloric (when temperature has a constant value) and modern Entropy[24]:

> Communicated by L. Rosenfeld
> Abstract. A simple connection is pointed out between the theory of heat formulated in Sadi Carnot's: "Réflexions sur la puissance motrice du feu" (1824) and the later Kelvin–Clausius thermodynamics. In both theories two well–defined quantities, a heat function and a work function, exist and can be calculated by integrating along a reversible path. In thermodynamics the work function (energy) is conserved, whereas the heat function (entropy) increases by irreversible processes. In Carnot's theory the heat function is conserved, whereas the work function decreases, so that in this theory the irreversible process is characterized by a *loss of work*.[25]

With his analogy, Lervig (1972) presented his outline of the problem:

> *A List of Questions.* A satisfactory understanding of CARNOT'S theory should enable a few key problems to be solved, as for instance
> 1. How can the following statements be consistent with each other:

[23]Other specific analogies–studies are: Renaud (1937), Brønsted (1955), Pisano (2003), pp 327–348, (2004), pp 203–211.

[24]On historical and epistemological studies of entropy–energy, one can see: Scott (1971), Falk (1985), pp 108–115, Fuchs (1987), pp 215–216, Hoyer (1976), pp 221–228.

[25]Lervig (1972), p 222, line 3; author's *italics.*

(a) One of the axioms of CARNOT'S theory is that a *perpetuum mobile* (of the first kind) is impossible.
(b) CARNOT'S theory was given up precisely because it was inconsistent with the energy principle (first law).
2. Is it true, as several authors seem to have felt, that many of CARNOT'S results are virtually correct in a modern sense if "calorique" is replaced by "entropy"? If so, why?
3. In *precisely which sense* is CARNOT'S theory a forerunner of the correct thermodynamics?
[Lervig 1972, 223–224, line 39. *Italic style* and capital letters belongs to the author].
[...] Carnot's Axioms. Calorique and Chaleur
The introduction can be summarized in *two theses*:
I. The *fundamental assumption* of CARNOT is that a *perpetuum mobile* (of the first kind) is impossible.
II. To CARNOT, *heat is conserved.*
These theses will be the foundation for the rest of the paper [Lervig 1972]. If II is admitted, it follows that in CARNOT'S *book the words* calorique *and* chaleur *are both used to describe heat as one and the same conserved quantity.*[26]

In another article (Lervig 1976) he also interpreted Sadi Carnot's principles and found them very similar to those of modern thermodynamics:

> As Professor Mendoza has taught us [Mendoza 1960], it may be reasonable to suppose that on the day when the *Réflexions sur la puissance motrice du feu* (Ref. 2) appeared in the bookshops, its author, Sadi Carrot had finally convinced himself thought that its contents could not be true. I have [...] And wrong it was [...]; in the sense that it was based upon the erroneous assumption of conservation of heat. [...] I intend to show in some detail – and this is my point – that Carnot's theory had precisely the structure needed to make possible an adequate theory of heat. [...] However, what I intend to explain is that with this strange theory by the young French engineer S. Carnot, the whole logical framework of thermodynamics was in principle complete. (Lervig 1976, p 200, line 5). [...] It is a fascinating fact that Carnot's theory, which was to be the source of this enormous wealth of scientific development, was in a certain sense wrong; that is to say, wrong in the sense that it was based upon the erroneous assumption of conservation of heat. In fact, I believe that this was an essential part of Carnot's feat: to take seriously the theory of heat as a substance; to develop it into a logical structure, describing the interconnection between heat and mechanical motion, and to develop this theory to the point where it was possible to understand its weaknesses. This had never been done before. [...] What I intend to show today is that Sadi Carnot in fact came extremely close to the correct theory.[27]

Thus, in our opinion, it is interesting to try to understand whether the principles of the modern theory can be reduced to those Philip Lervig interprets as being Sadi Carnot's only two principles.

First, we should note that Sadi Carnot himself expressed his misgivings with regard to the foundations of the theory through DNSs, particularly concerning the role played by *calorique*, *chaleur* at the end of his book.

[26] *Ivi,* 227, line 16. (Author's *italics* and capital letters).
[27] Lervig (1976), p 200, line 18.

Below, we list only a few of them:

> L'on a souvent agité la question [= we do not know] de savoir si la puissance motrice de la chaleur est limitée ou si elle est sans bornes [≠ we are able to say that it has bounds] [...].[28]
> La phénomène de la production du mouvement par la chaleur n'a pas été considéré sous un point de vue assez général. [= not dependent from the details]. [≠ it has been considered an independent manner from the details].[29]
> Pour envisager dans toute sa généralité le principe de la production du mouvement par la chaleur, il faut le concevoir in dépendamment [≠ separately] d'aucun mécanisme, d'aucun agent particulier [...].[30]
> [Doubts on caloric theory]. La loi fondamentale que nous avions en vue de confirmer [$Q = 0$] nous semblerait exiger cependant, pour être mise hors de doute [≠ it is considered as valid law], des vérifications nouvelles; [...].[31]
> [Follow by DNS 61]. [...] Elle [$Q = 0$] est assise sur la théorie de la chaleur telle qu'on la conçoit aujourd'hui, et, il faut l'avouer, cette base ne nous parait pas d'une solidité in ébranlable [≠ it is questionable].[32]

From a historical point of view, Lervig (by another method with respect to ours) also discussed some topics related to them:

> Whether Carnot really believed in the conservation of heat is a matter which has been much discussed. I find that this question can be answered from a study of the above [on principles] argumentation. Of course, in a certain respect he did not believe in the conservation of heat, as is apparent from his notes, presumably written at the same time. However, it must be stressed that the assumption was nevertheless necessary for his *argument.*[33]

With regard to closing Carnot's work function and a related analogy between Carnot's theory and modern thermodynamics, Lervig argued:

> Carnot would have needed to explain this [work function in cycle] more complicated situation in detail since the structure he described was a logical structure. I therefore claim that the assumptions made by Carnot are as follows:
> *Carnot's theory*
> First Law: A perpetuum mobile is impossible
> Second Law: Heat is conserved
> I prefer this formulation of Carnot's theory in terms of a first law and a second law, because I find that it clearly demonstrates the analogy between his theory and later thermodynamics.
> *Thermodynamics.*
> First Law: the energy principle
> Second Law: A perpetuum mobile of second kind is impossible
> I admit that now I am in a sense leaving the history of science proper, and enter what might be called a comparative study of physical theories. It is interesting to compare the two sets of axioms, one set for Carnot's theory, one for thermodynamics, as given above.

[28] Carnot (1978), DNS 3, p 6, line 18.

[29] *Ivi*, DNS 4, p 7, line 11

[30] *Ivi*, DNS 5, p 8, line 1.

[31] *Ivi*, DNS 61, p 89, line 5.

[32] *Ivi*, DNS 62, line 8.

[33] Lervig (1972), p 201, line 17. (Author's *italics*).

> They have much in common, and in particular the two first laws look very similar. Are they different at all? It is often claimed that the two statements (that energy is conserved, and that a perpetuum mobile is impossible) are identical, but they are not! It is easy to show that in Carnot's theory, based as it is upon the assumption of the impossibility of a perpetuum mobile, there is certainly no conservation of energy.[34]

Moreover, Lervig (1976) identified one more principle in Sadi Carnot's theory: W_{max} *is a function of state* based on his observation that in the mathematical footnote *Réflexions sur la puissance motrice du feu* included an absolute "temperature ϑ" which induced him to note two more analogies. The first one occurs between the absolute temperature T (in modern thermodynamics) and temperature t (in Carnot's mathematical footnote). Hence, Lervig states the existence of the new function: *work A as a state function.*

> The work function A in Carnot's theory is then defined as a function of state, closely analogous to the energy function in thermodynamics.[35]

The second analogy concerns the state function as work A (attributed to Sadi Carnot) and the energy function U in modern thermodynamics. We can observe that through calculations of Carnot's cycle, Ferdnand Reech (1805–1880; Reech 1853, pp 357–378; Drago and Pisano 2008; Chapter 10 below) previously obtained an energy function (and potentially the relationships of Maxwell as well). We should also remark that Reech (see Chapter 10 below) followed Clapeyron's idea of arranging two systems of curves in a PV diagram as two isotherms and two adiabatic equations:

$$\varphi(V, p) = t\psi(V, P) = u$$

Reech's premise is important in order to recognize an implicit principle of Sadi Carnot's theory: *every point of PV diagram, for any gas, is obtained by intersection between one adiabatic and one isotherm.*

When it comes to Lervig, by means of different reasoning, he seems to concur with the first part of Reech's work[36] (Reech 1853, pp 357–363). Let us also note that Clausius (Mendoza 1960) and Kelvin studied, for over a year, how to reconcile Carnot's theory with Joule's experiment regarding the equivalence of heat and work: they had to reconcile the two aforementioned sentences on heat and work in the two theories, respectively. They reconciled them by transforming the logical organization of the theory into an entirely deductive AO, by laying out a first

---

[34]Lervig (1976), p 202, line 2. (Author's *italics*)

[35]Lervig (1976), p 205, line 7.

[36]Lervig did not mention Reech's work.

principle $U = W + Q$, although this sentence is an idealized expression (because in this formula, Q can be entirely converted into work; then into the second principle which is necessary for returning to represent reality). Now, by considering Lervig's assumptions and logical arguments discussed in the previous section, one should take into account that, generally speaking, within the Latin and French languages, a double negation is a common grammatical construction, particularly in discursive French. In fact, *Réflexions sur la puissance motrice du feu* was written in a discursive style; nevertheless, we should specify that – in this case – the book's content is also scientific.

> It is hardly necessary to insist at this point on the dangers inherent in the discussion of precedents in the history of science. Such discussion can lead us all too easily to underestimate the originality of a discovery or innovation and to overlook the gulf that normally exists between a precedent and its resurrection is some modified form or new synthesis.[37]

Therefore, a clarification of its semantics is essential to understanding the epistemological objective. In his works *la Logique* (Condillac [1780] 1821) and *la Langue des calculs* (Condillac 1798), Etienne Bonnot Abbé de Condillac (1714–1780) had already emphasized the importance of language in logical reasoning, stressing the need for a scientific language for a scientific theory. The following is an example of the importance of language in a scientific theory at Sadi Carnot's time:

> SECONDE PARTIE. L'analyse considérée dans ses moyens et dans ses effets, ou l'art de raisonner réduit à une lange bien faite. [...].[38]
> CHAPITRE V. *Considérations sur les idées abstraites et générales ou comment l'art de raisonner se réduit a une langue bien faite.* [...].[39]
> CHAPITRE VII. *Combien le raisonnement est simple quand la langue est simple elle–même,* [...]. Erreurs de ceux qui préfèrent la synthèse à l'analyse. Toutes les sciences seraient exactes si elles parlaient toutes une langue fort simple. Problème qui le prouve. L'évidence d'un raisonnement consiste uniquement dans l'identité qui se montre d'un jugement à l'autre. Les sciences peu exactes sont celles dont les langues sont mal faites. L'algèbre n'est proprement qu'une langue.[40]
> [...] Tout confirme donc ce que nous avons déjà prouvé, que les langues sont autant de méthodes analytiques; que le raisonnement ne se perfectionne qu'autant qu'elles se perfectionnent elles–mêmes; et que l'art de raisonner, réduit à sa plus grande simplicité, ne peut être qu'une langue bien faite.[41]

Antoine Laurent de Lavosier (1743–1794) also wrote about the same ideas in the *Preface* of his revolutionary *Traité élémentaire de chimie* in the eighteenth century.

> Il [Condillac] y établit que nous ne pensons qu'avec le secours des mots [...].[42]

---

[37]Robert Fox in Carnot (1986), p 22, line 38.

[38]Condillac (1821), p 222, line 18. (Author's *italics* and capital letter).

[39]*Ivi,* p 224, line 11. Please, see also p 126. (Author's *italics* and capital letters).

[40]*Ivi,* p 225, line 13. Please see also p 147. (Author's *italics* and capital letters).

[41]*Ivi,* p 162, line 7.

[42]Lavoisier ([1789] 1937, I, p XXV, line 9).

[Condillac] enfin que l'art de raisonner se réduit à une langue bien faite.[43]

[. . .] De supposition fausse, en supposition fausse nous nous sommes égarés parmi une multitude d'erreurs[44];

[. . .] Alors nous n'avons su raisonner que d'après les mauvaises habitudes que nous avions contractées [. . .].[45]

7.6 Which and How Many Principles did Sadi Carnot Express in his *Réflexions sur la puissance motrice du feu?*

Below (Table 7.4), we list both the sentences and the scientific contents, expressed concisely, of all the DNSs which show the methodological principles in thermodynamic theory (Truesdell and Bharatha 1977) of *Réflexions sur la puissance motrice du feu*. It is evident that more than one original scientific sentence may concern a modern principle didactically written in a concise manner.

Table 7.4 The five methodological principles in Sadi Carnot's *Reflections*

The original sentences in *Réflexions sur la puissance motrice du feu* which express the groups of principles in the theory (Carnot 1978)	The scientific content of five principles in the *Réflexions sur la puissance motrice du feu*
A set of sentences (Drago and Pisano 2001, DNSs 8; 47≈*48**)[a] La production de la puissance motrice est donc due, dans les machines à vapeur, non à une consommation réelle [≠ to conservation] du calorique, mais à son transport d'un corps chaud à un corps froid [. . .] [that is a formal manner of its conservation] (Carnot 1978, p 10, line 20). [Suppositions: quantities of caloric given and absorbed are equal]. Ce fait n'a jamais été révoqué en doute [≠ it is sure]; [. . .]. (Carnot 1978, p 37, ft 1, line 8). $Q = 0$ in a cycle]. Le nier [to negate caloric theory], ce serait renverser[≠ to affirm it would mean to maintain = so theory is correct] toute la théorie de la chaleur, à laquelle il sert de base]. (Carnot 1978, p 37, ft 1, line 9).	1. In the machines $W > 0$, does not consume *Q, that is* $Q = cost.$
B set sentences (Drago and Pisano 2001, DNSs 7, 9, *10**) La vapeur n'est ici qu'un [≠ it is] moyen de transporter le calorique (Carnot 1978, p 10, line 2).	2. $W > 0$ only when there exists $\Delta t \neq 0$

(continued)

[43]*Ibidem*, line 13.

[44]*Ivi*, XXXVIII, line 2.

[45]*Ibidem*, line 6.

Table 7.4 (continued)

[Production of engine power and importance of thermostats for different temperatures]. [. . .] sans lui [the cold], la chaleur serait inutile [≠ through it is always useful]. (Carnot 1978, p 11, line 8).	
[About vapour] O*ù* la placerait–on une fois qu'elle aurait pris naissance? Il ne faudrait pas croire que l'on put [indefinitely] [≠you can blow off it], ainsi que cela se pratique dans certains machines, la rejeter dans l'atmosphère: l'atmosphère ne la recevrait pas. (Carnot 1978, p 11, line 11).	
C set sentences (Drago and Pisano 2001, DNS 12)	3. *Q* causes both motion and $\Delta V > 0$
La chaleur ne peut évidemment que [≠ it is] être une cause de mouvement qu'en vertu des changements de volume ou de forme qu'elle fait subir aux corps. (Carnot 1978, p 14, line 6)	
D set sentences (Drago and Pisano 2001, DNS's $22^* = 23^* = 29^* = 30^* \approx 27 = 28$).	4. Impossibility of perpetual motion
[Return of caloric from B body to A body]. [. . .] De recommencer une opération entièrement semblable à la première [caloric from A body to B body] et ainsi de suite: ce serait là, non seulement le mouvement perpétuel, mais une création indéfinie de force motrice sans consommation [≠ maintaining constant] ni de calorique ni de quelque autre agent que ce soit. (Carnot 1978, p 21, line 1).	
[Une semblable création est tout–à–fait contraire [≠ motive power creation is according to] aux idées reçues jusqu'à présent, aux lois de la mécanique et de la saine physique; [but this creation] elle est inadmissible. [≠ the conservancy is admissible]. (Carnot 1978, p 21, line 5).	
[Perpetual motion and electricity] Ne sait–on pas d'ailleurs *à posteriori* que toutes les tentatives faites pour produire le mouvement perpétuel par quelque moyen que ce soit ont été infructueuses? [= perpetual motion was not realized ≠ every motion has an end] [≠ we know it] (Carnot 1978, p 21, ft 1, line 6).	
Que l'on n'est jamais parvenu à produire un mouvement véritablement perpétuel [≠ we were able to produce motion that has a sure end], c'est–à–dire un mouvement qui se continuat toujours sans altération [≠ constant] dans les corps mis en œuvre pour le réaliser? [≠ we are] (Carnot 1978, p 21, ft 1, line 8).	
Si elle était possible [motive power creation], il serait inutile [≠ conservation makes useful] de chercher dans les courans d'eau et d'air, dans les combustible, cette puissance motrice [. . .]. (Carnot 1978, p 22, ft 1, line 11).	
[If there is motive power creation]. [. . .] Nous eu aurions à notre disposition une source intarissable [≠ conservation leads exhaustion] o*ù* nous pourrions puiser à volonté. (Carnot 1978, p 22, ft 1, line 13).	

(continued)

Table 7.4 (continued)

The original sentences in *Réflexions sur la puissance motrice du feu* which express the groups of principles in the theory (Carnot 1978)	The scientific content of five principles in the *Réflexions sur la puissance motrice du feu*
E set sentences (Drago and Pisano 2001, DNSs 39) On peut le regarder [Δt] comme nul en théorie, sans que pour cela les raisonnements perdent rien de leur exactitude. [$\neq$ they continue to be exact results] (Carnot 1978, p 25, line 10).	5. Validity of all reasoning even when $\Delta t \to 0$

Legend: the scientific content of Sadi Carnot's principles are numbered from 1 to 5 in the last column on the right

[a]In the list, by means of " * ", we indicate that the sentences include an *ad absurdum* proof. When a DNS is in a footnote, its number is written in *italics*; when an asterisk is added the DNS is included in an ad *absurdum proof*

There are five methodological principles in Sadi Carnot's theory – three more than Lervig counted – and all of them are characterized by DNSs. Below, Table 7.5 summarizes the methodological principles and the DNSs' serial numbers:

Table 7.5 Methodological principles and corresponding DNSs in *Réflexions sur la puissance motrice du feu*

Methodological principles	DNS's serial number
In heat machines L > 0 does not consume Q	8; 47 ≈ 48*
$W > 0$ when there exists $\Delta t \neq 0$, only	9
Q causes both motion and $\Delta V > 0$	12
Impossibility of perpetual motion	22*(=23* = 29* = 30*) ≈ 27 (=28)
Validity of all reasoning even when $\Delta t \to 0$	39

Legend: DNS's serial numbers refer to the original sentences (Drago and Pisano 2001)

Finally, Table 7.6 compares the principles of Sadi Carnot's theory (reduced to two) with the two expressed by Lervig and with the two of the modern mechanical theory of heat.

The first thing we notice in Table 7.6 is that Lervig (for his own reasons) has reversed the order of Sadi Carnot's principles (also with respect to modern thermodynamics); that is, he has set Sadi Carnot's (methodological) principle of the impossibility of perpetual motion as the first principle. If we restore the original order, Table 7.6 presents us with a more natural comparison. Another significant fact is that in every–case–conserved quantities (it is the first principle of every theory) are claimed:

(a) In the first column (Sadi Carnot) that quantity is caloric.
(b) In the second column (Sadi Carnot according to Lervig with the principles conveniently inverted) that quantity is heat.
(c) In the third (mechanical theory of heat) that quantity is energy.

Table 7.6 Comparison of the principles of Sadi Carnot, Philip Lervis' reasonings, and the mechanical theory of heat

Logical organization of theory	Synthesis of Sadi Carnot's contents by his DNSs (from Table 7.4)	S. Carnot's principles interpreted by Philip Lervig	(Modern) Mechanical theory of heat
Thermodynamics expressed in a PO theory by means of DNSs	In order to obtain work, heat does not decrease. (*A* set DNSs)	*Perpetual motion is possible*	*It is not true that heat is not work*
	A perpetual motion is impossible. (*D* set DNSs)	*There is no consumption of heat (caloric)*	*Perpetual motion of the second kind is impossible*
Thermodynamics expressed in an AO theory without DNSs	*In a cycle caloric fluid is conserved*	A perpetual motion is impossible	Energy conservation
	A maximum efficiency does exist	Heat is conserved (?)	Entropy definition $\Delta S = \frac{\Delta Q}{T}$

Legend: in order to better distinguish the contents we wrote in a different format. Use of Italics indicates material added by myself. Negated words are underlined

Finally, the second principle itself undergoes an evolution, though merely from the qualitative (methodological) to the quantitative (axiom) while the nature of the principle does not change (that is granted that) Sadi Carnot implicitly intended the impossibility of perpetual motion of the second kind.

7.7 DNSs and Mathematics in Sadi Carnot's Theory

Previous studies (Drago and Pisano's works, Pisano's works) on the logical foundation in Sadi Carnot's reasoning show that some DNSs express the crucial correlations between both the work function (for a fixed quantity of heat Q, i.e. the quantity that expresses the *input* of a heat machine) $W > 0$ (which expresses the relevant *output* of a heat machine) and the single variables t, V, A (gas at issue) in a heat machine. Let us notice that due to the difficulty of the theory on unknown gases at the time, in order to be thorough we should also list other parameters unknown by him. We can synthesize them with ξ. Thus, one can write all variables necessary for the theory t, V, A, ξ. In this sense the whole first part of *Réflexions sur la puissance motrice du feu* includes a subordinate thread of arguments on one more crucial problem: *is W/Q a mathematical function?* In other words: *is $W/Q = f(t, V, A, \xi)$ a valid mathematical notion?* (See below Chapters 9 and 11).

Let us also notice that the pressure is not included because Sadi Carnot has plans to use a gas as an agent and he had already shown (in Carnot 1978, pp 223–234) that for a gas, a state equation on three variables p, V, t holds true. Subsequently, by means of an *ad absurdum proof* (DNS 48*), Sadi Carnot presents his famous theorem (DNS 49); it states the independence of W_{max}/Q from the kind of gas; hence, it characterizes the ratio W/Q as an abstract concept from all possible bodies; therefore, it may be correctly considered as a true mathematical function (although it is obvious on which variables it depends). Early in the first part of *Réflexions sur la puissance motrice du feu* Carnot expressed the idea that to obtain work, it is necessary to have a temperature difference ($\Delta t \neq 0$), (early in his exposition he had already presented it through a methodological principle of the theory, DNS 9). Then, Sadi Carnot goes on to state (by means of some DNSs) that, in a heat machine, W_{max}/Q has to depend on t only (DNS 49). In other words, he operatively defines the physical quantity W_{max}/Q, as a mathematical function of a variable, only $W_{max}/Q = f(t)$. In fact, in the second part of *Réflexions sur la puissance motrice du feu* – i.e. the part regarding gas theory in a long mathematical footnote (Carnot 1978, pp 73–79) –Sadi Carnot reasons mathematically, with the goal of finding function f and its dependence with respect to temperature. Following this point of view, logic would not only be an aspect of mathematical science, but would acquire a cultural approach for a positive reception of the *logic of physics and its relation to mathematics*. Below, in Fig. 7.6, we summarize the role played by DNSs to introduce mathematics in *Réflexions sur la puissance motrice du feu*:

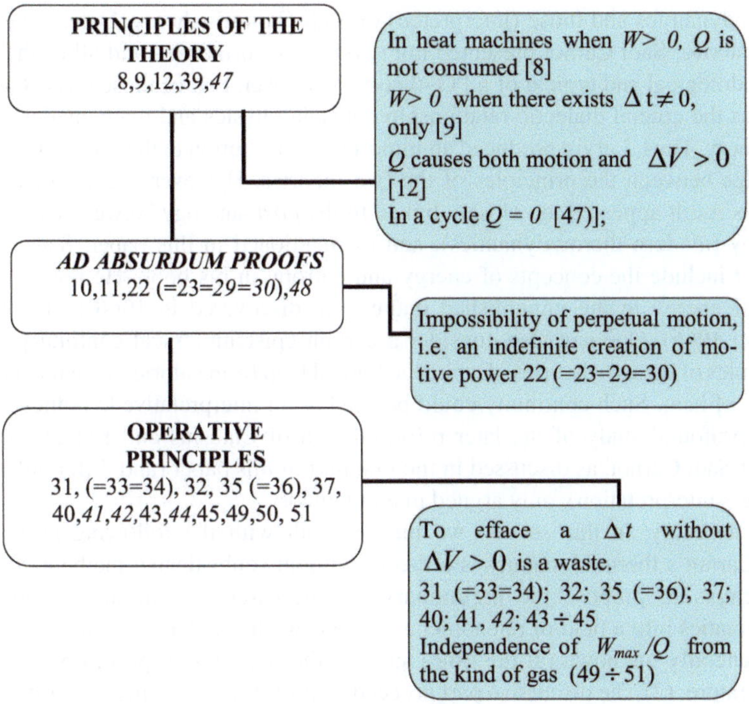

Fig. 7.6 The relationship between mathematics and logic expressed by DNSs

7.8 Final Remarks

Sadi Carnot developed the entire first part of *Réflexions sur la puissance motrice du feu* without mathematics. Despite the fact that Sadi Carnot was a former student of *École Polytechnique de Paris*, he did not start his work by assuming sophisticated mathematical concepts. On the contrary, the book is entirely discursive and essentially based on his famous cycle.[46] Relevant mathematical reasoning is only relayed by a footnote (Carnot 1978, pp 73–79; Pisano 2000; Drago and Pisano 2005). In fact, several DNSs (including those reported in this paper) show that Sadi Carnot, rather than using mathematics, used logic to establish his celebrated reasoning. For almost a century, engineers have developed this practice without any leading scientific principle.

Philip Lervig's interpretation of Sadi Carnot's thermodynamics is essentially based on the search for a link between the concepts of entropy and caloric. He finds it within an analogy between the terms of the two principles of modern

[46] As previously mentioned, the cycle in the *PV* diagram was firstly introduced by Clapeyron (1834) after Carnot's publication of his book in 1824.

thermodynamics and those (interpreted) of Sadi Carnot's theory. Instead, as illustrated above, Sadi Carnot presented not two but five principles and all of them are methodological and typical of a PO–theory. Moreover, two of those (sets *B* and *E*) address the crucial dialectic relationship between physics and mathematics within the theory. Thus, Lervig produced an important contribution to the effort of creating a bridge between the principles of the two theories. However, we should remark that his result appeared too closely linked to the *post*–analogy between caloric and entropy (modern thermodynamics); and as mentioned in this paper, Sadi Carnot did not include the concepts of energy and entropy in his book. He presented his idea of energy in the unpublished manuscript discovered in 1966 (Gabbey and Herivel 1966). One can also consider a certain epistemological continuity in the principles of the two theories, despite the logical leap from caloric to the mechanical theory of heat. Such continuity could be used as an interpretative hypothesis for a more profound study of the later reformulation of Clausius and Kelvin's theory even if Sadi Carnot, as discussed in the first part of this paper (and differently from Lervig's interpretation), only argued in a PO–theory.

In summary, in this section we mainly dealt with the following points: (1) Sadi Carnot's thermodynamics is a theory without sophisticated mathematics and it attempts to propose the preliminary and necessary conditions to introduce mathematics into a field of phenomena where new physical magnitudes were *fuzzy* and generally unknown (at the time) such as the relationship played by heat and temperature. (2) The discursive part proceeds with DNSs following a rigorous path. (3) The theory includes five methodological principles considered an exemplary instance of the problematic organization of the physical theory. (4) These results also explain why during such a great *timespan* his booklet was impenetrable to historical interpretations. Since the author was unaware of this logical figure, the DNSs illustrate (5) what the author's presentation implicitly presupposes. In other works (Drago and Pisano 2001, 2007), Carnot's fundamental hypotheses, beyond his discursive exposition of the theory, are revealed. Furthermore, DNSs are so great in number (more than 60) that they are sufficient to summarize the completely scientific content of Carnot's theory. (6) In the end, the historical instance of inductive thinking is even more interesting for one more reason: *Carnot obtained the theory by arguing about the practical application of the most utilitarian tools, e.g., heat engines*.

Moreover, we also discussed DNSs as a specific aspect of non–classical logic. It would then seem that when we use the negation or double negation, we open the door to a multiplicity of different meanings both in natural language and in mathematically formalized logic: the comprehension of the meaning of a sentence entails great intellectual difficulty. That kind of investigation could suggest that the difficulty may lie in not having clarified a whole system of reasoning rules (a system that is based on classical or non-classical logic). However, we have also tried to show that DNSs can play a role in the scientific structure of a theory. They could also contribute to intellectual clarification, not only of the way of reasoning, but (also) of the language normally used. In fact, beginning with the study of these types of scientific ambiguities and of natural language abuses, it would be possible to begin the study of thinking rules to justify the need for a *formalization* that would precisely crystallize the different semantics of double negation into an entirely different logical system. In this way, the theorization would not appear as an abstract formalism, but would be strongly motivated and therefore the different formalizations could (possibly) come forth naturally. PO–theories contain at least one DNS not corresponding to its positive sentences. In this sense, they belong to non-classical logic. Thus, the choice for the kind of organization of the theory is characterized by the kind of logic employed. This fact thoroughly distinguishes PO–theories and AO–theories. It is possible to view Carnot's logical exposition, not as an attempt to translate qualitative data into quantitative data (i.e. a mathematical formulation) but rather, as an effort[47] to clarify the fundamental method that provides a means of introducing mathematics into thermodynamics.

Sadi Carnot's doubts on the nature of heat (*chaleur, calorique*) and related uncertainty, e.g., Clément–Desormes law in his *Réflexions sur la puissance motrice du feu* are naturally justified. Moreover, the effort to reconcile[48] the conservation of caloric–theory and the novelty of the conservation of energy was valued by science in France at the time. We know that the *conservation law* would only appear later, mainly thanks to results of Joule (Fox1969; Joule 1965, pp 277–281, 1847, pp 173–176), Clausius and Thomson (Clausius 1850, 1865a, b; Thomson 1848, 1851a, b).

Finally, in agreement with some historical theses[49] – confirmed by the epistemological point of view in that section – *Réflexions sur la puissance motrice du feu* lacks a certain mathematical sophistication that characterized other works of physics at the time, and for this reason it would have been preferable to eliminate certain doubtful or marginal arguments. Nevertheless, the effort to distinguish his logical method of research by trying to introduce mathematics into new topics of physics is an evident trace of a meticulous process of composition, typical of genius worthy of his father.

---

[47]An effort, which appears similar to that made by Leonardo da Vinci (1452–1519) and later by Galileo Galilei, to introduce mathematics in a mechanical theory: the deformability (Pisano 2010, pp 34–50) of beams for the first and the resistance (Pisano and Capecchi 2010; Capecchi and Pisano 2010a, b; Pisano 2009a, b, c, d, 2013) of cantilevers for the second.

[48]The role played by "conservation" and "chaleur spécifique à pression et à volume constants" related with Delaroche and Bérard's (Carnot 1978, pp 56–58, 70–72) Clément and Desormes (*Ivi*, 59), Gay–Lussac and Welter's results (Laplace 1822, p 268) are very important in Sadi's book. As mentioned, he also wrote a long footnote (*Ivi*, pp 73–79; Fox 1988, p 292; Pisano 2000; Drago and Pisano 2005) to better explain these facts mathematically. (Carnot 1978, pp 122–126; Delaroche and Bérard 1813; Desormes and Clément 1819; Fox 1988, pp 290–300). (See below Chapters 9 and 11).

[49]Fox (1988), p 298.

Chapter 8
What are the Scientific Roots of Sadi Carnot's Cycle?

Ciò che pressuppor possiamo, e che convengono tutti i fisici, è che il fluido elettrico molta analogia ha con il fuoco [...]

(Volta 1918–1929, IV, p 353, line 14; s. transl. below)

In this section we investigate the possible origin of the idea of the cycle in Sadi Carnot's work, following the hypothesis of an analogy with the electric circuit in Alessandro Volta's battery. First we will present a comparison from the standpoint of the fundamental concepts between Sadi Carnot's theory of thermodynamics and the theory of electricity. Secondly we will propose an analogy between Carnot's cycle and the cyclic path of the I current in Volta's battery, whose current (between two potentials) corresponds to the heat flux which flows between the two thermostats of two reversible heat engines paired by Sadi Carnot. Additionally, we will report and comment on some analogies between electrostatic, electric phenomena and heat engines. In conclusion we present a global vision of every possible connection and will discuss its compatibility.

8.1 An Outline of the Problem

Sometimes, not always,[1] in the history of science, a process of scientific knowledge advances with help from the assumptions that one has previously acquired from another domain. In our case, one can consider the important analogy Sadi Carnot (following his father, see Chapter 11) discovered between water and heat, which respectively performed mechanical work and thermal work: the former by using falling water, the latter by a decrease in temperature (see Chapters 9 and 11).

[1]Let us note, for example, the role played by artisans and practical science with respect to the birth of modern (and more mathematical) science; or generally speaking, *the role played by mechanical science in engineers' and architects' designs during the Renaissance*. (Pisano 2008).

C.C. Gillispie and R. Pisano, *Lazare and Sadi Carnot: A Scientific and Filial Relationship*, History of Mechanism and Machine Science 19, DOI 10.1007/978-94-017-8011-7_8,

Another assumption regards the principle of the *impossibility of perpetual motion* for its maximum independence of the working substance in a cyclical heat machine. Generally speaking, in the history of science, by considering the crucial role played by mechanics in the development of scientific knowledge as compared, e.g., with all other physical and mathematical domains, it is not "[. . .] surprising that people attempted to apply mechanics conceptions wherever possible".[2] It is quick and easy to think of Newton's laws of universal gravitation and Coulomb's electrostatic law. Nevertheless, the history of science itself teaches us that this approach is not always advantageous (i.e., *normal science and scientific anomalies . . .*). Let us think about, e.g., the differences among the foundations of mechanics and the foundations of chemistry (Pisano 2007b), the foundations of thermodynamics (Pisano 2010), and the relationship between physics and mathematics in the history of science (Pisano 2011a), e.g., Descartes 'coordinates' coordinates system and Clapeyron's diagram, space and time as physical and mathematical magnitudes, etc. (previously discussed in Chapters 6 and 7; see also Bussotti and Pisano 2013).

It is well known[3] that the concept of the cycle expressed by Sadi Carnot in *Réflexions sur la puissance motrice du feu* had no significant precedents in theoretical physics until the nineteenth century. It is also true that his father Lazare discusses the state of a mechanical system that returns to its initial state with the potential internal springs brought back to the initial tension (Scott 1971, pp 82–103, 212–241); Poisson also talks about a cycle in his *Traité de mécanique* (Poisson 1823b, pp 5–16, 1833 II, pp 552–554; see also Truesdell 1980, pp 36–38, 1976, 208–235). However, these are previous cases that do not seem to have guided Sadi Carnot in his book. Therefore, it is natural to wonder: *where could the theory of the cycle have come from?* On the other hand, as far as we know, previous historical reflections on the history of heat machines and related arguments, had not considered engines from a cyclic point of view as well. For example, Watt proposed a cycle of working operations of the entire engine but not of the working substance, for which he only proposed a prolongation of the isothermal expansion with a phase of adiabatic expansion – these are just two of the four phases of Carnot's cycle. Thus one may see the cycle in electricity and thermodynamics domains by

> [. . .] bringing into parallelism a domain of facts with another more familiar domain. Rather we must to understand by this word [theory] a complete and systematic representation of the [scientific, historical] facts.[4]

E.g., *could the concept of the cycle in Sadi Carnot's theory come from the typically cyclic operation of the electric current in Alessandro Volta's battery*[5] *(1745–1827)?*

---

[2]Mach ([1896] 1986), p 314, line 12.

[3]An overview of this chapter was published in: Pisano (2003), pp 327–348.

[4]Mach ([1896] 1896), p 415, line 5. That is typically regarding the history of the foundations of physics or, the long–term comprehension of the development of scientific knowledge in historical epistemology (see, e.g., Chapters 6 and 9).

[5]Count Alessandro Giuseppe Antonio Anastasio Volta ([1968] 1998). Alessandro Volta invented the battery when he was 54 years old and already one of the most famous European physicists. In 1778 he was appointed Regius Professor of Experimental Physics at Pavia University in Italy

In his *Simultaneous discovery* (Kuhn 1959), Kuhn investigated the history of the formulation of the law of conservation of energy and claimed:

> The availability of conversion process resulted principally from the stream of discoveries that flowed from Volta's invention of the battery in 1800.[6]

Further ahead he deduces that new formulations are made possible thanks to the sometimes *simultaneous* contribution of scientists and studies from different cultural and scientific backgrounds. To this end, Kuhn cites Mary Sommerville (1780–1872) in *On the Connexion of the Physical Science* from 1832 and writes:

> "The progress of modern science" she said in her preface, "especially within the last five years, has been remarkable for a tendency to [. . .] unite detached branches [of science, so that today] [. . .] there exists such a bond of union, that proficiency cannot be attained in any one branch without a knowledge of others". Mrs. Sommerville's remake isolates the "new look" that physical science had acquired between 1800 and 1835.[7]

Below, we list five facts – and some reflections – which reasonably suggest an analogy between the electric phenomena in Volta's battery and the thermodynamic phenomena in Sadi Carnot 's cycle:

1. Sadi Carnot cites Volta's battery in a lengthy note in *Réflexions sur la puissance motrice du feu* (Carnot 1978). In a footnote (Carnot 1978, pp 21–22, ft 1) he affirms the impossibility of perpetual motion and feels it is necessary to consider it valid for electric phenomena as well. This discussion was of general interest because when Volta invented his battery, he frequently maintained that he had found perpetual motion (he could do so since the Volta effect, origin of the electromotive force, or most commonly *e.m.f.*, was explained more than a century after the invention of the battery) (Fig. 8.1).

(Pancaldi). The *physics department* was named after him; *idem* for the unit of measurement of potential, which Volta called *tension* (later named the *volt.*) Nowadays more than 100 pieces of apparatus invented or used by Volta were conserved at Pavia department of physics and digitalized *Le Opere di Alessandro Volta* composed by 15 volumes for a national edition. See also the international and very important project *European Cultural Heritage Online Project* (ECHO) by *Max Planck Institute for the History of Science* in Berlin and other institutions. For details see: http://echo.mpiwg-berlin.mpg.de/home and http://echo.mpiwg-berlin.mpg.de/content/electricity.

[6]Kuhn (1959), p 323, line 24.

[7]*Ivi*, p 324, line 27. (Author's quotation marks).

(21)

sure de recommencer une opération entièrement semblable à la première et ainsi de suite: ce serait là, non seulement le mouvement perpétuel, mais une création indéfinie de force motrice sans consommation ni de calorique ni de quelque autre agent que ce soit. Une semblable création est tout-à-fait contraire aux idées reçues jusqu'à présent, aux lois de la mécanique et de la saine physique; elle est inadmissible (1). On doit donc conclure que *le*

(1) On objectera peut-être ici que le mouvement perpétuel, démontré impossible par les seules actions mécaniques, ne l'est peut-être pas lorsqu'on emploie l'influence soit de la chaleur, soit de l'électricité; mais peut-on concevoir les phénomènes de la chaleur et de l'électricité comme dus à autre chose qu'à des mouvemens quelconques de corps, et comme tels ne doivent-ils pas être soumis aux lois générales de la mécanique? Ne sait-on pas d'ailleurs *à posteriori* que toutes les tentatives faites pour produire le mouvement perpétuel par quelque moyen que ce soit ont été infructueuses? Que l'on n'est jamais parvenu à produire un mouvement véritablement perpétuel, c'est-à-dire un mouvement qui se continuât toujours sans altération dans les corps mis en œuvre pour le réaliser?

L'on a regardé quelquefois l'appareil électromoteur (la pile de Volta) comme capable de produire le mouvement perpétuel; on a cherché à réaliser cette idée en construisant des piles sèches, prétendues inaltérables. Mais, quoi que l'on ait pu faire, l'appareil a toujours

(22)

maximum de puissance motrice résultant de l'emploi de la vapeur est aussi le maximum de puissance motrice réalisable par quelque moyen que ce soit. Nous donnerons, au reste, bientôt une seconde démonstration plus rigoureuse de ce théorème. Celle-ci ne doit être considérée que comme un aperçu. (V. pag. 29).

On est en droit de nous faire, au sujet de la proposition qui vient d'être énoncée, la question suivante: Quel est ici le sens du mot

éprouvé des détériorations sensibles, lorsque son action a été soutenue pendant un certain temps avec quelque énergie.

L'acception générale et philosophique des mots *mouvement perpétuel* doit comprendre, non pas seulement un mouvement susceptible de se prolonger indéfiniment après une première impulsion reçue, mais l'action d'un appareil, d'un assemblage quelconque, capable de créer la puissance motrice en quantité illimitée, capable de tirer successivement du repos tous les corps de la nature, s'ils s'y trouvaient plongés, de détruire en eux le principe de l'inertie, capable enfin de puiser en lui-même les forces nécessaires pour mouvoir l'univers tout entier, pour prolonger, pour accélérer incessamment son mouvement. Telle serait une véritable création de puissance motrice. Si elle était possible, il serait inutile de chercher dans les courans d'eau et d'air, dans les combustibles, cette puissance motrice; nous en aurions à notre disposition une source intarrissable où nous pourrions puiser à volonté.

Fig. 8.1 Sadi Carnot cites Volta's Pile (Carnot 1978, pp 21–22, ft 1)

2. Since it was in use at the time, Sadi Carnot called his invention *appareil électromoteur* (Carnot 1978, p 21, ft 1, line 11). Regarding this point, Volta wrote:

> And above all, concerning the sense of tact: if, [...] I establish on one hand a good communication using an extremity of my *elettro–motore* device (it is necessary to give new names to new devices, not only for the shape but also for the effects or for the principle on which they depend) [...].[8]

It should be noted that during Sadi Carnot's lifetime, it is probable that this was the only type of *engine* that was known besides the first heat engine. Generally speaking, currently, we know that the battery is a device that converts chemical

[8] Volta ([1968] 1988), pp 526–527, line 36. (The translation is ours: RP). Cfr.: Volta (1796), (1800a, b), (1987).

energy into electric energy, just as Sadi Carnot's heat machine is a device that transforms heat (thermal energy) into (thermal) work.[9]

3. The invention of Volta's battery achieved a scientific change, as well as a socio–technological one. It had given birth[10] to the electric current then seen as a fluid analogous with caloric fluid for thermal phenomena. In the following listed and cited works, *Lettera Settima, Saggio teorico e sperimentale di elettricità* and in *Del modo di render sensibilissima la più debole Elettricità sia Naturale, sia Artificiale* (the latter read on March 14, 1782) Alessandro Volta emphasized his interest in science within scientific domains, focusing on the analogy between the two fluids:

> I cannot avoid making note of an analogy, whose effects so much so, between electric fluid and caloric fluid [. . .].[11]
>
> [*On the electric fluid, and its spreading in others domains*]. [. . .]. Paragraph 83. That which we can presuppose, and all physicists agree with, is that electric fluid is very much analogous with fire, and produces various effects similar to those of fire; and electric fire equally to electric fluid as all of them agreed to call it, [. . .] and we distinguish it from common fire. [. . .].[12]
>
> Who would not be struck by such a beautiful analogy in which electricity sheds light on the new doctrine of heat, and in turn receives it? I speak of the heat doctrine of *latent heat orspecific heat* [. . .] in which Black and Wil[c]ke sowed the seeds for a later proposal by Dr. Crawford [his work *Animal heat*] after the experiences of Dr. Irwine.[13]

In addition, Sadi Carnot seems to echo Volta's words, when, at the last moment – and without plausible motives – he deletes the word "Chaleur" from the title of his work and substitutes it with the *old–fashioned* "feu" (Carnot 1878a, folio 1; see above Chapter 6).

4. Furthermore, Sadi Carnot's *Réflexions sur la puissance motrice du feu* are based on the concept of *re–establishment of equilibrium of caloric*; in his theory on mechanical machines, his father Lazare Carnot had mainly focused on general theory of machines and principles of equilibrium and of motion.[14] It should be noted that the concept of heat equilibrium was introduced much earlier than Sadi Carnot used in his thermodynamics theory. For example, in the eighteenth-century, (1728–1799) during his professorship of medicine and investigations on the calorimeter (Black) in Glasgow, based on his experiments

---

[9]In this chapter we will use the term "work" as Lazare and Sadi perceived it at that time. For our purposes, at times, such as a meta reflection in modern terms, will refer it as "quantity" and without explicit references Coriolis.

[10]Let us also note the previous invention of the *Leyda bottle* by Peter Mussheenbroek (1692–1761) in 1745.

[11]Volta (1816), p 245, line 1. (The translation is ours: RP).

[12]Volta (1918–1929), IV, p 353, line 14. (The translation is ours: RP). He had quite a good knowledge of French and English experiments and studies as he frequently referenced them by names and descriptions (Volta 1782, pp 237–280; see also selected references in: Mottelay, pp 248–249).

[13]Volta (1782), pp 278–279, line 28. (The translation is ours: RP).

[14]Carnot (1786), pp iv–v, 11–12; see also Carnot (1778), §§ 27–79, (1780), § 102, §§ 133–141; Gillispie (1971), Appendix C, § 102, pp 301–303, §§ 133–141, pp 317–321. For the comparison and relationship between the two Carnots on that argument see below Chapter 11.

on transformations from liquid to vapor, heat's effect (latent) as the main cause of the change from one phase to another, etc., Joseph Black maintained that if two bodies were in contact (today we say in thermal contact) the heat flows (out) of the body in which the intensity (temperature) is highest, to the body in which the intensity is lowest. The flux stops when the heat has the same intensity in both bodies, that is to say, when both bodies are at the same temperature. In other words, Black inductively clarified these concepts, establishing that under certain experimental conditions, the phases of a compound can coexist in a particular state called the *state of equilibrium of heat,* in our case, between the two bodies. With respect to these crucial arguments of physics and the history of physics, we list some original quotes concerning *Lectures on the Elements of Chemistry delivered in the University of Edinburgh by the late Joseph Black* (Black) during his professorship of medicine and chemistry (Partington; Ramsey):

> 2. *Of the Distribution of Heat.* A second improvement in our knowledge of heat, which has been attained by the use of thermometers, is the more distinct notion we have now than formerly, of the *Distribution* of heat among different bodies. I remarked formerly, that, even without the help of thermometers, we can perceive a tendency of heat to diffuse itself from any hotter body to the cooler around, until it be distributed among them, in such a manner that none of them are disposed to take any more heat from the rest. The heat is thus brought into a state of equilibrium. This equilibrium is somewhat curious. We find that when all mutual action is ended, a thermometer, applied to any one of the bodies, acquires the same degree of expansion: Therefore the temperature of them all is the same, and the equilibrium is universal. No previous acquaintance with the peculiar reaction of each to heat could have assured us of this, and we owe the discovery entirely to the thermometer. We must therefore adopt, as one of the most general laws of heat, that "all bodies communicating freely with each other, and exposed to no inequality of external action, acquire the same temperature, as indicated by a thermometer." All acquire the temperature of the surrounding medium. By the use of these instruments we have learned, that if we take 1000, or more, different kinds of matter, such as metals, stones, salts, woods, cork, feathers, wool, water and a variety of other fluids, although they be all at first of different heats, let them be placed together in the same room without a fire, and into which the sun does not shine, the heat will be communicated from the hotter of these bodies to the colder, during some hours perhaps, or the course of a day, at the end of which time, if we apply a thermometer to them all in succession, after the first to which it is applied has reduced the instrument to its own temperature, none of the rest are disposed to increase or diminish the quantity of heat which that first one left in it. This is what has been commonly called an equal heat, or the equality of heat among different bodies; I call it the *equilibrium of heat*. The nature of this equilibrium was not well understood, until I pointed out a method of investigating it. Dr. Boerhaave imagined, that when it obtains, there is an equal quantity of heat in every equal measure of space, however filled up with different bodies; and Professor Muschenbroeck expresses his opinion to the same purpose: "Est enim ignis aequaliter per omnia, non admodum magna, distributus, ita ut in pede cubico auri et aëris et plumarum, par ignis fit quantitas.[+]" The reason they give for this opinion is, that to whichever of those bodies the thermometer be applied, it points to the same degree. But this is taking a very hasty view of the subject. It is confounding the quantity of heat in different bodies with its general strength or intensity, though it is plain that these are two different things, and should always be distinguished, when we are thinking of the distribution of heat.[15]

[15]Black, I, pp 76–78, line 14. (Author's quotations marks and *italics*).

In regard to logic, it is also interesting to note that in his argument, Sadi Carnot uses DNSs belonging to non–classical logic.[16] Such a high number of double negative sentences suggests that Sadi Carnot wants to express a concept that he considers fundamental for the development of his theory. In particular, various times in his book,[17] Sadi Carnot maintains that the work of the heat machine depends on the presence of $\Delta T \neq 0$. To achieve work, it is necessary to *re–establish the equilibrium of the caloric* (Carnot 1978, p 23). Before Volta, Luigi Galvani (1737–1798) asserted that an intrinsic electricity was already present in a dead animal (the famous frog). In the state of *disequilibrium,* the external conductors induced contractions in the frog's muscles, therefore allowing for the flux of this electricity even inside the animal; which according to Galvani, had accumulated principally in the frog's muscles. Galvani did not use his name when he published *Dell'uso e dell'attività dell'arco conduttore nelle contrazioni dei muscoli* (Galvani 1794a, [1841] 1998, pp 430–431 and footnotes; see also 1794b) in which he argued with Volta regarding Galvanic phenomena within animals. In particular, he also described one of the most important experiments of electrophysiology nowadays so called: muscular reactions caused by the contact of nerves and corresponding muscle. In other words, according to Galvani, the animal electricity is already in a state of equilibrium and it is ready to move if opportunely stimulated by external or internal actions. Thus, Galvani thought it plausible to claim:

> This disequilibrium in the animal must either be natural, or be introduced by artifices. Of course, if it will be in accordance to confess that the animal has[a] particular machine for producing this disequilibrium ["disequilibrio"] [then] it will be reasonable to call animal this kind of electricity in order to denote, not any electricity, but a particular one, [that is] applied to a particular machine: if instead the disequilibrium is inducted by artificial things, [then] to consider the animal as any humid body, as Mister Volta thinks, will be sufficient.[18]

Following this idea, it also seems plausible that Sadi Carnot could have connected the previous concept of disequilibrium belonging to Galvani's theory, to that of caloric disequilibrium in his theory on heat machines. In fact, both Carnot's ideal heat machine and Volta's battery are devices made up of a *fluid* that circulates between two potentials with different values. Therefore, the comparison between the two types of engines could be proposed.

5. We will add a general interpretive hypothesis to explain historically how the concept of the cycle was born in *Réflexions sur la puissance motrice du feu.* Sadi Carnot developed his work on heat engines in an incredibly short amount of time: less than 3 years. It is therefore reasonable to presume that it was actually

---

[16]The DNSs are discussed in previous Chapters 6 and 7. The DNSs that concern Sadi Carnot's reasonings on equilibrium are: 8, 15, 31, 45 (Carnot 1978, pp 1–112) and they are also listed below in the Appendix. In addition, please see also: p 9, line 10; p. 10, line 7; p 10, line 10; p 12, line 3; p 14 line 2; p 14, line 15; p 16, line 17; p 17, line 1; p 23, line 5; p 26 ft 1, line 2.

[17]DNSs 9, 31(=33 = 34), 32, 35 (=36), 37, 40, 41, 42, 43, 44, 45. (See below Appendix).

[18]Galvani ([1841] 1988), p 201, line 27; see also Galvani (1794a). (Translation is ours: RP). On Galvani one can also consult: Galvani (1791, 1797).

his father, Lazare, who conceived the ambitious project of generalizing his theory about mechanical machines (see below Chapter 11), extending it to thermal machines as well. He also wrote a rough draft on thermodynamics which he passed on to his son, when Sadi Carnot went to visit his father in Magdeburg in 1820. Therefore, Sadi Carnot provided a continuation to his father's work, albeit with extraordinary results. In fact, the most reasonable correlation between mechanics and electrical magnitudes is based on Lazare's mechanics and not on Newtonian paradigm mechanics (Table 8.1).

Table 8.1 Mechanics and electricity. Some general analogies

Lazare Carnot's mechanics	Electric theory
Quantity of motion Ip_i	Electric charge Q
Flux of quantity of motion Ip_i	Flux of electriccharge of current I_Q
Velocity v_i	Electric potential V
Viscosity η	Electric conductance σ
Mechanical resistance R_p	Electric resistance R_{Qr}
Mass of a body, m	Capacity of a charged body, C
Mechanical Inductance *1/k* (k of a spring)	Electric inductance I

In this sequence of reasoning, we can assume that it was Lazare Carnot, rather than Sadi Carnot, who first had the idea of the cycle for heat machines. This idea could have been conceived at the moment in which Lazare Carnot and other French academic scientists (see Figs. 8.2 and 8.3 below) witnessed Alessandro Volta's

Fig. 8.2 Experiences by Alessandro Volta in front of Napoleon (1801) ("Esperienza di Alessandro Volta al cospetto di Napoleone (Parigi 1801). Affresco di Nicola Cianfanelli nella tribuna Galileo a Firenze (1841)" Original Photo credit: Saulo Bambi, *Museo di Fisica e Scienze Naturali di Firenze,* Italy. With permission of *Museo di Fisica e Scienze Naturali di Firenze*, *sezione Zoologica.* University of Firenze, Italy). (1841. Nicola Cianfanelli (1793–1848))

Fig. 8.3 Alessandro Volta presents the battery to Napoleon Buonaparte (1801) ("Alessandro Volta espone la pila a Napoleone Buonaparte (Parigi 1801). Autore Giuseppe Bertini (1897))". Plate from the original, with permission of *Volta Temple* in Como. (1897. Giuseppe Bertini (1825–1898))

experiments on the electric battery (which involves a cyclic path of the electric current) that he presented on 7 November 1801 at Napoleon's court. In fact, Lazare Carnot always attended the presentations of new technological inventions and was a great supporter of them (Reinhard). He may have been so impressed that he then emphasized the importance of that apparatus to his son, thereby giving him the suggestion for the first cycles in *Réflexions sur la puissance motrice du feu.*

Volta invented the first exemplar of the electric battery which came to be known as *Volta's Pile* (or also voltaic battery) in late 1799 (Volta 1799; Grattan-Guinness 1990a, b, p 491); and formally in a publication in 1800 (Volta 1800a, b). In fact, on March 20, 1800, he wrote to Sir Joseph Banks (1743–1820), President of *The Royal Society of London* describing the technique for producing an electric current using his voltaic battery (*Ibidem*; see Fig. 8.4). In the end the French expedition to promote his invention from Paris to *Consulta* at Lyon covered ca. 8 months, from 1 September 1801 to 22 April 1802. This may have been enough time for French scholars to appreciate his invention.

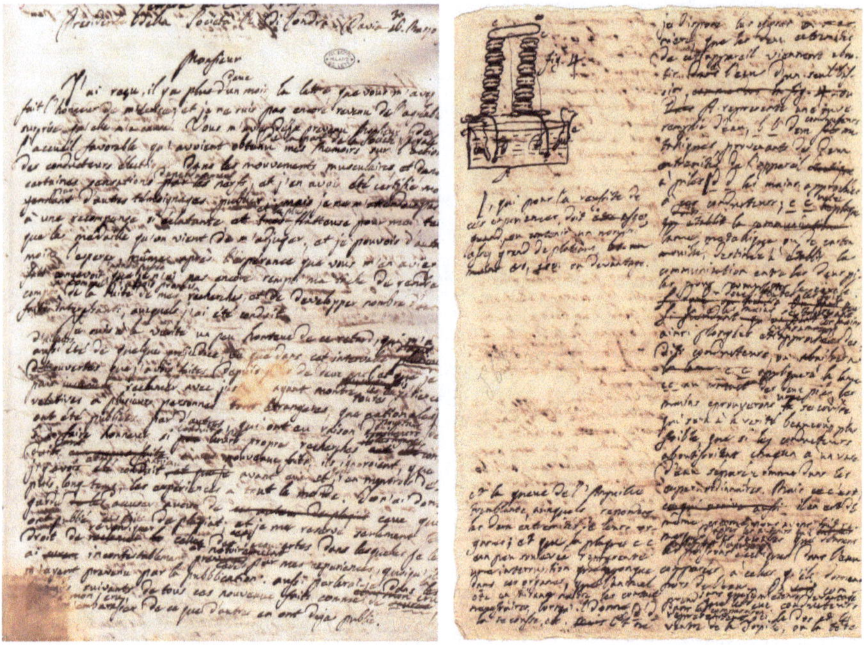

Fig. 8.4 Plates from Volta's manuscript sent to The Royal Society of London (1800). Plates from the original. With permission of *Manuscript Archive of Istituto Lombardo Accademia di Scienze e lettere*: "Volta E27" (on the *left*) and "Carteggio Voltiano J68" (on the *right*)

8.2 A Comparative Analysis: Volta's Battery and Sadi Carnot's Heat Machine

The idea, therefore, is to study which analogies with electric phenomena could have guided Sadi Carnot in forming the idea of the cycle in his book. Here, we attempt a comparative analysis between the cyclic operation (path) of Volta's battery and the (cyclic) operation of Sadi Carnot's heat machine and therefore between the physical quantities (and actions) that intervene in the thermal cycle and Volta's battery (Volta 1800a, b).

8.2.1 An Analogy Between Heat Fluid and Electric Fluid

The following figure is performance and usual diagram of a heat machine (that can work cyclically) (Fig. 8.5):

As previously stated, in *Réflexions sur la puissance motrice du feu,* Sadi Carnot writes that he wants to reestablish the equilibrium of caloric and asserts that he can achieve it in every heat cycle that makes the heat pass from temperature T_1 to T_2. According to the theory of caloric, Sadi Carnot's heat machine, whose fluid Q flows

Fig. 8.5 Simplified performance model of a simple heat machine

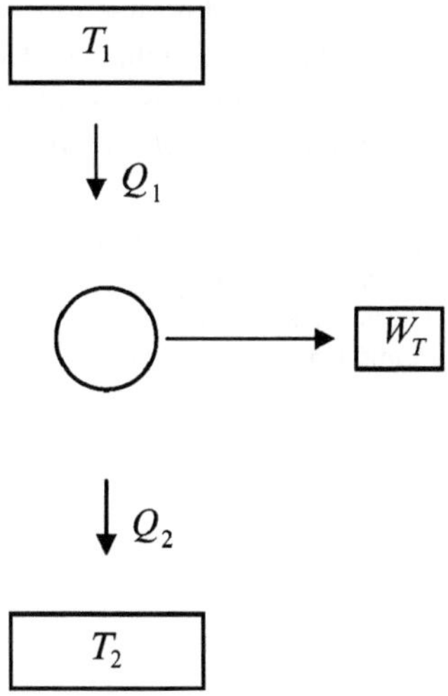

from temperature T_1 to temperature T_2 ($T_1 > T_2$), has a peculiar characteristic: at the end of the production (or absorption) work process, the fluid returns to its initial temperature; therefore it is possible to reinitiate a *heat disequilibrium* which brings the caloric back to T_1 and will be followed by a production of work that restarts from the source at temperature T_1. Even without the caloric hypothesis it can be seen that in order to have caloric circulation, it must be brought back to temperature T_1, and this can be done with a reversible heat machine between the same temperatures; additionally, heat is not lost in the environment. Therefore two paired, reversible heat machines allow for the heat circle between the two sources practically to infinity. This fact corresponds to the cyclic motion of the electric fluid in the battery circuit (both external and internal). The correspondence leads to the connection of the *fem* with the ΔT and the heat Q with the electric charge in motion q.

The Voltaic battery[19] is a set of cells placed in a series. In the late 1800s Volta learned that when two metals and brine–soaked cloth or cardboard are arranged in a circuit they produce an electric current. In 1800, Volta stacked several pairs of alternating copper (or silver) and zinc discs (electrodes) separated by cloth or cardboard soaked in brine (electrolyte) to increase the electrolyte conductivity. In detail, Volta used pairs of copper and zinc disks alternating with a wool cloth soaked in salted or acidulous water (electrolyte H_2SO_4) following the sequence

[19] In a modern performance of the Voltaic battery all of the electric potential differences which occur between every ionic pair that flow in the solution (before they reach the respective electrodes) should be studied. However, assuming the exemplifying hypothesis that there are only two ionic species of opposing charges in the solution (and that, in the solution, they move methodically), we can consider a single constant electric potential difference (such as a hypothesis) for every couple (of the two species) of charges found in the solution. It is also necessary to hypothesize that in the acidulous solution, other disruptive phenomena such as those of overvoltage do not occur.

copper–zinc–solution–copper . . . We can consider the Voltaic battery as an electric engine that works with two electric potentials, between which an electric fluid flows. This fluid completes a cyclical path made up of an external tract ($Cu|Zn$) and an internal one ($Zn|Sol. + Sol. + Sol.|Cu$).

Fig. 8.6 Volta's Pile at *Volta Temple* in Como, Italy (Original photo. With permission of *Volta Temple* in Como)

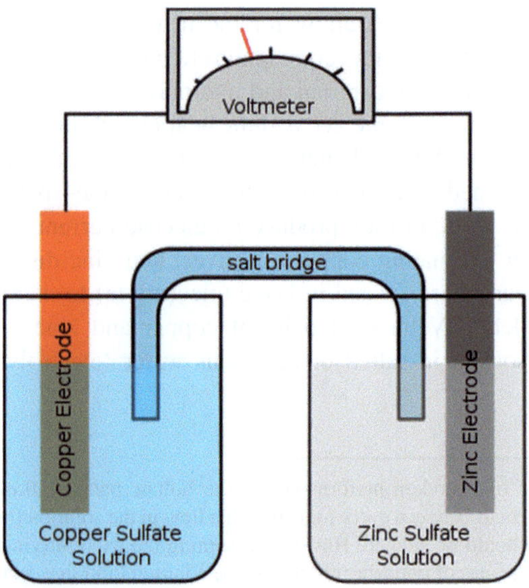

Fig. 8.7 Simplified performance model of voltaic battery (One can consider the motion from a potential V_1 to another one V_2 ($V_1 > V_2$) which (by also modernly considering chemical–physical phenomena in parallel) produces electric work W_E. Source: http://en.wikipedia.org/wiki/File:Galvanic_Cell.svg)

In addition to Volta's battery (intended as an electric motor) we can *adjust* the second principle of thermodynamics according to Kelvin where *no process is possible in which the sole result is the absorption of heat from a reservoir and its complete conversion into work.*

8.2.2 *An Analogy Between the Cycle of Electric Fluid and the Cycle on Heat States*

The heat engine and the cycle achieved by the state of the thermal system is composed of four physical processes. In the voltaic battery, there are also four passages but they are contemporaneous. In order to have a thermal analogue operation, it is necessary to consider an engine which is not single–cylinder but is composed of four–cylinders: here all four phases occur simultaneously, just as the electric current flows simultaneously in the external and internal circuits and also in the transformations to the electrodes. It is well known that Sadi Carnot's heat cycle is made up of four physical transformations, alternating two by two. In a voltaic battery four passages could be thought of (we know now they are in fact present) which are chemical–physical actions (production and annihilation of ions and motions in the solution and the wire conductor) due to the passage of the electric charges (Table 8.2).

Table 8.2 Paths and analogies between Volta's Pile and Sadi Carnot's machine

Phases	Physical actions in electric cycle during motion of current between external electrodes	Kind of path	Analogies between heat and electric phases
1	The charges q move from *Zn*	Metal–solution	Adiabatic
2	The charges q move the solution	An internal path within solution, but external to electrodes	Isothermal
3	The charges q move to *Cu*	Solution–metal	Adiabatic
4	The charges q move from *Cu* to *Zn*	An external path between to electrodes	Isothermal

Clausius's principle can also be repeated in the same way: *no process is possible whose sole result is the transfer of heat from a body of lower temperature to a body of higher temperature.*

8.2.3 *An Analogy Between Heat Work and Electric Work*

As previously mentioned, Sadi Carnot maintained that in order to obtain non–null work, (in addition to a caloric fluid) a $\Delta T \neq 0$ (among thermostats) is necessary.

In Volta's battery as well, in order to obtain non–null work, a $\Delta V \neq 0$ is needed (between reservoirs–potentials). The work produced by S. Carnot's thermal machine W_T (which uses gas as an working substance) is given as:

$$W = Fs \tag{8.1}$$

Equation 8.1 can be written as:

$$F_T \Delta s = PA\Delta s = W_T = p\Delta V \tag{8.2}$$

in which P is the pressure on the piston and A is its area, W_T is the thermal work, s is the displacement and F_T to mention an *ad hoc* force for thermal facts. Equation 8.2 has an analogous formula in the work produced by an electric field W_E in Volta's battery, that is to say moving the charges from one potential to another: $\Delta V = V_2 - V_1$ (see Figs. 8.6 and 8.7):

$$W_E = q\Delta V = \Delta V I t \tag{8.3}$$

where W_E is the electric work. In the cycle of Sadi Carnot's heat machine, heat work is caused by the contributions of work calculated on just two isotherms. Sadi Carnot fortuitously concluded (it is currently the result of the equivalence in first principle of thermodynamics) that the work of the two adiabatic transformations should not be considered (see below Chapters 9 and 11) since they are equal and opposite and therefore (in the cycle) they annul each other (Pisano 2001; Drago and Pisano 2005). After collecting every element of this analogy, it is reasonable to construct a theoretical parallel as shown in the following Table 8.3 which summarizes the corresponding quantities between Sadi Carnot's heat machine in thermodynamics and Volta's battery in electricity:

Table 8.3 Thermodynamic magnitudes and their correspondence with electric magnitudes in Volta's battery

Sadi Carnot's thermodynamics theory	Volta's electric theory
Heat machine: Reservoirs–Potentials ($T_1 > T_2$), heat, four complete phases	Volta's battery: Electrodes–Potentials ($V_1 > V_2$), electric charges, four complete phases
$\Delta T = T_1 - T_2$ between Reservoir T_1 and Reservoir T_2	$Fem = \Delta V = V_1 - V_2$ between "*Zn* and *Cu*" plus "*Zn* and solution" plus "solution–and–solution" and "solution and *Cu*"
$W_T = p\Delta V$	$W_E = q\Delta V = \Delta V I t$
Heat (*Chaleur–Calorique*)	Current I (electric charges in motion)
A full four–phases in heat cycle:	A full four–phases in electric cyclical path:
1. An adiabatic transformation	1. The charges move from *Zn* (internal path)
2. An isotherm transformation	2. The charges move within solution (internal path)
3. An adiabatic transformation	3. The charges move toward *Cu* (internal path)
4. An isotherm transformation	4. The charges move from *Cu* to *Zn* (external path)
In order to obtain W_T, one needs:	In order to obtain W_E one needs:
1. A $\Delta T \neq 0$	1. A difference of potential between electrodes

(continued)

Table 8.3 (continued)

Sadi Carnot's thermodynamics theory	Volta's electric theory
2. The heat is transferred from one of the two reservoirs	2. An internal path done by ionic dissociations on contact–surfaces between solution and two metals (within solution)
The absorption of heat from the other reservoir	The absorption of charges from a metal

8.3 On the Analogy Between Thermodynamics and Electrostatics According to Mach's Investigations

One might hypothesize that the analogy presented here could have been useful to Sadi Carnot, at least for a general direction in his scientific reasoning. In particular on the development of the thermodynamics laws in history and their analogies, Mach claimed:

> 6. So far we have spoken principally of the *agreement* with one another of the forms of energy. The *differences* of these forms from one another must not be overlooked, and heat especially offers differences as compared with other forms. If we take the standpoint mentioned above we see that: (1) The mere *exact* knowledge of the conservation of energy is sufficient to obtain the theorem of Carnot and Clausius: and (2) Because this theorem holds for the different forms of energy, a special position for heat is not conditioned by this theorem. 7. As regards the first point. We have already mentioned that a full insight into the conservation of energy not only allows us to recognize the transformation of one kind of energy *A* into another *B* but also the fall of potential of *A* and the rise of potential of *B* which is necessarily connected with this transformation. That the two properties of the transformation of energy have been formulated in two different theorems – the two "laws of thermodynamics" – is only due to the historical accident that a period of twenty years lies between our knowledge of them. Only after ten years more were both properties expressed in the theorem of Carnot and Clausius which is simply a more complete expression of the fact of which the first law shows only one slide*. In the history of physics this is not without an analogue. [...] 8. With respect to the second point, the following is clear. Analogy is not identity; heat may therefore have other properties which are peculiar to itself. In fact, it has such properties; but this particularity lies in circumstances independent of the theorem of Carnot and Clausius. Every transformation of a kind of energy A is connected, in the case of heat as well, with a fall of potential of the kind of energy considered. But while, for other kinds of energy, with the fall of potential a transformation is inversely connected – and consequently a loss in energy of the kind which falls in potential –, heat behaves in another way. Heat *may* Buffer a fall in potential without experiencing a loss of energy – at least according to the usual way of measuring it. If a weight sinks, it must necessarily generate kinetic energy or heat or some other energy. An electric charge, too, cannot undergo a fall of potential without a loss of energy, that is to say without transformation. Heat, on the other hand, may be transferred with a fall of temperature to a body of greater capacity, though the energy of heat remains the same. This is what besides its property of energy, gives heat the character of a material substance*.[20]

However, we should note that official and shared and full theorization on the phenomena concerning the *battery* and *electricity* was lacking at that time (Pisano

[20] Mach ([1896] 1986), pp 310–311, line 20. (" *" = Author's footnote).

2012b). In fact, a comprehension–path from electric phenomena to electric–magnetic phenomena was not immediate. For example, in the following table a concise historical panorama[21] of the scientific studies around the time of Volta's invention are presented (Table 8.4):

Table 8.4 Electricity and magnetism laws

Year	Electricity	Magnetism
1785	Coulomb's law $F = K\frac{q_1 q_2}{r^2}$	
1800	Volta's battery	
1813	Poisson's equation electrostatic potential $\nabla^2 V = -4\pi\rho$	
1820	Ørsted's experiments	
1820	Biot–Savart's law: $dB = id\vec{l} \times \vec{r}/r^3$	
1821		Faraday's lines of forces
1825	Ampère's mathematization for the interaction between magnetism and electricity (motion of particles)	
1825–1827	Ohm's law: $V = Ri$	
1831	Faraday's law[a]: $emf = -\frac{d\Phi(\vec{B})}{dt}$	
1832	Faraday's law for electrolytic conduction: $m = \frac{Mq}{Fz}$	
1834		Lenz's rule

[a]In 1831 Michael Faraday (1791–1867) demonstrated (without mathematics) the reciprocal effect, in which a moving magnet in the vicinity of a coil of wire produced an electric current. Ørsted's experiment on a magnetic needle led Faraday to conceive the notion of a magnetic field (Pisano 2012b)

Where briefly (for our aims),

- *1785: Coulomb…* attraction and an irreversible advancement toward new phenomena to explain the repulsion and try not to lose the Newtonian physical mathematical approach (1785).

[21]Another concise list might be (for our purpose): 1745: Grummert studies electric light in *vacuo*. 1745: Miles reads at *The Royal Society* (March 7) a paper concerning phosphorus, electricity and the role played by conducting bodies. Pivati writes *Lettere della elettricità medica*. 1753: Beccaria, scientific works on electricity are produced. 1756: Le Chevalier Jacques CF de la Perrierc de Roiffe write *Mécanismes de l'électricité et de l'univers*. 1757; Wilcke studies the production of electricity by means of melting electrical substances (following Stephen Grey's studies). 1769: Volta addresses his *de attractiva ignis electrici* to Beccaria. 1775: Cavallo studies relationships between electricity and atmosphere and invents a small electroscope and a condenser of electricity. 1775: Volta produces quite important experiments on his inventions making them known by letters, e.g., the electrophorus, a sort of perpetual reservoir of electricity. 1759–1778: Benjamin produces important essays on electricity. 1781: Lavoisier proposes (see also Volta and Laplace) that electricity is developed when solid or fluid bodies pass into the gaseous state. 1781: Kirwan, *President of the Dublin Society and of the Royal Irish Academy* produces quite important works on magnetism and electricity and receives from the *English Royal Society* its *Gold Copley medal*. 1790: Vassalli publishes his views concerning the electricity of bodies, electricity of water and ice. 1793: Fontana works on animal electricity. (Cfr.: Mottelay).

- *1799–1800: ... but Volta's* experimental (only) innovations: electrophorus produced an electric charge without a mechanical Coulomb balance; electric circuits; dynamics without inertia and mechanical forces; cyclical dynamics. *Newtonian cause–forces* were substituted by *interaction. The chimera of perpetual motion.* A new theory advanced without a basic (mechanical) model and without sure scientific dialectic among foundations and without relationship among known structures.
- *1820: Ørsted* performed an important experiment which showed that there was a connection between electricity and magnetism. When a current was switched on through a wire, it made a compass needle turn so that it was at right angles to the wire. The current had produced a magnetic field strong enough to cause the compass needle to turn (Ørsted 1822, 1823).
- *1820: Ampère* (on 11 September), produced specific works based on Ørsted's discoveries on the magnetic needle.
- *1820: Ampère* (on 18 September), mathematically (*parallelogram rule*) demonstrated that parallel wires carrying currents attract or repel each other, depending on whether currents are in the same (attraction) or in opposite directions (repulsion). Ampère's theoretical hypothesis: with respect to Ørsted's experiments, he proposed a set of circular current: "*... not the conductor crossed by current become a magnetic ..., but the magnetic object is a complex system of currents*" (Ampère 1822a, 1822b, 1826, 1827).
- *1820: The law of Biot and Savart,* formulated by Jean–Baptiste Biot and Félix Savart around 1820, is the fundamental law of magnetic force, analogous to Coulomb's law for electric force. It specifies the direction and strength of the magnetic field next to a conductor carrying an electric current. Like Coulomb's law, there is an inverse–square dependence on distance, but the behavior of magnetic fields is much more complicated than that of electric fields.
- *1825: Ohm* examines the decrease in the electromagnetic force produced by a wire as the length of the wire increased. The paper deduced mathematical relationships based purely on the experimental evidence that Ohm had tabulated.
- *1826: Ohm* conduction in circuits modelled on Fourier's study on propagation of heat.
- *1827: Ohm's law* appeared in the famous book *Die galvanische Kette, mathematisch bearbeitet* (*The Galvanic Circuit Investigated Mathematically*) in which he gave his complete mathematical theory of electricity with respect to a non–mathematical approach to those physical phenomena at that time.
- *1831–1845: Faraday's experiments* and innovations: *e.m.f.* and magnetic field, vectors without use of a mathematical approach.

In other words, one might wonder if the analogy could carry on in light of modern results regarding electric phenomena. We will now see a few studies on the subject.

In 1896, Mach in his book on the historical development of the *Principles of the theory of heat* (Mach [1896] 1986) wrote several pages generalizing Carnot's cycle claiming that

> The reversible cyclic process which performs work is not limited to thermal processes. There is no difficulty in imagining analogous cyclic processes for any other events whatever electrical, for instance. Footnote:* For the last twenty years it has been my habit to discuss such examples in my lectures.[22]

In the end a question might be: *was the cycle a concept specific to the thermodynamics of 1800 or can there be equivalents of Sadi Carnot's cycle in other scientific theories? Moreover, can that technique of arguing be adopted for another scientific theory?* With regard to the possibility of generalizing Sadi Carnot's cycle on a meta–theoretical level, Mach argued the fact that if a part W' of a kind of energy, e.g., $W + W'$ at potential V' transforms itself into one or more kinds of energies, then the remaining part W decreases to a potential V.

Thus one can write:

$$\frac{(W + W')}{V'} - \frac{W}{V} = 0.$$

From that equation – according to Mach – other related operations follow (Mach [1896] 1986, pp 309–310). In particular, the aforementioned formula (*Ivi*, p 309) expresses, in our historical epistemology interpretation of Sadi Carnot's thermodynamics, the formula for Carnot's cycle[23]:

$$\frac{(Q_2 + q)}{T_1} - \frac{Q_2}{T_2} = \frac{Q_1}{T_1} - \frac{Q_2}{T_2} = \Delta S = 0 \tag{8.4}$$

where it should be noted that he uses the first principle of thermodynamics, since the work is given by the disappearance of Q' (W'). Therefore, he maintains that through Sadi Carnot's cycle, other physical systems, which come from very different physical theories, can be interpreted: mechanics or electrostatics for example. In fact, Mach describes two rather significant and direct *analogies* regarding Sadi Carnot's heat cycle:

1. The first one, cited by Mach, belonged to Zeuner ([1860] 1866) and discussed thermal processes and mechanical processes in order to shed light on the meaning of term *Q/T* which he referred to as "weight of heat (Wärmegewicht)" in an attempt to compare Sadi Carnot's heat cycle process to mechanical process using heavy masses. (Mach [1896] 1986, pp 275–276, 306–309).
2. The second one was proposed by Mach himself and concerns an electrostatics system which expands and contracts. It is mainly composed of a body having a great capacity of charge between two given potentials (Mach [1896] 1986, pp 308–309).

---

[22] Mach ([1896] 1986), p 308, line 10; see also pp 142–145.

[23] For the sake of consistency, at letters *W* and *V* in the quotation from Mach we respectively substituted the letters *Q* and *T*.

Let us examine Mach's analogy. In the following an electrostatic mechanics cycle see (Fig. 8.8) by means of an expandable metallic sphere which dilates and contracts between two capacitors at potentials V_1 and V_2 ($V_1 > V_2$) is represented.

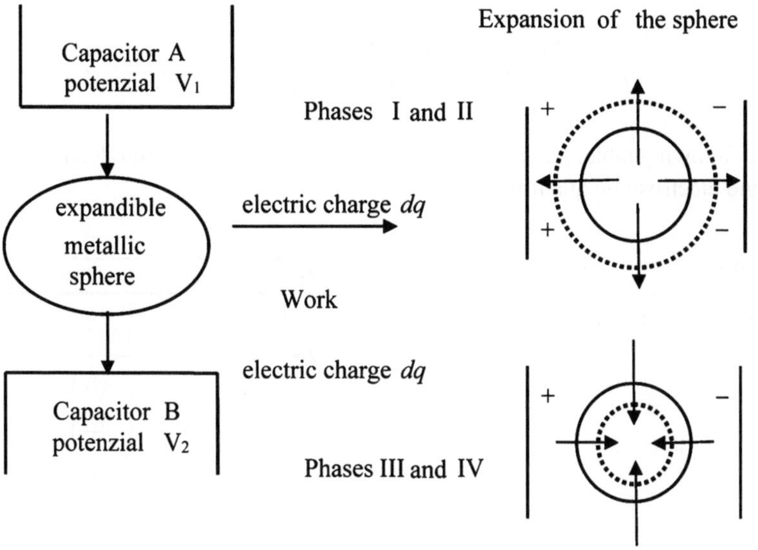

Fig. 8.8 A simplified performance model based on Mach's arguments (We tried to report and adapt Mach's discursive reasoning in a performance model (Mach [1896] 1986, pp 308–309))

Below, we also provide a comparison between the electrostatic phases described in the previous figure and the phases in Carnot's cycle (Table 8.5):

Table 8.5 Electrostatic paths and Sadi Carnot's cycle

Phases in an electrostatic cycle	Phases in Sadi Carnot's heat cycle
1st phase: from *A*, which is at constant potential V_1, the metallic sphere receives electric charge Δq and expands due to electric repulsions.	Isothermal transformation
2nd phase: after the sphere is isolated and it continues its expansion until potential V_2.	Adiabatic transformation
3rd phase: by entering into *B*, the sphere contracts at constant potential V_2, transferring Δq to it.	Isothermal transformation
4th phase: after the sphere is isolated, and it continues its contraction until it reaches potential V_1.	Adiabatic transformation

Therefore, for the cycle in the metallic sphere and in Sadi Carnot's cycle the following are valid (for both cases, and keeping in mind the first principle):

$$\frac{\left(\frac{\Delta q V_2}{2}\right)}{V_2} = \frac{\left(\frac{\Delta q V_1}{2}\right)}{V_1} \approx \frac{Q_2}{T_2} = \frac{Q_1}{T_1}, \tag{8.5}$$

and in the end the efficiency is:

$$\eta = \frac{\frac{\Delta q(V_1 - V_2)}{2}}{\frac{\Delta q V_1}{2}} = \frac{(V_1 - V_2)}{V_1} \approx \frac{W}{Q_2} = \frac{(T_1 - T_2)}{T_1}. \tag{8.6}$$

In the following table, we list the correspondences between thermodynamic quantities and electrostatic quantities:

Table 8.6 Correspondence between the physical thermodynamic quantities and the quantities of an electrostatic cycle

Thermodynamics	Electrostatic machine
T_1, T_2	V_1, V_2
$2\frac{Q}{T} = 2\Delta S$	$\frac{\left(\frac{\Delta q V}{2}\right)}{V}$
Q_1	$\left(\frac{\Delta q V_1}{2}\right)$
Q_2	$\left(\frac{\Delta q V_2}{2}\right)$
W	$\frac{\Delta q(V_1 - V_2)}{2}$
$\eta_T = \frac{(T_1 - T_2)}{T_1}$	$\eta_{E_{Mach}} = \frac{(V_1 - V_2)}{V_1}$

In conclusion, he proposed a strong connection between the magnitudes of the two theories. However, in Table 8.6, the two quantities of electric charge indicated by Mach are different and the quantities of heat yielded and absorbed are also different. This contradicts caloric theory and is consistent with modern heat theory. Therefore, the previous analogy would have been useful for Sadi Carnot if only he (according to the doubt expressed in his footnote (Carnot 1978, pp 73–79)) had definitively abandoned caloric theory. In addition, it must be noted that Mach's contribution concerns electrostatics, whereas for the purpose of this document, the link between thermodynamics and the electric current is needed. However, while Sadi Carnot used the heat entering and exiting the machine in thermodynamics, Mach treated a cycle achieved by quantities of electric charges entering and exiting the sphere; in fact, in Mach's description, charge Δq first exits and then enters the sphere, then falls at potential V_2; therefore the charge transfers from A to B (without considering time and space). This transfer can be thought of as a current analogous to that of Volta's battery. Therefore, Mach's analogy represents the dynamics of electric charges just as Sadi Carnot's cycle represents heat dynamics. Therefore the analogy between Carnot's heat machine and Volta's battery could advance by placing Q in place of heat, the (general) quantity $\frac{\Delta q V}{2}$, until entropy is reached, but diverging from caloric theory.

8.4 Other Kinds of Analogies

Let us examine some other analogies and attempt to determine whether they were feasible for Sadi Carnot. We now conclude the section dedicated to Mach and begin a new section by discussing Mach's interesting point of view on the crucial role played by scientific analogies in the history of science:

> The free fall of bodies first becomes familiar to us. The conceptions of force, mass, and work are transferred, with proper modifications, to electric and magnetic phenomena. A stream of water is said to have furnished to Fourier his first intuitive picture of a thermal stream. A particular case of the vibration of strings investigated by Brook Taylor explained to Fourier a particular case of the conduction of heat. Just as Daniel Bernoulli and Euler compounded the most various vibrations of a string out of Taylor's cases, so Fourier compounded the most various distributions of temperature out of simple ones analogous to Taylor's particular solution; and this method has spread over the whole of physics. Ohm modelled his notion of electric currents after Fourier's of thermal currents; then came Fick's theory of diffusion; and in an analogous way a notion of magnetic currents was developed. All kinds of stationary currents display in fact, common features; and even the state of complete equilibrium in an extensive medium shares these features with the dynamical state of equilibrium. Things so far apart as magnetic lines of force of an electric current and the stream–lines of a frictionless fluid vortex enter thus into a relation of similarity. The conception of potential, which was originally set up in a narrowly bounded domain, assumed a wide applicability. Things so unlike as pressure, temperature, and electromotive force show agreement in their relations to the conceptions derived from them in a definite way: slope of pressure, slope of temperature, slope of potential and likewise intensity of a fluid, thermal or electric current. Such a relation of systems of conceptions in which both the dissimilarity of any two homologous conceptions and the agreement in logical respects of any homologous pair of ideas corner to clear consciousness, we are accustomed to call an "analogy". It is an efficient means of subduing heterogeneous domains of facts by taking a single view. Here is clearly indicated the way in which we shall at last get a general physical phenomenology which embraces all domains and would be an exposition of physics which is free from hypotheses. The theorem of Carnot and Clausius, which was originally borrowed from a resemblance in the behavior of heat to that of a heavy liquid, may be transferred by paying attention to such analogies, to all domains of physics [...].[24]

8.4.1 On the Birth of Electricity and Thermodynamic Theories

An interesting *surrealistic story* was proposed (Fuchs 1986, 1987, 1996) as an analogy on the origins of the theory of electricity. It emphasizes that the new theory may have been conceived according to the same historical stages as the birth of thermodynamics. In this sense, the author is very clever in letting the analogies between the magnitudes in the two theories emerge naturally. Therefore, the scientists also change historical placement in the story. The research mainly deals with the working substances on a parallel–plate–capacitor.[25] We will examine parts of Fuch's *story* (Fuchs 1986), including and discussing his main *corpus* to see if some of his analogies and elements might be of interest for our purposes.

---

[24]Mach ([1896] 1986), pp 368–369, line 8. (Author's quotation marks).

[25]Let us note that the electric phenomena examined in this section are naturally different from the electrostatic phenomena presented in previous section.

Fuchs' analogy concerns internal energy E_i of an electric system that can change by adding (or removing) electric charges, and by achieving (or absorbing) *mechanical* work W.

An equation can be written:

$$dE_i = dK + dW. \tag{8.7}$$

This establishes the content of the first principle of thermodynamics. Therefore, in that surrealistically story, "Clausius" (who notably studied thermodynamics but in Fuchs' tale appears to be a scholar of electro–technique) began to study electric machines (engines) implementing calculations for an ideal parallel–plate–capacitor. Below Eqs. 8.8 and 8.9 are obtained.[26] The following is the equation of state for a parallel–plate–capacitor:

$$Fx^2 = \frac{1}{2}\varepsilon_0 A V^2 \tag{8.8}$$

In the story Eq. 8.8 conceptually corresponds to the ideal gas equation $PV = nRT$. Furthermore, in the story "Clausius" concluded that if the capacitor is no longer connected to the battery, a force F between the plates does not depend on distance x (provided that it is a short distance): distance and potential vary in such a way that F is left unaltered (Fuchs 1986, p 907). Next, according to Fuchs, "Clausius" also obtained (directly from the aforementioned observation) the expression of internal energy E_i from a parallel–plate–capacitor:

$$E_i = Fx = \frac{1}{2}\varepsilon_0 A\left(\frac{V^2}{x}\right). \tag{8.9}$$

Thus, "Clausius" determined the value of the work of a particular closed cycle of this machine whose plates can be moved short distances. The cycle was achieved through two isovoltaic phases *2* and *4* (at constant potential, for which F and x vary along a square hyperbola) and two anelectric phases *1* and *3* (in an anelectric phase, $\Delta K = 0$ and $F = const.$). In Fig. 8.9 phases *1* and *3* are anelectric, phase *2* is isovoltaic at potential V_2, and phase *4* is isovoltaic at potential V_4. The work produced by this machine, by definition as force–displacement, can by calculated as the area represented in the diagram *F–x,* force and distance (see Fig. 8.9); for Eq. 8.9 this diagram has the same function as Clapeyron's *P–V* diagram in thermodynamics.

It should be noted that from the first principle with one transfer to differential Eq. 8.9, the following expression can be obtained:

$$dE_i = \frac{\partial E}{\partial V}dV + \frac{\partial E_i}{\partial x}dx = \varepsilon_0 A\frac{V}{x}dV - \frac{1}{2}\varepsilon_0 A\frac{V^2}{x^2}dx. \tag{8.10}$$

[26]In the Eqs. 8.8 and 8.9 we exchanged the letter "*P*" used by Fuchs with the letter "*V*".

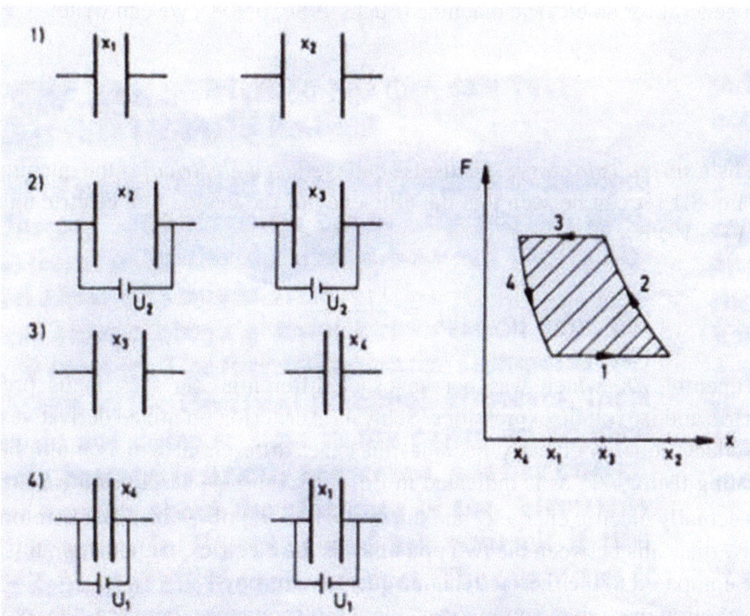

Fig. 8.9 Simplified performance model of electric cycle machine composed by two anelectric (*1* and *3*) and two isovoltaic (*2* and *4*) phases proposed by Fuchs (1986, p 908)

Moreover, it must be noted that if we include, in the second addendum of Fuchs's calculations (see Eq. 8.10), the term $dW = Fdx = \frac{1}{2}\varepsilon_0 A\frac{V^2}{x^2}dx$ (with a positive symbol) Eq. 8.7 will be:

$$\begin{aligned} dK &= \varepsilon_0 A\frac{V}{x}dV - \frac{1}{2}\varepsilon_0 A\frac{V^2}{x^2}dx - \frac{1}{2}\varepsilon_0 A\frac{V^2}{x^2}dx \\ &= \varepsilon_0 A\frac{V}{x}dV - \varepsilon_0 A\frac{V^2}{x^2}dx \end{aligned} \tag{8.11}$$

From which the following equations suggested in the story are obtained (Fuchs 1986, p 908, Eqs. 4 and 6):

For anelectric transformations: $dK = 0 \Rightarrow V \approx x$
For isovoltaic transformations: $dV = 0 \Rightarrow dK = -\varepsilon_0\frac{AV^2}{x^2}dx$

At this point, if we rethink and include Mach's reasoning (from the previous paragraph) with Fuchs's *surrealist story*, we obtain the following result for Carnot's cycle:

$$\frac{\Delta K_1}{V_1} + \frac{\Delta K_2}{V_2} = 0; \tag{8.12}$$

and in general for an electric machine (Fuchs 1986, p 908), we can write:

$$\Delta\sigma = \oint \frac{\Delta K}{V} = 0. \tag{8.13}$$

In Fuchs's *surrealistic story*, "Clausius" calls *reduced electricities* the quantity $\frac{dK}{V}$. From Eq. 8.13 it can be seen that the efficiency of the cycle of an electric machine (with $V_2 > V_1$) is:

$$\eta_E = \frac{V_2 - V_1}{V_2}. \tag{8.14}$$

Consequently, dK which was not an exact differential (in fact, in its previous anelectric and isovoltaic expressions, Schwarz's criterion on mixed derivatives does not produce equality) clearly becomes an exact differential if it is multiplied by integrating factor *1/V*, as is indicated in Eq. 8.13; thus $\frac{dK}{V}$ has a primitive function that is actually electric charge Q and the cyclic integral of $\frac{dK}{V}$ is null. Therefore, the quantity that falls between the two potentials is the electric charge that, in this case, can be compared to the thermodynamic quantity entropy $\frac{dQ}{T}$.

The following table summarizes the correspondence between the physical quantities of thermodynamics and of electricity according to Fuchs' story.

Table 8.7 The thermodynamic physical magnitudes and their correspondence in electrostatics according to Fuchs' story

Thermodynamics	Electrostatics
T_1, T_2	V_1, V_2
$\Delta S = \oint \frac{dQ}{T}$ (Entropy)	$\Delta\sigma = \oint \frac{dK}{V}$ ("reduced electricity")
$W_T = p\Delta V$	$W_E = \frac{1}{2}\varepsilon_0 E_0^2 Ax$
$W_T \Leftrightarrow Q$ (equivalence work–heat)	$W_T \Leftrightarrow K$ (equivalence work–electricity)
$dU = dQ + dW$	$dE_i = dK + dW$
$PV = nRT$	$Fx^2 = \frac{1}{2}\varepsilon_0 AV^2$
$\eta_T = \frac{W}{Q} = \left(\frac{T_2 - T_1}{T_2}\right)$ (Efficiency < *100%*)	$\eta_E = \left(\frac{V_2 - V_1}{V_2}\right)$ (Efficiency < *100%*)
Reversibility of heat machines	Reversibility of electric circuits
Irreversibility of heat machines	No

It should be noted that in both of the previous cases we saw a physical object that had an expansive force (or for example repulsion between charges), which is regulated by the entry of material (electric charges); just as in thermodynamics, where the physical system is characterized by an expansive force of gas, which is regulated by the entry of heat. The comparison of Tables 8.6 and 8.7 allow for several reflections on their affinity, e.g.:

- How could a person have conceived the analogy between heat Q and $\frac{\Delta qV}{2}$ during Sadi Carnot's lifetime?

- Would it be necessary to understand whether or not Mach's formulas came before or after Sadi Carnot's work?
- Furthermore, did they come before or after the invention of the parallel–plane capacitor?

8.4.2 On the Caloric and Entropy

We will now examine another analogy (Schmid 1984). The formulation of thermodynamics proposed by Schmid is based on an analogy between the concepts of entropy and concept of caloric. In an article from 1911, Callendar claimed:

> Finally, in 1865, when its importance was more fully recognised, Clausius [...] gave it the name of "entropy," and defined it as the integral of dQ/T. Such a definition appeals to the mathematician only. In justice to Carnot, it should be called caloric, and defined directly by his equation $W = AQ(T - T_0)$, which any schoolboy could understand. Even the mathematician would gain by thinking of caloric as a fluid, like electricity, capable of being generated by friction or other irreversible processes.[27]

The analogy is obtained by reformulating all of thermodynamics as a consistent theory of two fluids: *S* (entropy) and *E* (energy) (Schmid 1984, pp 794–799; see also Falk et al. 1983; Falk 1985). Below, we provide the schematization of the electric machine according to Schmid's ideas. This is a global vision of the entering and exiting of energy fluids and entropy from a heat machine (Fig. 8.10). It is clear that in this way, the time variable *t*, ignored by Sadi Carnot, is introduced as an essential element:

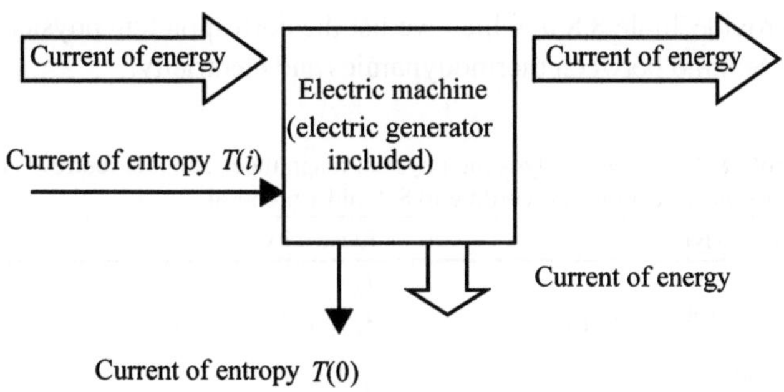

Fig. 8.10 Simplified performance model of Carnot's cycle by Schmid (Adapted by Schmid 1984 p 794)

Here, Schmid uses entropy (seen, in his opinion, as Sadi Carnot's caloric at $T = const.$) for the calculation of efficiency η, and in doing so includes the first principle of thermodynamics in his theory. He argues that the most important quantities of quantum mechanics are energy, quantity of motion, angular momentum, the electric charge and the cell: these are different from the concepts of the principles in

[27] Callendar (1910), p 1. (Author's quotation marks).

mechanical paradigm in the history, whose mass and quantity describe the trajectory and are instead similar to the extensive quantities of thermodynamics.[28] These quantities, according to Schmid, have something in common conceptually; each of them can be thought of as being both contained in a physical system, and in the flux from one system to another. Therefore each of them can be imagined as a type of "fluid". This is the reason that pushed Schmid to define these quantities as "substance–like" (Schmid 1984, p 794). Among them, entropy is considered the typical quantity of thermodynamics. On this subject, Schmid wrote:

> The idea that a physical quantity is substance–like first appeared historically in connection with quantities like mass and electric charge. In the phenomena of heat, two substance–like quantities are manifest: energy and entropy. Each is contained in a material body. It even makes sense to speak about the amount of energy E or entropy S contained within an arbitrary region of space. If a body or in general, a region of space filled homogeneously with energy and entropy is divided in two each contains half the original total amount of E or S. Energy obeys a general conservation law (the First Law of thermodynamics). Entropy obeys a general law forbidding its destruction (the Second Law of thermodynamics). These two statements mean (1) the amount of energy contained within a body or, in general, within a region of space can only increase (decrease) if energy flows into (out of) the body or region in question; and (2) the amount of entropy contained within a body or, in general, within a region of space can only increase (decrease) if entropy flows out of the body or region. The amount of energy can increase, however, either via the inflow of entropy into the body or region in question or by the local creation of entropy. When we say a quantity A is substance–like, we mean several things at once, most of which are generally familiar from the traditional description of electric charge. (1) A spatial density ρ_A exists such that the amount of A contained in an arbitrary region R of space is given by the volume integral $A = \int_R \rho_A dV$ [...].[29]

Below, we provide Table 8.8 in which we list the corresponding physical quantities proposed by Schmid between thermodynamics and electricity:

Table 8.8 The thermodynamic physical magnitudes and their correspondence in electrostatics according to Schmid's proposal

Thermodynamics	Electricity
Q	I_E
$\Delta S = \oint \frac{dQ}{T}$ (Entropy)	$\frac{d}{dt}A + I_A = \sum_i A_i$
Energy E	$\frac{d}{dt}A + I_A = 0$
Reversibility of heat machines	Reversibility of electric circuits
Irreversibility of heat machines	No

Where:

[28]It should also be noted that energy, quantity of motion and angular motion also appear in mechanics; but in this case only as convenient instruments for calculation, never as fundamental concepts; in fact in this theory the fundamental quantities are trajectory, velocity, mass and force; where the quantity of motion is just another name to indicate the product of velocity times mass and energy as a certain constant of motion.

[29]Schmid (1984), pp 794–795 (*Author's italics,* numbers in running text and capital letters).

(a) ρ_A = the *spatial density* of a quantity A, within an arbitrary *region R* of space. A is given by the *volume integral*, $A = \int_R \rho_A dV$. dV is the *volume element* of R.[30]
(b) I_A = *flux of A* (the intensity of the current of A) through an oriented surface F is given by the integral of the *current density* j_A on the entire surface $I_A = \int\limits_F j_A df$. df is the *volume element* of F.[31]
(c) $\frac{d}{dt}A + I_A = \sum_i A_i =$ is the rate at which the quantity A and is created (*negative creation* = *destruction*) in a given region R. It is a continuity equation and expresses the conservation of the quantity of charge A. It comes from the equation $\frac{d}{dt}\rho_A + \nabla J_A = \sigma_A$ where σ_A is the *source density* of the quantity A.[32]

For our purposes, let us see the validity of the concept of reversibility in Schmid's analogy. In fact, while reversibility is a crucial *discovery* in Sadi Carnot's theory, Schmid in his scientific teaching aims, *a priori* presupposed reversibility and (correctly) wrote the following formula

$$I_E = TI_S$$[33]

Here the *entropy current* I_S is always accompanied by an *energy current* I_E. In this formula, the term T is the absolute quantity temperature. It concerns how much energy is transported by *entropy current* I_S, or how much *energy current* is that which is charged to the *entropy current*. The equation $I_E = TI_S$ assumes a more familiar appearance and we consider currents I_E and I_S in terms of variations *dE/dt* and *dS/dt* which express how the quantities of energy E and entropy S are contained within a system which change in time. In agreement with previously listed (a), (b), (c) and with Schmid's formula (Schmid 1984, formula (5), p 795), the following quantities can be written:

$$\begin{aligned} I_E &= -\frac{dE}{dt} \\ I_S &= -\frac{dS}{dt}. \end{aligned} \tag{8.15}$$

And revisiting Schmid's analogy, they can be also written as:

$$dE = TdS. \tag{8.16}$$

They are better known as Josiah Willard Gibbs's (1839–1903) fundamental formula for a system in which all of the extensive variables (firstly V) are kept constant, except S . By using Eq. 8.15 Schmid uses it as if it expresses the content of the first principle. With it, he then attempts to construct an analogy between electricity and the mechanical theory of heat. It should be noted, however, that Eq. 8.16 (proposed

---

[30] Schmid (1984), p 795.

[31] *Ibidem.*

[32] *Ibidem,* respectively formulas (3) and (4).

[33] *Ibidem*, line 6, formula (5).

by Schmid as the expression of the analogous first principle) partially includes the content of the first principle of thermodynamics since $W = pdV$ *is* lacking.[34] Therefore, as announced at the beginning of the paragraph, it (such as a derivative) indicates at most a local reversibility. In other words, from a physics standpoint, it can only express local information on the variation in a point of a determined phenomenon; and not information on the global variation of a given physical system which expresses the reversibility of heat (or mechanical) *machines en général*. Thus, in the comparison between caloric and entropy (if we truly have historical facts to establish that in *Réflexions sur la puissance motrice du feu*, caloric at a constant temperature might be defined as sort of Carnot's entropy), T and Q/T appear as quantities given *a priori*. Instead, Sadi Carnot, especially in the discursive part of *Réflexions sur la puissance motrice du feu* (see Chapter 7) works to define them. In addition, in this analogy, entropy is linked to energy, which Sadi Carnot (1978) does not consider; at most, W_{max} is considered.

In the end, for his reasonable scientific teaching aims, Schmid attempts to mathematically quantify his arguments by adopting *a priori* the first principle of thermodynamics, which inevitably leads him to the obvious physical–mathematical consequences that emerge when the concept (evenonly intuitively) of caloric is adopted in classical thermodynamics. In this sense, with respect to Mach's and Fuchs' analogies, the introduction of energy and entropy concepts in Sadi Carnot's work appear historically inadequate; however, they are very interesting for better defining its foundations.

8.5 Final Remarks

The analyses proposed in the first paragraphs and suggested by Mach, Fuchs and Schmid, explicitly adopt the equivalence of work and heat and address the analogy between entropy and electric charges. Now, if we shift our focus to producing history, we can note that the first analogy proposed may have been thought of (maybe?) by Sadi Carnot as:

$$\oint \frac{dQ}{T} = \Delta S \tag{8.17}$$

while the second (in hindsight) could have been seen as

$$dq = Idt. \tag{8.18}$$

Returning to our way of producing history, it must be stressed that Eq. 8.18, however, is not part of Sadi Carnot's scientific *corpus*.

Below, we provide Table 8.9, which compares analogies proposed by the authors cited in this section:

[34] In Schmid's correct interpretation and calculus that term was not strictly necessary. Instead here we try to take into consideration what is interesting for us.

Table 8.9 Comparative analysis between thermodynamics, electrostatics and electricity

Concepts	Mach (Es) 1896	Schmid (Ey) 1984	Fuchs (Es) 1986	Pisano (Ey) 2003, 2012b
	Interpretation of SCc in Es	Analogies: E and MTH	Analogies: Es and Thm	Analogies: A cyclical path in S. Carnot and in Volta
Q Heat	$\Delta q/2$ Es charges	$I_E(i), I_E(o), A$ FCE "substance–like"	E (Es charges)	$I = dq/dt$ Current in voltaic battery
Reservoirs T_1, T_2	V_1, V_2	$T(i)$, $T(o)$	V_1, V_2	V_1, V_2
$W = p\Delta V$ Work	$W = \frac{\Delta q(V_2 - V_1)}{2}$	$I_E(i) - I_E(o) =$ $= \frac{T(i)-T(o)}{T(i)} T(i) I_S(i) +$ $- T(o) I_S(produced)$	$W = \frac{1}{2}\varepsilon_0 E_0^2 Ax$	$W = q\Delta V =$ $= \Delta VIt$
Heat Machine	Metallic spheres	Electric engine	Parallel–plate *capacitor*	Voltaic battery
Efficiency of heat machine	$\eta_{Mach} = \frac{V_1 - V_2}{V_1}$	$\eta_{Schmid} = \frac{T(i)-T(o)}{T(i)} +$ $- \frac{T(o) I_S(produced)}{T(i) I_S(i)}$	$\eta_{Fuchs} = \frac{V_2 - V_1}{V_2}$	$\eta = \frac{W_{max}}{Q}$ [?]
$\Delta S = \oint \frac{dQ}{T}$ Entropy	$\Delta qV/2$	$\frac{d}{dt}A + I_A = \sum_i A_i$	$\Delta\sigma = \oint \frac{dK}{V}$ "Reduced electricity"	?
Energy	No	$\frac{d}{dt}A + I_A = 0$	$dE = dK + dW$	No
I–PT	Yes	Yes	Yes	$\approx$ Yes
II–PT	Yes	Yes	Yes	Yes
Cycle	Yes	No	Yes	Yes

Legend: *Ey* electricity, *Es* electrostatics, *SCc* Sadi Carnot's cycle, *MTH* mechanical theory of heat, *FCE* flux of current of energy, *Thm* thermodynamics, *I–PT* use of the first principle of thermodynamics, *II–PT* use of second principle of thermodynamics

Thus, a disparity in the correspondence between thermodynamic quantities and various electric quantities should be noted.

In conclusion, one can also claim, that Sadi Carnot, given the weakness of (formalized) knowledge of electric phenomena at his time, might not have been fully able to apply his (and his father's) idea of a comparison between states in the electric cycle of the voltaic battery and the four phases in a heat cycle, while trying to stress a possible analogy between two different domains of physics. On the other hand, the analogy might have provided him with input to use the concept of the cycle as both an instrument of intellectual reasoning and an instrument of calculus alternative to differential equations. However, it might not have allowed him to draw any *direct* conclusion at all, e.g., regarding the independence of the working substance or more generally, the upper bound of the efficiency related to heat machines. However, an extensive and specific historical investigation (1799–1824) on Volta–Sadi Carnot is not yet complete (Pisano 2012b).

Chapter 9
A Historical Epistemology of Thermodynamics. The Mathematics in Sadi Carnot's Theory

Si l'on admettait la constance de la chaleur spécifique d'un gaz lorsque son volume ne change pas, mais que sa température varie, l'analyses pourrait conduire à une relation entre puissance motrice et le degré thermométrique. Nous allons faire voir de quelle manière, cela nous donnera d'ailleurs occasion de montrer comment quelques unes des propositions établies ci–dessus doivent être énoncés en langage algébrique.

(Carnot 1978, pp 73–74, ft 1, line 1; s. transl. below)

The mathematical footnote in *Réflexions sur la puissance motrice du feu* can be considered the highest mathematical level presented by Sadi Carnot in his book. It also represents his way of mathematically interpreting physical phenomena and the relationship between physics and mathematics in thermodynamics. Without proposing new concepts or sophisticated functions in trying to explain *the modernity* of the result obtained by Sadi Carnot, here we will present an historical analysis of the footnote. Moreover, strictly based on primary sources and documents, we will provide an epistemological interpretation of the arguments that Sadi Carnot used to calculate the formula of the efficiency of a heat machine; and centring on the role of shared knowledge, of challenging objects, and of knowledge reorganization. The idea is presenting an integrated history and epistemology of scientific methods which combine epistemological and historical approaches to identify significant historical hypotheses. Such epistemological interpretations are subjected to historical facts of scientific activity and original documents in order to trace their historical development. In this case the discussion regards Carnot's mathematical foundations, considering both the relationship between physics and mathematics and the mathematical (infinitesimal) analysis of the nineteenth century.

C.C. Gillispie and R. Pisano, *Lazare and Sadi Carnot: A Scientific and Filial Relationship*, History of Mechanism and Machine Science 19, DOI 10.1007/978-94-017-8011-7_9,

9.1 An Outline of the Problem

The mathematical footnote[1] in *Réflexions sur la puissance motrice du feu* (Carnot 1978, ft 1, pp 73–79) has been a subject of inquiry for the leading Carnot historians. Some of them sought very sophisticated concepts and functions explaining *the modernity* of the result obtained by Sadi Carnot. A discussion related to the use of the two isotherms without considering the two adiabatic transformations in his complete cycle continues to be a crucial point of his theory. However, *why did Sadi Carnot disregard the two adiabatics in his reversible cycle without explanation?*

> In p. 30, some conclusions drawn by Carnot from his general reasoning were noticed; according to which it appears, that if the value of μ for any temperature is known, certain information may be derived with reference to the saturated vapor of any liquid whatever, and, with reference to any gaseous mass, without the necessity of experimenting upon the specific medium considered. Nothing in the whole range of Natural Philosophy is more remarkable than the establishment of general laws by such a process of reasoning.[2]

It should be noted that in order to obtain η, the calculation of work *W* and heat *Q* on each of the four thermodynamics processes of his cycle (in order to get W_{tot}/Q_{tot}) would not have led him to the result, as his adiabatic equation was wrong. According to a method derived from his father, Lazare Carnot (1813, pp 217–253; see also Drago and Pisano 2007), the *synthetic method*, it is conceivable that he legitimized the *elimination of the adiabatic* within the calculation of the efficiency in his theory. In another work one of us (Drago and Pisano 2005) showed that Sadi Carnot intended the adiabatic as auxiliary variables to "supprimer" (Carnot 1978 p 39, line 9; *ibidem*, line 15). In *Réflexions sur la puissance motrice du feu* Sadi Carnot only uses mathematical calculus of the 19th century within the mathematical footnote (*Ivi*, ft 1, pp 73–79). The footnote summarizes a considerable part of the discursive reasoning in the first part of Sadi Carnot's book.

Finally, we will outline another important difficulty. It regards the role played by experimental data[3] (sometimes uncertain) at a time during which Sadi Carnot used his mathematical reasoning (Carnot 1978, ft 1, pp 73–79). We mainly mention Delaroche and Bérard's studies (Delaroche and Bérard 1813), Clément and Desormes' experiments (Clément and Desormes [1819] 1905), and Gay–Lussac and Welter's observations (Laplace (1822), p 268; see also Laplace 1822, pp 266–268; Poisson 1823a, b). In detail we can propose the following resume:

Delaroche and Bérard: observations on specific heat ("chaleur").
Sadi Carnot's errors–doubts: (Carnot 1978, p 58, 68, pp 72–79).

---

[1]The research presented is based on Drago and Pisano (2000, 2005, 2007), Pisano (2001, 2012a).

[2]Thomson read it on 30 April, 1849. The content is discussed in an Appendix by the author (Thomson [1890] 1943, p 179, line 3; see also: *Id.*, 1852, XXXV, 1882–1911, I, pp 113–155; et al. editions).

[3]A meaningful study on the experimental data used by Sadi Carnot is available in Fox (1988), pp 288–299.

Clément and Desormes[4]: experiments on full heat ("chaleur totale")
Sadi Carnot's errors–doubts (*Réflexions sur la puissance motrice du feu*: Carnot 1878a, ff 65–66; Carnot 1978, p 59, pp 123–124;).
Gay–Lussac and Welter: observations on the constancy[5] of $\gamma = C_p/C_V$.
Sadi Carnot errors–doubts: he considered $\gamma = \left(C_p - C_V\right)$ *which would have produced non–constancy of caloric* (Carnot 1978, pp 42–53, p 59).
*In the manuscript Notes sur les mathématiques, la physique et autres sujets presented (*1876b*) by Hyppolite at Académie des sciences Sadi Carnot suggested the conservation of energy – "puissance mécanique"* – (Carnot 1878a, folio 7v; Carnot 1878b,[6] p 95; Carnot 1953, pp 134–135; Carnot 1978, p 248; Picard pp 81–82).

9.2 A Book of Physics Without Mathematics?

Carnot's friends (although they were students at *École polytechnique, Paris*) considered his book a difficult work to understand. In fact, it contained astonishing news: the idea of the cycle, a limit to the efficiency of heat machines, a theorem demonstrated by an *ad absurdum* proof, and new gas laws. The work seemed even more difficult to interpret due to Sadi Carnot's surprising obtainment of results which were almost entirely accurate, despite his relying alternatively either on caloric theory or on heat theory. Moreover,

> The book is particularly rich in what have been generally supposed to be the discoveries of later writers, and in enunciations of principles now recognized as those forming the base and the supporting framework of that latest of the sciences. As stated by Tait[7] in his history of Thermodynamics, the "two grand things" which Carnot originated and introduced were his idea of a "cycle" and the notion of its "reversibility," when perfect. "Without this work of Carnot, the modern theory of energy, and especially that branch of it which is at present by far the most important in practice, the dynamical theory of heat, could not have attained its now enormous development". These conceptions, original with our author, have been, in the hands of his successors, Clausius and other Continental writers, particularly, most fruitful of interesting and important results; and Clapeyron's happy thought of so employing the Watt diagram of energy as to render them easy of comprehension has proved a valuable aid in this direction.[8]

[4]In 1827 César–Mansuète Despretz (1798?–1863) showed the uncertainness of the Clément and Desormes law (Fox 1988, ft 26, p 301).

[5]Laplace resumed Gay–Lussac and Welter's experiments in a part of his work. In particular, he properly focused on the constancy of γ (Laplace 1822, ft 18, p 268).

[6]We let note that the manuscripts edited by Gauthier–Villars (Carnot 1878b) is not always integrally reproduced. For any complete English consulting of the Sadi Carnot's manuscripts, please see Fox (1986).

[7]Tait (1877).

[8]Carnot (1943), p 6, line 27. (Author's wording and quotation marks).

Réflexions sur la puissance motrice du feu and the last consideration concerning the theory of gases (Carnot 1978, pp 223–225) appeared in a brief time frame, less than 3 years. Some reviews considered Sadi Carnot's thermodynamics too practical and therefore less valuable for theoretical physics in general; that was also the opinion of Sadi Carnot's contemporaries, who were accustomed to infinitesimal analysis. Carnot's book was not sufficiently comprehensive because it lacked the basic concepts for a nineteenth century scientific theory: e.g., absolute space and time, cause–force, differential equations, material–point. Moreover, in 1873 it was revealed that in some *posthumous notes*, Carnot was in favor of the mechanical theory of heat (Callen 1974, pp 423–40; Callendar 1910) and maintained that he had even obtained the value of the mechanical equivalent of heat several years before Mayer's work[9] (Mayer 1842a, b). In spite of all this, Carnot's theory has been underrated by science historians for over a century, essentially because his theory was accused of being based on caloric theory, which is considered incorrect in modern times. Other studies, on the other hand, considered the young French scientist's thermodynamics fascinating, for the simple kind of language used as well as for the air of *mystery* surrounding the lack of mathematical calculations in this work.

> This helps to explain why the argument has none of the elaborate mathematics that Carnot's cotemporaries were accustomed to finding in the work of such men as Laplace, Poisson and Fourier, all of whom wrote on the theory of heat, as mathematical physicists, between 1800 and the 1830s. [...]. And yet the simplicity of the book is snare. For although this *Réflexions* lacks the mathematical sophistication of much of the physics of the period, its unassuming presentation conceals an argument of extraordinary depth and, in parts, of baffling complexity. It is for this reason that a lengthy discussion text seems indispensable.[10]

We should note that Sadi Carnot also tried to provide a general theory that could be applied to any machine and could work with any fluid. The theoretical physics content in such a theory is actually very high, going far beyond the engineering study of machines. Sadi Carnot built a theoretical model based on state functions, which helped him reach some important theoretical conclusions on gas laws, known at the time only through empirical facts. In the 1970s and 1980s, Truesdell[11] and Hoyer[12] tried to interpret the foundations of thermodynamics with this new historical–foundational approach.

---

[9]The estimate value proposed by Sadi Carnot in his *Notes sur les mathématiques, la physique et autres sujets* (folio 7v; see also Carnot 1878b, pp 94–95; Fox 1986, pp 191–192) for the mechanical equivalent of heat of 370 Kg/cal is not so much farther from the correct accepted value (427) than Mayer's later determination (365). The difference is only 11%. (See below Chapter 10; Hoyer 1975; Edmunds 1902, p 127).

[10]Robert Fox in Carnot (1986), p 2, line 9. (Author's *italic*).

[11]Truesdell (1970), *Id.*, (1980), pp 208–235.

[12]Hoyer in Taton (1976), pp 221–228. He showed, by using the cycles, that S. Carnot's caloric theory and the modern theory, coincide for an infinitesimal process: that is $\Delta t \rightarrow 0$. That is correct in the first and second approximation in thermodynamic theory.

As just mentioned (see above Chapter 7), Philip Lervig (1972, 1976) produced an interesting contribution to the debate and interpretation of Sadi Carnot's work. Particularly, regarding mathematics and Sadi Carnot's footnote, Lervig argued that, to make Sadi Carnot's footnote more understandable, it is necessary to examine it in general terms, considering the footnote in the context of a Maxwell relation on state variables (see below). In regard to Sadi Carnot's mathematical footnote, we will also indicate where Lervig[13] provides a personal interpretation. Moreover, Lervig performs the calculations of the footnote (without discussing the role of the cycle and the "suppression" of the two adiabatics) by these two functions (ϑ and A) in an attempt to imitate modern thermodynamics, given as an inevitable end point. He does in fact find, as a test, an example of one of Maxwell's relations for thermodynamics. He applies it to the calculation of c_V at the end of the footnote, reaching the same result obtained by Sadi Carnot. However, this does not seem to reveal hidden aspects of Carnot's thoughts (he was already working with state functions). We should also observe that Reech[14] (1853, XVIII, pp 357–378) had already obtained, by the calculations of the cycles, the internal energy function (and therefore potentially the Maxwell relations; see Chapter 10; Pisano 2012b). In 1977, Truesdell and Bharatha (1977) proposed the distinction between general and special axioms in Carnot's theory. Regarding the mathematical footnote, they discuss the non–linearity of the efficiency function in Q:

$$W = G(\theta^+, \theta^-, Q) \stackrel{?}{=} W = G(\theta^+, \theta^-)Q.$$

9.3 The Calculations and Historical Interpretation of the Mathematical Footnote

An historical epistemology reconstruction of the mathematical footnote (Carnot 1978, ft 1, pp 73–79) is presented (Fig. 9.1). Without listing all of the original mathematical passages expressed by Sadi Carnot in his footnote, we summarize their content in eight crucial steps assembled in the following Tables 9.1 and 9.2.

---

[13]We will use a modern notation in comparison to that used by Lervig.

[14]We will also discuss in the following chapters the details proposed by Reech on Carnot's theory.

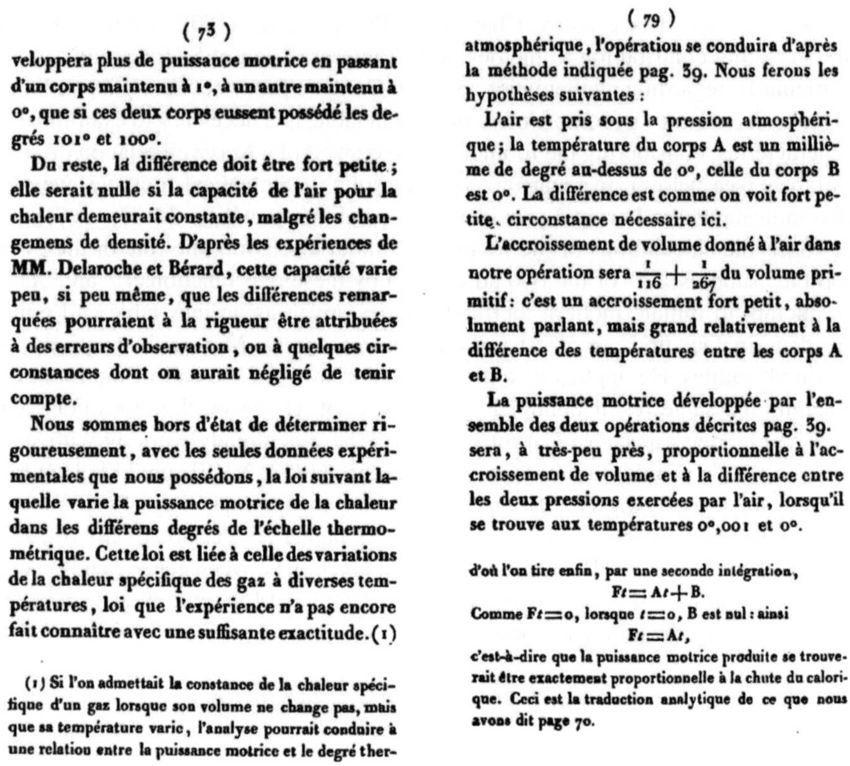

(73)

veloppera plus de puissance motrice en passant d'un corps maintenu à 1°, à un autre maintenu à 0°, que si ces deux corps eussent possédé les degrés 101° et 100°.

Du reste, la différence doit être fort petite; elle serait nulle si la capacité de l'air pour la chaleur demeurait constante, malgré les changemens de densité. D'après les expériences de MM. Delaroche et Bérard, cette capacité varie peu, si peu même, que les différences remarquées pourraient à la rigueur être attribuées à des erreurs d'observation, ou à quelques circonstances dont on aurait négligé de tenir compte.

Nous sommes hors d'état de déterminer rigoureusement, avec les seules données expérimentales que nous possédons, la loi suivant laquelle varie la puissance motrice de la chaleur dans les différens degrés de l'échelle thermométrique. Cette loi est liée à celle des variations de la chaleur spécifique des gaz à diverses températures, loi que l'expérience n'a pas encore fait connaître avec une suffisante exactitude. (1)

(1) Si l'on admettait la constance de la chaleur spécifique d'un gaz lorsque son volume ne change pas, mais que sa température varie, l'analyse pourrait conduire à une relation entre la puissance motrice et le degré ther-

(79)

atmosphérique, l'opération se conduira d'après la méthode indiquée pag. 39. Nous ferons les hypothèses suivantes :

L'air est pris sous la pression atmosphérique; la température du corps A est un millième de degré au-dessus de 0°, celle du corps B est 0°. La différence est comme on voit fort petite, circonstance nécessaire ici.

L'accroissement de volume donné à l'air dans notre opération sera $\frac{1}{116} + \frac{1}{267}$ du volume primitif: c'est un accroissement fort petit, absolument parlant, mais grand relativement à la différence des températures entre les corps A et B.

La puissance motrice développée par l'ensemble des deux opérations décrites pag. 39. sera, à très-peu près, proportionnelle à l'accroissement de volume et à la différence entre les deux pressions exercées par l'air, lorsqu'il se trouve aux températures 0°,001 et 0°.

d'où l'on tire enfin, par une seconde intégration,

$$Ft = At + B.$$

Comme $Ft = 0$, lorsque $t = 0$, B est nul : ainsi

$$Ft = At,$$

c'est-à-dire que la puissance motrice produite se trouverait être exactement proportionnelle à la chute du calorique. Ceci est la traduction analytique de ce que nous avons dit page 70.

Fig. 9.1 The first and last page of the mathematical footnote (Carnot 1824, p 73, 79, ft 1. *Idem* pages are presented in all of editions)

In order to explain mathematical arguments, we adopt modern notation. Our additional and summarized interpretative notes are in italics or in square brackets:

Table 9.1 Sadi Carnot's notation and their correspondences in modern theory

Sadi Carnot's main statement (Carnot 1978)	Carnot's notation	Modern notation
"$N = p/267$ [...] constant" (*Ivi*, p 74)	N	R Constant of gas
"[...] la quantité du puissance motrice produite [...]" (*Ibidem*)	r	W work
"$u = F(t)$ [...] la puissance motrice développée [...]" (*Ivi*, p 75)	u	η efficiency
"[...] quantité de chaleur[a] [...]" (*Ivi*, p 76)	e	Q heat

[a]Please take into account our *errate corrige* for the use of *chaleur* and *calorique* as reported in Chapter 7

Table 9.2 The main calculations in Sadi Carnot's mathematical footnote

Step	Contents and calculus	References (Carnot 1978)
1	*Aim*: "If we were to assume that the specific heat of a gas remains unchanged when its volume is kept constant [$c_v(t) = const.$] [...] it would be possible to derive a relationship between motive power and the temperature" (The efficiency of a reversible heat machine): $\eta = W_{[max]}/Q = f(t_1, t_2)]$	*Ivi*, p 73, line 1
2	Let Q be a quantity of heat on isotherm t, the calculations of the work W between unitary volume and V be: $W_t = N\,(t + 267)\log V + c$ [and also $Q_t = A_t + B_t \log V$]	*Ivi*, pp 74–75
3	Calculations of W on isotherm $t + dt$, between unitary volume and V: $W_{t+dt} = N\,(t + 267 + dt)\log V + c$ $\left[\text{and also } Q_{t+dt} = A_{t+dt} + B_{t+dt} log V\right]$; [*By compensating the two infinitesimal adiabatics by means of the synthetic method*] we have: $dW_{max} = R log V dt$	*Ivi*, p 75
4	*On his Theorem:* let is *Q (=1)*, that (*by means of calorique theory*) propagates between two isotherms t and $t + dt$, we have: $d\eta = \eta_{t+dt} - \eta_t = F(t)dt$	*Ivi*, pp 75–76
5	[*Definition of* η *for a quantity of heat* Q ($\neq 1$) *between two isotherms* t *and* $t + dt$]. Equalizing $Q d\eta$ and dW_{max} [*factorization of* Q; see step 3], we have: $Q = \tau(t)\log V$ with $\tau(t) = N/F$	*Ivi*, p 76
6	Calculations of C_V by three processes which produce Q, K, q_0.[1° Sadi Carnot's inaccuracy: he writes $c_V(t) = dq/dt$ with *dq/dt, instead* of dK/dt]. [Wrong] formula for $c_V(t)$: $c_V(t) = dq/dt = \tau_t'(t)\log V_t + q'$ $[c_V(t) = (dK/dt)_V = K_t' - \tau_t'(t)\log V_t]$	*Ivi*, p 77
7	Hypothesis: $c_{V_t}(t) = q_t' - \tau_t'(t)\log V_t = \text{constant}$ for all temperatures; then: $\tau_t' = A_t$ with A_t constant in t. [*He makes calculations and obtains*]: $\eta = \frac{R}{c}\ln(A_t t + B_t) + c'$ and $\tau = A_t t + B_t$	*Ivi*, pp 77–78
8	[2° Sadi Carnot's inaccuracy: *he again confuses* dK/dt with $c_P(t) = dq/dt$]. By integrating: $\tau = c = const.$ From the definition of τ, we have: $d\eta = \frac{Ndt}{c} = \lambda dt$. Thus $\eta = \int \lambda dt = \lambda t + \text{b}.$ $\eta_t = A_t t''$ "In other words, the motive power produced is shown to be exactly proportional to the fall of the caloric[a]"	*Ivi*, pp 78; *Ivi*, p 79, line 5

[a]"C'est–à–dire que la puissance motrice produite se trouverait être exactement è proportionnelle à la chute du calorique" (Carnot 1978, p 79, ft 1, line 5)

In the following section, we provide an interpretation of the footnote by discussing the arguments as indicated in Table 9.2. Our additional interpretations and considerations on Sadi Carnot's procedure are in *italics*. Sadi Carnot's goal is to utilize what was at the time a bare hypothesis, the constancy of specific heat, to obtain a decisive result based on the efficiency function.

Step 1. The footnote begins with the following lines:

> If we were to assume that the specific heat of a gas remains unchanged when its volume is kept constant [$c_v(t) = \text{const.}$], but its temperature varies, it would be possible to derive a relationship between motive power and the temperature.[15]

For this reason, even the final result (Carnot 1978, p 79, ft 1) *is valid according to the aforementioned hypothesis. Therefore in the end he could not be sure of the result, however interesting, which we now know to be valid* (*Ivi* p 78, ft 1, line 10).

Step 2. Sadi Carnot calculates work W_t for a certain quantity Q of heat between V and $V = 1$ and for a specific value t of temperature. With regard to this process the work W appears to be the following function: $W_t = Wt(V)$.

Step 3. Sadi Carnot calculates the increment dW from isotherm t to isotherm $t + dt$. *We remark that the calculation makes sense if the two isotherms are intended as part of a cycle. Let us also note that if – as Sadi Carnot did and justified (Ivi,* pp 38–40*) – the two adiabatics are considered, one should be added and the other should be eliminated such as in an algebraic calculations between two opposites terms. Here an adiabatic process is the variable ε (added and then eliminated) of his father's synthetic method. Thus, for a complete cycle in four phases, between t and t + d, he can write:*

$$dW_{max} = Rlog\, V\, dt.$$

Step 4. Only by taking into account an infinitesimal cycle, whose work is done by the two isotherms alone is it possible to justify that in this step Sadi Carnot is able to consider the efficiency function η for a *unitary quantity Q of heat*

$$\left(\textit{which in general is} : \eta = \frac{W_{\max}}{Q} \right).$$

He then calculates the $d\eta$ for a reversible heat machine, operating between the isotherms t and $t + dt$. *This allows him to introduce his theorem*:

> The motive power of heat [chaleur] is independent of the working substances that are used to develop it; its quantity is determined exclusively by the temperatures of the bodies between which, at the end of the process, the passage caloric [calorique] takes place.[16]

---

[15]Carnot (1978), p 73, ft 1, line 1.

[16]Carnot (1978), p 38, line 4. (Author's *italic*).

which is fundamental for the entirety of the follow–up in the footnote: the efficiency of a reversible thermal machine is independent of the working substance used. It only depends on the thermal jump, e.g.,

$$\eta = F(t).$$

Step 5. Sadi Carnot *calculates the maximum work for a quantity of heat* $Q \neq 1$. *Here lies a further hypothesis, which is, from a strictly mathematical point of view, not justified: in the expression of* W_{max} *the function of state* Q *is factorizable*:

$$W_{\max} = g(Q, \Delta t) = Q\eta(\Delta t).$$

This aspect of the mathematical footnote was discussed by previous scholars; among them we note Truesdell and Bharatha (1977, pp 61–63), Montrbrial (1976 in Taton, p 335), Cropper (1987), and Dias & C (Cropper 1987) who did, however, provide justifications concerning previous problem. In our analysis, we hypothesize that Sadi Carnot utilized factorization in order to facilitate his physical reasoning as a theorist usually does. He begins his argument by proposing $Q = 1$. *Next, we hypothesize that the idea, taken from his father (Carnot 1780, p 153; see also 1786 and Gillispie 1971, Appendix C, pp 330–331), of the efficiency as a relation input (heat)–output (work) of a machine, naturally led him to the calculation of the relation* $\eta = W/Q$, *considered as a quantity independent of* Q. *Moreover, Sadi Carnot was reasoning on a cycle which brought all the state variables back to their initial states, included the quantity* Q *(which, according to him, was caloric).* W *was the unique variable not considered to be cyclic. Therefore, it was on this variable alone that he had to execute the calculations in relation to* Q *which was considered a state function. It is necessary to specify that he could have written* $\eta = F(\Delta t)$. *Nevertheless, he set the initial temperature to 0, so his notation F(t) does not provide problems.* Thus, Sadi Carnot calculates

$$\eta = \frac{W_{\max}}{Q}$$

for a non-unitary quantity of heat Q (*in fact he executes the same calculation indicated in the previous fourth step, in which he provided* $Q = 1$). He then calculates $d\eta$:

$$Qd\eta = Q\eta'(t)dt.$$

Strictly speaking, it is necessary to claim that he uses two cycles , one between the temperatures 0 and t; the other between $t + dt$ *and 0. We previously noted Joule 's experiments on the equivalency of work–heat. Let us also note that Joule's process spontaneously transformed mechanical work into heat, and thus it is an irreversible process. We also note that the first principle,* $\Delta U = Q - W$, *includes special quantities, internal energy* ΔU. *The principle is limited to a cyclical transformation where* $\Delta U = 0$, *that is* $Q = W$ *is valid. However, we know that something else* *is necessary: a heat machine that runs a power cycle and that transfers part of the heat from a high–temperature reservoir T_1 to a low–temperature reservoir T_2.*

Let Q_1 and Q_2. *be such quantities; a process can be also ideally reversible if*

$$\frac{Q_1}{T_1} = \frac{Q_2}{T_2}.$$

In other words, through his theorem, Sadi Carnot provided the preliminary operative conditions so that a heat–work process should be valid (as suggested by Clausius and Kelvin some years later). Thus, the role played by missing heat via a reversible to low–temperature reservoir is very important for establishing (also experimentally) the importance of having two different temperatures to produce thermal work for a complete cycle. In fact, only in this way can he write about an incremented efficiency. Nevertheless, it is possible to understand the calculation as accomplished on a cycle between 0 and t, while the higher temperature is incremented by a dt. Therefore, Sadi Carnot assumes that (for infinitesimal increments of the higher temperature in a cycle) the increment of η is linear. Moreover, we remark that now $d\eta$ is an exact differential, because η depends on one t–variable, only. Therefore, if one considers Q as a state function in the theory of caloric, then it also depends on dW_{max}. This aspect might epistemologically suggest that Sadi Carnot, in this case, only used state variables. We should also remark that from the definition of η, the following formula can be obtained:

$$d\eta = \frac{dW_{\max}}{Q} - W_{\max}\frac{dQ}{Q^2}.$$

Nevertheless, in the continuation of the footnote, Sadi Carnot does not consider the second addendum of the previous $d\eta$. Therefore, it seems to us that he does not consider η as a function of Q, but as a function of t, only. That is, as if $W_{\max} = F(Q,t)$ was factorizable in Q, i.e. $W_{\max} = Q\eta(t)$. This exclusion is correct only if we reason in terms of caloric. The term $W_{\max}\frac{dQ}{Q^2}$ is zero because (in the theory of caloric) in a cycle, $dQ = 0$. Now, equalizing *$Q d\eta$ to the expression of $dW_{\max}$, calculated in three steps*, we have:

$$Q\eta'(t)dt = R\log V dt$$

and dividing both members by dt, we obtain

$$Q = \frac{R\log V}{\eta'(t)};$$

and indicating $\frac{R}{\eta'}$ by $\tau(t)$ we have *(in the cycle and for the theory of caloric)*

$$Q_{t+dt} = Q_t = Q = \tau(t)\log V.$$

Step 6. Sadi Carnot calculates the specific heat at constant volume $c_V(t)$. *Here, he introduces* the quantities of heat *q*, *Q*, and *K on three processes, represented below. An isotherm* $Q(V)_t$ *useful (under certain conditions) for calculating the work; an isochor K(t), useful for defining the specific heat at constant volume* $c_{V_t}(t)$*; and a process which gives q(V, t), the difference of* $Q(V)_t$ *and K(t), believing that in a cycle the sum of the quantities of heat was null. That is to say, here he unequivocally applies the theory of caloric. Therefore the following calculation is erroneous from a modern point of view* (Fig. 9.2).

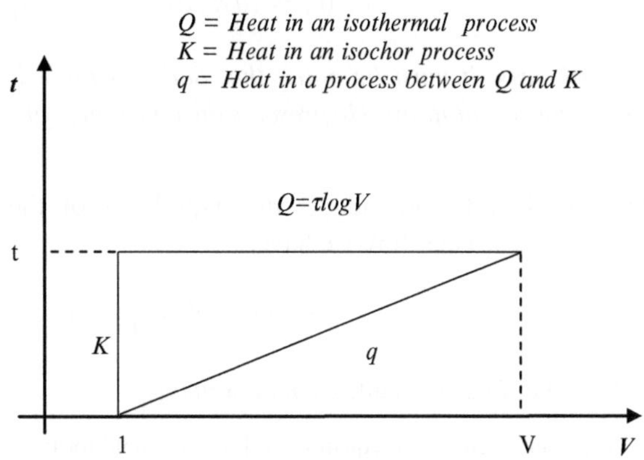

Fig. 9.2 Simplified performance of quantities *Q*, *q* and *K* set by Sadi Carnot in his mathematical footnote (Carnot 1978, pp 77–78, ft 1)

Given the conservation of the caloric in a cycle, Sadi Carnot proposes the following relation:

$$q = Q + K = \tau \log V + K.$$

He then commits an inaccuracy. He confuses K with q. In fact, he writes (the modern notations are in square brackets):

> If we take *s* [*q*] the quantity to represent the amount of heat [chaleur] required to bring the air we have been considering from a volume *1* and a temperature 0° to a volume *v* [*V*] and a temperature *t*, the difference between *s* [*q*] and *e* [*Q*] will be the amount of heat required to rise the temperature of the air, at the constant volume of *1* from 0° to *t*°. This amount of heat depends on *t* alone; we will call it *U* [*K*]: standing for some unknown function of *t*. it. [...]. If this equation is differentiated [see previous equation] with respect to *t* alone, and if the differentials of $T[\tau]$ and $U[K]$ are represented by differentials coefficients $T'[\tau']$ and $U'[K']$, it follows that
>
> $$ds/dt = T' \log v + U' \quad (5)$$

> $\left[\left(\frac{\partial q}{\partial t}\right)_V = \frac{d\tau}{dt}\log V_t + \frac{dK}{dt} = \tau'_t \log V_t + k'_t\right]$ [and] *ds/dt* is equal to the specific heat [chaleur] of the gas at constant volume and our equation (5) is the analytical version of the law stated on page 58.[17]

In effect we know that $c_V(t) = \frac{dK}{dt}$. *We potentially suggest that ds/dt (in modern terms: dq/dt), for a particular and unexpected length of the isotherm could coincide with* $c_P(t)$*: that is, the specific heat at constant pressure. Consequently, for a determined value of t, a correct expression for the specific heat at constant volume is:*

$$c_{V_t}(t) = (dK/dt)_V = q'_t - \tau'_t \log V_t \tag{a}$$

Nevertheless, it must be noted that Sadi Carnot's inaccuracy does not affect his following calculations. He presents the constancy of τ' *for the accuracy of previous Equation (a).*

Step 7. Sadi Carnot adopts the hypothesis of the constancy of $c_V(t)$ for all temperatures. Therefore we have:

$$c_V(t) = q'_t - \tau'_t \log V_t = c; \tag{b}$$

with c_V being a constant. *Then he writes:*

> [...] and therefore, if equation (5) [(a)] is to hold for two particular [separate] values of v [V_t], it will be necessary that T' [τ'_t] and U'[K'_t] must also be independent of t[18];

Therefore it should be:

$$\tau'_t = A_t,$$

$$q'_t = B_t.$$

If we follow Carnot's reasoning (Ivi, pp 77–78, ft 1) or that proposed by Lervig (1972) *presented in previous Chapter 7 (except the use of statement* $\vartheta = \tau$*)* the following expressions of τ and η are obtained only with $\tau' = A_t$:

$$\tau = A_t t + B_t$$

$$\eta = \frac{R}{c}\ln(A_t t + B_t) + c' \tag{c}$$

Step 8. *Even here Sadi Carnot maintains the previous inaccuracy. He writes*:

> In equation (5), as we have seen, the first term [*ds/dt*] represents the specific heat [chaleur spécifique] of the air occupying a volume v [V].[19]

---

[17]Carnot (1978), p 77, ft 1, line 1.

[18]*Ivi*, p 77, ft 1, line 15.

[19]Carnot (1978), p 78, ft 1, line 13.

At this point, Sadi Carnot introduces an experimental fact of the almost constancy of $c_V(t)$ for large changes in volume (*Ivi*, p 78, ft 1, line 16). Therefore the same equation has it that the coefficient τ' of *log V is* null or negligible. If $\frac{d\tau}{dt} = \tau' = 0$, then $\tau = B_t = const.$; therefore from the definition of $\tau = R/\eta'$ we have $d\eta = \frac{Rdt}{B_t} = \lambda dt$; and, integrating, Sadi Carnot obtains:

$$\eta_t = \int \lambda dt = \lambda t + b,$$

or, even though for t = 0 $\eta_t = 0$, *we can write:*

$$\eta_t = \lambda t$$

which is the efficiency function of an ideal heat machine, which Sadi Carnot achieves at the end of the footnote:

> [...] by a second integration, we arrive finally, [to the following equation], $F(t) = At + Bt$. Since $F(t) = 0$ when $t = 0$ [also], B is zero; and so $F(t) = At$, in other words, the motive power produced is shown to be exactly proportional to the fall of the caloric.[20]

By analyzing this statement, many have also tried to explain why S. Carnot's final result effectively appears to be *fortuitous*, since in modern thermodynamics the work of the two (infinitesimal) adiabatics are mutually cancelled out. Let us see.

9.4 The "Suppression" of Adiabatic Curves in Sadi Carnot's Cycle

9.4.1 Why does Sadi Carnot "Supprime" the Adiabatic Curves?

The issue of the suppression of the two adiabatics, both in the arguments and in the calculations regarding the cycle of the four thermodynamic phases, is quite delicate and strictly related to the closing cycle:

> The Carnot Cycle! What memories this name must stir in every engineer! To some it will recall early struggles to comprehend the nature of heat engines; to others it represents the ideal of performance sought after in design and operation. Many will recall the keen delight that came from a thorough understanding of the Carnot cycle which forms one of the principal foundations of thermodynamics. Every student of this subject can now write the elementary principles of this cycle in our modern terms of heat and thermodynamics. How many have given a thought to the conception and development of the idea when many

[20]Carnot (1978), p 79, line 1, op. cit.

> of the laws of physics as now stated were either imperfectly known or not yet discovered? What could be more interesting than to read the thoughts of the originator of the Carnot cycle and to follow his reasonings that led to his great discovery?[21]

Based on our previous discussion, Sadi Carnot seemed to be confident that the equivalence between the contributions of the work of the two adiabatic processes should be mathematically valid, but only from a methodological point of view.

> [Necessary condition: $\Delta t \rightarrow 0$]. Then the movements of the piston being so slight during stages 3 and 5 that these stages could have been suppressed [ignored] without significantly affecting the amount of the of motive power produced [ces périodes eussent pu être supprimées sans influence sensible sur la production de la puissance motrice]. Only a very small change [fort petit] in volume is required, in fact, in order to cause a correspondingly small change in temperature; such as small change in volume is negligible in comparison with those occurring in stages 4 and 6, which can be of unlimited extent.[22]

Therefore he reasoned on the formula of the efficiency, either without executing the calculation of the work on the two adiabatic processes; or, if he executed it, he did not attribute importance to it (maybe) because his formula of the adiabatic was *tentative*:

> The vapor of water so made may then be treated as a permanent gas; and subject to all [gas] laws. Consequently the formula [of adiabatic equation] $t = \frac{(A+B\ \log[\ln]v)}{(A'+B'\ \log[\ln]v)}$[23] should be applicable to it [statured steam], and be consistent with the table of pressures based on the experiments of M. Dalton. We can in fact confirm that our formula, very approximately represents the experimental evidence, once the arbitrary constants have been appropriately determined. The minor anomalies that appear are small enough to be plausibly attributed to observational error.[24]

Therefore *what is the reasoning, never revealed by Sadi Carnot in Réflexions sur la puissance motrice du feu, that pushed him to calculate the efficiency function* $\eta = W_{\max}/Q$ *without considering the two adiabatics?*

It is well known that, for a given physical system, an adiabatic transformation is a thermodynamic process in which the net heat transfer to or from the working substance is zero; or, a process without heat exchanging with the other systems. Here, we are studying thermodynamic systems which essentially differ from the mechanical systems due to heat, which is a phenomenon not reducible to mechanical work: *what happens in a thermodynamic process if we prevent the passages of heat? Is it certain that we will obtain a mechanical situation?* We have no proof of this. This uncertainty could be typically expressed by a double negation (see above Chapter 7; Drago and Pisano 2000; Pisano and Gaudiello 2009b): a

[21] Carnot (1943), "Foreword" [by Thurston] line 1.

[22] Carnot (1978), p 39, line 7.

[23] Poisson proposed another formula.

[24] Carnot (1978), p 66, line 7.

scientific sentence that is not equivalent to the corresponding affirmative sentence. *Therefore, is an adiabatic process formalizable by means of a double negation of the mechanical work?* Let us see.

Modernly, it is clear that by considering the validity of the first law[25] for a cyclical process,[26] and by taking dU as an infinitesimal change in internal energy, we cannot state the following affirmative expression from an experimental point of view:

$$dU = 0 \Rightarrow \delta Q = \delta W$$

(*The heat input is equal to the work output over any cycle*)

without discussing the upper bound proposed by the second law. For example, it is impossible that in a cyclical running machine ($dU = 0$) all heat spontaneously flows from a unique reservoir (at a given temperature) for producing work. Thus, in this case and under the first law, one can experimentally observe that it not true that all of the heat can be completely transformed into work. Another reservoir of temperature is necessary. Moreover, there is an important distinction between heat and work that is not evident in the first law.[27] This ambiguity is frequently represented by the following words: *it is* not *true that heat is* not *work.*

On the other hand, by considering a given $p \neq 0$, we can banally claim that:

$$\Delta V \neq 0 \Rightarrow W = p\Delta V.$$

However, we also know that:

$a\Delta V$ *is valid even in the case of an adiabatic process* ($Q = 0$).

Therefore everything also very much depends on the modality by which the adiabatic process occurs.

It should be also noted that Sadi Carnot evaluated the reversibility (Carnot 1978, pp 10–14, 23–35) of his thermodynamic processes by ΔV in $W = p\Delta V$. Moreover, if we explicitly refer to the theory of the equivalence heat–work and consider a typical molecular hypothesis, the latter can also co–justify the previous epistemological discussion. In the interaction of the molecules (collision and forces) the net heat transfer is zero; *can we claim that within a system, only a mechanical interaction exists? That is to say, without, e.g., friction and viscosity?* In other words,

[25]The first law is usually formulated by saying that the change in the internal energy of a closed thermodynamic system is equal to the difference between the heat supplied to the system and the amount of work performed by the system on its surroundings.

[26]Briefly, if a physical system received a certain amount of heat, the same amount of work has been done by the system. Thus the total variation of the internal energy after a cycle is null.

[27]The first law of thermodynamics makes no distinction between processes that occur spontaneously and those that do not. The second law of thermodynamics establishes which processes do and which do not occur.

is the adiabatic process a compression by means of an elastic expansion as was supposed at the time? We cannot be sure of this and we certainly cannot decide it *a priori*. Thus, it is worth indicating the experimental ambiguity of an adiabatic process using the following double negative expression:

$$(Q = 0) = \neg\neg W = \textit{Adiabatic is } \underline{\textit{non}}\ \underline{\textit{non}}\ \textit{Work}.$$

This fact is of considerable importance, since the adiabatic is a crucial process in Carnot's heat cycle, a DNS could be intended by Sadi Carnot as the additional variable "ε" of the *synthetic method* (see below). Thus, in Sadi Carnot's cycle, the mysterious suppression of two adiabatic processes corresponds to the suppression of the variable due to the application of his father's *synthetic method*.

9.4.2 A Possible Solution. Lazare Carnot's Synthetic Method

Generally speaking the *synthetic method* (Drago and Pisano 2005) and the *analytic method* have their historical origins in the ancient period with Pappus[28] from Alexandria (290–350 b.C.), and later assumed different meanings even for a single author, e.g. Descartes (1897–1913). With regard to Lazare Carnot, the *synthetic method*[29] also explains the nature and the frame of mind behind his research against the metaphysical conception – which prevailed at the time – of infinitesimals, as he declares at the beginning of his famous book, *Réflexions sur la métaphysique du calcul infinitésimal*[30]:

> It is my object to ascertain in what the true spirit of the Infinitesimal Analysis consists.[31]

"The Renaissance of the synthetic method at the École Polytechnique" (Schubring, pp 295–308) reached its pinnacle with a new conception of the metaphysics of infinitesimals, re–elaborated by Lazare Carnot as based on exclusively operative or effective algorithms (Schubring, pp 309–445), which, in mathematics, utilize the potential infinite only. Lazare Carnot pointed out that both the methods utilize auxiliary variables, *analytical quantities*[32] (êtres de raison: Carnot 1813, p 189,

---

[28]Pappus from Alexandria, *Mathematical collections* (Commandino 1565; see also: Pappus from Alexandria 1588).

[29]Carnot (1786), pp 33–35, 85; Carnot (1813), pp 12–21, 189, 200, 242–243, 217–253.

[30]Cfr.: Gillispie and Youschkevitch (1979), pp 251–298, § 13, p 256, ft 1, p 182.

[31]Carnot (1832), p 1. ["Je cherche à savoir en quoi consiste le véritable esprit de l'Analyse infinitesimal" (*Id.*, 1813, p 1, line 1)]. Let us note that the English translation of the 1832 version is differently organized in comparison to the original 1813 version. E.g., the number of paragraphs in the two versions does not correspond.

[32]See also its English translations "[...] creatures of reason [...]" (Carnot 1832, p 105, line 13).

line 1) for the analytic method, and *absolute quantities* (quantités absolues: *Ivi*, p 243, line 10) *for* the *synthetic method.* With regard to infinitesimal calculus for suppression, he claimed:

> *Fundamental Principle.* [...] Up to that point I regard my calculation as only a simple method of approximation. Those, who in order to reconcile the rigour of the calculation throughout the operation with the simplicity of my Algorithm [...].[33]
> *On the Differential Calculus.* [...] But if the two systems are supposed to approach one another as much as we please, the difference of the two values of the same variable may be also made as small as we please, and it will become what is called a differential, and will be nothing but the common difference simplified by the suppression of quantities which in their expression might be found to be infinitely small with respect to other terms of which it is composed. Such is the principle of Differentiation.[34]

Thus, these *quantité désignées et non désignées* help to generalize the problem. Below, we provide Lazare Carnot's famous words in the original French:

> [...] la source de cette application continuelle des géomètres à généraliser, vient de ce qu'ils sentent, que, presque toujours généraliser c'est simplifier [...].[35]

According to Lazare Carnot the variables have to be *simply implied to simplify the calculation*, if, at the end of the reasoning, once the solution to the problem has been obtained, we are able to make these variables disappear "[...] in order to arrive infallibly at new truths." (Carnot 1832, p 116, line 24; Carnot 1813, p 200, line 8). That is:

> To neglect quantities of this nature then is not only admissible in parallel cases, but it is necessary, and it is the only method of expressing exactly the conditions of the problem.[36] Hence the sign which announces that the desired compensation has taken place, is the absence of the arbitrary quantities which produced these errors, and hence to effect this operation, it remains but to eliminate these arbitrary quantities. Let us now attempt to give to these ideas the degree of precision and of generality which they merit.[37]

Therefore, according to Lazare Carnot, even the *synthetic method* can accomplish this cycle of operations of infinitesimal analysis, while staying anchored to effective quantities. In fact, Lazare Carnot presented a crucial problem related to the role played by infinitesimal quantities in the rigor of mathematics (and his mechanics). A similar but greater methodological (and currently epistemological) problem was

---

[33]Carnot (1832), p 34, line 6. (*Id.*, 1813, p 37, line 15). Author's *italic.*

[34]Carnot (1832), pp 55–56, line 19. (*Id.*, 1813, pp 65–66, line 22). Author's *italic.*

[35]Carnot (1785), § 100; see also Gillispie (1971), Appendix A, § 100, p 257, line 19.

[36]Carnot (1832), p 10, line 1. (*Id.*, 1813, p 12, line 4).

[37]Carnot (1832), p 15, line 1. (*Id.*, 1813, p 17, line 6).

previously proposed by George Berkeley (1685–1753) who memorably disapproved of infinitesimals as "the ghosts of departed quantities" as he claimed in one of his most frequently quoted passages:

> XXXV. [...] Whatever therefore is got by such Exponents and Proportions is to be ascribed to Fluxions: which must therefore be previously understood. And what are these Fluxions? The Velocities of evanescent Increments? And what are these same evanescent Increments? They are neither finite Quantities nor Quantities infinitely small, nor yet nothing. May we not call them the Ghosts of departed Quantities?[38]

Lazare Carnot seemed to echo Berkeley's conceptual streams (Carnot 1813, pp 12–15, 30–34) concerning the *théorie de la compensation des erreurs à la Berkeley*[39] for the infinitesimal calculus: *two errors* are made, which nevertheless are algebraically annulled in the end, and "[...] by a compensation of errors: which compensation, however, is a necessary and certain consequence of the operations of the Calculus". Carnot (1832, p 105, line 20, Carnot 1813, p 189, line 10).

> The inventor might then be led to his discovery by a very simple train of reasoning. If in the place of a proposed quantity, he might say, I employ in the calculation another quantity not equal to it, there will thence result some error or other: but if the difference of the quantities employed one instead of the other is arbitrary, and if I can make it as small as I please, this error will never be important: I might even make several similar errors at the same time without any bad result, so long as I should continue master of the degree of precision which I should wish to give to my results. Moreover, it might happen that these errors

[38]Berkley ([1734] 2002), sect. XXXV, p 18, line 9.

[39]Berkeley's lemma claims: "If with a View to demonstrate any Proposition, a certain Point is supposed, by virtue of which certain other Points are attained; and such supposed Point be itself afterwards destroyed or rejected by a contrary Supposition; in that case, all the other points, attained thereby and consequent thereupon, must also be destroyed and rejected, so as from thence forward to be no more supposed or applied in the Demonstration". (Berkeley 1734, sect. XII–XIV; See also: Berkeley [1734] 2002). Let us consider the function $f(x) = x^2$; let us pass to the calculation of its derivative: the increment of the variable is dx (which at that time was not distinguished by Δx) and the increment of the function is df (or Δf), that is equal to: $df = f(x + dx) - f(x) = (x + dx)^2 - x^2 = x^2 + 2xdx + dx^2 - x^2 = 2xdx + dx^2$. The last term is then suppressed, since it is much smaller compared to the others. Dividing df by dx we obtain $2x$, which is exactly the derivative the function of the function. Berkeley applied his lemma, pointing out that at the beginning of the calculation $dx \neq 0$ has been set; however, subsequently the calculation eliminates the infinitesimal of higher order at dx, considering it equal to zero. Therefore, either dx is zero, as stated at the end of the reasoning – but then the entire calculation is devoid of content – ; or it is not zero, as stated in the middle of the reasoning – but this goes against the aforementioned lemma. This criticism allows Berkeley to suggest an original interpretation of the efficacy of this type of calculation. At the beginning, $dx \neq 0$ is set and this, Berkeley claims, constitutes the first error because df as dx does not represent a specified number: nor does zero, nor a number different from zero. So there is no reason to include it in the only calculation that we know very well: elementary algebra. Continuing with the calculation, the last term is suppressed – which we know to be different from zero (otherwise we would have already eliminated it along with all of the dx) – and this clearly represents an algebraic error. Since the first result is correct, this second error cancels out the first one. (Cfr.: Drago 2007a, b; see also Parigi (2010)).

> mutually compensate for one another, and that thus my results become perfectly exact. But how to effect this compensation in all cases? Some reflection can alone discover this. Let us suppose, the inventor might say, that the compensation be made, and let us observe by what mark it ought to show itself in the result of the calculation. Now what naturally must occur is, that the quantities which had occasioned the errors having disappeared, the errors also have disappeared; for the quantities such as MZ, RL having by hypothesis arbitrary values, they ought no longer to enter into any formulas or results which are not so, and which, having become exact by supposition, depend alone not on the will of the calculator, but on the nature of those quantities, the ratio of which it was proposed to discover as expressed by these results. Hence the sign which announces that the desired compensation has taken place, is the absence of the arbitrary quantities which produced these errors, and hence to effect this operation, it remains but to eliminate these arbitrary quantities. Let us now attempt to give to these ideas the degree of precision and of generality which they merit. [...].[40]

Consequently the *dx* belonging to analysis is not something infinitely small; it is rather an auxiliary variable of any considered value. Most importantly, it could be useful for simplifying the search for a solution to the problem, and at the end of the calculations it disappears (e.g. with the procedure of limit) leaving a perfect equivalence. In other words, by introducing the auxiliary variables, Lazare Carnot accepts the power of analysis, avoiding some metaphysical concepts, but he does not accept the way in which the latter introduces it. In fact, he deprives the infinitesimals of any metaphysical value. For that

> This approach is essentially what distinguishes analysis and synthesis. [...].[41] The synthesis does not involve absolute quantities [quantités absolues] and immediately executable operations which may satisfy the arguments proposed; while the analysis considers all algebraic forms that can satisfy equations found.[42]

In agreement with previous investigations (Gillispie 1971, pp 90–100, 1976, pp 23–34; see above Chapter 6) according to which Sadi Carnot's theory could be parentage from his father's theory, we claim that Lazar Carnot's *synthetic method* is present in Sadi Carnot's theory (1978, pp 18–19, 36, 38–39, 73–79, ft 1). On the other hand, in the first discursive part of the *Réflexions sur la puissance motrice du feu*, Sadi Carnot introduces an infinitesimal cycle, the fundamental concepts of infinitesimal Δt and a cycle of reasoning (see Chapter 7); these elements occur in the application of the *synthetic method*. He reasoned introducing and then "supprimant" the auxiliary variable. Sadi Carnot argues that this is possible when using infinitesimal calculus. In regard to this crucial historical element, we provide his own wording (Fig. 9.3):

[40] Carnot (1832), pp 14–15. (*Id.*, 1813, pp 16–17, line 1). Author's original wording.

[41] Carnot (1813), "Note", p 242, line 7.

[42] *Ivi*, p 243, line 9.

Fig. 9.3 A relationship between mathematics and physics in Sadi Carnot's *Réflexion sur la puissance motrice du feu* (Carnot 1978), pp 18–19, ft 1)

(1) On s'étonnera peut-être ici que le corps B se trouvant à la même température que la vapeur puisse la condenser: sans doute cela n'est pas rigoureusement possible; mais la plus petite différence de température déterminera la condensation, ce qui suffit pour établir la justesse de notre raisonnement. C'est ainsi que, dans le calcul differentiel, il suffit que l'on puisse concevoir les quantités négligées indéfiniment réductibles par rapport aux quantités conservées dans les équations, pour acquérir la certitude du résultat définitif.

Le corps B condense la vapeur sans changer lui-même de température: cela résulte de notre supposition. Nous avons admis que ce corps était maintenu à une température constante. On lui enlève le calorique à mesure que la vapeur le lui fournit. C'est le cas où se trouve le métal du condenseur lorsque la liquéfaction de la vapeur s'exécute en appliquant l'eau froide extérieurement, chose pratiquée autrefois dans plusieurs machines. C'est ainsi que l'eau d'un réservoir pourrait être maintenue à un niveau constant, si le liquide s'écoulait d un côté tandis qu'il arrive de l'autre.

On pourrait même concevoir les corps A et B se maintenant d'eux-mêmes à une température constante, quoique pouvant perdre ou acquérir certaines quantités de chaleur. Si, par exemple, le corps A était une masse de vapeur prête à se liquéfier, et le corps B une masse de glace prête à se fondre, ces corps pourraient, comme on sait, fournir ou recevoir du calorique sans changer de degré thermometrique.

We may perhaps wonder here that the body B being at the same temperature as the steam is able to condense it. Doubtless this is not strictly possible, but the slightest difference of temperature will determine the condensation, which suffices to establish the justice of our reasoning. It is thus that, in the differential calculus, it is sufficient that we can conceive the neglected quantities indefinitely reducible in proportion to the quantities retained in the equations, to be certain of the exact result.

The body B condenses the steam without changing its own temperature this results from our supposition. We have admitted that this body may be maintained at a constant temperature. We take away the caloric as the steam furnishes it. This is the condition in which the metal of the condenser is found when the liquefaction of the steam is accomplished by applying cold water externally, as was formerly done in several engines. Similarly, the water of a reservoir can be maintained at a constant level if the liquid flows out at one side as it flows in at the other.

One could even conceive the bodies A and B maintaining the same temperature, although they might lose or gain certain quantities of heat. If, for example, the body A were a mass of steam ready to become liquid, and the body B a mass of ice ready to melt, these bodies might, as we know, furnish or receive caloric without any change in temperature.

This is what Lazare Carnot (1813, pp 213–257; Gillispie 1971, pp 31–32), George Berkeley (Grattan–Guinness 1970, pp 215–227) and "de notre immortel" (Carnot 1813, p 122, line 15) Joseph–Louis Lagrange[43] (1736–1813) had already done.

An immediate consequence of this (introduction of the adiabatic auxiliary variable in the *synthetic method*), following Lazare Carnot's wording, is the variable simplifying the solution to the problem: to find the efficiency of a heat machine. Therefore (within an ideal cycle) the contributions of the work on the two infinitesimal adiabatic processes mutually cancel each other out. The *suppression of the adiabatic* in Sadi Carnot's cycle may indicate precisely that operation, typical in the *synthetic method*, of the elimination of the auxiliary variable ε. In Lazare Carnot's reasoning:

FUNDAMENTAL PRINCIPLE
24. *Two quantities which are not–arbitrary, cannot differ from one another by any other than a quantity which is not arbitrary. Proof.* Since the two proposed quantities are never arbitrary, we change neither the one [n]or the other: hence we cannot change their difference: *hence this difference is never arbitrary.*

COROLLARY I
25. *Two quantities that are not–arbitrary are exactly equal to one another, whenever their supposed difference may be supposed as small as we please.* Let *P* and *Q* be the two proposed quantities: we have just seen that their difference cannot be arbitrary: it cannot therefore be supposed as small as we wish; which is contrary to [the] hypothesis. Hence this supposed difference does not exist. Consequently *P* and *Q* are rigorously equal.

COROLLARY II
26. *To be sure that two assigned quantities are rigorously equal, sufficiently to be proved that their difference is an unassigned quantity.* In effect, assigned quantities are not–arbitrary: hence their difference cannot be arbitrary; their difference therefore is necessarily an assigned quantity: Hence, in order to prove that this difference does not exist, and that consequently the quantities are equal, it is sufficient to prove, that if it existed it could only be an assigned quantity.

COROLLARY III
27. *Every value that we may make as near as we please to the real quantity which it represents, without being at the same time compelled to change either the one or the other, is of necessity rigorously correct.* Since there is no occasion to change either the proposed quantity or its value in order to render the latter as near an approximation to the former as we wish: i. e. since we may regard both of them as determined, and consequently non–arbitrary, in order that they may differ as little as we please from one another. Therefore they fall under the case of Corollary II. Thus they are of necessity and rigorously equal.

COROLLARY IV
28. *Every quantity that we may suppose as small as we please may be overlooked as if it were absolutely zero, in comparison with every other quantity which cannot like the former be supposed to be as small as we please; without the errors (possibly arising hence in the course of the calculation) affecting the result, so soon as all the arbitrary quantities shall*

[43]Lagrange (1797, 1806). Very interesting Lazare Carnot's quotation on Lagrange's *Théorie des Fonctions Analytiques*: Carnot (1813), p 47.

have been eliminated from it. Whilst neglecting as absolutely zero the quantities which may be taken as small as we please, since they are added to or taken from others which do not admit of being taken as small as we please, it is clear that the errors which can arise in the course of the calculation, may similarly be considered as small as we please; there will remain then in the result some arbitrary quantity: but this is opposed to the hypothesis, since all arbitrary quantities are supposed to be entirely eliminated.

COROLLARY V
29. *Every quantity, the relation of which to another quantity may be supposed as small as we please, may be neglected as absolutely zero in comparison with this last, without the errors (to which that would give rise in the course of the calculation) affecting the result so soon as all arbitrary quantities have been eliminated from it.* This corollary is but an extension of the former. In corollary IV, it was supposed, that those quantities in comparison of which we might overlook the others, could not themselves be supposed as small as we please: but in corollary V, we suppose, that both the one and the other may be made as small as we please, but that the ratio of one to the other is also susceptible of being taken as small as we please: consequently, of what nature so ever both of the quantities may be, we may neglect, in comparison with the others, those quantities, the ratios of which to these last we may suppose as small as we please. And the demonstration is the same as that for corollary IV; for it is evident, that if some errors arose from suppressing these quantities, we might always diminish them as much as we please, either in the course of the calculation, or in the result. Not that that can take place as respects this last case, for then some arbitrary quantity would be introduced there; and this is contrary to hypothesis, it being provided, that all arbitrary quantities should be eliminated.[44]

In fact, by means of elimination, in Sadi Carnot's cycle as well, we return to the initial system with the solution to the initial problem. That is, the thermodynamic system goes back to the *initial state* with the solution to the question of how much the accomplished work is for a certain quantity Q. Motive power is the matter in question for Sadi Carnot; in the first of the cited sentences,[45] "without significantly affecting the amount of motive power produced" (Carnot 1978, p 39, line 9). Therefore, Sadi Carnot may have thought that it was not necessary to know the correct equation of the adiabatic, because he relied on his father Lazare's scientific method. He performed the calculation of the infinitesimal reversible cycle by adding and subtracting the variable ε (the adiabatic) – as is done in such a method – that is to say, in his own words, "être supprimées" (*Ibidem*). Below, we provide a performance model of the application of the compensation of the error based on Lazare Carnot's arguments (Fig. 9.4):

[44]Carnot (1813), pp 30–34, line 20. (*Id.*, 1832, pp 27–31, line 18). Author's capital letters, wording and *italics*. Please note that the use of wording, *italic* and quotations marks by the author (or publisher) are not the same in the two (1813 and 1832) editions cited.

[45]Please refer to Lazare Carnot's quotations. The wording seems identical.

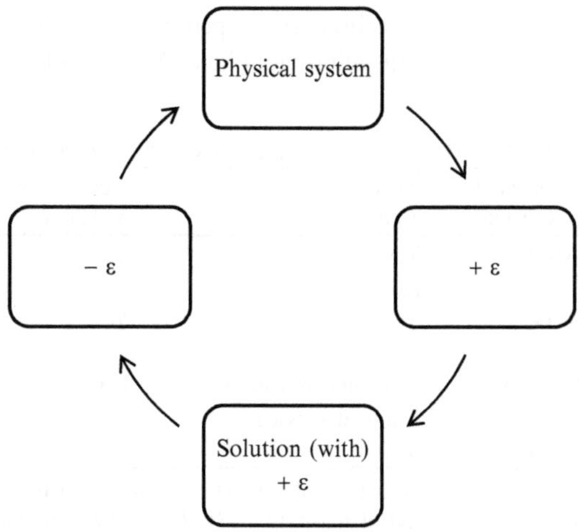

Fig. 9.4 Simplified performance model of the application of the *synthetic method*

These other states of the system are then, properly speaking, only auxiliary systems, which are interposed in order to facilitate the comparison between the parts of the former. The differences of quantities, which correspond in all these systems, may consequently be supposed as small as we please, without affecting those which compose the former, and the ratio of which is the object of our enquiry. So that these differences have the nature of what we call Infinitely small Quantities [*infiniment petites*]: since they are considered as continually decreasing, and as admitting of being made as small as we please, without affecting the value of the sought unknown quantities. Unity divided by a quantity infinitely small is called an Infinite, or an Infinitely great Quantity.[46]

[...] hence as this last never admits Infinitesimal Quantities, it is quite necessary that such as [can] be admitted into the former always conclude with being eliminated.[47]

In the first [system], the infinitesimal quantities are only auxiliary quantities, which must necessarily be eliminated to achieve the results looked for; in the second [system], the isolated negative quantities, being imaginary, are no more than of auxiliary use, and as tools which become absolutely extraneous to the building, once it is erected.[48]

9.4.3 *The Synthetic Method in Réflexions sur la puissance motrice du feu*

Below we present Sadi Carnot's reasonings in *Réflexions sur la puissance motrice du feu* regarding the thermodynamic cycle through the application of the *synthetic method.* Particular attention is paid to Sadi Carnot's reasoning process while

[46]Carnot (1832), pp 17–18, line 21. (*Id.*, 1813, p 20, line 1). Author's *italics* and capital letters.

[47]Carnot (1832), p 19, line 13 (*Id.*, 1813, p 21, line 25). Author's capital letters.

[48]Carnot (1813), p 252, line 18.

theorizing from a three–phase primitive cycle (or composed cycles), until he creates the idea of a four–alternate–phase cycle (Table 9.3).

Table 9.3 The *synthetic method* in the *Réflexions sur la puissance motrice du feu*

Choices of S. Carnot	2 cycles by 3 phases	2 cycles by 3 phases	2 cycles by 4 phases	In footnote
ASM	Yes	Yes	Yes	Yes
UIA	No	No	No	Yes
SMR	Yes	Yes	Yes	Yes (suppression of two adiabatics)
SMAIA	None in application But he discussed it	No	No	Yes
Reference	p 18, ft 1	p 19	p 36	p 75, ft 1; p 78, ft 1

Legend: *ASM* application of synthetic method, *UIA* use of infinitesimal analyses, *SMR* synthetic method as method of reasoning, *SMAIA* synthetic method as method of calculus alternative to infinitesimal analyses (ca. eighteenth to nineteenth centuries)

Generally speaking in the first discursive part of *Réflexions sur la puissance motrice du feu* (Carnot 1978, pp 1–73) Sadi Carnot applied, for the first time, his father's *synthetic method* to the cycle using vapor such as a working substance. The vapor is added and then subtracted from the cylinder, as is usually done with the auxiliary variable in the *synthetic method.* Finally a solution to the problem is obtained: the machine produces work:

> The vapor [steam] serves as means of transporting the caloric [. . .].[49]

Then Sadi Carnot introduces an infinitesimal cycle, and, additionally, the fundamental concept of a Δt (Carnot 1978, pp 1–73). At this point he reduces the range of temperature (Δt) *with continuity* until it is annulled. In regard to this, he argues:

> In order for caloric to pass from one body to another, there must to be a difference of temperature between them. However, we may suppose the difference [of temperature] to be as small as we like. For the purposes of our theory, it [Δt] may be taken to be zero [that is, ideally], without the argument being undetermined in any way.[50]

In fact, based on previous discussion, he first introduces the auxiliary variable Δt and he then "supprime" it. Sadi Carnot had previously argued that this is possible because it is the current procedure adopted in the differential calculus.

> It may be perhaps surprising here that the body B being at the same temperature as the vapor [steam] is capable of condensing it: doubtless this is not strictly possible; but the most minute difference in temperature is enough to bring about condensation, and, is just in our

[49] Carnot (1978), p 10, line 2.

[50] *Ivi*, p 25, line 7.

reasoning. It is just as in differential calculus, where we achieve rigorously demonstrated proofs by ignoring quantities, that are infinitesimally small relative to the ones that we retain in our equations, to make certain of the exact result.[51]

Finally, Sadi Carnot proposes two main cycles:

1. The cycle as composed by two isotherms (as all the cycles he suggested in *Réflexions sur a puissance motrice du feu*), alternated with two isochors, having $W = 0$.[52]
2. The cycle as composed by two isotherms alternating with two adiabatic, $Q = 0$.[53]

When Sadi Carnot conceives the cycle with isochors, he reasons correctly since he uses an infinitesimal range of temperature, which eliminates the non–reversibility of the isochors. Then, he passes from the infinitesimal isochors to the finite isochor (Carnot 1978, pp 39–40). Therefore, he commits an imprecision since he cannot extend the reversibility to the finite isochor processes (this fact is commonly pointed out). Let us examine this in detail.

It must be noted that Sadi Carnot did not define reversibility in modern terms, but in terms of $\Delta V > 0$ when Q produces a gap, $\Delta t > 0$. His definition of reversibility was a criterion capable of distinguishing only the case study $\Delta V > 0$ (positive work), from the other case study $\Delta V = 0$ (zero work). Therefore, it is reasonable to suppose that he could not distinguish the infinitesimal situation from the finite situation. Since, according to Sadi Carnot, the case study of a process where ΔV is infinitesimal for an infinitesimal Δt is valid, we can also hypothesize that he considered the case study of finite processes as a generalization which was supported by the general method (i.e., the *synthetic method*) that he applied in *Réflexions sur a puissance motrice du feu*. Therefore, we might also claim that Sadi Carnot inaccurately applied the *synthetic method*, which is applicable only when a condition of continuity exists.

Sadi Carnot introduced the definitive cycle with two isotherms and two alternate adiabatics only at the end of a series of cycles. Let us note the adiabatics are indeed intended as the two *new* auxiliary variables. In this case, as previously mentioned, he does not commit an error when turning to the finite transformations, since, in modern terms, we know that the finite adiabatic process does not destroy the reversibility of the infinitesimal thermodynamic transformation.

Essentially, we think that this trust in the *synthetic method* is the only reason that later led Sadi Carnot to fundamentally build two cycles:

1. A first cycle consisting of, in addition to the two isotherms (which are present in all the cycles he suggested in *Réflexions sur la puissance motrice du feu*), two isochors.
2. A second cycle with two adiabatics.

---

[51]*Ivi*, 18, ft 1, line 1.

[52]Carnot (1978), pp 39–40.

[53]*Ivi*, pp 29–38.

It should be noted that the two previous case–studies respectively represent:

$$W = 0 \text{ and } \neg\neg W = 0$$
$$\textit{Two case} - \textit{studies based on his efficiency calculated using a determined value of } Q$$

Therefore, it was only thanks to the application of the *synthetic method* that he thought that infinitesimal works dW and dW' of the two adiabatics did not provide any contribution, since $\Delta dW = dW'$. Consequently, in the application of the *synthetic method* Carnot obtained:

- *The cycleshould not be constituted by three transformations (as would be sufficient to build a cycle), but rather by four transformations.*
- *Among the four transformations, two should necessarily be isotherms.*
- *The two remaining adiabatictransformations, if infinitesimal, give contributions that mutually cancel (supprimer) each other out, therefore they can be (mathematically) ignored in the cycle.*

In that sense, an explanation has been attempted to account for the closure of the cycle which perplexed Klein (Klein in Taton 1976, pp 213–219) since such a closure did not attribute any special role to the initial state – we also explained that Carnot's final result appears to be fortuitous since in modern thermodynamics, the work of the adiabatics mutually cancel each other out.

> Carnot sets up the cycle in what seems to be a very awkward way, a way not followed by any of his successors or commentators, so far as I know.[54]

9.5 The Mathematical Character of the Footnote

The variables on which the function $W/Q = f(t, Q)$ depends are all state variables (even Q is a state function in the theory of caloric), which vary in reversible processes. Therefore we encounter the following problem: *was Sadi Carnot capable of calculating the maximum of W/Q studying the differential d(W/Q) by means of infinitesimal analysis?*

Since ca. the seventeenth to eighteenth centuries, the derivative[55] has been seen as an important operation for calculus. It was heavily relied on along with new

---

[54] Martin J. Klein in Taton (1976), p 214, line 17.

[55] Very briefly we mention the ancient *Moscow Mathematical Papyrus* (also called the *Golenischev Mathematical Papyrus* at *Pushkin State Museum of Fine Arts* in Moscow), where it remains today demonstrating knowledge of a formula for the volume of a pyramidal frustum. Greek Eudoxus (ca. 370 BC), for areas and volumes by breaking them up into an infinite number of shapes for which the area or volume was known. Further advancements were made in the early seventeenth century by Isaac Barrow (1630–1677) and by Torricelli (Pisano and Capecchi 2010; Capecchi and Pisano 2010a) who provided hints of a connection between integration and differentiation. One of the first Barrow proofs of the fundamental theorem of calculus was provided by Barrow and it is known that Wallis generalized *Cavalieri's method* (Cavalieri 1635).

physical quantities during the birth of modern science, currently also referred to as *derived quantities*, e.g.: velocity and acceleration. For instance, both physical and mathematical quantities appear in the laws of motion. Without derivatives such as mathematical–physical operations, the above cited *physical quantities* would lack in mechanics. However, the next significant purely mathematical advancements in integral calculus appeared in ca. the sixteenth century, e.g., with Bonaventura Cavalieri (1598–1647) and later the use of integration, together with differentiation started to become one of the main operations in calculus applied to physical phenomena (Capecchi and Pisano 2007, 2008) and in geometry applied to the *science of weights* (Capecchi and Pisano 2010b; Pisano 2008). At this time, the work of Cavalieri with his *method of indivisibles* (Cavalieri 1635) and the works of Pierre de Fermat (c. 1601–1665) (Mahoney) began to lay the foundations of modern calculus. Thus, integration as a global operation was considered for the calculation of the area instead of the calculation of the tangent.[56] Thus, the use of integration for mechanical machines and thermodynamic machines could be acceptable due to their physical complexity: a global mathematical interpretation for running machines appeared reasonable and preferential: a working substance, volume, two reservoirs an input and an output, work.

In fact, Lazare Carnot's theory of mechanical machines utilizes, as a typical mathematical operation (in its fundamental equations, see above Chapters 2, 3 and below Chapter 11), the sum of all the parts that compose the physical system. By combining a sequence of elements using addition, the result is their sum. Generally speaking, without other explicit details, a sum may be expressed (under certain conditions) as a definite integral.

As a matter of fact, Lazare Carnot's theory of mechanical machines utilizes, as a typical mathematical operation (in its fundamental equations, see above Chapters 2, 3 and below Chapter 11), the summation on all the parts that compose the physical system. By combining a sequence of elements using addition, the result is their sum. Generally speaking, without explicit other details, a sum may be expressed (under certain conditions) as a definite integral

$$\sum_a^b f = \int_a^b f dx.$$

Even Sadi Carnot began his theory by considering global variables such as P, V, and t. Such globalist behavior from both father and son suggested to Sadi Carnot that he should theorize according to more ancient mathematical tradition, that is

[56]Nevertheless, we agree with the current secondary literature of the history of mathematics that a major advance in integration appeared in ca. the seventeenth century: an independent discovery of the fundamental theorem of calculus by Newton and Leibniz. Here, a strictly connection between integration and differentiation was proposed. Of course, for our aim, and equal in historical importance, is the comprehensive mathematical framework that both Newton and Leibniz developed, since we mainly take in account mathematical analysis calculus in the history until XIX century.

to first propose the global mathematical operation, the integration. In fact, in the *unpublished manuscript* (Carnot 1978, pp 223–225), Sadi Carnot attempted to calculate the integral

$$\oint \frac{dW_{\max}}{Q}$$

for steam in three phases only, without the correct closing of the cycle and reversibility (Carnot 1978, p 228, line 18). Therefore, in *Réflexions sur la puissance motrice du feu*, it was possible to study a cyclic integral $\oint d\left(\frac{W_{\max}}{Q}\right)$rather than the differential $d(W_{max}/Q)$. Nevertheless, *a priori* this integral depends on the path of integration; it essentially depends on the kind of machine being used. Sadi Carnot could have solved the problem by searching for the multiplicative factor of the integrand function which makes the cyclic integral null (see above); and then he could have obtained a primitive function which, as a new physical quantity, could have provided the correct *dynamics* of the process. We remark that this strongly abstract approach did not relate to physics. Instead, since it was dimensioned, the integrating factor would have changed the quantity η into another unknown quantity. In fact, passing from an operative mathematics in the unpublished *Recherche d'une formule propre à représenter la puissance motrice de la vapeur d'eau* (Carnot 1986, pp 167–180; see significant pages reported below) to the exposition in *Réflexions sur la puissance motrice du feu,* Sadi Carnot abandoned this calculation following an advanced mathematics. In fact, he utilized the mathematics of his time (Carnot 1978, pp 73–79, ft 1) to calculate the differential of the efficiency of a heat machine (*Ivi*, p 75, ft 1; see also below Chapter 9), operating on a cycle composed by two isotherms (t and $t + dt$) and two adiabatics (which, according to him, did not give any contribution and therefore were not taken into account) (Fig. 9.5).

As previously mentioned, in the first part of *Réflexions sur la puissance motrice du feu*, Sadi Carnot underlined the idea that, in order to obtain work it was necessary to have, first of all, a $\Delta t \neq 0$. He later stated that, in a heat machine, W_{max}/Q also depends on Q, thought of in terms of caloric (see footnote above). In this sense, Q and Δt appear to be the only important variables for Carnot's reversible heat machine. There could have been many more variables involved; for instance, the volume V and the pressure of a working substance. Nevertheless, it can be hypothesized that, by following Lazare Carnot's idea of studying the heat machines *en général*

> We draw a distinction here between the vapor [steam] engine and heat ["à feu"] engine[57] in general ["*en général*"]. The latter may be employed with any working substance, steam or anything else, in order to develop the motive power of heat.[58]

[57]On the first page of the manuscript presented by Hippolyte Carnot to the *Académie des sciences* on 16 December 1878, one can see (Carnot 1878a, folio 1) Sadi Carnot avoided the ancient use of the word "feu" (fire) in his book (see details Chapters 6 and 7 above). Nevertheless something remained in the book, e.g.: "[...] machine à vapeur de la machine à feu en général [...]". (Carnot 1978, p 8, ft 1, line 1).

[58]Carnot (1978), p 8, ft 1, line 1.

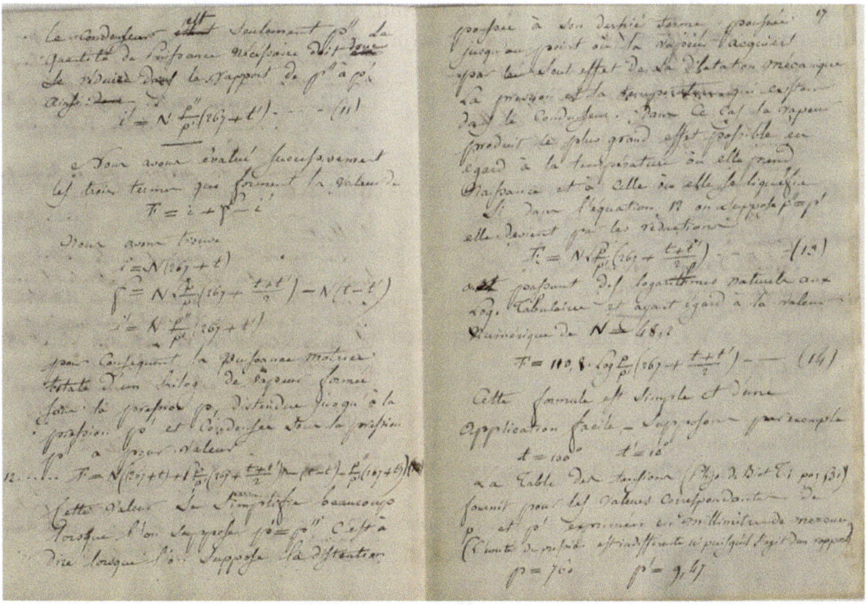

Fig. 9.5 Sadi Carnot's state equation on gas theory in the *unpublished manuscript* (Carnot S–EP, folia 16–17, © *Collections archives de la bibliothèque de l'École polytechnique de Paris* (see also Gabbey and Herivel; Fox 1986, pp 178–179))

In this sense, he decided to compare the *input* and the *output* of a machine by means of a performance coefficient. Consequently he theorized the functioning of a heat machine using the quantities that can be *seen* from the outside: the input Q, the output W, and the Δt of the two external reservoirs. In this way, he obtained the new quantity W_{max}/Q as a mathematical function of the variable Δt alone:

$$\eta = dW_{\max} = f(\Delta t).$$

Subsequently (as already discussed in reference to the footnote interpretation in step 5), Sadi Carnot calculated the *maximum* work for a quantity Q using a hypothesis that might seem arbitrary from a mathematical point of view: Sadi Carnot thought that in the expression of W_{max} the state function Q was *factorizable* (Carnot 1978, p 76, line 8, *Ivi, 1db*):

$$W_{\max} = g(Q, \Delta t) = Q\eta(\Delta t).$$

As previously stated from a strictly mathematical point of view, this hypothesis is not justified. Sadi Carnot then tried to physically establish that the efficiency of a heat machine is independent from the working substance used (Carnot 1978, p 38, line 4). In addition, since Sadi Carnot set (Carnot 1978, p 75, line 11) the initial temperature to *0*, he could simplify the formula to $\eta = f(t)$. Thus, $d\eta$ is an exact differential because η depends exclusively on one variable (Carnot 1978, p 75, line 17, p 76, line 7).

9.6 How did Sadi Carnot use his Adiabatic Equation? A Possible Calculation

As already mentioned, the fact that Sadi Carnot calculated a cycle between only two isotherms, t and $t + dt$ can leave us perplexed because the (final) result is still valid from a modern point of view.

> Carnot now goes on to calculate the motive power obtained in the cycle [4 phases] for air, using the convenient assumption that the "fall" of temperature [between two isotherms] is so small that the effect of the adiabatic can be ignored.[59]

Nevertheless, nothing proves that this is also valid for a finite adiabatic. Let us see the possible calculus. Sadi Carnot utilized the following adiabatic formula[60]:

$$t(V) = \frac{K + \tau \log [\ln] V}{K' + \tau' \log [\ln] V}. \tag{9.1}$$

Experimentally, the coefficient τ' of *log V* is a very small quantity, therefore, in order to simplify the calculations, as Sadi Carnot also suggested (see above) at the end of the mathematical footnote, $\tau' = 0$ can be introduced. In this case, Eq. 9.1 can be expressed in the following way:

$$t(V) = \frac{K}{K'} + \frac{\tau \log V}{K'},$$

from which, if we consider

$$\frac{K}{K'} = M \text{ and } \frac{\tau}{K'} = \Omega \tag{9.2}$$

it follows that

$$t(V) = M + \Omega \log V. \tag{9.3}$$

[59]Robert Fox in Carnot (1986), p 147, En. 78, line 1. (Authors' quotations).

[60]Carnot (1978), p 66, line 10.

Let us see consider Sadi Carnot's *ad hoc* cycle:

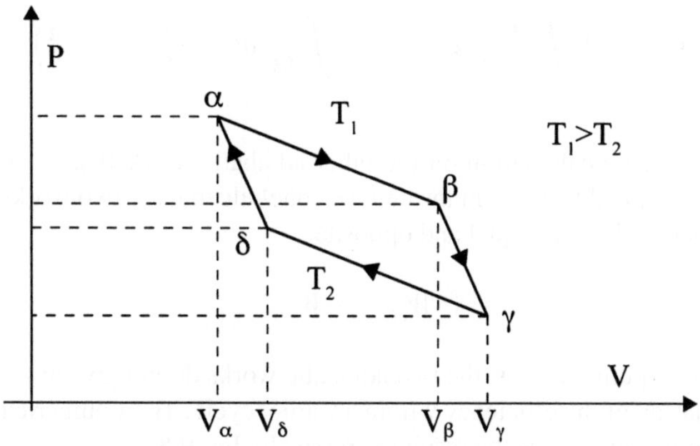

Fig. 9.6 An *ad hoc* simplified performance diagram of Sadi Carnot's processes

Let us calculate the work $W_{\gamma\beta}$ on the adiabatic expansion of an ideal gas (see Fig. 9.6.)

$$W_{\gamma\beta} = \int_{\beta}^{\gamma} pdV. \tag{9.4}$$

From Sadi Carnot's state equation of ideal gas $pV = Nt$ (Carnot S–EP folia 5–6; Carnot 1986, pp 172–173; see also below Chapter 9) we obtain $p = N\frac{t}{V}$, which substituted in Eq. 9.4, gives:

$$W_{\gamma\beta} = N \int_{\beta}^{\gamma} \frac{t(V)}{V} dV. \tag{9.5}$$

From Eq. 9.3 we can write

$$\log V = \frac{t(V) - M}{\Omega},$$

obtaining the following expression for volumes V and dV, which depend on t–variable temperature only:

$$V = e^{\frac{t-M}{\Omega}} = e^{\frac{t}{\Omega} - \frac{M}{\Omega}} = e^{-\frac{M}{\Omega}} e^{\frac{t}{\Omega}} = Be^{\frac{t}{\Omega}}$$

$$dV = e^{\frac{t}{\Omega}} \cdot \frac{B}{\Omega} dt \tag{9.6}$$

Thus, by substituting Eq. 9.6 in Eq. 9.5 we have:

$$W_{\gamma\beta} = N \int_{\beta}^{\gamma} \frac{t}{e^{\frac{t}{\Omega}}} \frac{1}{\Omega} e^{\frac{t}{\Omega}} dt = N \int_{\beta}^{\gamma} \frac{t}{\Omega} t\, dt = \frac{N}{2\Omega} \left(t_{\gamma}^{2} - t_{\beta}^{2} \right). \tag{9.7}$$

Executing the same calculation on the other adiabatic–work $W_{\delta\alpha} = \frac{N}{2\Omega}\left(t_{\delta}^{2} - t_{\alpha}^{2}\right)$, and since $t_{\gamma} = t_{\delta}$ and $t_{\beta} = t_{\alpha}$, in the end we conclude that the two works differ only by a sign; that is, they are equal and opposite:

$$W_{\gamma\beta} = -W_{\delta\alpha} \tag{9.8}$$

From a modern point of view, the two adiabatic works do not give any contribution to the calculation of the efficiency within a Carnot cycle. Their numerical values are equal and opposite; such as the result expressed by Eq. 9.8.

9.7 A Reflection on the Modern Calculation of Efficiency

Generally speaking the first principle of thermodynamics establishes that:

$$\oint dU = \oint dQ + \oint pdV.$$

Let us consider a physical system Γ and we will suppose that it is adiabatically isolated, that is, $\oint dQ = 0$. The adiabatic work accomplished on the system (chosen as positive) in bringing it from a state *1* to a state *2* is, for the first principle:

$$W = \int_{1}^{2} dU = U_2 - U_1.$$

Let us calculate, for an ideal gas, ($\Delta U = c_V \Delta t$ and $\Delta U = c_V \Delta t$) the contribution of the work on the isotherms and on the adiabatic in Sadi Carnot's reversible cycle.

Isothermal expansion $A \rightarrow B$ at temperature T_2. With regard to the diagram of the Carnot' cycle, for the first principle we have $\int dU = 0$, consequently we have $-Q = W$, that is:

$$-Q_2 = W_{AB} = \int_{A}^{B} pdV = nRT_2 \int_{A}^{B} \frac{dV}{V} = nRT_2 \log \frac{V_B}{V_A},$$

having $Q < 0$, since it is absorbed, and $W > 0$ because $\Delta V > 0$.

Adiabatic expansion $B \rightarrow C$ between the temperatures T_2 and T_1. For the first principle we have $\int dQ = 0$, therefore $\Delta U = W$:

$$\Delta U = W_{BC} = nC_V(T_2 - T_1),$$

having $W > 0$ because $\Delta V > 0$.

Isothermal compression $C \rightarrow D$ at temperature T_1. For the first principle, we have $\int dU = 0$, therefore $Q = -W$:

$$Q_1 = -W_{CD} = -\int_C^D p dV = -nRT_1 \int_C^D \frac{dV}{V} = -nRT_1 \log \frac{V_D}{V_C},$$

having $W < 0$ because $\Delta V < 0$.

Adiabatic compression $D \rightarrow A$ between temperatures T_1 and T_2. For the first principle we have $\int dQ = 0$, therefore $\Delta U = -W$:

$$\Delta U = -W_{DA} = -nC_V(T_2 - T_1),$$

having $W < 0$ because $\Delta V < 0$. Finally we arrive at the calculation of the work on the entire cycle. It is simple to verify that it is necessary to calculate the area which is enclosed by the four thermodynamic curves.

Let us point out that, for the whole cycle, $W_{BC} + W_{DA} = 0$; i.e., the contribution of the work on the two adiabatics, is zero. Therefore, Sadi Carnot was right to disregard it. However, in modern thermodynamics the two adiabatics depend on the nature of the gas through the coefficient γ. Therefore the cycle efficiency may be considered as independent from the nature of the gas only if the adiabatics do not give any contribution to the work. In fact, applying the first principle to the whole cycle we have:

$$\oint dU = 0$$

$$Q_{ciclo} = Q_2 + Q_1 = W_{AB} + W_{CD} = W_{ciclo}$$

$$W_{ciclo} = nR(T_2 - T_1) \log \frac{V_B}{V_A} + nR(T_2 - T_1) \log \frac{V_C}{V_D}$$

which is the area sought under the curve of the heat work performed. Now we consider the relation between heat Q_1, obtained by isothermal expansion and heat Q_2, obtained by isothermal compression:

$$\frac{Q_1}{Q_2} = \frac{T_1 \log \frac{V_D}{V_C}}{T_2 \log \frac{V_B}{V_A}}. \tag{9.9}$$

It is interesting to show that in modern thermodynamics this relation is independent from the *ratio of compression* V_D/V_C and V_B/V_A. In fact, applying the adiabatic equation to the reversible transformations *AD* and *BC,* we have:

$$V_B T_2^{\frac{1}{\gamma-1}} = V_C T_1^{\frac{1}{\gamma-1}}$$
$$V_A T_2^{\frac{1}{\gamma-1}} = V_D T_1^{\frac{1}{\gamma-1}}$$

and dividing member by member, we obtain equality $V_B/V_A = V_C/V_D$, which, when substituted in (9.9), gives the following result:

$$\frac{Q_1}{Q_2} = \frac{T_1}{T_2}.$$

In conclusion, the reversible efficiency of a heat machine in a Carnot cycle is expressed as follows:

$$\eta_{rev} = \frac{Q_2 - Q_1}{Q_2} = 1 - \frac{Q_1}{Q_2} = 1 - \frac{T_1}{T_2.}$$

9.8 Final Remarks

> I believe that analysis of Carnot's description raises questions whose answers have both physical and historical interest.[61]

We hope that this historical epistemology of thermodynamics focused on Sadi Carnot's footnote in *Réflexions sur la puissance motrice du feu* discussed by means of his original level of mathematics and the application of his father's *synthetic method* can contribute to clarifying an enigmatic and crucial aspect of the birth and development of both Sadi Carnot's theory and the historical knowledge of thermodynamics. This is an aspect whose importance has been underestimated, since the possibility that Sadi Carnot could have applied his father's *method* was ignored. This cultural parentage strengthens the hypothesis of a strict scientific and filial relationship between Sadi Carnot and his father's science, especially in regard to mathematics (see Chapter 11).

We should consider that even from a modern point of view, Sadi Carnot's result on two adiabatics could have been obtained by using the *synthetic method*, that is, neglecting the works of two adiabatics. Therefore, even though only in thermodynamics, the *synthetic method* appears to have such a general and simple nature, that it can be independent from the theory to which it is applied.

[61]Martin J. Klein in Taton (1976), p 214, line 19.

> A theory is more impressive the greater simplicity of its premises is, the more different kinds of things it relates, and the more extended its area of applicability. Therefore the deep impression which classical thermodynamics made upon me. It is the only physical theory of universal content concerning which I am convinced that, within the framework of the applicability of its basic concepts, it will never be over thrown.[62]

In the end, it is possible to consider that Sadi Carnot's exposition is not an attempt to translate qualitative data into quantitative data, e.g. a mathematical formulation. Instead, it represents an effort to clarify the fundamental method that allows us to introduce mathematics into thermodynamics. An effort that might to appear similar to that of Galileo Galilei (1564–1642) which tried to introduce mathematics in his mechanics presented in *Le Mecaniche*[63] (1634) and in *Discorsi e dimostrazioni matematiche sopra a due nuove scienze*[64] (1638).

---

[62]Einstein (1949 [1970]) vol. VII, p 33; also quoted in: Klein (1967), pp 509–516.

[63]Galileo (1890–1909), II, pp 147–191. For a recent review of *Le Mecaniche* see Romano Gatto's work (Galileo 2002) and, of course indispensable Stillmann Drake' works on the Galileo (e.g., Drake 1973, 1985, 2000; Drake and Drabkin (1969)). See also an interesting French version, of the *les mécaniques de Galilée* by Egidio Festa and Sophie Roux (forthcoming). In regard to *Le Meccaniche,* Gatto comments both *versione breve* (4 manuscripts, Galileo 2002, pp 3–42) that *versione lunga* (14 manuscripts, *Ivi*, pp 43–154). The manuscript edited by Antonio Favaro (1847–1922) in *Opere nazionali di Galileo Galilei* is the *versione lunga*. It is composed of 10 manuscripts at his time known (Galileo 1890–1909, II, 155–190). In the *versione lunga* one can see the effort cited in the running text. For recent works concerning Galilei's method and his mathematical approach one can consult: Pisano (2008, 2009a, b, c; Pisano and Bussotti (2012)).

[64]Galileo (1890–1909), VIII, pp 39–448; see also Galileo (1954).

Chapter 10
Studies and Heritage of *Réflexions sur la puissance motrice du feu*

Une chose digne de remarque c'est le rapport que les Réflexions sur la puissance motrice du feu de SADI CARNOT ont avec les considérations renfermées dans un petit ouvrage de son illustre père, le Général CARNOT, publié en 1783, sous le titre d'Essai sur les machines en général. Ce que le père a fait pour la mécanique ordinaire, le fils l'a fait pour La mécanique de la chaleur.

(Saint–Robert 1870b, II, p 447, line 24; s. transl. below)

In order to describe a possible scientific and historical panorama of the role played by the physical foundations of *Réflexions sur la puissance motrice du feu* when it was first published in this section, we've selected scientific studies and historical studies regarding Sadi Carnot's theory. As we have presented in previous chapters, this theory was essentially based on three *anomalies* of the paradigm at the time: the novelty of the cycle both as a device of reasoning and an, *ad absurdum* proof in calculus and the use and doubts regarding caloric. Therefore, for our purposes, we focus on: the contemporary birth of the theory of heat and its developments (Fourier, Lamè), the scientific studies on cycles (Hoyer), and the developments of thermodynamics devoted to the physics and mathematics of the nineteenth century regarding the dynamical effects of heat based on Sadi Carnot and Clapeyron's arguments (Reech). Finally, we examine a thermodynamics textbook (Saint–Robert 1865, 1870c) used at the end of the nineteenth century which was fundamentally devoted to Lazare and Sadi's typical scientific approach.

10.1 An Outline of the Problem

First, we briefly present historical reflections on Gabriel Lamé's *Leçons sur la théorie analytique de la chaleur* (Lamé 1861b) and Joseph Fourier's *Théorie analytique de la chaleur* (Fourier 1822), focusing the physical–mathematical relationship

C.C. Gillispie and R. Pisano, *Lazare and Sadi Carnot: A Scientific and Filial Relationship*, History of Mechanism and Machine Science 19, DOI 10.1007/978-94-017-8011-7_10,

in the theory of heat. In the second part, we address Sadi Carnot's theory in terms of the characteristic cyclic process with the aim of obtaining results independent from the *chaleur* and *calorique*, in order to discuss the equivalence of heat and work. Therefore, we present Reech's *Théorie générale des effets dynamiques de la chaleur,* 1853 (Reech 1853, pp 357–378; see also Reech 1851, pp 567–571) in which he adopted and generalized Sadi Carnot's (and Clapeyron's) reasonings in order to obtain a general formula from which each of the two theories (on caloric and on heat) can be derived under the right conditions. In this sense, more recently than Reech, Hoyer (1975, 1976) demonstrated that the modern theory of the equivalence of heat and work and caloric theory in Sadi Carnot's work coincide by means of an infinitesimal process. For our part, we suggest that there is a way to reconsider Reech's general result according to Hoyer's modern suggestion, thereby obtaining the modern theory of the equivalence and so explaining one of Sadi Carnot's surprising results. In the third part, we present Paul de Saint–Robert, an Italian scholar who published one of the first textbooks on the subject that included scientific studies, as well as historical and biographical notes on Sadi Carnot. The book is entitled, *Principes de thermodynamique* (Saint–Robert 1865, 1870c). With regard to the historical period from 1850 to 1870, we provide a scientific and historical study of the foundations of the first edition (Saint–Robert 1865) to examine his way of discussing the second principle. In particular, we compare it to its original version in *Réflexions sur la puissance motrice du feu* published 41 years before.

10.2 *La théorie analytique de la chaleur*. Reflections on Fourier and Lamé

10.2.1 On Fourier's théorie analytique de la chaleur (1822)

On December 21st, 1807, Jean Baptiste Joseph Fourier[1] (1768–1830) (Grattan-Guinness 1972; Grattan–Guinness in Gillispie 1970–1980)[2] presented his work on heat entitled, *Théorie de la propagation de la chaleur dans les solides*, at the *Académie des sciences* in Paris (Fourier 1807, I, pp 112–116, Fourier 1888–1890). He focused on heat transmission between discrete masses and gave special case studies of continuous bodies (e.g. ring, sphere, cylinder, rectangular prism and

[1]The content of this chapter is based on Pisano (2001, 2005), Drago and Pisano (2008), Pisano and Capecchi (2009). (The discussion on Sadi Carnot Reech and Hoyer was also previously arranged by one of us in his lauream dissertation in Physics).

[2]See also next 1990 edition.

cube). A committee composed, among others, by Joseph–Louis Lagrange (1736–1813) and Pierre Simon Laplace (1749–1827) who rejected it, judged his work. According to Grattan–Guinness:

> The significance of the contributions made to the development of mathematical physics by Joseph Fourier (1768–1830) has been made difficult to evaluate by the controversies, which they stirred amongst his colleagues and rivals. One result of this situation was that his major paper, submitted to the Académie des sciences at Paris in 1807, was never published. The manuscript is still extant, however, and its major features are discussed in the light both of contemporary results and also of immediately succeeding work which it did so much to inspire. An edition of this and other Fourier manuscripts on mathematical physics is in preparation.[3]

In 1811[4] in occasion of a prize competition for the propagation of heat in solid bodies, Fourier improved his work and presented it again (Fourier 1829, VIII, pp 581–622, Fourier 1888–1890, pp 595–614). A crucial mathematical problem was the propagation (one dimension) of heat in an infinite domain. He won the competition but his work was not given the possibility of being published in the *Mémoires de l'Académie*. In 1822, he published a monographic work, *Théorie analytique de la chaleur* (Fourier 1822, pp i–xxij; Fourier 1888–1890, I) which integrally contained the *Mémoire* of 1811 in its first part. Gabriel Lamé wrote about Fourier in his *Discours préliminaire,* contained within *Leçons sur la théorie analytique de la chaleur*:

> Taking into account the history [...]. Fourier's original work on the analytic theory of heat, only considers the case–study of the propagation of heat, for homogeneous solid bodes. The accepted principle leads to this corollary, that is the flux of the heat, which moves within plane–body is expressed by a variation, normal to surface, of the function, and that function is the temperature. This corollary quickly establishes the second–order partial differential equation, which prevails within bodies, [...].[5]

In formulating heat conduction in terms of a partial differential equation and developing methods to solve it, Fourier introduces many innovations because the science of heat and the theory of differential equations were both in an early stage of development at his time, as we have just observed. One of the most important aspects of Fourier's work was the method of solution. However, his physical insights were also quite relevant. In 1822, in his *Discours préliminaire* (Fourier 1822, pp i–xxij), Fourier also discussed the problems of his theory (Fig. 10.1):

[3]Grattan-Guinness (1969), p 230.

[4]The prize was only conferred in 1812.

[5]Lamé (1861a), p XII, line 21.

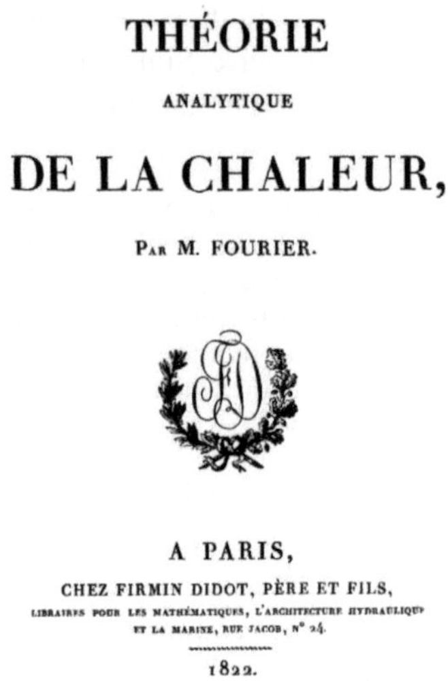
THÉORIE
ANALYTIQUE
DE LA CHALEUR,
PAR M. FOURIER.
A PARIS,
CHEZ FIRMIN DIDOT, PÈRE ET FILS,
LIBRAIRES POUR LES MATHÉMATIQUES, L'ARCHITECTURE HYDRAULIQUE ET LA MARINE, RUE JACOB, N° 24.
1822.

Fig. 10.1 Fourier's preliminary arguments (Fourier 1822)

> Why do terrestrial temperatures cease to be variable at a depth so small with respect to the radius of the earth?[6] Why do not the annual changes alone in the distance of the sun from the earth, produce at the surface of the earth very considerable changes in the temperatures?[7] [. . .] how shall we be able to determine that constant value of *the temperature of space*, and deduce from it the temperature which belongs to each planet?[8]

He also dealt with other problems and the social importance of solving them, since in his opinion they remained unresolved at the time. Among the various problems, he emphasized those typically associated with early heat machines,[9] which began to be widespread at the time.[10] In the first pages of his *Discours préliminaire,* Fourier had already noted the great results obtained by rational mechanics (Fourier 1822, pp i–ij): he included the principles, which regulated the motion of the planets, the equilibrium and oscillation of the seas, the propagation of light and the capillary phenomena. Moreover, he claimed heat phenomena were not referable to mechanical phenomena, which already had a consolidated scientific theory (Fourier 1822, pp ij–iv). In this sense, a theoretical and satisfactory theory for heat phenomena still seemed distant–at least until Sadi Carnot's scientific work appeared in 1824 (Carnot 1978).

---

[6] Fourier (1878), p 4, line 27. (*Id.*, 1822, p vij, line 25). We listed the references of both the French edition (Fourier 1822) and the English edition (Fourier 1878).

[7] Fourier (1878), p 4, line 35. (*Id.,*1822, p viij, line 12).

[8] Fourier (1878), p 5, line 15. (*Id.,*1822, p ix, line 10). Author's italics.

[9] E.g.,: Denis Papin (1691), Thomas Savery (1698), *Boulton & Watt company* in Soho, near Birmingham (1776–1800), Richard Trevithick (1801) with the first steam engine called "Catch me who can"; Stephenson (1813).

[10] Singer ([1954–58] 1993), IV, pp 152–221, Dickinson (1939), chaps. 1 and 2, Tredgold (1838), II; Fox and Weisz (2009).

> But whatever may be the range of mechanical theories, they do not apply to the effects of heat. These make up a special order of phenomena, which cannot be explained by the principles of motion and equilibrium. We have for a long time been in possession of ingenious instruments adapted to measure many of these effects; valuable observations have been collected; but in this manner partial results only have become known, and not the mathematical demonstration of the laws which include them all.[11] [...] To found the theory, it was in the first place necessary to distinguish and define with precision the elementary properties which determine the action of heat. I then perceived that all the phenomena which depend on this action resolve themselves into a very small number of general and simple facts; whereby every physical problem of this kind is brought back to an investigation of mathematical analysis. From these general facts I have concluded that to determine numerically the most varied movements of heat, it is sufficient to submit each substance to three fundamental observations.[12]

The basic proprieties assumed by Fourier concern the capacity of bodies to absorb and transfer heat. A small group of *primary facts* deduces the *principles* of the theory, as one can see in rational mechanics:

> The principles of the theory are derived, as are those of rational mechanics, from a very small number of primary facts, the causes of which are not considered by geometers, but which they admit as the results of common observations confirmed by all experiment. The differential equations of the propagation of heat express the most general conditions, and reduce the physical questions to problems of pure analysis, and this is the proper object of theory. They are not less rigorously established than the general equations of equilibrium and motion.[13]

The main principle adopted by Fourier for the heating transmission is the following:

> SECTION III. *Principle of the communication of heat.*
> 57. We now proceed to examine what experiments teach us concerning the communication of heat. If two equal molecules are formed of the same substance and have the same temperature, each of them receives from the other as much heat as it gives up to it; their mutual action may then be regarded as null, since the result of this action can bring about no change in the state of the molecules. If, on the contrary, the first is hotter than the second, it sends to it more heat than it receives from it; the result of the mutual action is the difference of these two quantities of heat. In all cases we make abstraction of the two equal quantities of heat which any two material points reciprocally give up; we conceive that the point most heated acts only on the other, and that, in virtue of this action, the first loses a certain quantity of heat which is acquired by the second. Thus the action of two molecules, or the quantity of heat which the hottest communicates to the other, is the difference of the two quantities which they give up to each other.[14]

The mathematical side emerges from the physical one, both for experimental phenomena and for hypotheses and the reasoning necessary to introduce them. In the end, one can only discuss the solutions of equations and their physical–mathematical limits.

---

[11]Fourier (1878), p 2, line 3. (*Id.*, 1822, p ij, line 24).

[12]Fourier (1878), p 2, line 16. (*Id.*, 1822, p iij, line 14).

[13]Fourier (1878), p 6, line 11. (*Id.*, 1822, p xj, line 4).

[14]Fourier (1878), p 41, line 3. (*Id.*, 1822, p 39 line 23). Author's italics.

59. Denoting by v and v' the temperatures of two equal molecules m and n, by p, their extremely small distance, and by dt, the infinitely small duration of the instant, the quantity of heat which m receives from n during this instant will be expressed by $(v' - v)\varphi(p).dt$. We denote by $\varphi(p)$a certain function of the distance p which, in solid bodies and in liquids, becomes nothing when p has a sensible magnitude. The function is the same for every point of the same given substance; it varies with the nature of the substance.[15]

Fourier is indifferent to the problem of the nature of heat also because, from a mathematical point of view, it does not modify any reasoning regarding the partial derivative of the new analytical theory of heat. He explained this in the first page regarding the preliminary concepts of the propagation of heat.

SECTION II. *Preliminary definitions and general notions.*
22. Of the nature of heat uncertain hypotheses only could be formed, but the knowledge of the mathematical laws to which its effects are subject is independent of all hypothesis; it requires only an attentive examination of the chief facts which common observations have indicated, and which have been confirmed by exact experiments.[16]

The propagation of heat in solids is interpreted by functions in time and space and differential equations:

142. THEOREM IV. From the preceding Theorems it is easy to deduce the general equations of the propagation of heat. *Suppose the different points of a homogeneous solid of any form whatever, to have received initial temperatures which vary successively by the effect of the mutual action of the molecules, and suppose the equation* $v = f(x, y, z, t)$ *to represent the successive states of the solid, it may now be shown that v a function of four variables necessarily satisfies the equation*

$$\frac{dv}{dt} = \frac{K}{CD}\left(\frac{d^2v}{dx^2} + \frac{d^2v}{dy^2} + \frac{d^2v}{dz^2}\right)^{17}.$$

Nevertheless, at the time, Fourier's analytic theory of heat was too premature to be involved in a definitive mathematical and mechanical scheme (Grattan-Guinness 1969, pp 230–253). In fact, "Part of its impact was that it did not fit into the scheme of rational and celestial mechanics".[18] In any case, in the first half of the 1800s, one can remark a crucial change in physical theorization because central forces were found to be ineffective for the production of heat phenomena. Despite Fourier's introduction of mathematical techniques used in Newtonian physics (differential equation), he understood[19] that not all of his results guaranteed an evident *continuum* of the previous theory of heat (and heat machines).

In summary, although the theory had earned the use of advanced mathematics, the theoretical advancement proposed by Fourier relied too heavily on a Newtonian

[15]Fourier (1878), p 42, line 34. (*Id.*,1822, p 42, line 11).

[16]Fourier (1878), p 26, line 11. (*Id.*,1822, p 18, line 10).

[17]Fourier (1878), p 112, line 15 (*Id.*, 1822, pp 134–135, line 19). Author's italics.

[18]David Keston, *Joseph Fourier, Politician & Scientist*, also available via: http://www.todayinsci.com/F/Fourier_JBJ/FourierPoliticianScientistBio.htm

[19]"[...] most complex effects of all the natural forces, and thus has the thought of Newton been confirmed: *quod tam paucis tam multa proestet geometria gloriatur*. But whatever may be the range of mechanical theories, they do not apply to the effects of heat. These make up a special order of phenomena, which cannot be explained by the principles of motion and equilibrium." (Fourier 1878, p 1–2, line 27, 1822, pp ij–iij, line 20). Author's italics.

paradigm; it was also too narrow without clear perspectives for the further development of a general theory of heat.

10.2.2 On Lamé's Leçons sur la théorie analytique de la chaleur *(1861)*

In 1861 Gabriel Lamé (1795–1870) wrote his textbook, *Leçons sur la théorie analytique de la chaleur* (Lamé 1861b), focusing on the mathematical role played by the transmission of heat, independently of both physical laws and the nature of heat. In a previous work, *Leçons sur la théorie mathématique de l'élasticité des corps solides* (Lamé 1852), he had already advanced his preliminary ideas on the role played by mathematics in a physical theory:

> The Physics mathematics itself, is an entirely modern conception, which belongs exclusively to the Geometers of our century. Today, the science only includes three chapters, one each other differently thought, which are rationally conceived; that is they are based on principles or uncontestable laws. These chapters are: the theory of static electricity studied on the surface of the conductors–bodies; the analytic theory of heat; [and] within mathematical theory of elasticity of solid bodies. The latter is the most difficult, the less complete; and it is also the most useful was, during a time in which we appreciate the importance of a mathematical theory since results which it can immediately give to the industrial practice.[20]

Therefore, in this sense see the following image (Fig. 10.2)

LEÇONS

SUR LA

THÉORIE MATHÉMATIQUE

DE

L'ÉLASTICITÉ DES CORPS SOLIDES,

PAR M. G. LAMÉ,

MEMBRE DE L'INSTITUT.

PARIS,

BACHELIER, IMPRIMEUR-LIBRAIRE

DU BUREAU DES LONGITUDES ET DE L'ÉCOLE POLYTECHNIQUE,

QUAI DES AUGUSTINS, 55.

1852

Fig. 10.2 His preliminary concepts on physics mathematics

[20]Lamé (1852), p V, line 11.

> The Analysis should be including soon, no doubt, to embrace other parts of the General Physics, such as the theory of light, and the electrodynamics phenomena. [. . .]. But, we cannot repeat all the time, the truth of physics mathematical is a so rigorous science, so accurate such as rational mechanics. It differs, consequently, for all applications on given principles, on free or convenient hypotheses, on empirical formulas; most are the numerical calculations for a synthetically classification. [. . .] it is uncontestable that all these empirical and partial theories are non-definitive sciences.[21] [. . .] Their reign is mainly provisory. They will survive in time if rational Physics invade their domain. Then, they will have more than a historical importance.[22]

Moreover, as was common at the time, in *Leçons sur la théorie analytique de la chaleur*, Lamé was influenced by the positivistic environment of French science at the time (Gillispie 1980, 2004; Fox 1995, Fox and Guagnini 1993). In order to better understand his analytical approach, certain lines from his previous work (Lamé 1852) must be noted. In this work, he focused a great deal of his scientific thoughts on the role played by mathematics in respect to physical phenomena, moving towards a *rational physics* called *physique rationnelle* (Guinness 1990a, b). Gabriel Lamé's concepts (e.g. velocity, propagation) with regard to heat were supported by advanced mathematics, differential equations and integral calculus. This mathematical aspect led him to a rational theory, which demonstrates its true power when it is included in a larger and consolidated theory and interpreted mathematically. Particularly, this approach was based on methods and techniques, which were quite different from Lavoisier and Sadi Carnot's theories as previously demonstrated.

Generally speaking, Lamé is not interested in experimental procedures of measurements, hypotheses on nature of heat, the physical theory of heat or specific heat, as one can read in his theoretical premise of the theory in his *Discours préliminaire* of the *Leçons sur la théorie analytique de la chaleur* (Fig. 10.3):

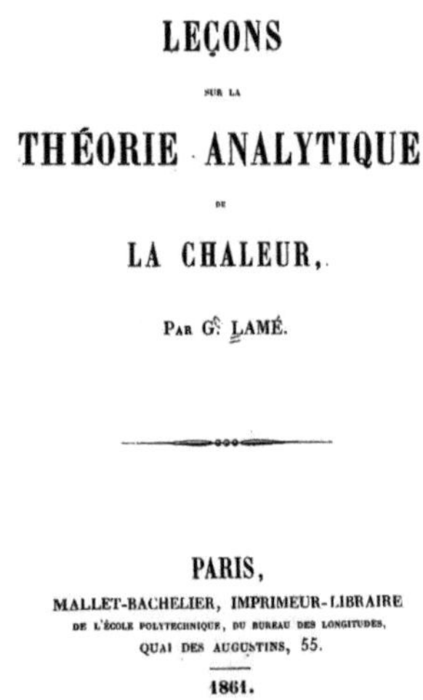
LEÇONS

SUR LA

THÉORIE ANALYTIQUE

DE

LA CHALEUR,

PAR G. LAMÉ.

PARIS,

MALLET-BACHELIER, IMPRIMEUR-LIBRAIRE

DE L'ÉCOLE POLYTECHNIQUE, DU BUREAU DES LONGITUDES,

QUAI DES AUGUSTINS, 55.

1861.

Fig. 10.3 His preliminary teaching concepts (In regard to this matter, one of Lamé's previous works in his *Cours de physique* at *École polytechnique* (Lamé 1836) is very interesting. It includes crucial aspects of his main idea concerning the relationship between mathematics and physics (Pisano 2011a))

[21]Lamé (1852), p VI, line 4.

[22]*Ibidem*, line 24.

> The Course that I am going to do nowadays is mainly focused on the purpose to establish the [contents of] analytical theory of heat, without using any hypothetical principle concerning the internal constitution of the matter, without using presupposing the laws on exchange of heat, or particular reasoning, without adopting any restriction on conductibility around a given point. I am more convinced that by avoiding in this way, any preconceptions about the natural laws, [. . .] the Theory of Elasticity, completely free of any hypothetical principles, can rigorously demonstrate, based on the facts, in diaphanous environments, the ponderable vibrations of *lights*.[23]

Thus, he removes any physical considerations. He also introduces temperature as a mathematical function:

> The laws, which rule internal part of homogeneous solid, or functions of three– coordinates, such as variables for static conditions only, and the time, such as a fourth variable for dynamic state are proposed to determinate.[24] [. . .] It is then necessary to modify or adopt the principle already presented [. . .]: I° the temperature is a continuous function of time–coordinates; 2° the quantity of heat transferred, within very short time, from hotter first point to a colder second point, very near [infinitesimal] one each other, [. . .] is proportional to very small difference [infinitesimal] of temperature of these two points.[25]

In detail, we have:

1. The temperature is a function of the space and time $T = f(x, t)$ (continuous cases only).
2. Constant orientations–transfer of heat into a material object.
3. For $T = f(x, y, z)$ with three variables, he defines a static state.
4. For $T = f(x, y, z, t)$ with four variables, he defines a dynamic state.

He wrote:

> The corresponding flow to any plane–elements
>
> $$(23)\ \omega dt\left(-q\frac{dV}{dn}\right);$$
>
> *dn* is increment of the normal on element, and *q* is constant conductibility. Thus, in a non–crystalline homogeneous solid each elementary flow includes only a normal variation of the temperature, and the coefficient of that variation is the same for all plane–elements, whatever is their orientation. The expression (23) is the same expression deduced by Fourier but starting from a physical law, considered as point of departure. In the current theory, that expression is rigorously established; and we clearly see that it must not include the tangential–variations–of–temperature [. . .].[26]

---

[23] Lamé (1861b), p. V, line 1. Author's italics.

[24] Lamé (1861b), p VI, line 26.

[25] Lamé (1861b), p 2, line 8.

[26] Lamé (1861b), p. 16, line 8. Since it is a difficult original print, we changed some letters in the formula "(23)". Nevertheless, the mathematical content does not change.

In the following section we avoid focusing attention on the intimate nature of heat, which is somewhat uncertain in Fourier's work[27] and more decisive in Lamé's work, perhaps also due to the temporal distance between the two scientists. *Leçons sur la théorie analytique de la chaleur* echoes Fourier's theory while adding some improvements (Lamé 1861b, p 1), especially for mathematical techniques. In particular: it introduces elliptical coordinates, non-isotropic solids and more complex geometrical forms of bodies.

> 15. It still remained to compare the facts with theory. [...] These experiments confirm the principle from which we started, and which is adopted by all physicists in spite of the diversity of their hypotheses on the nature of heat.[28]

10.2.3 Final Remarks on Fourier and Lamé's Theories

Observed from a historical epistemological point of view, the initial research on heat seemed to echo the scientific programs in Laplace's *Traité de mécanique céleste* and *Exposition du système du monde* (Laplace 1805, [1836] 1984), focusing on central forces and offering the possibility of realizing differential equations. The exaltation of the importance of the mathematical theory of heat occurs when it is included in a new theory, e.g., generally speaking, propagation, velocity applied to heat, etc. as occurs in the following flow chart (Fig. 10.4).

This foundational aspect mainly began thanks to Fourier first and Lamé later. In these types of theories there are various mathematical aspects so the theory appeared entirely mathematical, e.g. the analytic theories. In fact, the content was advanced by advanced mathematics (differential calculus by partial derivatives, integral calculus, series, etc.) to interpret each field of phenomena. Therefore, a physical theory in origin is absorbed by mathematics, producing two crucial consequences: the birth of a new theory (e.g., *Théorie analytique de la chaleur*) and, at the same time, the scientific–cultural demise of the previous classical physical theory of heat.

In the end, the birth and developments of analytical theories of heat also based on results and concepts developed between the eighteenth and nineteenth centuries, seemed uninterested in the revolutionary debate of the nineteenth century caused, for example, by the birth of thermodynamics. Their development progressed independently from *chaleur*, *calorique* and heat machines. Since in general they were methodologically based on, a mechanical view of nature (mathematical space and time, velocity, differential calculus, etc.), they contributed to the success of the triumphant *mécanique céleste*: one of the most relevant theories of the nineteenth century absorbed by mechanics. The latter, may be the most important

[27] Generally speaking the analytical theory of heat is not interested in the intimate nature of heat.

[28] Fourier (1878), p 22, line 17; *Id.*, 1822, p 12 line 16. See also Fourier (1822), pp 4–6.

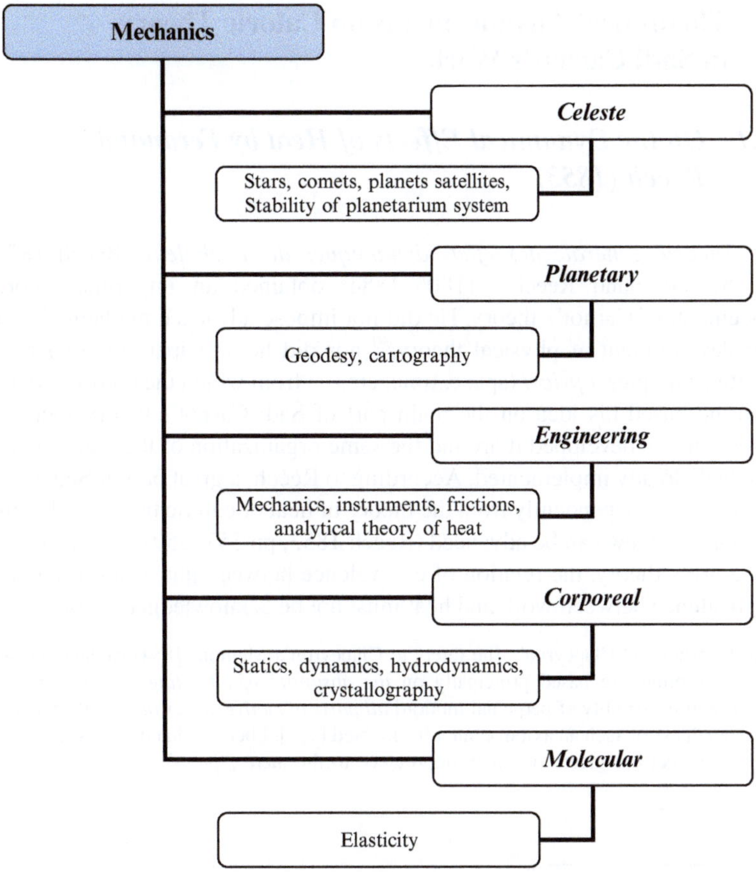

Fig. 10.4 An example of some theories included in the general domain of mechanics (Let us note that the inclusion concerns the foundations (mathematics, physics) and the mechanical approach (Pisano 2011a))

scientific system of knowledge but it is also science's most conservative and difficult approach. We must wait for the first two decades of the twentieth century to see a radical change in the direction of both foundations and methods (2009d). This latter hypothesis could also suggest why after Lamé, the *Théorie analytique de la chaleur* proved to be too specialized; it appeared too focused on the needs of advanced mathematics, further specified within an idealistic and difficult paradigm (Pisano 2012a).

10.3 Theoretical Advancements on Caloric Theory in Sadi Carnot's Work

10.3.1 On the Dynamical Effects of Heat by Ferdnand Reech (1853)

In his *Théorie générale des effets dynamiques de la chaleur* (Reech 1853, pp 357–378), Ferdnand Reech[29] (1805–1884) obtained an important theoretical advancement, of Carnot's theory. He did not impose advanced mathematical ideas on the development of physical theory,[30] nor did he anticipate the experimental data with principles *a priori* (space, time, etc.... from which the theory is derived). Instead, he based his idea on the main part of Sadi Carnot's theory focusing on the cycle and he developed it around the same organization of the theory that Sadi Carnot had already implemented. According to Reech, a great part of Sadi Carnot's theory is valid independently from the theory of heat or caloric theory and therefore a new point of view can be advanced (Reech 1853, pp 357–359). That is to say, for Sadi Carnot's theory, the relation of equivalence between quantities q' and q, and the equivalence between work and heat must not be acknowledged *a priori.*

> [Sadi] Carnot and Clapeyron's [he refers to Clapeyron's Memoir, 1834] calculations and their reasonings are based principally on *the absurdity of admitting the possibility of creating* [impossibility of perpetual motion] *all parts of motive power or heat* ["*chaleur*"]. The soundness of such an axiom cannot be doubted [...]. I believe that most research done on this subject has given too much importance to the mere ["pures"] hypotheses, losing

---

[29]Ferdnand or Ferdinand or Frederick or Frédéric Reech, French marine engineer, entered the *École du génie maritime* in Brest after graduating from *École polytechnique* (1825). Upon the completion of his naval training, he was promoted to the rank of lieutenant in 1829. He soon devoted himself exclusively to teaching and research in geometry, mechanics, thermodynamics, and hydrodynamics. Reech was director of the *École du génie maritime* (1831–1870) and in 1854 was transferred back to Paris, where he devoted himself to teaching and also became a professor at Université Sorbonne, Paris. Reech published ten memoirs, ranging from a brochure supporting his unsuccessful candidacy for membership at the *Académie des sciences*: e.g., notes on mechanics, mathematics physics applied to mechanics and topics of applied thermodynamics. Most remember him for his contribution to the first formulation of the hydraulic–model law of gravitational similitude: the necessary proportionality between the velocity of a ship and the square root of its length (more generally known as Froude's law of similarity). This law was included in his lectures as early as 1831, but it is first mentioned in his memoir from 1844 and then discussed in detail in his course notes from 1852. In 1870, Reech retired from active duty and returned to Lorient in Brittany, a fascinating region in the North West of France. He was also a member of the *Association française pour l'avancement des sciences* (1875–1883). At his request, he was buried in Soultz–sous–Forêts, close to the village of his birth. Truesdell included Reech in his summary of thermodynamic views (Truesdell and Bharatha 1977; see also Drago and Pisano 2008).

[30]He rebuilt the definition of integral, which, in a global conception of differential calculus concerning the calculus of an area for a given function, we can consider is the first mathematical operation.

> sight of the logical filial ["filiation logique"] of Mr. [Sadi] Carnot's reasoning, which was not affected, in my opinion, by objection [to caloric], and only needs to be completed from a new point of view.[31]

His reasoning concerns Carnot's cycles (which he supposes exist for every working substance), that is Carnot's result which was utilized by everyone who related Sadi Carnot's theory to modern thermodynamics. Therefore, without being familiar with Lazare Carnot's synthetic method, Reech emphasized the part of Sadi Carnot's theory which, as we have seen, depended on it: Carnot's reversible heat machine operating between two sources *A* and *A'*. Let us examine Reechs' main hypotheses[32] and reasoning (Drago and Pisano 2008) (Figs. 10.5, 10.6 and 10.7).

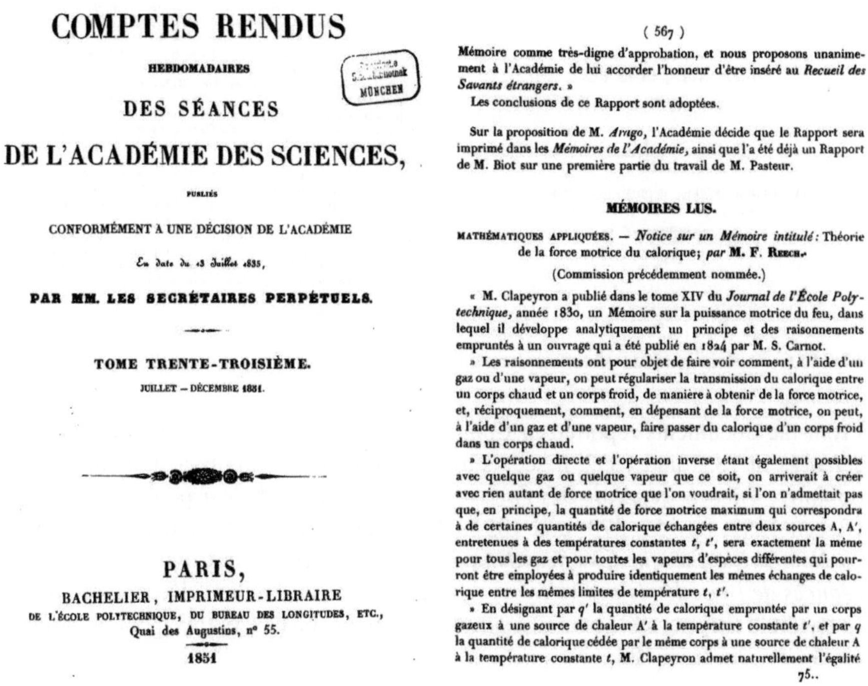

COMPTES RENDUS

HEBDOMADAIRES

DES SÉANCES

DE L'ACADÉMIE DES SCIENCES,

PUBLIÉS

CONFORMÉMENT A UNE DÉCISION DE L'ACADÉMIE

En date du 13 Juillet 1835,

PAR MM. LES SECRÉTAIRES PERPÉTUELS.

TOME TRENTE-TROISIÈME.

JUILLET — DÉCEMBRE 1851.

PARIS,

BACHELIER, IMPRIMEUR-LIBRAIRE

DE L'ÉCOLE POLYTECHNIQUE, DU BUREAU DES LONGITUDES, ETC.,

Quai des Augustins, nº 55.

1851

(567)

Mémoire comme très-digne d'approbation, et nous proposons unanimement à l'Académie de lui accorder l'honneur d'être inséré au *Recueil des Savants étrangers.* »

Les conclusions de ce Rapport sont adoptées.

Sur la proposition de M. *Arago*, l'Académie décide que le Rapport sera imprimé dans les *Mémoires de l'Académie*, ainsi que l'a été déjà un Rapport de M. Biot sur une première partie du travail de M. Pasteur.

MÉMOIRES LUS.

MATHÉMATIQUES APPLIQUÉES. — *Notice sur un Mémoire intitulé:* Théorie de la force motrice du calorique; *par* M. F. REECH.

(Commission précédemment nommée.)

« M. Clapeyron a publié dans le tome XIV du *Journal de l'École Polytechnique*, année 1830, un Mémoire sur la puissance motrice du feu, dans lequel il développe analytiquement un principe et des raisonnements empruntés à un ouvrage qui a été publié en 1824 par M. S. Carnot.

» Les raisonnements ont pour objet de faire voir comment, à l'aide d'un gaz ou d'une vapeur, on peut régulariser la transmission du calorique entre un corps chaud et un corps froid, de manière à obtenir de la force motrice, et, réciproquement, comment, en dépensant de la force motrice, on peut, à l'aide d'un gaz et d'une vapeur, faire passer du calorique d'un corps froid dans un corps chaud.

» L'opération directe et l'opération inverse étant également possibles avec quelque gaz ou quelque vapeur que ce soit, on arriverait à créer avec rien autant de force motrice que l'on voudrait, si l'on n'admettait pas que, en principe, la quantité de force motrice maximum qui correspondra à de certaines quantités de calorique échangées entre deux sources A, A', entretenues à des températures constantes *t*, *t'*, sera exactement la même pour tous les gaz et pour toutes les vapeurs d'espèces différentes qui pourront être employées à produire identiquement les mêmes échanges de calorique entre les mêmes limites de température *t*, *t'*.

» En désignant par *q'* la quantité de calorique empruntée par un corps gazeux à une source de chaleur A' à la température constante *t'*, et par *q* la quantité de calorique cédée par le même corps à une source de chaleur A à la température constante *t*, M. Clapeyron admet naturellement l'égalité

75..

Fig. 10.5 Theory on motive force of caloric (1851) (Reech 1851, pp 567–571)

[31]Reech (1853), p 357, line 14. (Translation is ours: RP; Author's italic).

[32]Reech explicitly stated four main hypotheses–premises for this reasoning on Sadi Carnot's theory.

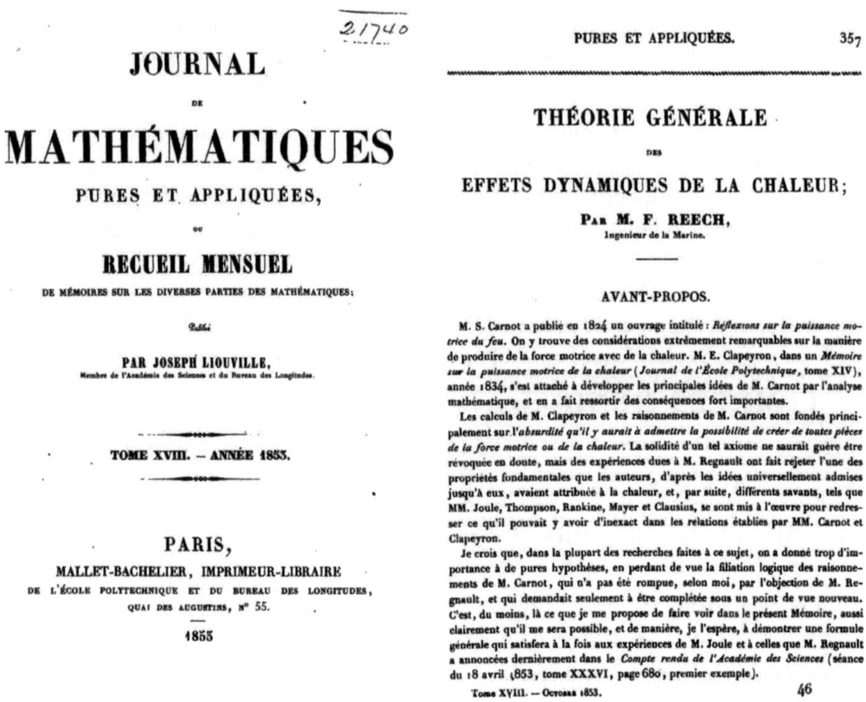

JOURNAL

DE

MATHÉMATIQUES

PURES ET APPLIQUÉES,

OU

RECUEIL MENSUEL

DE MÉMOIRES SUR LES DIVERSES PARTIES DES MATHÉMATIQUES;

Publié

PAR JOSEPH LIOUVILLE,

Membre de l'Académie des Sciences et du Bureau des Longitudes.

TOME XVIII. – ANNÉE 1853.

PARIS,

MALLET-BACHELIER, IMPRIMEUR-LIBRAIRE

DE L'ÉCOLE POLYTECHNIQUE ET DU BUREAU DES LONGITUDES,

QUAI DES AUGUSTINS, N° 55.

1853

PURES ET APPLIQUÉES. 357

THÉORIE GÉNÉRALE

DES

EFFETS DYNAMIQUES DE LA CHALEUR;

PAR M. F. REECH,

Ingenieur de la Marine.

AVANT-PROPOS.

M. S. Carnot a publié en 1824 un ouvrage intitulé : *Réflexions sur la puissance motrice du feu*. On y trouve des considérations extrêmement remarquables sur la manière de produire de la force motrice avec de la chaleur. M. E. Clapeyron, dans un *Mémoire sur la puissance motrice de la chaleur* (*Journal de l'École Polytechnique*, tome XIV), année 1834, s'est attaché à développer les principales idées de M. Carnot par l'analyse mathématique, et en a fait ressortir des conséquences fort importantes.

Les calculs de M. Clapeyron et les raisonnements de M. Carnot sont fondés principalement sur l'*absurdité qu'il y aurait à admettre la possibilité de créer de toutes pièces de la force motrice ou de la chaleur*. La solidité d'un tel axiome ne saurait guère être révoquée en doute, mais des expériences dues à M. Regnault ont fait rejeter l'une des propriétés fondamentales que les auteurs, d'après les idées universellement admises jusqu'à eux, avaient attribuée à la chaleur, et, par suite, différents savants, tels que MM. Joule, Thompson, Rankine, Mayer et Clausius, se sont mis à l'œuvre pour redresser ce qu'il pouvait y avoir d'inexact dans les relations établies par MM. Carnot et Clapeyron.

Je crois que, dans la plupart des recherches faites à ce sujet, on a donné trop d'importance à de pures hypothèses, en perdant de vue la filiation logique des raisonnements de M. Carnot, qui n'a pas été rompue, selon moi, par l'objection de M. Regnault, et qui demandait seulement à être complétée sous un point de vue nouveau. C'est, du moins, là ce que je me propose de faire voir dans le présent Mémoire, aussi clairement qu'il me sera possible, et de manière, je l'espère, à démontrer une formule générale qui satisfera à la fois aux expériences de M. Joule et à celles que M. Regnault a annoncées dernièrement dans le *Compte rendu de l'Académie des Sciences* (séance du 18 avril 1853, tome XXXVI, page 680, premier exemple).

Tome XVIII. – Octobre 1853. 46

Fig. 10.6 Théorie générale des effets dynamiques de la chaleur (1853) (Reech 1853, pp 357–378)

From the documents reported in *Comptes rendus des séances de l'Académie des sciences* in 1851 and 1852, it seems to us that a part or simply a summary of his definitive *Théorie générale des effets dynamiques de la chaleur* published in 1853 (Reech 1853, pp 357–378) and 1854 (Reech 1854) would have been anticipated by *Note sur la théorie des effets dynamiques de la chaleur* in *Comptes rendus des séances de l'Académie des sciences* during "Séance du lundi 1er Décembre 1851". In fact, it is only possible to read a simple remark on the lack of this memoir in these cited *Comptes* from 1851:

> APPLIED MATHEMATICS – *Note on theory of dynamics effects of heat*; *by* Mr. Reech. [...] the author discusses Mr Thomson and Mr Clausius' theories on a subject just presented in his previous communication [Reech 1851, pp 567–571], [this *Notes* is] postponed for the examination of the committee which will be named for the next session on the 17th of this month.[33]

[33]Reech (1851), p 602. (Capital letters and *italics* from the original; the translation is ours: RP).

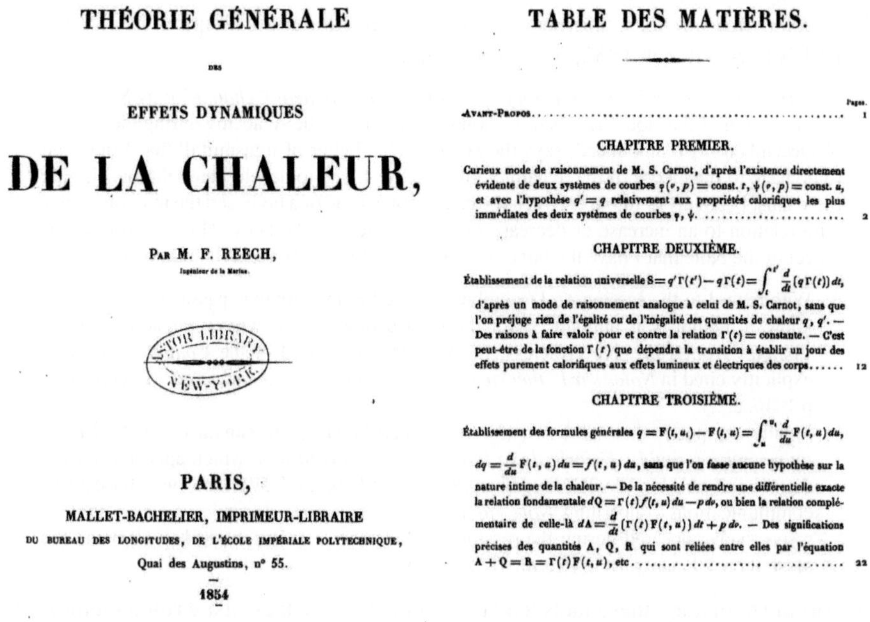

THÉORIE GÉNÉRALE

DES

EFFETS DYNAMIQUES

DE LA CHALEUR,

PAR M. F. REECH,

Ingénieur de la Marine.

PARIS,

MALLET-BACHELIER, IMPRIMEUR-LIBRAIRE

DU BUREAU DES LONGITUDES, DE L'ÉCOLE IMPÉRIALE POLYTECHNIQUE,

Quai des Augustins, n° 55.

1854

TABLE DES MATIÈRES.

Fig. 10.7 Théorie générale des effets dynamiques de la chaleur (1854) (Reech published his *Théorie générale des effets dynamiques de la chaleur* both in *Journal de mathématiques pures et appliquées* (Reech 1853, pp 357–378) and, extended but using *idem* title, in a publication by Bachelier in 1854 (Reech 1854). The focus on Sadi Carnot and Clapeyron continues to be present as does the use of the physics and mathematics of his time)

Therefore:

(a) From the comment on his announced memoir, *Note sur la théorie des effets dynamiques de la chaleur*, it seems to us that Reech's aim did not strictly regard Sadi Carnot and Clapeyron's formulas, such as instead it appears in his following memoires in 1853 (Reech 1853, pp 357–378) and 1854 (Reech 1854).

(b) The aforementioned session on 17 December 1851 lacks in these *Comptes*. In fact, when turning pages, one first sees a session on 15 December 1851, something related to 8 December, and immediately after read the session on 22 December 1851 and so on.

(c) The above cited memoir in the original passage concerns Reech's work titled, *Théorie de la force motrice du calorique* (Reech 1851, pp 567–571) reported in *Comptes rendus des séances de l'Académie des sciences*. Twenty-seven pages before, one can read that this memoir is presented by Poncelet as "[...] un travail très–considérable [...]" (Reech 1851, p 540, line 35). Its board of judges is composed by Dupin, Poncelet, Regnault and Lamé (*Ivi*).

Following history, in 1852 in another issue of *Comptes rendus des séances de l'Académie des sciences*, another communication – under the chapter "MÉMOIRES PRÉSENTES" (Reech 1852, p 18) – appears:

> MECHANICS. *Second Note on theory of dynamic effects of heat* ["*chaleur*"]; by M. Reech. In the closing of the Note that I had the honor of reading to the Academy during the session of December 1st, I announced, says the author in his Letter of transmittal, that I intended to go back to the theory of the dynamic effects of heat, and to develop the relationship that there must be, in my opinion, between the change of caloric of a body and the integral $\int p dv$ with relation to an increase or decrease of the volume v of the body. This is certainly the object of the Note that I have the honor of submitting for the Academy's judgment today.[34]

(d) With regard to the session on December 1st cited in the previous passage, as far as we know, a memoir was announced for the second time, yet its written version lacks the *Comptes* from 1852. Nevertheless, in the index at the end of the *Comptes*, Reech is explicitly cited in *Notes sur la théorie des effets dynamiques de la chaleur* (Reech 1852, p 1046, cl 2).

(e) Finally we specify that in the two aforementioned *Comptes* one can only read *Notice sur un mémoire intitulé : Théorie de la force motrice du calorique* which appeared during the session on Monday 24 November (Reech 1851, pp 567–571) and two consecutive communications concerning *Note sur la théorie des effets dynamiques de la chaleur* respectively appeared during the session on December 1st (Reech 1851, p 602) and the session on 5 January 1852 (Reech 1852, p 21).

According to history, these facts lead us to think that the lack of a written memoir on *théorie des effets dynamiques de la chaleur* in the *Comptes* from 1851 (and 1852) might be related to the lack of an official session on 17 December 1851 for which we presently have no comments. Alternatively, it was too similar to his previous *Théorie de la force motrice du calorique*. The reflections could also provide some details regarding its subsequent absence in *Comptes* from 1852. Moreover, one can simply and more probably suppose that its complete content was not yet ready for publication. In fact, presently, we know that a definitive work on the dynamic effects of heat by Reech appeared in 1853 (Reech 1853, pp 357–378) and in an extended book in 1854 (Reech 1854). On the other hand, based on our studies, it is probable that the lacking memoires did not include crucial discoveries regarding Sadi Carnot with respect to the two publications from 1853 to 1854. However, we cannot strictly confirm these suggestions as *historical hypotheses*.

We will now address Reech's work and refer the reader to one of Reech's previous studies entitled *Théorie de la force motrice du calorique* (Reech 1851, pp 567–571), which, however, we prefer not to include here for the sake of brevity. In this work, Reech includes various topics later addressed in his lengthy work from 1853. Therefore, the following pages will focus on *Théorie générale des effets dynamiques de la chaleur* (Reech 1853, pp 357–378).

Reech (1853, p 357) primarily followed Clapeyron's example of transferring Sadi Carnot's reasoning to a P–V diagram in order to state the existence of two systems of

[34] Reech (1852), p 21, line 23. (Capital letters and italics from the original; the translation is ours: RP).

curves on the P–V plane, the isotherms and the adiabatics, respectively of equations (*Reech's first hypothesis*):

$$\varphi(V, p) = t$$

$$\psi(V, p) = u.$$

Reech's premise is important for recognizing implicit assumptions in Sadi Carnot's theory, summarized here below:

- Each point of the plane (V, p), for any gas, is the intersection of an adiabatic and an isotherm distinct amongst themselves.
- From this assertion, Carathéodory's axiom follows as a consequence in its usual quotation: *in the neighborhood of any initial state, there are states which cannot be approached arbitrarily close through adiabatic changes of state.*
- It is not necessary to take these curves as a strict system of coordinates (*Reech's second hypothesis*) since a metric lacks in Clapeyron's diagram.

Reech's calculation of a double infinitesimal cycle in adiabatic coordinates and isotherms is also of little relevance. It was extraneous to the original spirit of Sadi Carnot's work. It is, however, important for our aim to emphasize when Reech discusses the reversibility of a thermodynamic system.

Let q be the quantity of heat[35] from which temperature t' in source A' passes to lower temperature t in source A. Reech observes that, in determined temperatures t and t', W as the maximum motive power possible (*Reech's third hypothesis*) must increase in proportion to q (*Reech's fourth hypothesis*). Thus, under these conditions, Sadi Carnot's thesis (Carnot 1978, pp 38–39) can effectively express a mathematical relationship for any kind of working substance adopted (*Reech's fifth hypothesis*):

$$W = qf(t, t') \tag{10.1}$$

where the function $f(t, t')$ is the efficiency which is presently referred to as η in most thermodynamic textbooks. If then, by means of the same machine, n identical cyclical operations are carried out, in the end, a total amount of motive power called nW must be the same as that which one would obtain using a single cyclical operation that uses an amount of heat equal to $nq = Q$. In short, by *ad absurdum* proof, we can write

$$W = nW_n = nqf(t, t') = Qf(t, t') \tag{10.2}$$

where the function f is – such as in Carnot's theorem – independent from the gas employed (Carnot 1978, pp 23–38).

[35] We have kept Reech's original symbols. Consequently, the letter q indicates the quantity of heat, which is indicated by the letter Q in Sadi Carnot's work.

Reech proposes a thermodynamic system made up of two machines: one direct and the other inverse,[36] which operate cyclically between two thermostats and have a difference of temperature which, according to Sadi Carnot,

> [...] we may suppose the difference [of temperature] to be as small as we like ["comme nul"]. For the purposes of our theory, it [Δt] may be taken to be zero [that is, ideally], without the argument being in any way undetermined.[37]

Let "*C*" be a particular function depending on temperature *t*, and valid for any kind of working substance. Thus, the following expression can be written (Reech 1853, p 365):

$$W = Q \int_{t}^{t'} \frac{dt}{C}, \tag{10.3}$$

If we consider the process after *n* cycles in the first and n_1 cycles in the second, we have a total work:

$$W_t = nW - n_1 W_1. \tag{10.4}$$

It is obtained for the heat absorbed ($q'_t = nq' - n'_1 q'_1$) and for the heat given off ($q_t = nq - n_1 q_1$). In practice, the same result would be obtained with another source A'' at temperature t'' different from t. Thus, by means of calculations (Reech 1853, pp 365–368), which we will omit for the sake of brevity, the efficiency of a reversible heat machine is given by an efficiency of heat conversions $\Gamma(t')$which is a generalization of the $\eta = W/Q$. It is valid in the theory of the equivalence of heat and work:

$$\Gamma(t') = \frac{W_t}{q'_t} = \frac{\dfrac{W}{q} - \dfrac{W_1}{q_1}}{\dfrac{q'}{q} - \dfrac{q'_1}{q_1}}. \tag{10.5}$$

In Carnot's theorem, Γ represents a certain function of the temperature that should be the same for any kind of gas. At this point, (Reech 1853, pp 369–371) by using the same reasoning but removing the higher temperature source that takes into account an n and an n_1 to have $q'_t = 0$, Reech obtains:

$$n = \frac{N}{q'}, \; n_1 = \frac{N}{q'_t} \tag{10.6}$$

[36] Sadi Carnot used a single machine to operate first in a direct manner and then in an inverse manner.

[37] Carnot (1978), p 25, line 10.

$$W_t = N\left(\frac{W}{q'} - \frac{W_1}{q'_t}\right) \tag{10.7}$$

$$q_t = N\left(\frac{q_1}{q'_1} - \frac{q}{q'}\right). \tag{10.8}$$

$$\eta = \frac{W_t}{q_t} = \frac{\dfrac{W}{q'} - \dfrac{W_1}{q'_1}}{\dfrac{q_1}{q'_1} - \dfrac{q}{q'}} = \Gamma(t) \tag{10.9}$$

From Eqs. 10.5 and 10.9, we arrive at the following relations:

$$\frac{\dfrac{W}{q} - \dfrac{W_1}{q_1}}{\dfrac{q'}{q} - \dfrac{q'_1}{q_1}} = \frac{\dfrac{W}{q'} - \dfrac{W_1}{q'_1}}{\dfrac{q_1}{q'_1} - \dfrac{q}{q'}} = \Gamma(t)$$

$$\frac{\dfrac{q_1W - qW_1}{qq_1}}{\dfrac{q'q_1 - q'_1q}{qq_1}} = \frac{q_1W - qW_1}{q'q_1 - q'_1q} = \frac{\dfrac{q'_1W - q'W_1}{q'q'_1}}{\dfrac{q'q_1 - q'_1q}{q'q'_1}} = \frac{q'_1W - q'W_1}{q'q_1 - q'_1q} = \Gamma(t).$$

That is,

$$\begin{aligned} q_1W - qW_1 &= \Gamma(t)(q'q_1 - q'_1q) \\ q'_1W - q'W_1 &= \Gamma(t)(q'q_1 - q'_1q) \end{aligned} \tag{10.10}$$

By considering the system composed by Eq. 10.10, adding and subtracting, we have:

$$W_1(qq'_1 - q_1q') = \Gamma(t)(q_1 - q'_1)(q'q_1 - q'_1q).$$

It follows that:

$$\begin{aligned} W_1 &= \Gamma(t)\frac{q_1 - q'_1}{qq'_1 - q_1q'}(q'q_1 - q'_1q) = \\ &= -\Gamma(t)(q_1 - q'_1) = q'_1\Gamma(t') - q_1\Gamma(t). \end{aligned} \tag{10.11}$$

In the same way, one can obtain:

$$W = q'\Gamma(t') - q_1\Gamma(t). \tag{10.12}$$

From Reech's equations (Reech 1853, p 371, Eq. (3)), therefore also from Sadi Carnot and (later) Clapeyron's formula, one can write the following general expression (*Ibidem*):

$$W = q'\Gamma(t') - q\Gamma(t) = \int_t^{t'} \frac{d}{dt}[q\Gamma(t)]dt, \tag{10.13}$$

which reduces itself to Reech's Eq. (3) (Reech 1853, p 371) $q' = q$ is possible to find in an experiment. In this case, one can obtain (Reech 1853, p 371):

$$\Gamma(t) = \text{constant} = G$$

and thus we have

$$W = G\left(q' - q\right).$$

In the end, Ferdnand Reech summarizes Eqs. 10.12 and 10.13 representing the two cases – caloric and heat – using the same formula for obtaining the work in a cycle:

$$W = q'\Gamma(t') - q\Gamma(t) = (q' - q)\Gamma(t') + q[\Gamma(t') - \Gamma(t)]. \tag{10.14}$$

Therefore he obtains the following mathematical relations:

$$\textit{Theory of caloric}: (q' = q) \Rightarrow W = q[(\Gamma(t') - \Gamma(t)], \tag{10.15}$$

$$\textit{Theory of work} - \textit{heat equivalence}: \ \{\Gamma(t) = \cos t\} \Rightarrow W = J(q' - q). \tag{10.15a}$$

For this Reech clearly acknowledged his debt to Sadi Carnot (Fig. 10.8).

The aforementioned generalization is of great importance because it provides a general theoretical result independent from the nature of heat. It allows for mathematical reasoning in thermodynamics. In addition, Eq. 10.15 also suggests why Sadi Carnot, who was familiar with the mathematics of his time, differentiated the efficiency on temperature (Carnot 1978, pp 76–77). It can be hypothesized that the expression $q[(\Gamma(t') - \Gamma(t)]$ induced him to do this. Overall, Reech, reasoning only on his cycles, effectively continues Sadi Carnot's heritage, allowing him to obtain the efficiency of heat machines independently from working substances. These two decisive choices are also reminiscent of the role played by the concept of the cycle in other domains of physics (see Chapter 8 and the discussion on Mach).

PURES ET APPLIQUÉES. 371

devra être remplacée par la formule un peu plus générale

$$(3) \qquad = q'\Gamma(t') - q\Gamma(t) = \int_t^{t'} \frac{d}{dt}(q\Gamma(t))\,dt,$$

qui se réduirait précisément à l'autre, si l'expérience nous faisait trouver $q' = q$, et de laquelle, au contraire, nous tirerions

$$S = G(q' - q),$$

si l'expérience ou le raisonnement nous faisaient trouver

$$\Gamma(t) = \text{constante } G.$$

Il y a à faire remarquer que je suis parvenu à la formule (3) sans faire aucune hypothèse sur la nature intime de la chaleur dans le volume des fluides élastiques. La formule (3) sera donc le pivot fondamental de la théorie des effets dynamiques de la chaleur; ce sera une relation supérieure qui dominera à la fois la théorie de M. Carnot et les théories plus récentes de quelques autres physiciens.

Par le mode de raisonnement que j'ai suivi, en me laissant aller au cours naturel des idées dans la voie ouverte par M. Carnot, la formule (3) est devenue du premier coup aussi complète que la formule (a), après trois raisonnements consécutifs; car, en supposant que l'on veuille imaginer maintenant une série de sources auxiliaires à des températures quelconques $t_0, t_1, t_2, t_3, \ldots, t_{n-1}, t_n$, perdant et recevant successivement les unes après les autres les quantités de chaleur $q_0, q_1, q_2, q_3, \ldots, q_{n-1}, q_n$, sans qu'il reste ni excédant ni déficit dans aucune des sources intermédiaires

$$A_1, \quad A_2, \quad A_3, \ldots, \quad A_{n-1},$$

l'on trouvera, par la formule (3), les quantités de force motrice partielles

$$\begin{gathered} q_1\Gamma(t_1) - q_0\Gamma(t_0), \\ q_2\Gamma(t_2) - q_1\Gamma(t_1), \\ q_3\Gamma(t_3) - q_2\Gamma(t_2), \\ \cdots\cdots\cdots\cdots\cdots \\ q_n\Gamma(t_n) - q_{n-1}\Gamma(t_{n-1}), \end{gathered}$$

dont la somme se réduira précisément à la différence

$$q_n\Gamma(t_n) - q_0\Gamma(t_0),$$

que la même formule (3) fera trouver pour une seule opération de t_0 à t_n, ou inversement de t_n à t_0. Cette propriété vient de ce que le second membre de la formule (3) a pris de suite la forme d'une intégrale définie de t à t'.

On a encore identiquement

$$(3\ bis) \quad \begin{cases} S = q'\Gamma(t') - q\Gamma(t) = q'(\Gamma(t') - \Gamma(t)) + (q' - q)\Gamma(t) \\ \qquad = (q' - q)\Gamma(t') + q(\Gamma(t') - \Gamma(t)), \end{cases}$$

Fig. 10.8 Reech's result and his quotation on Sadi Carnot (Reech 1853, p 371)

10.3.2 On Cycles and Equivalent Mechanic of Heat

Like Gillispie, Fox, and Lervig here cited, Ulrich Hoyer is another of the speakers at the Paris International Conference in honor of *Sadi Carnot et l'essor de la thermodynamique*. Below, we will discuss his contribution to the conference proceedings (Taton 1976).

Ulrich Hoyer studied *Notes sur les mathématiques, la physique et autres sujets* by Sadi Carnot (1878b, pp 89–102) and wrote an interesting and original reasoning entitled *La théorie de Carnot première et seconde approximations de la thermodynamique* (Hoyer 1976, pp 221–228; see also Hoyer 1975). According to Hoyer (1976, p 222), in order to calculate the value of the mechanical equivalent of heat (and in lack of experimental data), Sadi Carnot applied the principle of the conservation of caloric (Carnot 1878b, p 24) to a triangular cyclic process, considering a small thermal excursion equal to $\Delta t = 1°C$. The following is Hoyer's performance model (Fig. 10.9):

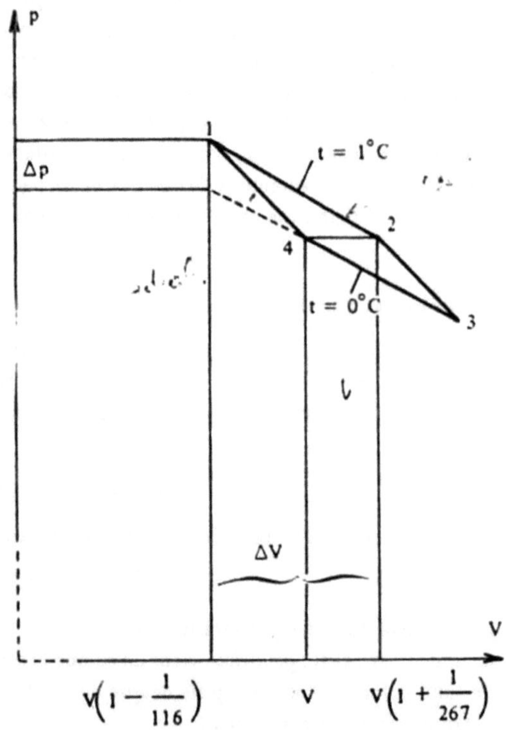

Fig. 10.9 Simplified performance model of Carnot's cycle by Hoyer (1976, p 222)

Hoyer showed that the modern theory of the equivalence of heat and work and Carnot's caloric theory in *Notes sur les mathématiques, la physique et autres sujets* (Carnot 1878b, pp 89–102; see also Carnot 1986, pp 181–206) coincide in infinitesimal processes.

For our part, here we will try to combine Reech's previous reasoning with that of Hoyer. Thus, based on Reech's previously presented reasoning, our remarks consist of showing that there is a way to reconsider Reech's general result (see above) according to Hoyer's reasonings and thereby obtain a modern theory of the equivalence. Our reasoning should also explain Sadi Carnot's surprising results (in respect to Mayer[38]) which were not presented by Sadi Carnot in *Réflexions sur la puissance motrice du feu* (Carnot 1978). Let us see.

Following Hoyer's reasoning (Hoyer 1976, pp 222–223; see also Fig. 10.9) he obtained:

$$\int_1^2 dq = \int_1^4 dq + \int_4^2 dq \tag{10.16}$$

At that point, on adiabatic *4–1* (see Fig. 10.9) one can easily write that

$$\int_1^4 dq = 0. \tag{10.17}$$

Following Sadi Carnot's reasoning, Hoyer obtained:

$$\int_1^2 dq = \int_4^2 dq. \tag{10.18}$$

It is easy to see that the Carnot caloric absorbed during isothermal expansion coincides with the caloric necessary to increase (at constant pressure) the air temperature from *0*°C to *1*°C (Hoyer 1976, p 223).

On the other hand, Sadi Carnot calculated a motive power ΔA, where A is mechanical work produced (Carnot 1878b, p 45). This fact suggested an agreement with the corresponding values for a water cycle (Carnot 1878b, p 47) and an alcohol cycle (Carnot 1878b, p 49). This consideration also led Sadi Carnot to consider the hypothesis of the equivalence of heat and mechanical work and therefore to determine his value of the mechanical equivalent of heat. Therefore, according to Hoyer (1976, p 223) under this hypothesis, the heat absorbed in isothermal expansion will be equal to the thermal work produced by the isothermal transformation. Therefore, the following expression should be valid:

$$\Delta q = \int_1^2 p dV = p\Delta V.$$

[38]The result was presented by Sadi Carnot in *Notes sur les mathématiques, la physique et autres sujets* (III.Folio 7v; see also Carnot 1878b, p 95, 1896, p 191).

At this point Hoyer obtained (Hoyer 1976, p 223 and Eqs. (7) and (9)) a good value of the approximation of the mechanical equivalent of heat:

$$1000\,kg = 2.697Kcal.$$

It is possible to compare the specific heat at constant[39] pressure with the thermal work given off by the isothermal expansion. By using calculations, which we omit for the sake of brevity, Hoyer proposed an important question for our aim:

> *How* [Sadi] *could Carnot's theory, although based on the conservation of heat, in principle, provide a satisfactory value of the mechanical equivalent of heat?*[40]

It is known that, according to the modern first principle, the internal energy is conserved during a cycle. By considering *1–2–4* process (see Fig. 10.9)

$$0 = U_{12} + U_{24} + U_{41} = Q_{12} - W_{12} + Q_{24} - W_{24} - W_{41}$$

and being $\oint dU = 0$, according to Hoyer (1976, p 224) the following expression can be written:

$$\int_1^2 dq + \int_2^4 dq = \oint pdV.$$

Alternatively, one can write

$$\int_1^2 dq = \int_4^2 dq + \oint pdV \qquad (10.19)$$

where

$$\oint pdV = \frac{1}{2}\Delta p \Delta V \qquad (10.20)$$

that is to say, $\oint pdV$ represents a second-order term compared to Δp (or with regard to ΔV).

---

[39] As Hoyer also remarked, it should be noted that in *Réflexions sur la puissance motrice du feu* (Carnot 1978, p 46) Sadi Carnot cites Delaroche and Bérard's experiments, which both showed that specific heat at constant pressure was equal to $c_p = \Delta q = \int_4^2 dq = 0.267kcal$; *(m = 1 kg).*

[40] Hoyer (1976), p 223, line 22. Author's italics.

On the other hand, for the *1–2–3–4* path (see Fig. 10.9) while still considering the first principle, we obtain:

$$\oint dU = \oint dQ + \oint pdV,$$

therefore

$$0 = U_{12} + U_{23} + U_{34} + U_{41} = Q_{12} - W_{12} - W_{23} + Q_{34} - W_{34} - W_{41},$$

from which

$$\int_1^2 dq - \int_1^2 dW - \int_2^3 dW + \int_3^4 dq - \int_3^4 dW - \int_4^1 dW = 0$$

$$\int_1^2 dq + \int_3^4 dq = \oint dW. \tag{10.21}$$

However, for Eq. 10.19 we can write:

$$\oint_{1234} dW = \oint_{124} dW + \oint_{234} dW = \frac{1}{2}\Delta p \Delta V + \frac{1}{2}\Delta p \Delta V.$$

By substituting it in Eq. 10.21, we obtain:

$$\int_1^2 dq + \int_3^4 dq = \Delta p \Delta V. \tag{10.22}$$

Now, by applying the second principle we reach the conclusion that the variation of entropy must only be calculated for isothermal processes in the *1–2–3–4* cycle (see Fig. 10.9) since, in a reversible cycle that produces $\Delta p \Delta V$, the variation is zero. Therefore, disregarding superior-order terms, in accordance with Hoyer (1976, p 224) we have:

$$\frac{1}{T}\int_1^2 dq = \frac{1}{T - \Delta T}\int_4^3 dq \tag{10.23}$$

and by series-expansion of the second member, it is possible to write:

$$\frac{1}{T}\int_1^2 dq = \frac{1}{T}\left(1 + \frac{\Delta T}{T} + \ldots\ldots\right)\int_4^3 dq \tag{10.24}$$

from which it follows that:

$$\frac{1}{T}\int_{1}^{2} dq = \frac{1}{T}\int_{4}^{3} dq + \frac{1}{T}\frac{\Delta T}{T}\int_{4}^{3} dq + \dots . \qquad (10.24bis)$$

If we disregard quantities superior to the second order, one can write:

$$\int_{1}^{2} dq = \int_{3}^{4} dq + \frac{\Delta T}{T}\int_{4}^{3} dq.$$

By using Eq. 10.22 one obtains:

$$\Delta A = p\Delta V = \frac{1}{T}\Delta T \Delta q \qquad (10.25)$$

which is in complete agreement with the mechanical work per unit of heat. Thus, if we assume our hypothesis is factorizable (see Chapter 9) in accordance with Hoyer, we can state that:

> *The theory of the mechanical work produced in the Carnot engine expressed by analytical equation* [see Eq. 10.22], *is a second approximation of the two principles of thermodynamics.* From two previous results follows: [1] *Carnot's theory* [of caloric] *is absolutely correct in the sense of thermodynamics provided one considers the infinitesimal reversible.* [2] *Carnot's theory* [of caloric] *is an example of the progress of science by developing approximation.*[41]

By considering our previous suggestions in Hoyer's reasonings and, as previously announced, let us return to Reech's general result (see above).

By accepting Hoyer's suggestion of focusing on first order of expansion approximations, the modern theory of the equivalence can be reached. In fact, from Reech's conclusion, we immediately obtain:

$$\frac{dW}{dq} = \Gamma + dq\frac{\Gamma}{dq},$$

envisioning it as a development in a series, analogously to how Hoyer operated. If we stop at the first term of expansion (intending an infinitesimal process such as a constant) we obtain, combining Hoyer and Reech on Sadi Carnot's reasoning, the modern theory

$$\frac{dW}{dq} = \Gamma = const.$$

[41] Hoyer (1976), p 226, line 22. *Italic style* from the author.

10.4 On a Comparative Analysis Between Paul de Saint–Robert's *Principes de thermodynamique* (1865) and Sadi Carnot's *Réflexions sur la puissance motrice du feu* (1824)

10.4.1 On Paul de Saint–Robert

The Italian scholar Paolo Ballada (1815–1888), also called Paul de Saint–Robert or Paul comte de Saint–Robert, is not very well known in respect to other distinguished thermodynamicists and scholars of his time. Nevertheless, he was one of the XL members of the *Società Italiana delle Scienze*, and a member of *R. Accademia dei Lincei*. He loved poetry, mountaineering, geography, botany and entomology[42] and he was the founder and president of the *Club Alpino Italiano*. Saint–Robert was a Lieutenant colonel in the army (until age 42), and a ballistics teacher at the *Scuola di applicazione di artiglieria e genio* in Torino. He dedicated himself to the study of ballistics, applying physical–mathematical techniques where in general, rough calculations were used and became an internationally renowned specialist. In this sense, his theorems are well known. His collection of work also includes contributions to hypsometry (Saint–Robert 1872–1874). His scientific production took place during a particular period in the second half of the nineteenth century. He was finally commemorated by the *Regia Accademia delle Science di Torino* which dedicated notes to him in its *Atti* (Bassa 1888–1889, pp 235–244). A few other institutions also commemorated him. (Gianelli 1888, pp 387–389) (Fig. 10.10).

[42]In the Reference section of our book, we attempted to compile a list of Saint–Robert's works. (Cfr.: Drago 1993b; Pisano 2005). Recently the first popular and scientific St Robert's celebration (after those *by Accademia di Torino* in '70s of the past century) was organized by the *Comune di Castagnole delle Lanze* (Piemonte, Italy) and *Associazione Culturale Conte Paolo Ballada di Saint Robert (Ibidem)*. The event has celebrated both the man and scientific works. It has been patronaged by *European Society for the History of Science* and scientifically-historically supported by one of us (RP).

Fig. 10.10 Paul de Saint–Robert (Plate from the incision (10 × 16.5) cm. With permission of the *Club Alpino Italiano National Library*; Gianelli 1888)

It is not easy to find an exhaustive work on Paolo Ballada such as an official scientific biography. He was (maybe) the first to spread the news of Sadi Carnot's biography[43] at the *Accademia delle Scienze* of Torino in 1868. In Italy he was referenced by a project (Redondi 1974–1975, pp 281–318) that reevaluated Italian scientists.[44] In biographical literatures (Ghisalberti 1963), one can see that he is given just a brief profile and is referred to by his last name, Ballada, unlike in the scientific papers which refer to him as Saint–Robert. He is also remembered for a long series of literary, scientific and military interests. Concerning science, Saint–Robert wrote, in addition to various cultural and technological topics, almost exclusively about thermodynamics and mechanics (see the following Table 10.1).

[43]Saint–Robert (1869), IV, pp 151–170, 1870c, pp 431–450.

[44]His physical mathematical approach to thermodynamics concerning the role played by physics and mathematics of the nineteenth century was studied by Drago (1993a, b) and one of us (Pisano 2005). We also remember Cerruti (1999, pp 95–178) for a work in which he dedicated enough space to Ballada even though his topic was not strictly related to Saint–Robert.

Table 10.1 A list of Saint–Robert's works on heat and thermodynamics[a]

Year	Saint–Robert 's thermodynamics works
1863	Lettre concernant la théorie mécanique de la chaleur. *Cosmos* XXII: 200
1865	*Principes de thermodynamique*, 1st edn. Cassone, Torino
1866	Remarques à l'occasion d'une note de M. Clausius. *Archives des sciences physiques et naturelles* XXV: 34–41
1865–1866	*Note sur le travail mécanique dépensé dans la compression et sur le travail restitué par la détente d'un gaz permanent* Atti della Reale Accademia delle scienze di Torino. Tome I, pp 283–296
1867–1868	*Sopra un'opera del prof. Cavallero intitolata: sulle macchine motrici*, Atti Regia Accademia delle Scienze di Torino, Classe Scienze Fisiche e Matematiche III, pp 16–20
1868–1869	*Notice biographique sur Sadi Carnot*, Atti Regia Accademia delle Scienze di Torino, Classe Scienze Fisiche e Matematiche IV, pp 151–170 [also reported in the following 1870 edition, pp 431–450]
1870	*Principes de thermodynamique*, 2nd edn. Loecher, Turin et Florence (see also idem edition printed by Gauthier – Villars, Paris)
1870	*Jules Robert Mayer – Notice biographique.* Leipzig [reported in 1870 edition, pp 451–470]
1872	*Qu'est–ce que la force?*[b]. La revue scientifique de la France et de l'étranger 2s/II:985–993
1875	*Del calore attuale contenuto ne' corpi.* Rendiconti dello Reale Instituto Lombardo di Scienze e Lettere,VIII, pp 876–879
1876	*Intorno al calore che deve prodursi nell' esperienza immaginata da Galileo per misurare la forza di percossa.* Rendiconti dello Reale Instituto Lombardo di Scienze e Lettere, IX, pp 174–179

[a] A list of Saint–Robert's works on scientific topics is listed in the Reference section. Please also note others works listed below on the plate from the 1870 edition printed by Teubner, Leipzig (Saint–Robert 1870c, pp 485–486)

[b] In this work (Saint–Robert 1872, *Qu'est–ce que la force?*) the nature of heat (chaleur) in relationship with concept of force is discussed (*Ivi*, pp 990–993)

Table 10.1 (continued)

Du même Auteur:

Della fabbricazione della polvere da fuoco. Considerazioni e Proposte. In 8. 61 pag. Torino 1852.

Del moto de' projetti ne' mezzi resistenti. In 4. 105 pag. Torino 1855.

Nuovo projetto e nuova arma da fuoco. In 8. 39 pag. con 1 tavola. Torino 1857.

Del tiro. In 8. 83 pag. con 1 tavola. Torino 1857.

Des effets de la rotation de la terre sur le mouvement des projectiles. In 8. 62 pag. avec 1 planche. Paris 1858.

Études sur la trajectoire que décrivent les projectiles oblongs. 1ère Partie in 8. 36 pag. — 2me Partie in 8. 67 pag. Paris 1859. 1860.

Considérations sur le tir des armes à feu rayées dans leur état actuel. Proposition d'un nouveau système de projectiles et d'armes à feu. In 8. pag. VI. 45. Paris 1860.

Note sur le volume d'une embrasure. In 8. 7 pag. Paris 1860.

Sur l'analyse du Charbon destiné à la fabrication de la poudre. In 8. 16 pag. Paris 1860.

Del nitrato di soda in vece del nitrato di potassa nella polvere da fuoco. In 8. 11 pag. Torino 1860. (Rivista militare, anno IV. Vol. 4.)

Du mouvement des projectiles lancés par les armes à feu rayées. In 8. 56 pag. avec 3 planches. Paris 1861.

Teorema sulla similitudine delle trajettorie descritte dai projetti ne' mezzi resistenti. Applicazioni al tiro delle armi da fuoco. In 8. 20 pag. Pisa 1861

Théorie du Compresseur à colonne d'eau de MM. Grandis, Grattoni et Sommeiller. In 8. 60 pag. avec 1 planche. Paris 1863.

Sur la mesure des hauteurs à l'aide du baromètre. In 8. 15 pag. Paris 1864.

Barometrical formula resulting from the observations made by Mr. James Glaisher in eight balloon-ascents in 1862. In 8. 8 pag. (Philosophical Magazine for February 1864.)

On the measurements of heights by the barometer, and on atmospheric refraction, having regard to the constitution of the atmosphere, resulting from Mr. James Glaisher's observations. In 8. 25 pag. London 1864. (Philosophical Magazine for June 1864.)

Intorno alla formola barometrica, ed alla rifrazione atmosferica. In 8. 12 pag. Torino 1866. (Atti dell' accademia delle scienze di Torino. Vol. I. 1865. 1866.)

Sulla Saxifraga florulenta, Moretti. In 8. 3 pag. Torino 1866. (Atti dell' accademia delle scienze di Torino. Vol. I. 1865. 1866.)

Note sur le travail mécanique dépensé dans la compression et sur le travail restitué par la détente d'un gaz permanent. In 8. 16 pag. Turin 1866. (Atti dell' accademia delle scienze di Torino. Vol. I. 1865. 1866.)

Résultats d'expériences faites à diverses hauteurs touchant la durée de combustion de la matière de la poudre. In 8. 15 pag. avec 2 planches. Turin 1866. (Atti dell' accademia delle scienze di Torino. Vol. I. 1865. 1866.)

Sul vero significato di una terzina di Dantè. In 8. 15 pag. con una tavola. Torino 1866. (Atti dell' accademia delle scienze di Torino. Vol I. 1865. 1866.)

Table hypsométrique pour déterminer rapidement, sur place, la différence de niveau de deux stations et pour réduire les indications du baromètre, dans une station, à ce qu'elles seraient dans une autre. In 8. 15 pag. avec un tableau. Turin 1867 (Atti dell' accademia delle scienze di Torino. Vol. II. 1866. 1867.)

De la résolution de certaines équations à trois variables par le moyen d'une règle glissante. Caractère auquel on reconnaît qu'une telle résolution est possible. Graduation de la règle. In 4. 12 pag. Turin 1867. (Memorie dell' accademia delle scienze di Torino. Serie II. T. XXV.)

Nouvelles tables hypsométriques. In 4. 19 pag. avec 12 pag. de tables. Turin 1867. (Memorie dell' accademia delle scienze di Torino Serie II. T. XXV.)

Lettre de Mr. le Comte Paul de Saint-Robert au Directeur du Spectateur militaire. In 8. 7 pag. Paris 1867. (Spectateur militaire, livr. du 15. Avril 1867.)

Gita al Monte Ciamarella nelle Alpi Graie. In 8. 23 pag. Torino 1867. (Bollettino trimestrale del Club Alpino Italiano. No. 10 e 11.)

Tableau graphique donnant à vue l'altitude d'une station au moyen de la seule observation du baromètre et du thermomètre à cette station. In 8. 11 pag. avec 3 planches. Turin 1867. (Atti dell' accademia delle scienze di Torino. Vol.III. 1867. 1868.)

Des changements de température produits dans les corps solides de forme prismatique par une traction longitudinale. In 8. 22 pag. Turin 1868. (Atti dell' accademia delle scienze di Torino. Vol III. 1867 1868.)

Notice biographique sur Sadi Carnot. In 8. 22 pag. Turin 1868. (Atti dell' accademia delle scienze di Torino. Vol. IV. 1868. 1869.)

Leipsic: Imprimerie de B. G. Teubner.

Among these writings, the two editions of the thermodynamic textbooks from 1865 to 1870 should be noted. They are a significant sign of great pedagogical capability in a new (at the time) specific physical subject. In addition to his capability, a great attention to the history of thermodynamics must be added, both for the biography on Sadi Carnot and for a pamphlet on Mayer (he likely obtained the information directly from the author). In addition, Hyppolyte's son (Sadi's nephew[45]) compiled a biographical sketch which Saint–Robert (1868–1869b, pp 151–170) published.[46]

In Saint–Robert's studies, he summarized Sadi Carnot's theory and declared it compatible with the foundations of the mechanical theory of heat, in which, however, the principle of the conservation of energy is not emphasized. This is important for our aim since the formulation of the impossibility of perpetual motion used as

[45]Marie François Sadi Carnot (1837–1894) became the fourth President of the Third French Republic. On June 24, 1894, Sante Geronimo Caserio (1873–1894), an Italian anarchist, fatally stabbed President Carnot after his public banquet, to avenge Auguste Vaillant (1861–1894) and Emile Henry (1872–1894). Thus, Caserio assassinated the president who died after midnight on 25 June. The Board of Pardons decided against all appeals for clemency on August 14. Caserio was executed by guillotine in Lyon at 5 am, August 16, 1894. President Carnot was honored with an elaborate funeral ceremony in the Panthéon, Paris.

[46]Cfr.: Carnot (1986), pp 216–217.

a basic principle, the problem of the efficiency of machines and its solution greatly resembled Sadi Carnot's book. Moreover, he may have been the only scholar of his time to stress and cite the scientific relationship between Lazare and Sadi Carnot.

> One thing that is worth to be remarked is the relationship between the *Réflexions sur la puissance motrice du feu* by Sadi Carnot and the considerations claimed in a short book by his distinguished father. The General Carnot, published it in 1783 with title *Essai sur les machines en général*. What his father made for ordinary ["ordinaire"] the son made for the mechanics of heat ["mécanique de la chaleur"].[47] The principle of equivalence is what lacks to Sadi Carnot.[48]

In particular, the previous quotation concerns an interesting passage on two theories, including some ideas on the logical basis[49] of the theory (Saint–Robert 1870c, p 448). Below, we provide Saint–Robert 's original words on Lazare and Sadi Carnot (Fig. 10.11).

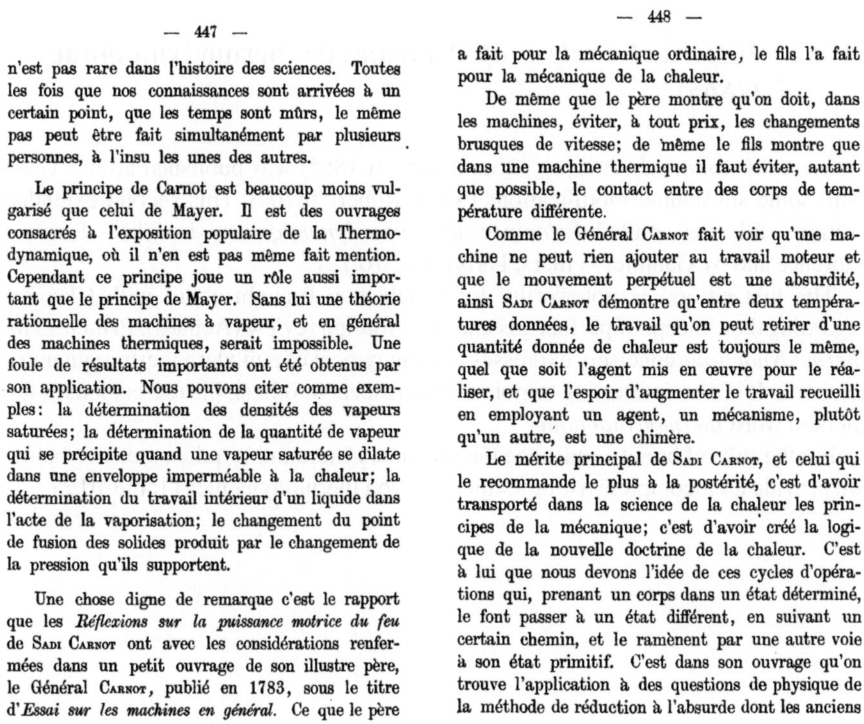

— 447 —

n'est pas rare dans l'histoire des sciences. Toutes les fois que nos connaissances sont arrivées à un certain point, que les temps sont mûrs, le même pas peut être fait simultanément par plusieurs personnes, à l'insu les unes des autres.

Le principe de Carnot est beaucoup moins vulgarisé que celui de Mayer. Il est des ouvrages consacrés à l'exposition populaire de la Thermodynamique, où il n'en est pas même fait mention. Cependant ce principe joue un rôle aussi important que le principe de Mayer. Sans lui une théorie rationnelle des machines à vapeur, et en général des machines thermiques, serait impossible. Une foule de résultats importants ont été obtenus par son application. Nous pouvons citer comme exemples: la détermination des densités des vapeurs saturées; la détermination de la quantité de vapeur qui se précipite quand une vapeur saturée se dilate dans une enveloppe imperméable à la chaleur; la détermination du travail intérieur d'un liquide dans l'acte de la vaporisation; le changement du point de fusion des solides produit par le changement de la pression qu'ils supportent.

Une chose digne de remarque c'est le rapport que les *Réflexions sur la puissance motrice du feu* de Sadi Carnot ont avec les considérations renfermées dans un petit ouvrage de son illustre père, le Général Carnot, publié en 1783, sous le titre d'*Essai sur les machines en général*. Ce que le père

— 448 —

a fait pour la mécanique ordinaire, le fils l'a fait pour la mécanique de la chaleur.

De même que le père montre qu'on doit, dans les machines, éviter, à tout prix, les changements brusques de vitesse; de même le fils montre que dans une machine thermique il faut éviter, autant que possible, le contact entre des corps de température différente.

Comme le Général Carnot fait voir qu'une machine ne peut rien ajouter au travail moteur et que le mouvement perpétuel est une absurdité, ainsi Sadi Carnot démontre qu'entre deux températures données, le travail qu'on peut retirer d'une quantité donnée de chaleur est toujours le même, quel que soit l'agent mis en œuvre pour le réaliser, et que l'espoir d'augmenter le travail recueilli en employant un agent, un mécanisme, plutôt qu'un autre, est une chimère.

Le mérite principal de Sadi Carnot, et celui qui le recommande le plus à la postérité, c'est d'avoir transporté dans la science de la chaleur les principes de la mécanique; c'est d'avoir créé la logique de la nouvelle doctrine de la chaleur. C'est à lui que nous devons l'idée de ces cycles d'opérations qui, prenant un corps dans un état déterminé, le font passer à un état différent, en suivant un certain chemin, et le ramènent par une autre voie à son état primitif. C'est dans son ouvrage qu'on trouve l'application à des questions de physique de la méthode de réduction à l'absurde dont les anciens

Fig. 10.11 Saint–Robert cites Lazare and Sadi Carnot (Saint–Robert 1870c, pp 447–448.)

[47]Saint–Robert (1870c), pp 447–448, line 24.

[48]Saint–Robert (1870c), p 449, line 7.

[49]See Chapters 6 and 7.

In our opinion, it seems relevant to mention his *Qu'est–ce que la force?* which is a paper written in *La revue scientifique de la France et de l'étranger* (Saint–Robert 1872, *Qu'est–ce que la force?*, pp 985–993). *Qu'est–ce que la force?* discusses the role played by the physical concept of force and criticizes several definitions in history. Lazare Carnot's *Principes fondamentaux de l'équilibre et du mouvement* is also quoted (*Ivi*, p 986) on cause and motion (Carnot 1803a, p xii) and his heritage is evident. In fact, Saint–Robert's approach and criticism concerning the use of measurable physical quantities, mechanics, Descartes, Leibniz, *inertia*, Newton's second and third principles, Galilean principles, the difference between force and pressure, Huygens, Jean Bernoulli, et al. is very interesting. For instance, his (modern) criticism of the use of force as a cause of motion is truly significant, such as for example, when he quotes Poisson's debatable definition (Poisson 1833, I, p 2; see also Poisson 1829).

10.4.2 Historical Analysis of Principes de thermodynamique *(1865)*

Principes de thermodynamique[50] (Saint–Robert 1865) was published after 5 years with some substantial modifications (Saint–Robert 1870c). This second extended edition, still was entitled *Principes de thermodynamique* by Loecher, Turin et Florence and by Gauthier–Villars, Paris) (Fig. 10.12).

The first edition is noteworthy for a debate it brought about with Clausius regarding the concept of perfect gas and for a foreign review (Achard). *Principes de thermodynamique* concisely expresses content typical of a modern thermodynamics textbook. It is also interesting to note that in respect to others, his textbook uses the modern word *thermodynamique*.

For the sake of brevity and since the secondary literature is quite extensive below we provide an incomplete list of textbooks of/on thermodynamics (Table 10.2).

[50]The book was also reviewed in *Les mondes Revue hebdomadaire des sciences et de leurs applications aux arts et l'industrie* (Moigno 1865, VIII, pp 514–515). In the secondary literature, a book by Saint–Robert entitled *Traité de thermodynamique* seemed to have been circulated. Its existence is doubtful since any reference–frontispieces–editions seems to be lacking.

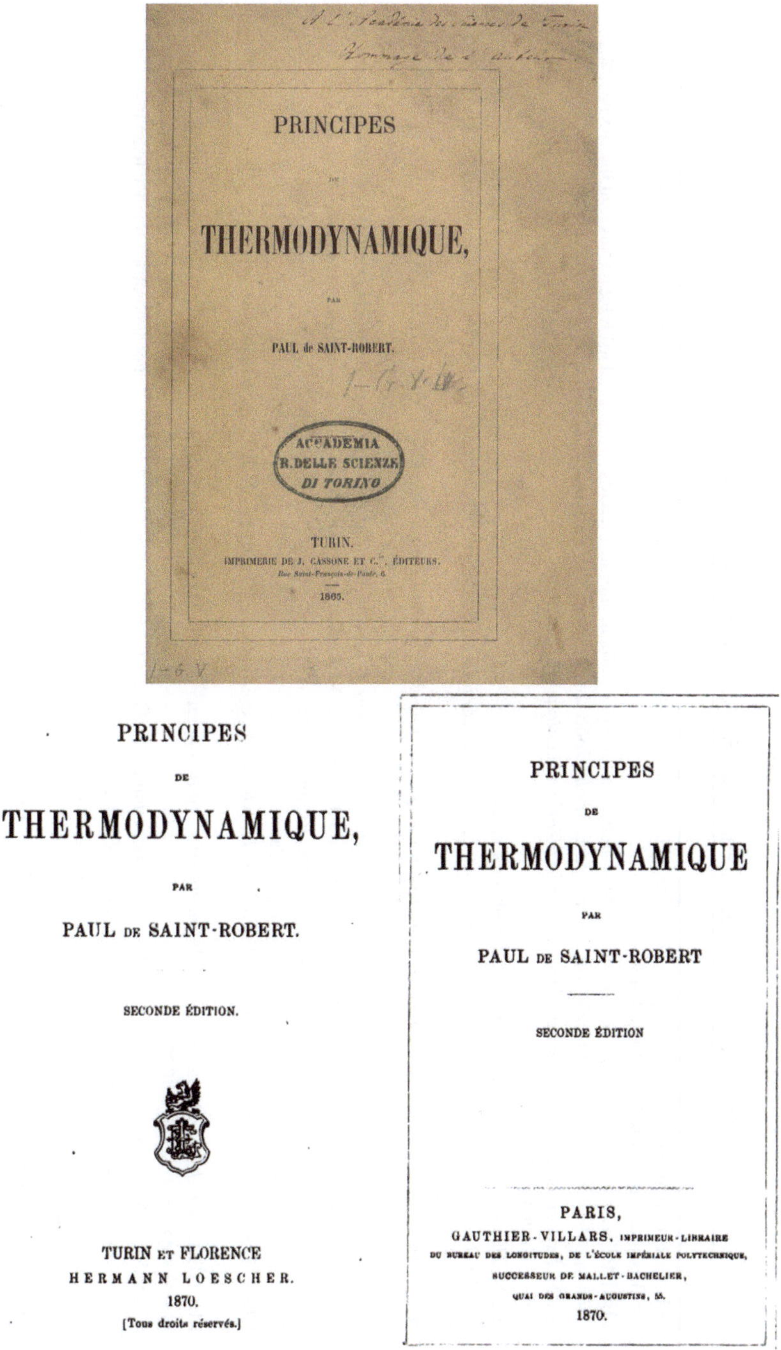

PRINCIPES

DE

THERMODYNAMIQUE,

PAR

PAUL de SAINT-ROBERT.

TURIN,
IMPRIMERIE DE J. CASSONE ET C.ie, ÉDITEURS.
Rue Saint-François-de-Paule, 6.

1865.

PRINCIPES

DE

THERMODYNAMIQUE,

PAR

PAUL DE SAINT-ROBERT.

SECONDE ÉDITION.

TURIN ET FLORENCE
HERMANN LOESCHER.
1870.
[Tous droits réservés.]

PRINCIPES

DE

THERMODYNAMIQUE

PAR

PAUL DE SAINT-ROBERT

SECONDE ÉDITION

PARIS,
GAUTHIER-VILLARS, IMPRIMEUR-LIBRAIRE
DU BUREAU DES LONGITUDES, DE L'ÉCOLE IMPÉRIALE POLYTECHNIQUE,
SUCCESSEUR DE MALLET-BACHELIER,
QUAI DES GRANDS-AUGUSTINS, 55.
1870.

Fig. 10.12 Two main editions of Thermodynamics by Saint–Robert

Table 10.2 An incomplete list of books also used for teaching thermodynamics and heat theories at Saint–Robert's time

Year	Author	Notes
1860	ZEUNER, *Grundzüge der mechanischen Wärmetheorie mit besonderer Rücksicht auf das Verhalten des Wasserdampfes*. Freiberg, Engelhardt.	For engineering students
1862	VERDET, *Leçons sur la théorie mécanique de la chaleur, professés en 1862 devant la société Chimique de Paris le 7 et le 21 février* [In *Œuvres de E. Verdet 1868–1872*, Tome VII, Paris].	Two long lectures on the Society of Chemistry are added.
	HIRN, *Exposition analytique et expérimentale de la théorie mécanique de la chaleur*, Mallet–Bachelier, Paris.	
	HIRN, *Théorie mécanique de la chaleur*, Tramblay, Leiber	
	ZEUNER, *Exposition analytique et expérimentale de la théorie mécanique de la chaleur,* Translated from German edition.	It is included in Hirn 1862 and regard to Zeuner 1860.
1863	HIRN, *Théorie mécanique de la chaleur. Confirmation expérimentale de la seconde proposition de la théorie*, Tramblay, Leiber	
	COMBES, *Exposé des principes de la théorie mécanique de la chaleur et ses applications principales*,[a] Bulletin de la Société d'Encouragement pour l'Industrie Nationale, Janvier, Paris	Based on principle of sufficient reason. And August 1864
1864	CLAUSIUS, *Die mechanische wärmetheorie*, Vieweg, Braunschweig, vol. I	
1865	SAINT–ROBERT, *Principes de thermodynamique*, Cassone, Torino.	
	HIRN, *Théorie mécanique de la chaleur. Exposition analytique et expérimentale,* Gauthier–Villars, Paris.	First part
1866	ZEUNER, *Grundzüge der mechanischen Wärmetheorie mit besonderer Rücksicht auf das Verhalten des Wasserdampfes*, 2nd edition.	Larger than previous edition
	STEWARD, *An Elementary treatise on Heat*, The Clarendon Press, Oxford.	For laboratory science teaching
1867	HIRN, *Mémoire sur la thermodynamique*. Gauthier–Villars, Paris.	Vol. I.
	JACQUIER, *Exposé élémentaire de théorie mécanique de la chaleur appliquée aux machines*, Gauthier–Villars, Paris.	
	CLAUSIUS, *Die mechanische wärmetheorie*, Vieweg, Braunschweig	Vol. II.
	COMBES, *Exposé des principes de la théorie mécanique de la chaleur et ses applications principales*, Bouchard–Huzard, Paris.	Based on principle of sufficient reason.

1868	HIRN, *Conséquences philosophiques et métaphysiques de la thermodynamique. Analyse élémentaire de l'univers*, Gauthier–Villars, Paris	
	VERDET, *Théorie mécanique de la chaleur,* Imprimerie Impériale, Paris.	Lectures at Sorbonne, 1884–1885
	JAMIN, *Cours de physique à l'École polytechnique*, Tome I–IV p. I–II, Gauthier–Villars, Paris	
	VERDET, re–edition in *Oeuvres de E. Verdet* 1868–1872 with new passages written by Proudhon and Violle, Masson, Paris.	
1868–1869	CLAUSIUS, *Theorie mécanique de la chaleur*, II vols., French édition, Translated by Folie, Lacroix éditeur, Paris.	And 1874
1869	BRIOT, *Théorie mécanique de la chaleur*, Gauthier–Villars, Paris.	
	DUPRÉ, *Théorie mécanique de la chaleur*, Gauthier–Villars, Paris.	
	REECH, *Théorie des machines motrices et des effets mécaniques de la chaleur*, Lacroix éditeur, Paris.	
	ZEUNER, *Théorie mécanique de la chaleur, avec des applications aux machines, 2 édition,* Translated from German by Arnthal Maurice ad Cazin Achille, Gauthier–Villars, Paris.	
1870	VERDET, re–edition in *Oeuvres de E. Verdet* 1868–1872.	Vol. II
	SAINT–ROBERT, *Principes de thermodynamique*, Loescher, Torino–Firenze.	
	SAINT–ROBERT, *Principes de thermodynamique*. Gauthier–Villars, Paris.	
	MAXWELL, *Theory of Heat*, Longmans, London.	
1886	JAMIN, *Cours de physique,* Tome I–IV p. I–II, Gauthier–Villars, Paris.	[1887, 1890, 1891]
1887	HIRN, *La thermodynamique et le travail chez les êtres vivants*, *Revue scientifique* 22: 673–684, [May, Paris, Bureau des revues,].	

[a] Saint–Robert listed this work in his final bibliography and added that "[. . .] is not concluded" (Saint–Robert 1865, p 206)

First, we immediately see that Saint–Robert's first book appears ca. 10 years after the confirmation of the second principle. That contrasts with both the great innovation that thermodynamics represented for physics and the great interest in motive power and engines at the time.[51] Of course, among the books listed in the previous table, we can note that those published[52] several times clearly have greater didactic influence, e.g., de Saint–Robert, Hirn (1815–1890), Clausius (1822–1888), Verdet (1824–1866), Zeuner (1829–1927).

Saint–Robert excludes traditional Newtonian mechanics in his thermodynamics – an aspect that we have already presented by using other elements (see Chapters 6 and 7):

> In 1686 Newton wrote, in the preface of *Mathematical Principles of Natural Philosophy*: "Until it is permitted to infer the other natural phenomena from the principles of mechanics with the same way of reasoning." [in Latin in the text]. This wish, as far as the theory of heat is concerned, was partially granted during our time. Thanks to Meyer's fortunate idea regarding the equivalence of heat to mechanical work, thanks to Sadi Carnot's previous ingenious reflections on the motive power of fire and on machines useful for developing this power, published in 1824, thanks to Joule's celebrated experiments, ultimately thanks to Thomson, Rankine and Clausius', etc. mathematical developments, a new science was achieved; it builds a bridge between mechanics on one side and physics and chemistry on the other, science of highly philosophical and practical capabilities: because this will allow us to one day penetrate the mystery of the internal composition of bodies and will allow us, from today forward, to create a precise idea of heat machines. The principles of the theory, although formulated over a dozen years ago, mainly by the work of W. Thomson, Rankine e Clausius, are still not well–known. There are also scientists who look at it mistrustfully. However, aside from pure mathematics, there are very few sciences that are based on more solid grounds.[53]

Before discussing his exposition of the second principle in detail, we believe that it is important to understand the historical reasoning that led Saint–Robert to his formulation and then compare it to the original version proposed by Sadi Carnot in *Réflexions sur la puissance motrice du feu*. Therefore, we concisely present the development of the foundations of Saint–Robert's theory, selected from the chapters most relevant for our aim: Chapters 1, 2, 3 and 8 (Table 10.3):

---

[51] We avoid any possible contribution to the discussion regarding the *steam engine, or* rather *motive power by heat* et al. in society, since this is not the main subject of our research. Historical foundations on physics and mathematics in the theory are considered here.

[52] We should note that Maxwell's (1831–1879) textbook was published in 1870, and accumulated many new editions.

[53] Saint–Robert (1865), pp V–VI, line 1. Author's italics. On Joule see: Joule (1844).

Table 10.3 A concise view of Saint–Robert's main concepts

Step	Saint–Robert's reasonings (Saint–Robert 1865, Chapter I, pp 1–24)
1.	Saint–Robert proposes his theoretical idea on heat theory (*Ivi*, p 1).
	He is only interested in the effects of heat and not in its intimate nature.[a]
2.	The effects of heat are composed by two theoretical parts:
	Analytical theory of heat, e.g., by Fourier. (*Ivi*, p 2)
	Definition of thermodynamics (*Ivi*, p 3)
	Dynamic theory of heat, on transformation of heat, (*Ivi*, p 3)
3.	Thermometry: definition of main magnitudes of theory (*Ivi*, p 3)
	Calorimetry: definition of main magnitudes of theory (*Ivi*, p 5)
4.	Thermal expansion of hard–bodies: $P = f(V, T)$ (*Ivi*, p 7)
5.	Relationship between *P, V, T*: $PV = R(\alpha + T)$ (*Ivi*, p 9)
6.	Mariotte's equation (*Ivi*, p 8)
	Gay–Lussac's equation: $V = V_0(1 + \alpha T)$ (*Ivi*, p 8)
	R depends on kind of gas used (*Ivi*, p 10)
7.	Equation of elasticity for a gas $PV = R$ (*Ivi*, p 11)

Legend: to indicate temperature, pressure and volume we use modern symbols, *T, P*, and *V* respectively, instead of "*t*" "*p*" and "*v*" which Saint–Robert uses, *SR* Saint–Robert

[a]In this sense, he is within a positivist panoramic view of his time as Lamé previously reported

1. We can see that Saint–Robert's initial premise on the theory's close nature, which he intends to examine. There are no speculations regarding the foundations of the theory of heat rather than caloric:

> The intimate ["intime"] nature of heat ["chaleur"] is not a known argument [...]. Therefore we will look at heat as a principle whose nature we do not want to know and we only work with its effects.[54]

2. Saint–Robert clearly explains that he will only discuss the theory's effects and describes two orders of phenomena: [...] *theory of propagation of heat* [...] and [...] *theory of transformation of heat*[55] (Saint–Robert 1865, pp 2–3, line 21). For this reason, from the first page on, he rids himself of the problem regarding the *unknown* nature of heat (Saint–Robert 1865, p 1). Only on page 51 does he discuss molecular aspects:

> These hypotheses [...] are not necessary in order to establish the theory. The attentive examination of the facts, provided to us through observation and experiments, aided by the powerful instrument of infinitesimal analysis is sufficient for establishing laws, to which their effects are subject [...].[56]

[54]Saint–Robert (1865), p 1, line 1.

[55]Author's italics.

[56]Saint–Robert (1865), p 1, line 6.

The depth of the preface is extraordinary. He argues, with extreme certainty, for the theoretical autonomy of thermodynamics as a *new theory* (Saint–Robert 1865, p V–VIII). In this sense, Saint–Robert's foundational attitude is also confirmed by the preface of the second edition of 1870:

> Thermodynamics is independent from every hypothesis on the cause of heat, even though usually the opposite is believed to be true. The idea that heat is only a movement of particles in bodies can be traced back to very distant past. However, this concept remained futile until the idea of comparing heat with the force sensitivity of a body, or with mechanical work was born. For the entire time in which we were limited to comparing heat to heat itself, it was of very little importance that it was considered as a fluid material or as a movement. The hypothesis that molecular movement makes regarding heat, quite differently than being used as a foundation of Thermodynamics, has received support from this that it could not have found elsewhere. The foundations of Thermodynamics are much more solid than any hypothesis on the nature of heat, a hypothesis whose direct verification will never be possible. Thermodynamics begins with an evident, indisputable principle: that it is impossible that in nature something comes from nothing. In other words, it is the principle of the impossibility of perpetual motion [. . .].[57]

3. (4., 5., 6. and 7.) Saint–Robert concludes by introducing a phenomenological section in which he defines the basic concepts of the theory without resorting to other physical theories.

Below, we provide an analysis of the second chapter of *Principes de thermodynamique* (which proves to be the most important chapter for this paper) in which Saint–Robert discusses the concept of the cycle and the second principle (see the following Table 10.4).

Table 10.4 A concise view of Saint–Robert's main concepts

Step	Saint–Robert's reasonings (Saint–Robert 1865, Chapter II, pp 25–50)
1.	Concept of state in *PV–diagram* (*Ivi*, p 26)
	Concept of heat and path: an infinitesimal succession of equilibrium–states between two initial states. (*Ivi*, p 26)
2.	Concept of mechanical work (*Ivi*, p 28)
	Assumption of equivalence theory of heat–work
	Definition of two heat reservoirs: "$Q > Q'$";
	1st reasoning based on an *absurdum proof* (1st AP) (*Ivi*, p 28)
	SR cites the erroneous theory of caloric referring to Sadi Carnot "$Q = Q'$" (*Ivi*, p 29)
	SR cites Hirn's experiences on vapor machines where "$Q > Q'$";
	2nd reasoning based on an *absurdum proof* (2nd AP) (*Ivi*, p 29)
3.	Concept of a reversible cycle (*Ivi*, p 30)
	3rd and 4th reasonings based on an *absurdum proof* (3rd and 4th AP) (*Ivi*, p 31; p 31)
4.	Proof of mechanical equivalent of calorimetry in a reversible cycle
	5th reasoning based on an *absurdum proof* (5th AP) (*Ivi*, p 32)

(continued)

[57] Saint–Robert (1870c), pp VI–VII, line 2. (Author's Capital letters).

Table 10.4 (continued)

Step	Saint–Robert's reasonings (Saint–Robert 1865, Chapter II, pp 25–50)
5.	Extension to *n* operations of the calculation of mechanical work of $nL = JnQ$
	6th reasoning based on an *absurdum proof* (6th AP) (*Ivi*, p 32)
	SR proves Sadi Carnot's theorem by cycles
	7th reasoning based on an *absurdum proof* (7th AP)
	SR, in this case, does not cite Sadi Carnot (*Ivi*, p 33)
6.	SR uses cycles to convert heat to (mechanical) *maximum* of work (*Ivi*, p 34)
7.	Assumption of equivalence theory of heat–work (Ivi, p 35)
	Definition of $1/J = A$ "équivalent calorifique du travail mécanique" (*Ivi*, p 35)
8.	Second law of thermodynamics (*Ivi*, p 37)
	SR assumes a different theoretical position in respect to Sadi Carnot (*Ivi*, p 37)
	Assumption of equivalence theory of heat–work (*Ivi*, p 37)
9.	Sadi Carnot's theorem. SR, in this case, does not cite Sadi Carnot (*Ivi*, p 40)
	SR generalizes, by means several heat reservoirs, a calculation on Sadi Carnot's theorem (*Ivi*, p 41)
	SR generalizes, by means equivalence theory of heat–work, a calculation on Sadi Carnot's theorem $mL = J(mQ_1 - mQ_0)$(*Ivi*, p 41)
	8th reasoning based on an *absurdum proof* (8th AP)
	SR generalizes, by means by means several heat reservoirs and using equivalence theory of heat–work, a calculation on Sadi Carnot's theorem (*Ivi*, p 41)
	Proof of second law by Thomson (*Ivi*, p 43)
10.	Proof of Sadi Carnot's theorem
	SR, in this case, does not cite Sadi Carnot (*Ivi*, p 44)
	Calculation of $F(T_1, T_0) = \frac{Q_1}{Q_0}$ (*Ivi*, p 44)

Legend: we used modern symbols *T, P, V*, for "*t* , "*p*" and "*v*"; we use the modern symbol *Q* for "*q*". Saint–Robert uses "*q*" until p 38, in particular, to indicate heat in the equation of the mechanical equivalent "*J* = *L/q*"; *AP ad absurdum* proof, we use the modern symbol *W* for "*T*", *SR* Saint–Robert

1. The second chapter is important for our aim since it establishes the theory of the second law. In the first step, Saint–Robert addresses the basic concepts of the theory. He specifies the concept of *state* and the path between states, using, at least conceptually, infinitesimal analysis, when dealing with an infinitesimal series of states: "A series of infinitely close, different states will construct a curve [...]" (Saint–Robert 1865, p 26, line 14).
2. Saint–Robert addresses the theory of the equivalence of heat and work and through *reductio ad absurdum*, including the impossibility of perpetual motion, demonstrates that in a cycle of an engine, the heat absorbed must be greater than that given off (Saint–Robert 1865 p 28, line 12). We immediately note two facts: a) Saint–Robert cites Sadi Carnot's caloric theory (Saint–Robert 1865, p 29, 37; see also p 57) and b) indicates his inaccurate formulation of the efficiency.
3. (and 4.) Saint–Robert, differently from Sadi Carnot, formulates the concept of the cycle as a closed path between two states in which the quantity of heat Q is not conserved, $Q > Q'$. In particular, Saint–Robert does this by introducing the mechanical equivalent of calorimetry $W / Q = J$ through the (reversible) cycles

of a heat machine. From here on, a different theoretical formulation compared to that of Sadi Carnot in *Réflexions sur la puissance motrice du feu* can be noted. In fact, from the beginning, Sadi Carnot constructed his theory according to cycles, first in three phases and then in four phases. Saint–Robert (obviously thanks to a better understanding of the phenomena), however, utilizes the mechanical equivalent of heat in order to construct his reasoning aimed at the formulation of the second law, which, instead, could be introduced even without the use of the equivalent. Nevertheless, it must be remembered that Saint–Robert's textbook is used for pedagogical purposes as well.

5. At this point, Saint–Robert once again utilizes the cycle method and the impossibility of perpetual motion, demonstrating Sadi Carnot's theorem: the efficiency is equal for all of the reversible heat machines (Saint–Robert 1865, p 33, line 8). It must be pointed out that Saint–Robert first constructs the concept of the cycle and then demonstrates Sadi Carnot's theorem. However, in the original version, Sadi Carnot firstly provides his theorem on the independence of working substances (Carnot 1978, p 38) and then proposes his reasoning regarding a complete four-phase-cycle (Carnot 1978, pp 29–38, line 6). Here, the pedagogical role played by its content is evident.
6. (and 7.) It is interesting to note that here Saint–Robert succeeds in converting heat into maximum motive power (Saint–Robert 1865, p 34, line 20). From here, he once again uses the equivalence to define the value $1/J = A$, which he will need in the following chapter to work out his definition of the various types of thermal states and then reach (Gibbs') potentials.
8. With regard to the erroneous caloric theory, we note that Saint–Robert declares his – previously announced – position (Saint–Robert 1865, p 37, line 26) to be different from that of Sadi Carnot. It is interesting to note that just after, Saint–Robert once again provides his calculations based on the equivalence of heat and work.
9. (and 10.) Here, Saint–Robert dedicates himself completely to Sadi Carnot's theorem (without, however, citing the author) and attempts to generalize it while considering other sources (as did Carnot). However, differently from the French author, he relies heavily on the equivalence of heat and work and infinitesimal analysis (we stress the presence of the seventh *reductio ad absurdum* that we encountered in Saint–Robert's book). In the end, he dedicates several pages to demonstrating Thomson's version of the second principle (Saint–Robert 1865, p 43, line 3), finally affirming that he prefers Clausius' statement (*Ivi*, p 50).

Below, we provide an analysis of the third chapter of *Principes de thermodynamique* in which Saint–Robert addresses the fundamental equations of the theory (Table 10.5).

Table 10.5 A concise view of Saint–Robert's main concepts

Step	Saint–Robert's reasonings (Saint–Robert 1865, Chap. III, pp 52–90)
1.	Reasoning on effects produced by heat expansion of bodies (*Ivi*, p 51): (1) An external work (e.g., by pressure); (2) An internal work (e.g., by molecules) Propagation of heat and definition of four main states (*Ivi*, p 52): (1) internal state of a body (*H*); (2) internal work of a body (*I*); (3) external work (*W*); (4) Work of living force [*vis viva*] (*F*).
2.	By means of equivalence work–heat he obtains: $W_{internal} + W_{esternal} = Work_{Total}$ (*Ivi*, p 52)
3.	Fundamental equation for the four states: $dQ = dH + AdI + AdW + \frac{1}{2}AdF \quad A = 1/J \quad (Ivi, p53).$
4.	dQ, in "new theory" is not an exact differential (*Ivi*, p 62)
5.	Equation used for the second law of thermodynamics $\frac{f(T)}{f'(T)} = C$ SR cites Sadi Carnot. (*Ivi*, p 63)
6.	Calculation of the inverse of the mechanical equivalent of calorimetry SR uses a method of reasoning opposite to that used in chap II. (*Ivi*, p 64) Calculation f specific heat at constant pressure (*Ivi*, p 65). SR uses infinitesimal analysis to prove the constancy of Carnot's function $C = T$. (*Ivi*, p 68) Introduction to entropy function attributed to Rankine (*Ivi*, p 69).

Legend: we use modern symbols *T*, *P*, *V*, for "*t*", "*p*" and "*v*"; we use the modern symbol *Q* for "*q*". Saint–Robert uses "*q*" until p 38, in particular, to indicate heat in the equation of the mechanical equivalent "*J* = *L/q*"; *AP ad absurdum* proof, we use modern symbol *W* for "*T*", *SR* Saint–Robert

1. (2. and 3.) Saint–Robert returns to his reasoning regarding the effects of heat which he began in Chapter 1. After having very precisely defined the behavior of a body exposed to a source of (temperature and) heat, he uses the infinitesimal calculation without hesitating: he introduces the thermodynamic potentials and illustrates their interchangeability in order to obtain the equations of state in thermodynamics.
4. Let us observe his mastery of the infinitesimal calculation and his connection to physical theory (the theory of equivalence) in his foundations when he emphasizes that in the theory of heat, dQ is not an exact differential.
5. (and 6.) Saint–Robert returns to the second law, now citing its original author (Saint–Robert 1865, p 63, line 11). We also remark that with a method that is inverse to that used in Chapter 2 (while still using infinitesimal analysis) he determines constant $A = 1/J$ which he will need for the calculation of constant C in Sadi Carnot's function. He follows Sadi Carnot's idea (and therefore Clapeyron's) which is decisive for quantifying the efficiency of a thermal process (Saint–Robert 1865, p 68, line 3). He obtains it, and then defines the entropy function, attributed to William John Macquorn Rankine (1820–1872), as a primitive function of an exact differential (Saint–Robert 1865, p 69, line 17).

Below, we end the analysis of *Principes de thermodynamique*, providing a summary of Chapter 8 in which Saint–Robert addresses general topics regarding heat machines (Table 10.6):

Table 10.6 A concise view of Saint–Robert's main concepts

Step	Saint–Robert's reasonings (Saint–Robert 1865, Chap. VIII pp 189–203)
1.	Analogy between a heat engine and an hydraulic engine. Reasonings on direct and inverse heat paths. (*Ivi*, p 190) Equivalence work–heat (*Ivi*, p 190) In a cycle one obtains: $W = J\left(\sum Q - \sum Q'\right)$ (*Ivi*, p 190)
2.	Efficiency of a heat machine $\eta = \frac{(\sum Q - \sum Q')}{\sum Q} = 1 - \frac{\sum Q'}{\sum Q}$ (*Ivi*, p 191)
3.	Efficiency of a simple heat machine $\eta = 1 - \frac{T'}{T} = \frac{T - T'}{T}$ (*Ivi*, p 192)
4.	Maximum work produced by any kind of a heat machine $\eta = 1 - \frac{\sum Q'}{Q}$ (*Ivi*, p 192)

Legend: we use modern symbols *T*, *P*, *V*, for "*t*", "*p*" and "*v*"; we use the modern symbol *Q* for "*q*". Saint–Robert uses "*q*" until p 38, in particular, to indicate heat in the equation of the mechanical equivalent "*J* = *L/q*"; *AP ad absurdum* proof, we use the modern symbol *W* for "*T*"; To indicate the efficiency of the work of a heat machine, I used the modern symbol η instead of "*E*" which Saint–Robert uses, *SR* Saint–Robert

In Chapter 8, Saint–Robert essentially dedicates himself to studying running heat machines.[58] First, he addresses a classical analogy, previously proposed by Sadi Carnot in *Réflexions sur la puissance motrice du feu*, (Carnot 1978, pp 16–17; see also Chapter 8) as well as an analogy between the cyclic path of heat in heat machines and that of water in hydraulic machines. Let us note that, in this case, Saint–Robert adopts the theory of equivalence, with which he formulates the modern efficiency of heat machines. He first discusses machines with only two sources and then expands his analysis to include machines with various thermostats for the calculation of maximum work (Saint–Robert 1865, p 192, line 1). The textbook does not ignore the applications (Saint–Robert 1865, pp 192–202), which are, for instance, elastic fluids, dilation, the flowing of fluids, the motion of projectiles and the calculation of the efficiency of heat machines.

In conclusion, Saint–Robert bases his arguments mainly on Sadi Carnot's theory, but he includes infinitesimal analysis and the equivalence of heat and work (as a constant of the mechanical equivalent of heat) in almost all of his calculations. However, as it was observed in the previous analysis, Saint–Robert's position concerning the two theories was initially neutral. In any case, the equivalence of heat and work appears to play a secondary role compared to the ideas of the second principle. In this sense, Sadi Carnot's heritage is evident. Internal energy is introduced only at the end of the textbook, and it is given as a particular thermodynamic potential. Saint–Robert used, since the 1865 edition, the thermodynamic potentials that Reech (see above) and Masson introduced, without creating a lasting effect. In 1873, Gibbs

[58]Let us note that this part of the theory is constructed pedagogically like that which appears today in polytechnic engineering institutes, usually known as technical physics.

(1839–1903) rediscovered them without, however, receiving too much consideration from the scientific community of the time (Truesdell 1980, p 300, 1970, pp 30–31).

The theory in *Principes de thermodynamique* shows a great mastery of the subject, a concrete understanding of thermodynamics and high level of logical and mathematical attention, as is typical in a pedagogical treatment. It may to represent an exception among the discussions and theories presented by Truesdell's historical outlines in the 1980s (Truesdell and Bharatha 1977; Truesdell 1980) where the authors discuss the level of *inaccuracy* in thermodynamic developments by using intelligent historical irony.

Saint–Robert's eloquence is very important since it precedes Tait's (1831–1901) work from 1868 by several years. It is quite probable that this work had an influence on Tait's contemporaries, thanks both to Tait's authority and to the controversy he elicited by diminishing Mayer's work. The textbook concludes with a bibliography of 32 essays specific to thermodynamics – this is yet another interesting aspect.

Finally, in regard to Sadi Carnot's heritage, Saint–Robert is certainly the most important Italian *thermodynamicist* of his time. Even abroad, he appeared to be one of the few who understood the theoretical and pedagogical modernity of this new theory. He exposed it to the maximum level, both theoretically and applicatorily (Fig. 10.13).

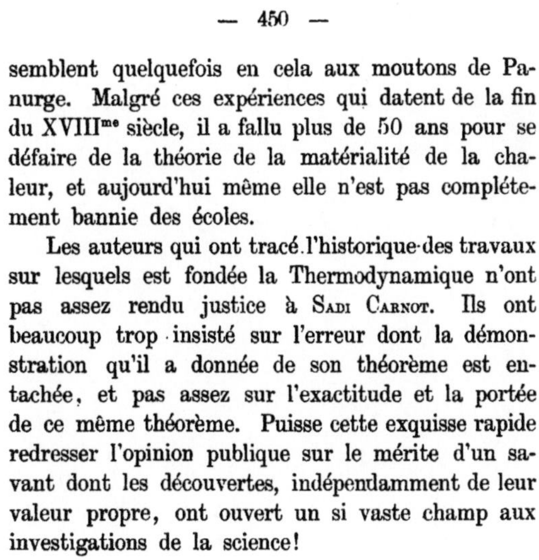
— 450 —

semblent quelquefois en cela aux moutons de Panurge. Malgré ces expériences qui datent de la fin du XVIIIme siècle, il a fallu plus de 50 ans pour se défaire de la théorie de la matérialité de la chaleur, et aujourd'hui même elle n'est pas complétement bannie des écoles.

Les auteurs qui ont tracé l'historique des travaux sur lesquels est fondée la Thermodynamique n'ont pas assez rendu justice à SADI CARNOT. Ils ont beaucoup trop insisté sur l'erreur dont la démonstration qu'il a donnée de son théorème est entachée, et pas assez sur l'exactitude et la portée de ce même théorème. Puisse cette exquisse rapide redresser l'opinion publique sur le mérite d'un savant dont les découvertes, indépendamment de leur valeur propre, ont ouvert un si vaste champ aux investigations de la science!

Fig. 10.13 Saint–Robert's conclusion on Sadi Carnot (Saint–Robert 1870c, p 450)

In this sense, he followed the change in the scientific attitudes of his time and recognized those of the past regarding the foundations of thermodynamic theory compared to those of mechanics.

Chapter 11
Final Remarks on the Scientific Relationship Between *Père et Fils*

Clearly, the relationship between the science of machines and thermodynamics was similar to that between Lazare and Sadi Carnot. It was one of parentage.

(Gillispie 1971, p 61, line 15).

Mainly based on the historical hypotheses announced by Charles Gillispie (1971), the Parisian workshop (Taton 1976) and finally on the historical and historical–epistemological investigations proposed in the previous chapters, we present here details and correlations on the scientific relationship between the two Carnots.

All evidence shows a common theoretical attitude to science and their sequence indicates that there is a great convergence of thought between the two theorists, at least as reflected together in one book.

11.1 An Outline of the Problem

As previously discussed, the thermodynamic theory Sadi Carnot presented in his *Réflexions sur la puissance motrice du feu* is essentially (and in a few parts) openly, based on the caloric hypothesis. We know that at that time, scholars considered a fluid capable of flowing from hotter to colder bodies; more importantly, this capability is conserved during phase changes. However, as we frequently discussed in previous chapters, Sadi Carnot correctly expressed doubts regarding the validity of caloric[1] (Carnot 1978, p 37, p 89) and in *Notes sur les mathématiques, la physique et autres sujets* he utilized in a modern way the hypothesis on the equivalence

[1]In particular, in *Notes sur les mathématiques, la physique et autres sujets* he explicitly wrote: "Perhaps at this point [after I dealt with caloric and friction] I may be allowed to offer a hypothesis concerning the nature of heat [chaleur]." (Carnot 1878a, folio 4 *1c*, pp 38–39; Picard 1927, pp 76–77); this is a problem that he was also surely aware of in *Réflexions sur la puissance motrice du feu* (Carnot 1978, pp 89–90) but he avoided addressing it explicitly.

C.C. Gillispie and R. Pisano, *Lazare and Sadi Carnot: A Scientific and Filial Relationship*, History of Mechanism and Machine Science 19, DOI 10.1007/978-94-017-8011-7_11,

of heat–work (see Fig. 11.1.). Here, he even calculated the constant J of the proportionality between a unit of heat and a unit of work: 1,000 kgm = 2,70 kcal where 1 kcal = 370 kgm.

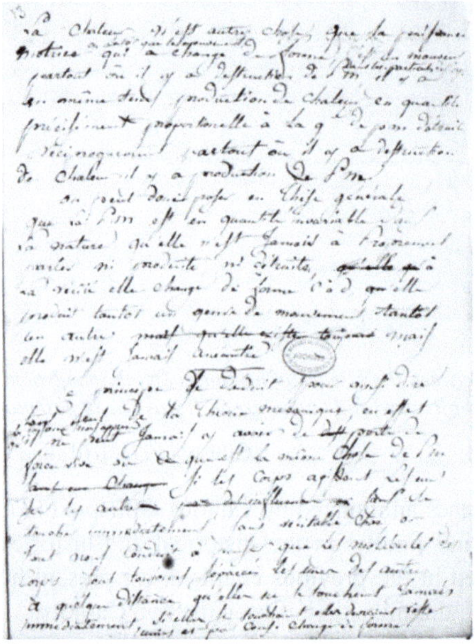

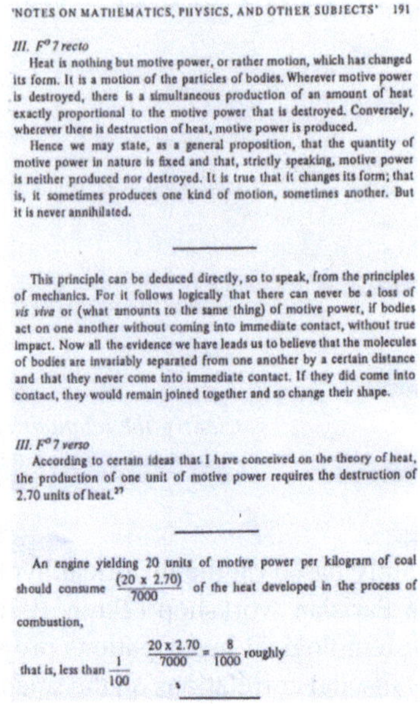

'NOTES ON MATHEMATICS, PHYSICS, AND OTHER SUBJECTS' 191

III. F° 7 recto

Heat is nothing but motive power, or rather motion, which has changed its form. It is a motion of the particles of bodies. Wherever motive power is destroyed, there is a simultaneous production of an amount of heat exactly proportional to the motive power that is destroyed. Conversely, wherever there is destruction of heat, motive power is produced.

Hence we may state, as a general proposition, that the quantity of motive power in nature is fixed and that, strictly speaking, motive power is neither produced nor destroyed. It is true that it changes its form; that is, it sometimes produces one kind of motion, sometimes another. But it is never annihilated.

This principle can be deduced directly, so to speak, from the principles of mechanics. For it follows logically that there can never be a loss of *vis viva* or (what amounts to the same thing) of motive power, if bodies act on one another without coming into immediate contact, without true impact. Now all the evidence we have leads us to believe that the molecules of bodies are invariably separated from one another by a certain distance and that they never come into immediate contact. If they did come into contact, they would remain joined together and so change their shape.

III. F° 7 verso

According to certain ideas that I have conceived on the theory of heat, the production of one unit of motive power requires the destruction of 2.70 units of heat.[27]

An engine yielding 20 units of motive power per kilogram of coal should consume $\frac{(20 \times 2.70)}{7000}$ of the heat developed in the process of combustion,

$$\frac{20 \times 2.70}{7000} = \frac{8}{1000} \text{ roughly}$$

that is, less than $\frac{1}{100}$

Fig. 11.1 Notes sur les mathématiques, la physique et autres sujets (*On the left*: Carnot 1878a folio 13. (2011 – with authorization of the *Académie des sciences*, *on the right*: an English language transcription by Robert fox (Carnot 1986, p 191). In French language see also Carnot 1953, pp 133–135; Picard 1927, pp 81–82, III–IV, Feuillet 7–9, pp 44–45)

We should also note two significant occurences:

1. Sadi Carnot wrote on equivalence of heat–work according to a modern definition of energy, which he referred to as *puissance motrice*: "[. . .] in nature motive power can never be either created or destroyed." (Carnot 1878a folio 6v; see also Picard, p 81 [III, feuillet 7, pp 44–45; Carnot 1986, p 118]; Blanchard, p 134).
2. The content on the equivalence of heat–work in *Notes sur les mathématiques, la physique et autres sujets* was surely written before 1832, so it preceded results of Julius *Robert* von Meyer (1814–1878) and James Prescott Joule (1818–1889)'s[2] (Challey 1971).

Moreover, the value of the *equivalent* which Sadi Carnot obtained is affected by only two percent compared to the mechanical *equivalent* (1 kcal = 365 kgm) subsequently calculated by Mayer between 1840 and 1842. What is even more surprising about this agreement is that it is possible to obtain the equivalency above based on numeric values, proportional to the air, already given by Sadi Carnot in *Réflexions sur la puissance motrice du feu* just as both Louis Philibert Décombe (1919) and Camille Raveau (1867–1953) (Raveau 1919) affirmed in 1919.

---

[2]E.g.: Joule 1844, 1845, 1847, 1965.

Below, we present the most important results (for our purposes) which help us stress the content of the two Carnots' work:

1. The long list of the most important *historical studies and commented documents* from 1834 to the second half of the past century on Sadi Carnot (see last tables below and the *References* at the end of this book).
2. The *historical and complete explanation* of *Lazare Carnot's mechanics* presented in the two *Mémoire sur les machines* (Carnot 1778, 1780), *Essai sur les machines en général* (*Id.*, 1786; 1808a, b), *Principes fondamentaux de l'équilibre et du mouvement* (*Id.*, 1803a) (Gillispie 1971; Gillispie and Youschkevitch 1979; see above Chapters 2, 3, 4, and 5).
3. Sadi Carnot's *Recherche d'une formule propre à représenter la puissance motrice de la vapeur d'eau–unpublished manuscript* date. The correct date of publication still lacks historical decisive evidence. Generally speaking, the most recent Carnot historians[3] agree (more or less) that it was written before *1824* (see Chapter 3, ft 44; see also Gillispie 1976, p 30, p 34) *or after April 1823* (Carnot 1986, p 168). On this point, Charles Gillispie suggested crucial reasons regarding the role played by the concepts of reversibility and incompleteness and completeness of a cycle (see Chapter 3; Gillispie 1976, pp 30–33). Robert Fox (Carnot 1986, pp 168–169) suggested his "tentative inclination to suppose" (*Ivi*, p 169) that the *unpublished manuscript* was mainly[4] written between November 1819 –when Clément (Lervig 1985) lectured as a professor at *Conservatoire des arts et métiers* in *Paris*– and 8 March 1827, when the latter acknowledged a "distinguished mathematician" (Carnot 1986, p 167) for information which added to his lecture. However, whether the composition of the *unpublished manuscript* was elaborated before or after (or during?) the composition of *Réflexions sur la puissance motrice du feu,* it should still be understood.
4. *Essai sur les machines en général* is "[...] in important respects, a precursor of the *Réflexions*" *sur la puissance motrice du feu* (Carnot 1986, p 13).
5. The *intellectual environment of Sadi Carnot* (Fox 1971a, 1995).
6. The *historical and complete explanation* of *caloric theory of gases* (Fox 1971b).
7. The role played by *new technologies on steam engines before 1824* (Fox 1976).
8. The *equation of state of ideal gases* by Sadi Carnot (Fox 1971b; Drago and Vitiello 1986; see below).
9. Lazare Carnot's mechanics are largely based on the *Principle of virtual work* (Gillispie 1971, Chapters 2 and 3; Drago and Manno 1989; Drago and Perno 2004; Drago 1993c, pp 69–80; Pisano and Drago 2013; Pisano and Capecchi 2013; see Chapters 2 and 3 and below).

---

[3]See Chapter 3; Gabbey and Herivel 1966; Challey 1971, p 80, p 83; Gillispie 1976, pp 30–33, p 34; Fox 1976, p 34, pp 149–168; Hornix 1982, p 403; Carnot 1986, pp 167–170; Drago and Vitiello 1986; Fox 1988, pp 294–297.

[4]For Robert Fox: (1) Sadi Carnot citing an *equation* in the *unpublished manuscript* (Carnot S–EP, folio 16; see also Carnot 1986, p 178, equation 12) from Clément's manuscript. (2) Sadi Carnot's doubts on Clément and Desormes's law. (3) The role played by unit *dyname* in the *unpublished manuscript* and its absence in *Réflexions sur la puissance motrice du feu* (Carnot 1986, pp 169–170).

10. *Machines en général* in the two Carnots' work (Pisano-forthcoming).
11. The *historical explanation* of the "*supprimer*" (Carnot 1978, p 39) *two adiabatics* for the complete cycle in *Réflexions sur la puissance motrice du feu* (Pisano 2001; Drago and Pisano 2005; Pisano 2010; see Chapters 7 and 9).
12. The hypothesis of the *scientific roots of Sadi Carnot's cycle* in *Réflexions sur la puissance motrice du feu* (Pisano 2003; Mach [1896] (1986); see Chapter 8).
13. *A model of organization* of a scientific theory based on resolving a general problem (PO) in *Réflexions sur la puissance motrice du feu* (Drago 1991; Drago and Pisano 2005, 2008; Pisano and Gaudiello 2009a, b; Pisano 2012a; see Chapters 6 and 7).
14. The *complete interpretation* of Sadi Carnot's famous *footnotes* (Carnot 1978, pp 73–79) written in *Réflexions sur la puissance motrice du feu* (Pisano 2001; Drago and Pisano 2005; see Chapter 9).
15. The finding of a logical base concerning *Double Negative Sentences* (DNSs) in *Réflexions sur la puissance motrice du feu* (Drago and Pisano 2000, 2002; Pisano 2004; see Chapters 6 and 7).
16. The *mathematical studies* regarding Sadi Carnot's thermodynamic theory and the *overcoming of caloric theory* (Pisano 2007a; Drago and Pisano 2008; Pisano and Capecchi 2009; see Chapter 10).
17. The novelty of a *relationship between mathematics and physics* in Sadi Carnot's *Réflexions sur la puissance motrice du feu* (Drago and Pisano 2007; Pisano 2011a; see Chapters 7 and 9).
18. Both Lazare and Sadi Carnot's *synthetic method* as a method of reasoning alternative to the infinitesimal analysis in use at the time (Drago and Pisano 2005, 2007; Pisano 2010, see Chapters 6, 7, and 9 and below).
19. The number of *principles* in thermodynamic theory presented by Sadi Carnot in *Réflexions sur la puissance motrice du feu* (Mach [1896] (1986); Drago and Pisano 2000; Pisano 2004, 2010; see Chapters 6 and 7 and below).
20. The recent study of the Lazare and Sadi's *documents concerning Leibniz* in *Sadi Carnot's archive* at *École polytechnique* in Paris (Pisano preprint).

11.2 A Hypothesis on Structures of the Scientific Parentage

The scientific relationship between the two Carnots is very complex since it doubtlessly begins with several of Lazare Carnot's manuscripts. In fact, it is possible that a very large program of research on the property of *machines* (mechanical and heat) *en général* involved *Père et Fils* for several years. Nevertheless, we currently have no historical proof that Sadi Carnot knew and/or read all of Lazare's works. It is quite probable, that he studied them, but – as we mentioned in previous chapters – all his work is lacking in evidence to support this supposition: e.g., he never cited his "father" or "Lazare Carnot". Therefore, we establish this scientific relationship based on parentage, on the content within both of their historical documents and writings; thus an historical epistemological hypothesis should be advanced. In this sense, we consider Lazare Carnot's most interesting books from Sadi Carnot's point of view.

Therefore, in order to establish a finite set of main elements on which a scientific relationship can be built, in this last section, we primarily list and summarize the main scientific concepts, occurrences and common details of the two Carnots and compare Lazare and Sadi Carnot's essential works (Carnot 1778, 1780, 1786, 1803a, 1813)[5] (Carnot 1878a, 1978, Carnot-EP) (Fig. 11.2).

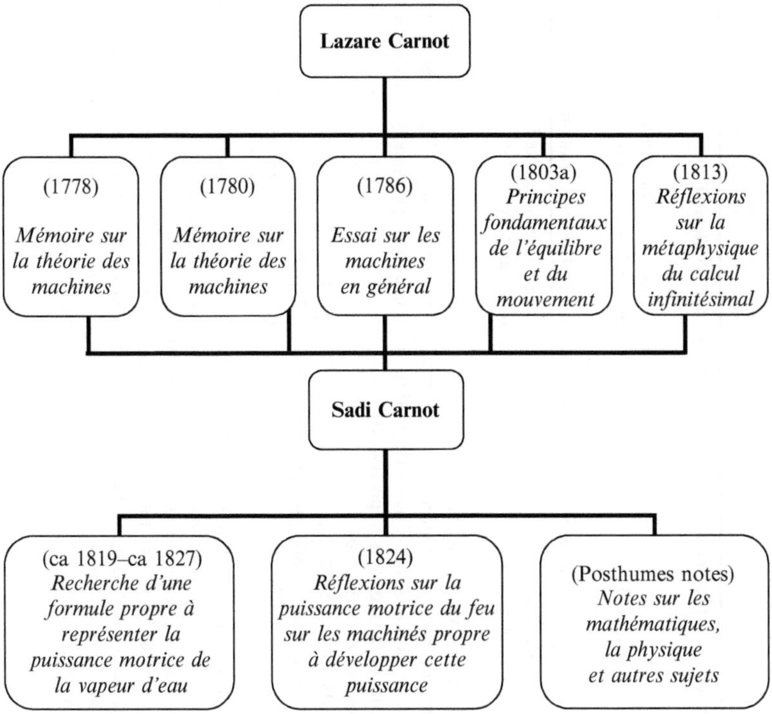

Fig. 11.2 A hypothesis on Lazare and Sadi Carnot's scientific parentage

Mainly based on previous historical–epistemological investigations (see above Chapters 2, 3, 6, 7 and 9), in this paragraph, a panoramic view of Lazare and Sadi Carnot's similar problematic organizations is presented.

> The difference is indicative of that between the two books: in the *Principes fondamentaux* Carnot set out in a didactic order what in the *Essai sur les machines* he had established in a problematic one. A consequence was that he himself began the process of obscuring the distinctiveness of his own approach while improving its clarity.[6]

Let us start this last *Carnots voyage* by listing the main common assumptions presented by the two Carnots in their respective general mechanical machines (Carnot 1778, 1780, 1786, 1803a, 1813) and general heat machines (Carnot 1978, 1878a, (and Carnot S–EP)), mainly focusing on the problems of the two theories. Some significant passages are added (see in the following Table 11.1):

[5]For some works, Carnot (1803b), as well.

[6]Gillispie 1971, p 87, line 21. (Author's *italic style*).

Table 11.1 On common and main assumptions in the theory[a]

Lazare Carnot (1778, 1780, 1786, 1803a, 1813)	Sadi Carnot ([1824] 1978, 1878a; Carnot S–EP)[b]
The use of the term "*Réflexion*" for the title of a book based on open problems[c]	The use of the term "*Réflexion*" for the title of a book based on open problems[d]
The cause of the motion of mechanical running machines[e]	The cause of the motion of heat running machines[f]
What is the best way of utilizing the *greatest possible effect produced* by a mechanical machine in motion?[g]	Is the motive power of a heat machine bounded?[h]
Lacking a complete theory of *impelling forces* and *resisting forces* in mechanics[i]	Lacking extensive and general laws on heat[j]

[a]Please note, we followed the order of the arguments presented by Sadi Carnot. Thus the order of the arguments presented by Lazare Carnot is changed and here opportunely follows Sadi's order. In this sense we suppose that Sadi Carnot (and his father (?)) advanced new theory while looking back at Lazare's scientific reasonings

[b]In this table, we mainly refer to the most significant reasonings. As we already made in previous sections of this book here we use *Réflexions sur la puissance motrice du feu* (Carnot 1978), *Notes sur les mathématiques, la physique et autres sujets* (Carnot 1878a, 1986; Picard 1927, and *Recherche d'une formule propre à représenter la puissance motrice de la vapeur d'eau* (Carnot S–EP). The corresponding references are opportunely cited in the footnotes

[c]Carnot (1813), frontispiece. On open discussions, see Carnot (1786), pp viij–x, pp 104–107

[d]Carnot (1978), frontispiece. On open discussions, see Carnot (1978), pp 1–9

[e]Carnot (1786), p vj, pp 13–14; see also Carnot (1780), § 103; Gillispie (1971), Appendix C, § 103, pp 301–302

[f]Carnot (1978), p 1, p 8; see also Carnot (1878a), folio 3rv(*Ib*); Carnot (1986), pp 185–186; Picard 1927, p 75

[g]Carnot (1786), pp ix–x, pp 89–94; see also Carnot (1780), §§ 149–160; Gillispie (1971), Appendix C, §§ 149–160, pp 327–340; Carnot (1803a), p xxj, pp 149, pp 247–250

[h]Carnot (1978), pp 6–7; see also Carnot (1878a), folio 1v(*Ia*); Carnot (1986), p 185; Picard 1927, p 73. It is interesting to note that here he added open problems on the improvements of heat engines (Carnot (1978), p 6) a topic which was greatly discussed in France at the time. On this matter, one can see: Carnot (1986), ft 1, pp 206–207; Fox (1992); Fox and Weisz (2009); see also Fox and Guagnini (1993)

[i]Carnot (1786), pp iv–v; Carnot (1780) § 129; Gillispie (1971) §129, p 316. Lazare Carnot was possibly one of the first to clarify (physically) the vector definition of *impelling forces* and *resisting forces*. In fact, he avoided using them as metaphysical causes of variation of motion. Generally speaking, he considered the (work) moment–of–activity "q", operated by *resisting forces*, as the *effect produced* by *impelling forces*. Instead, he considered the (work) moment–of–activity "Q", consumed by *impelling forces* at a given t–time. (Carnot 1786, §§ LII–LIII, pp 83–84, §§ LXIII–LXIV, pp 95–99). A full discussion is well documented by Gillispie (Chapters 2 and 3)

[j]Carnot (1978), pp 6–7.

Table 11.1 (continued)

Lazare Carnot (1778, 1780, 1786, 1803a, 1813)	Sadi Carnot ([1824] 1978, 1878a; Carnot S–EP)
Searching for a general theory of machines and principles of equilibrium and motion[k]	Searching for a general theory of heat machines[l]
Communication of motion and work[m]	Production of motion by heat and work; equilibrium of heat; communication[n]
Reducing the problems of mechanics to a practice–calculation and geometry[o]	Demonstrating how to express some of the propositions arrived at earlier in an algebraic language[p]
On the machine and its use[q]	On the machine and its use[r]
Absorbed motion and lost motion in a mechanical machine[s]	Carrying caloric from a hotter body to colder body[t]
On "momentum d'activité"[u]	On "puissance motrice"[v]
On the advantage[w]	On the advantage[x]

[k]Carnot (1786), p iv–v, pp 11–12; see also Carnot (1778) §§ 27–79; Carnot (1780), §102, §§133–141; Gillispie (1971), Appendix C, §102, pp 301–303, §§ 133–141, pp 317–321

[l]Carnot (1978), pp 7–8

[m]Carnot (1786), p iij–iv, p 44; Carnot (1803a), pp xiij–xvj; see also Carnot (1780), footnote " * ", § 148; Gillispie (1971), Appendix C, footnote " * ", § 148, p 309, pp 326–327

[n]Carnot (1978), pp 7–8, pp 23–35; see also Carnot (1878a), folio 4rv(*1c*); Carnot (1986), pp 186–187; Picard 1927, pp 76–77; Carnot (1878a), folio 5r; Carnot (1986), p 189; Picard 1927, pp 77–78; Carnot (1878a), folio 7v; Carnot (1986), pp 191–192; Picard 1927, pp 81–82; Carnot (1878a), folio 13r, pp 51–54; Picard 1927, p 86. We let note that here a difference in pagination was managed in Picard 1927's edition. However the latter adviced the reader about that

[o]Carnot (1786), p 12; see also Carnot (1780), § 113 and footnote " * "; Gillispie (1971), Appendix C, § 113 and footnote " * " , pp 308–309

[p]Carnot (1978), ft 1, p 74 (pp 73–79)

[q]Carnot (1786), p 19, pp 60–62; see also Carnot (1780), § 108; Gillispie (1971), Appendix C, § 108, p 303

[r]Carnot (1978), p 8; see also Carnot (1878a), folio 3rv(*Ib*); Carnot (1986), pp 185–186; Picard 1927, p 75

[s]Carnot (1786), pp 19–20; see also Carnot (1780), §§ 108–109; Gillispie (1971), Appendix C, §§ 108–109, pp 303–304

[t]Carnot (1978), pp 9–10; see also Carnot (1878a), folio 5r; Carnot (1986), pp 189–190; Picard 1927, pp 77–78

[u]Carnot (1786), p 88; see also Carnot (1780), § 129–132, § 149; Gillispie (1971), Appendix C, § 129–132, pp 316–317, § 149, p 327

[v]Carnot (1978), ft 1, p 6; see also Carnot (1878a), folio 4rv(*Ic*), Carnot (1986), pp 186–187; Picard 1927, pp 76–79; Carnot S–EP, ff 1–6

[w]Carnot (1786), p 85; see also Carnot (1780), § 151; Gillispie (1971), Appendix C, § 151, p 328

[x]"[...] *that superiority* [advantage of high–pressure engines in respect to machines with low–pressure engines] *lies essentially in their ability to utilize a greater fall of caloric* [*chute de calorique*]". (Carnot (1978), pp 97–98, line 20. Author's *italics*). He also remarked the origin his reasoning previously enunciated by Clément's law (Carnot (1978), p 98 and footnote). Finally, the last part of the book is mainly dedicated to single–acting heat engines in respect to double acting engines (Carnot (1978), pp 97–118)

Table 11.1 (continued)

Lazare Carnot (1778, 1780, 1786, 1803a, 1813)	Sadi Carnot ([1824] 1978, 1878a; Carnot S–EP)
Finding the actual ("réel") motion after interaction among bodies[y]	"Is heat ("chaleur") the result of a vibratory motion of molecules? [. . .] Can we find any instances of motive power being produced without an actual consumption of heat?"[z]
On the aim of running and general mechanical machines[aa]	On the aim of running and general heat machines[ab]
To obtain the *maximum effect produced*, no useless motions and interruptions have occurred[ac]	To obtain the *maximum motive power* ("puissance motrice"), no useless motions and interruptions have occurred[ad]
The operative conditions to *establish* the *maximum effect produced* for a hydraulic engine[ae]	The conditions for a *re–establishment of equilibrium of heat* for a heat engine[af]
Search for actual ("réelle") motion in mechanical machines[ag]	What exactly happens in a steam engine now in use ("en activité")? Search for production of motion of heat[ah]

[y]Carnot (1786), pp 21–24; see also Carnot (1878a) folio 2r(*Ia*), Carnot (1986), p 183; Picard 1927, p 73

[z]Carnot (1978), pp 9–10; Carnot (1878a), folio 5r; Carnot (1986), p 189; Picard 1927, pp 77–78. Here we let note that a difference in pagination was also object of attention by Robert Fox's in his English edition (Carnot 1986, pp 186–189)

[aa]Carnot (1786), pp 88–91; see also Carnot (1780), § 102, §§ 152–153; Gillispie (1971), Appendix C, § 102, pp 301–303, §§ 152–153, pp 328–332

[ab]Carnot (1978), pp 2–3, p 9; see also Carnot (1878a), folio 2r(*Ia)*; Carnot (1986), pp 183–184; Picard 1927, p 73

[ac]Carnot (1786), pp 89–91, pp 93–99. He searched for the *maximum* work. In this sense, he also proposed a famous *Corollary* on the equality "$Q=q$" (*Ivi*, Corollary V, § XLI, pp 75–76, pp 83–84; see also Carnot (1780), § 149; Gillispie (1971), Appendix C, § 149, pp 327–328). The *work* plays an important role in Lazare Carnot's mechanics of running machines. He was possibly one of the first to introduce the concept of *moment–of–activity* such as *FâĹŹds*. (Carnot 1786, pp 65–66, pp 96–97). In this regard, Gillispie develops a detailed discussion: above Chapters 2 and 3

[ad]Carnot (1978), pp 7–8, pp 21–23, pp 35–37 and footnotes

[ae]Carnot (1786), pp 89–94; see also Carnot (1780), §§ 149–152, §§ 155–157; Gillispie (1971), Appendix C, §§ 149–152, pp 327–330, §§ 155–157, pp 332–334

[af]Carnot (1978), p 10, pp 23–24; see also Carnot (1878a), folio 6v; Carnot (1986), pp 188–189 Picard 1927, p 77, p 80 (again pay careful attention to the page numbers). Let us note an important remark by Sadi Carnot on the equal situation of *invariability* ("immuable") between "quantity of power" and "[. . .] quantity of matter. In this case, there would be a fundamental difference between a direct restoration of the equilibrium of caloric and a restoration of its equilibrium accompanied by the production of motive power" (Carnot 1878a, folio 6v; Carnot 1986, pp 188–189; Picard 1927, p 78). On an experimental procedure to show the establishment of equilibrium, see also Carnot (1878a), folio 17r; Carnot (1986), p 202; Picard 1927, pp 91–92

[ag]Carnot (1786), pp 44–46; see also Carnot (1778) §§ 80–85; Carnot (1780), §106; Gillispie (1971), Appendix C, § 106, p 302

[ah]Carnot (1978), pp 8–11; see also Carnot (1878a), folio 4rv(*1c)*; Carnot (1986), pp 186–187; Picard 1927, pp 76–77; Carnot (1878a), folio 5r; Carnot (1986), pp 189–190; Picard 1927, p 78. In this case, Sadi Carnot also wrote on the fact that "[. . .] the production of motive power in a steam engine is due not to an actual ["réelle"] consumption of caloric [. . .]" thus it is an unnecessary condition for producing motive power (Carnot 1978, p 10, line 21). In fact it is due "[. . .] to its passage from a hot body to a cold one." (*Ibidem*)

Table 11.1 (continued)

Lazare Carnot (1778, 1780, 1786, 1803a, 1813)	Sadi Carnot ([1824] 1978, 1878a; Carnot S–EP)
For moving machines, what is lost in time or speed is always what is gained in force[ai] (*Golden rule*)	[aj]
The impossibility of perpetual motion[ak]	The impossibility of perpetual motion[al]
Ad abusurdum reasonings and proofs[am]	*Ad abusurdum* reasonings and proofs[an]
Geometric motions[ao]	Reversibility[ap]
On abstraction to study a machine[aq]	On abstraction to study a machine and true essence of bodies[ar]
On the science of (mechanical) machines[as]	On the science and social progress of (heat) machines[at]

[ai]Carnot (1786), pp iv–viii; see also Carnot (1780), § 153; Gillispie (1971), Appendix C, § 153, p 330

[aj]*For moving machines, what is lost in heat always what is gained in work.* The statement is not explicitly written in Sadi Carnot 's book (Carnot 1978). Lazare's statement is a general expression used in mechanics to argue on subjects in which something *lost* follows something *gained*. Nevertheless, generally speaking, one might easily adopt the statement for the thermodynamics of machines, e.g., *in order to produce–gain work, part of the heat is transferred–lost to a second reservoir*

[ak]Carnot (1786), p ix, pp 94–96; "[...] le mouvement perpétuel est une chose absolument impossible [...]" (*Ivi*, p 94, line 18); see also Carnot (1780), § 146, § 157; Gillispie (1971), Appendix C, § 146, pp 323–324, § 157, pp 333–337; Carnot (1803a), p xxi, pp 256–257

[al]Carnot (1978), pp 21–22 included the footnotes. "[...] this [an infinite creation of motive power] would be not only perpetual motion, but an unlimited creation of motive power without consumption either of caloric or of any other working substance whatsoever. Such a creation is entirely contrary to ideas now accepted, to the laws of mechanics and of sound physics. It is inadmissible" (Carnot 1978, p 21, line 5; see also Carnot 1878a, folio 4r(*Ic*); Carnot 1986, pp 186–187; Carnot 1878a folio 5v; Carnot 1986, p 190; Picard 1927, p 76, p 79)

[am]Carnot (1786), pp 28–36, p 107; see also Carnot (1780), §§ 113–114; Gillispie (1971), Appendix C, §§ 113–114, pp 308–310

[an]Carnot (1978), pp 11–15, pp 21–22, p 37 footnotes included

[ao]Carnot (1786), pp 28–34, pp 41–45; see also Carnot (1780), § 113; Gillispie (1971), Appendix C, § 113, pp 308–309

[ap]Carnot (1978), pp 23–35 included footnotes. Let us also note a special passage where Sadi also Carnot discussed an irreversible isochoric (Carnot 1978, pp 25–26) just after the statement of his theorem (Carnot 1978, pp 21–22)

[aq]Carnot (1786), pp 19–20, pp 60–63; see also Carnot (1780), § 108, §§ 116–118; Gillispie (1971), Appendix C, § 108, p 303, § 116–118, pp 312–313; Carnot (1803a), pp 256–257

[ar]Carnot (1978), p 5, p 24, pp 103–118; see also Carnot (1878a), folio 5r; Carnot (1986), pp 189–190; Picard 1927, p 77. Sadi Carnot's theory presents a high level of generality which is naturally and scientifically analogous to the machines in general proposed by his father

[as]Carnot (1786), p 21; see also Carnot (1780), § 107, §§ 109–111; Gillispie (1971), Appendix C, § 107, §§ 109–111, pp 303, pp 304–306

[at]Carnot (1978), p 1, pp 5–6; see also Carnot (1878a), folio 5r; Carnot (1986), pp 189–190; Picard 1927, p 77

Table 11.1 (continued)

Lazare Carnot (1778, 1780, 1786, 1803a, 1813)	Sadi Carnot ([1824] 1978, 1878a; Carnot S–EP)
The *effect produced* is always limited[au]	No creation of an "indéfinie" quantity of motive power[av]
On working substances[aw]	On working substances[ax]
Considering geometric motion independently of any dynamics rules[ay]	Considering the production of motion by heat independently of any particular mechanism or any particular working substance[az]
Reasonings by synthetic method[ba]	Reasonings by synthetic method[bb]
Produced work and consumed work[bc]	Produced work and consumed work[bd]
On the argument–hydraulic engine[be]	Analogy with hydraulic engine[bf]

[au]Carnot (1786), pp vij–ix, pp 86–87; see also Carnot (1780), §§ 151–152; Gillispie (1971), Appendix C, §§ 151–152, pp 328–329

[av]Carnot (1978), pp 21–22; see also Carnot (1878a), folio 5rv; Carnot (1986), pp 189–190; Picard 1927, pp 78–79

[aw]Carnot (1786), pp 86–87, pp 89–93; see also Carnot (1780), §§ 155–156; Gillispie (1971), Appendix C, §§ 155–156, pp 332–333. Carnot (1878a), folio 5r; Carnot (1986), pp 189–190; Picard 1927, pp 77–78

[ax]Carnot (1978), p 28, p 35, pp 37–39, p 112; see also Carnot (1878a), folio 2r(*Ia*); Carnot (1986), pp 183–184; Carnot (1878a), folio 3rv(*Ib*); Carnot (1986), pp 185–186; Carnot (1878a), folio 5rv; Carnot (1986), pp 189–190; Carnot (1878a), folio 6v; Carnot (1986), pp 188–189; Picard 1927, p 73, pp 76–79. (Again pay careful attention to the page numbers)

[ay]Carnot (1786), p iii; see also Carnot (1780), footnote "*"; Gillispie (1971), Appendix C, footnote "*", p 309

[az]Carnot (1978), p 8

[ba]Carnot (1786), pp 33–35; p 85; Carnot (1813), pp 12–21, p 189, p 200, pp 242–243, pp 217–253

[bb]Carnot (1978), pp 18–19, p 36, pp 38–39; pp 73–79, ft 1

[bc]Carnot (1786), p 66, p 85; see also Carnot (1780), §§ 129–132, §§ 153–154; Gillispie (1971), Appendix C, §§ 129–132, pp 316–317, §§ 153–154, pp 330–332

[bd]Carnot (1978), pp 20–21, ft 1, pp 73–79; see also Carnot (1878a), folio 4rv(*Ic*); Carnot (1986), pp 186–187; Carnot (1878a), folio 11r; Carnot (1986), p 194; Carnot (1878a), folio 14v; Carnot (1986), pp 199–200; Picard 1927, pp 76–77, pp 84–85, p 88

[be]Carnot (1786), pp ix–x, pp 88–81. Carnot (1803a), pp xxi, p 149, pp 247–250

[bf]Carnot (1978), pp 28–29, pp 35–36. In some previous pages Sadi Carnot noted (for the reader in footnote) an innovation with regard to "The matter here dealt with being entirely new, we are obliged to employ expressions not in use as of yet, and which perhaps are less clear than is desirable." (*Ivi*, ft 1, p 28). It was connected in the running text with "The motive power of a waterfall depends on its height and on the quantity of the liquid; the motive power of heat depends also on the quantity of caloric used, and on what may be termed, on what in fact we will call, the *height of its fall,** that is to say, the difference of temperature of the bodies between which the exchange of caloric is made." (*Ivi*, p 28)

Table 11.1 (continued)

Lazare Carnot (1778, 1780, 1786, 1803a, 1813)	Sadi Carnot ([1824] 1978, 1878a; Carnot S–EP)
On the role played by friction[bg]	On the role played by friction and creation of heat by motion[bh]
Percussions or brusque change. Impact between bodies and loss of *moment–of–activity*[bi]	Fall of caloric (or heat). Contact between parts of different temperatures should be avoided, since this leads to loss force vive, "what amounts to the same thing of motive power ("[...] ce qui est la même chose, de puissance motrice [...]")"[bj]
On the calculation of the *effect produced* for any machine; the initial conditions should be restored at the end of the process[bk]	Restoration of initial state on the argument–cycle[bl]

[bg]Carnot (1786), pp 43–44, pp 60–63, pp 94–95; see also Carnot (1778) §§ 1–26; Carnot (1780), §§ 1–100, § 160; Gillispie (1971), Appendix C, § 160, pp 337–340

[bh]Carnot (1978), ft 1, p 30; see also Carnot (1878a), folio 3rv(*1b*); Carnot (1986), pp 185–186; Picard 1927, p 75; Carnot (1878a), folio 8v; Carnot (1986), pp 192–193; Picard 1927, p 83

[bi]Carnot (1786), pp (45–48 and) 91–95; see also Carnot (1780), §§ 146–147, § 152, § 157; Gillispie (1971), Appendix C, §§ 146–147, pp 323–325, § 152, pp 328–329, § 157, p 333. Please note: "[...] qu'on appelle force ou puissance, dont la recherche est l'objet de la théorie des machines proprement dites" (Carnot 1786, p 62, line 29)

[bj]Carnot (1978), ft 1, p 6, p 16, pp 24–25, pp 22–26, pp 28–29. In *Notes sur les mathématiques, la physique et autres sujets* on *equivalent* heat–work (Carnot 1878a folio 7r; Carnot 1986, p 191; Picard 1927, p 81), he explicitly added that the principle just announced "[...] can be deducted directly [as an extended principle of conservation of mechanical energy] from principles of mechanics" (*Ibidem*). Moreover, he also dealt with the mechanical background of that principle comparing the theoretical roles played by force vive and motive power (*Ibidem*). On "fall"–term and its cultural background, please see: Cardwell (1965); Carnot (1986), ft 25, p 125. Particularly, the latter notes (Carnot 1986, ft 93, p 151) that on "[...] the changes in the temperature of solids or liquids occurring as an effect of compression and rarefaction would be but slight" (Carnot 1978, p 91, line 7). For his analogous adiabatic heating of a gas, Sadi Carnot "[...] almost certainly relied on [...] Berthollet's *Essai de statique chimique* [Berthollet 1803]" (Carnot 1986, ft 93, p 151). That is due to Berthollet's caloristic idea, which was circulating at the time, to associate the rise of temperature with a decrease in the volume of the solid body (Berthollet 1803, I, p 165, pp 248–250; Berthollet 1809, pp 441–448). Moreover, (1) in this passage from *Réflexions sur la puissance motrice du feu,* Sadi Carnot seems to accept (Carnot 1978, pp 90–92) Berthollet's argumentations; (2) In the *Notes sur les mathématiques, la physique et autres sujets* a contrast between the two passages is evident: here, he seems to reject (Carnot 1878a folio 8r; Carnot 1986, p 192; Picard 1927, pp 82–83) Berthollet's argumentations (Cfr.: Carnot 1986, ft 93, p 151, ft 30, p 209). For our part, we can only presume that there could be a mature and/or important convinction of the nature of caloric and related heat machines in *Notes sur les mathématiques, la physique et autres sujets*. On the other hand, only here does the *equivalent* of heat–work appear so clear; (3) Lazare Carnot's *Principes fondamentaux de l'équilibre et du mouvement* and Berthollet's *Essai de statique chimique* have the year of publication

[bk]Carnot (1803a), pp 259–261

[bl]Carnot (1978), pp 36–37

Table 11.1 (continued)

Lazare Carnot (1786)	Sadi Carnot (1824)
Preface. Although the theory here presented is applicable to all issues concerning the communication of motions, *Essay on machines in general* was given as the title of this pamphlet; first of all, because it is mainly the Machines that are considered as the most important argument of mechanics; secondly, because no particular machine is dealt with but we only deal with properties which are common to all of them	It is generally known that heat can be the cause of motion and that it possesses great motive power
	[. . .]. To heat also are due the vast motions which take place on the earth.
	[. . .]. Even earthquakes and volcanic eruptions are the result of heat
	From this immense reservoir, we may draw the moving force necessary for our purposes. [. . .]. To develop this power, to appropriate it to our uses, is the object of heat engines
This theory is based upon three main definitions; the first looks at some motions that I call *geometric*, because they can be only determined by the principles of geometry, and are absolutely independent of the rules of dynamics; I did not think that we would omit without creating obscurity in the statement of the main proportions, as, in particular, I let you see in the case of the principle of *Descartes*. By my second definition, I try to fix the meaning of the terms *impelling forces* and *resisting force*: it seems to me that without knowing a precise definition between these two different forces, we cannot clearly compare the causes and effects of the machines, without a well characterized distinction between these forces; and upon this distinction it seems to me that something vague and indeterminate was always left. Finally, my third definition is that by which I give the name of *moment of activity*[bm] of a force referring to a quantity which includes both a real force or an activity in motion that every instant employed by that force, that is to say, the time during which it acts. In any case, an agreement should be that this quantity, under whatever name one wishes to designate, to meet it in the	[. . .]. It appears that it must some day serve as a universal motor, and be substituted for animal power, water–falls, and air currents
	Over the first [animals] of these motors it has the advantage of economy, over the two others the inestimable advantage that it can be used at all times and places without interruption
	If, some day, the steam engine shall be so perfect that it can be set up and supplied with fuel at small cost, it will combine all desirable qualities, and will afford to the industrial arts a range the extent of which can scarcely be predicted
	[. . .]. There is almost as great a distance between the first apparatus in which the expansive force of steam was displayed and the existing machine as between the first raft that man ever made and the modern vessel
	[. . .]. Notwithstanding the work of all kinds done by steam engines, notwithstanding the satisfactory
	condition to which they have been brought to–day, their theory is very little understood, and the attempts to improve them are still directed almost by chance

[bm] We translated *puissance* with the general term *force* and, of course, *moment of activity* with *work*, even though here, Lazare Carnot, for the latter concept in the running text, refers to time and not to space. Let us also note that like many scholars of the seventeenth century, Lazare Carnot used, in general, the term "force", e.g., for *inertia*, *working substance* (that is *variation of quantity of motion in time*), *motive*, *acceleration*

Table 11.1 (continued)

Lazare Carnot (1786)	Sadi Carnot (1824)
analysis of Machines in motion is frequent. Using these definitions, I arrive at very simple propositions; I deduced them using the same fundamental equation. [. . .]. This equation is the most simple, generally extends to every conceivable case of equilibrium and motion, both this motion suddenly changes, that it change by insensible degrees; it also applies to all bodies, both hard [plastic[bn]] that they have any degree of elasticity; [. . .]. I easily obtain from this equation a general principle of equilibrium and motion for Machines properly so called; [. . .]. Everyone claims [a principle] that, for Machines in motion, what is gained in force is lost in time or speed; but after reading the best books of mechanics, where we should find proof and explanation of this principle, can we capture its importance and its true meaning? Since for most Readers its generality has irresistible evidence which must characterize mathematical truths. If Readers should find this guarantee striking, do not they see themselves as meccanici, educated by these works, [and] immediately abandon their chimerical projects? Do not they believe or at least suppose, in spite of everything one can tell them, that	The question has often been raised whether the motive power of heat* (* We use here the expression motive power to express the useful effect that a motor is capable of producing. This effect can always be likened to the elevation of a weight to a certain height. It has, as we know, as a measure, the product of the weight multiplied by the height to which it is raised) is unbounded, whether the possible improvements in steam–engines have an assignable limit, a limit which the nature of things will not allow to be passed by any means whatever; or whether, on the contrary, these improvements may be carried on indefinitely. [. . .]. We propose now to submit these questions to a deliberate examination. The phenomenon of the production of motion by heat has not been considered from a sufficiently general point of view. We have considered it only in machines the nature and mode of action of which have not allowed us to take in the whole extent of application of which it is susceptible. In such machines the phenomenon is, in a way, [interrupt and] incomplete. It becomes difficult to recognize its principles and study its laws. In order to consider in the most general way

[bn]Lazare Carnot, differently from tradition, used lost–quantity–of–motion. The latter point of view and the fact that the concept of geometrical form is not relevant in Lazare Carnot's mechanical theory (Carnot 1786), can emphasize that the *equivalent* between Carnot's *hard bodies* and mechanical *plastic bodies* seems epistemologically reasonable. Nevertheless, from a strictly mechanical point of view, we should also remark that the assumption cited is a *forcedequivalent*. In fact, Lazare Carnot's second fundamental law (Carnot 1786, p 22; see also Carnot 1803a) is related to hard bodies ("corps durs") or *perfectly hard* ("parfaitement durs & sans ressort") *à la* d'Alembert (d'Alembert, Lemme XI, pp 144–145), that is to say, bodies completely deprived of their elasticity (Carnot 1786, pp 22–23; see also Carnot 1803a, pp 8–10). Lazare Carnot justified his assumptions, introducing the important role played by experience in the theory. Next, he did not deal with elastic bodies or, to be precise, he assumed them to be a kind of limit–case of hard bodies: elastic bodies can be considered as composed of an infinity of hard bodies separated among them by elastic springs (*Ivi*, p 23; see also "corps *solides*" in: Carnot 1803a, p 8). Nevertheless, to consider *plastic bodies* as the most representative bodies for his collision theory is – as previously discussed – unfortunate, since plastic bodies are surely not hard bodies. On the other hand, we know that at that time the Newtonian tradition regarding hard bodies in which after collision, they conserve their forms but do not conserve the energy, was widely known. Thus, at first glance, we can assume that Lazare Carnot advanced these ideas to obtain a simplified performance model of *corps durs* to be interpreted by his mathematics

Table 11.1 (continued)

Lazare Carnot (1786)	Sadi Carnot (1824)
some magic is present in the Machines? The counter examples proposed are limited to simple Machines; they considered them not to be of such great effect; but none show them that it must be valid in every imaginable case; the case of two forces is considered only, and, for other cases, an analogy seems to be sufficient enough. [. . .]. The way to eradicate this error, is without doubt, [1] to fight its source, showing that, not only in all known Machines, but in all possible Machines, *loss in time or speed is always what is gained in force;* it is an inevitable law; and [2] to explain what this law clearly means; but in order to do that, one must move toward the greatest possible level of generality, and not study any Machine in particular, not adopt any analogy; in the end, it is necessary to propose a general proof, immediately and geometrically deduced by the first axioms of mechanics: that is exactly what I tried to do in this Essay.[bo] X. The Science of Machines in general is reduced to the following question: By *knowing the virtual motion of any system of bodies (that is to say that it would take each of these bodies, if it was free) to find the real motion which will be the next instant* [after the collision], *since there is mutual action of bodies, thus considering them as they are in nature, that is, having inertia common to all parts of matter*[bp]	the principle of the production of motion by heat, it must be considered independently of any mechanism or any particular working substance. It is necessary to establish principles applicable not only to steam engines* (* We distinguish here the steam–engine from the heat–engine in general. The latter may make use of any working substance whatever, of the vapor of water or of any other, to develop the motive power of heat) but to all imaginable heat engines, whatever the working substance and whatever the method by which it is operated. Machines which do not receive their motion from heat, those which have for a motor the force of men or of animals, a waterfall, an air current, etc., can be studied even to their smallest details by the mechanical theory. All cases are foreseen, all imaginable motions are referred to these general principles, firmly established, and applicable under all circumstances. This is the character of a complete theory. A similar theory is evidently needed for heat engines. We shall have it only when the laws of physics shall be extended enough, generalized enough, to make known beforehand all the effects of heat acting in a determined manner on any body. [. . .]. The production of motion in steam engines is always accompanied by a circumstance [. . . that] is the reestablishing of equilibrium in the caloric; that is, its passage from a body in which the temperature is more or less elevated, to another in which it is lower. What happens in fact in a steam–engine actually in motion? [. . .]. The production of motive power is then due in steam engines not to an actual consumption of caloric, *but to its transportation from a warm body to a cold body*, that is, to its re–establishment of equilibrium; an equilibrium considered as destroyed by any cause whatever, by chemical action, such as combustion, or by any other[bq]

[bo]Carnot (1786), pp iij–ix, line 1. (Author's *italics*)
[bp]Carnot (1786) § X, p 21, line 7. (Author's *italics*)
[bq]Carnot (1978), pp 1–11. (Author's *italics*)

Based on previous historical–epistemological studies on DNSs (see Chapters 6 and 7; Drago et al. 2001; Bellini et al. 2007), in order to better introduce all crucial aspects, we provide a concise reconstruction of Lazare and Sadi Carnot's main reasonings (Figs. 11.3 and 11.4):

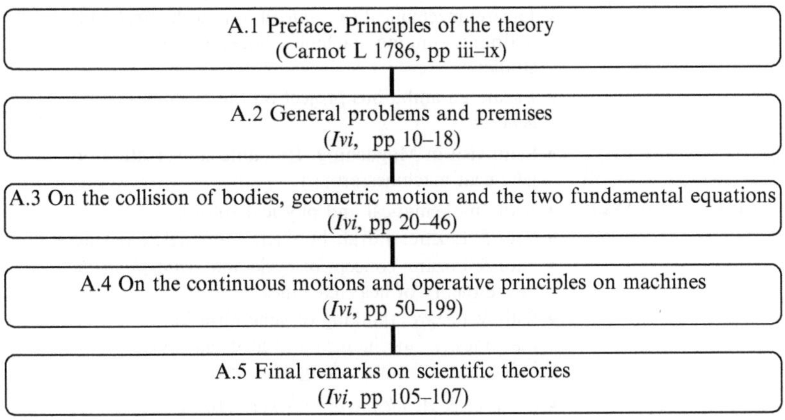

Fig. 11.3 Lazare Carnot's main reasonings (Carnot 1786)

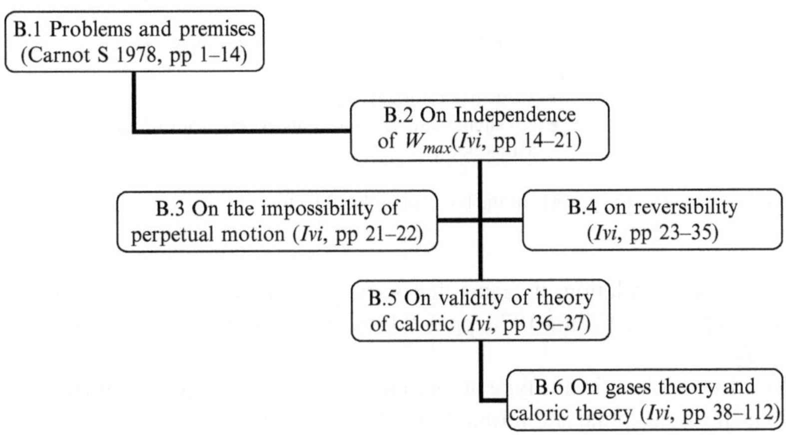

Fig. 11.4 Sadi Carnot 's main reasonings (Carnot 1978; Pisano 2010. For the reference and details, please see Chapters 6 and 7)

As stated in previous chapters there are several hypotheses on the relationship between the two Carnots: e.g., such as continuity of method and arguments focusing on mechanical machines and heat machines. Moreover, based on previous historical and epistemological open problems (Gillispie 1976, pp 23–33 see above Chapters 7 and 9) on Table 11.1 and (Fig. 11.5) we broaden and specify the common parts of the two main Carnot books (Carnot 1786, 1978). Below, we present a possible genesis of the *Réflexions sur la puissance motrice du feu* based on some crucial facts and events that occured at the time.

ca. 1778– ca. 1821	• Lazare: *letter on aerostat* (17 January 1784, lost). • Lazare: mechanical (*memoires*), on mechanical and hydraulic running machines, working substance studies. • Lazare an ambitious project on the machines *en general.* • Sadi: visit in Magdeburg. Probable conversation on new heat machines project.
ca. 1821– ca. 1823	• Sadi: mathematical and physical studies. • Sadi & Lazare: a draft booklet, many delete passages. • Sadi & Lazare: concept of cycle and *synthetic method* in the theory of heat machines. • Sadi & Lazare: thinking of generalization of theory of machines… mechanical and heat machines, independent from working substance. • [Fourier & *Académie des sciences* (18-11-1822)] • Lazare: dead on 2 August 1823. No commemoration... prudence.
1824	• Sadi: publication of *Réflexions sur la puissance motrice du feu* based on a theoretical general theory projected by(–with) Lazare. • Girard: publication of Sadi's topics in *Revue Encyclopédique*. • [Cauchy's rigorization]. • No significant reactions concerning printout of the Sadi's book.

Fig. 11.5 An idea on the genesis of the book and related factors

Based on our correlation, it seems that some part of *Réflexions sur la puissance motrice du feu* (e.g., Carnot 1978, pp 9–98) were prepared or written by four hands, *père et fils.*

What follows here is this type of interpretation with the objective of establishing a convincing conceptual and formal link between the two theories.

11.3 On the Lost Newtonian Paradigm

As previously discussed, like Lazare's Carnot's mechanics, Sadi's thermodynamics (and also modern thermodynamics) is also radically different from Newtonian theory. Consequently, the theoretical framework for studies on Sadi Carnot's heat could not be based on the Newtonian physical–mathematical paradigm; it is based on his father's mechanics, which comes precisely from an alternative organization, that is to say, an organization based on the *Principle of virtual work* (Capecchi 2012). Following this point of view, a correlation between the two theories on mechanical and heat machines is proposed according to a sequence of theoretical steps that range from basic ones to those which establish the mathematical formula of efficiency. Let us examine the details.

Lazare and Sadi Carnot's theoretical attitudes are independent from the (physical and mathematical) concept of absolute space typical of Newton (see above Chapter 6, Table 6.1). In fact, the theories of the two Carnots are independent from which positions the objects in question assume and which spatial motions (trajectories) they complete. For example, in Lazare Carnot's work, the solutions for the equations of motion are velocity and quantity of motion. In the same way, in Sadi Carnot's theory, the spatial paths, due to a hypothetical motion of heat, are not studied. The concept of space in the theories of the two Carnots reveals itself only with the finite volume of a system. Time is also different from Newtonian time. For example, in both Carnot theories, time is not absolute and does not have continual variations. It has only one dualistic variation: before and after. In this regard, it can also be noted that by avoiding the use of Newtonian absolute time and space, both theories also omit the use of physical quantities as non–finite mathematical variables, which in common theoretical physics are fundamental for the infinitesimal calculation of the variations of certain physical quantities. Therefore, from the very beginning, their theories did not contain abstract notions such as absolute space or force–cause. *Both authors considered mathematics as the result of common empirical knowledge and as an instrument to use when necessary.* Lazare Carnot's mechanical theory was limited to algebraic and trigonometric equations (because, in this theory, they are the types of equations of the invariants of motion which are to be solved with velocity only). In Sadi Carnot's theory, mathematics was also basic and, most importantly, it was only operative. Lazare Carnot made this attitude clear in his second book, *Principes fondamentaux de l'équilibre et du mouvement* (1803a) when he stated that all scientific (and mathematical) notions can only come from experiments (Table 11.2).

Table 11.2 On the common use of physical–mathematical quantities

Lazare Carnot (1786, 1803a, 1813)	Sadi Carnot (1824)
No Newtonian space and time	No Newtonian space and time
No mathematical space	No mathematical space
No mathematical time	No mathematical time
Independence from position in space	Independence from position in space
Velocity, Quantity of motion	Global space = Volume, Work, Efficiency
A finite mechanical Work	A finite heat Work
No local and infinitesimal variables	No local and infinitesimal variables
Importance of geometric motion	Importance of geometric motion

With "Newtonian" we intend the Newtonian science-paradigm in the history, particularly during 19th century (Pisano and Capecchi 2013)

In Lazare Carnot's words:

> Following this idea ["to avoid metaphysical notion of force" and... to use "the theory of communications of motions"[7]] we will soon see, as I previously mentioned, the necessity of turning to the experiment, and that is what I did, without neglecting to support myself with reasonings that can confirm it in the most plausible way, using or generalizing the results per induction. At times I even used the name of the force in the vague sense of which I spoke above [...].[8]
> [...] *Primitive ideas concerning the matter, the space, the time, the rest, the motion, etc.* 7. The first rule to establish in such delicate research on the laws of nature is to only admit notions so clear that they can comprise the bounds of our logic. We must therefore reject the definitions of *matter*, *time*, *space*, *rest*, and *motion* as expressions that are impossible to express with more clear terms, and the ideas that these expressions produce in us primitive ideas outside of which it is impossible to construct. But once these expressions are admitted, we will easily see that which is a body, speed, a motive force, etc. 8. The body is a given part of matter. 9. The apparent space that a body occupies is called its *volume*; the actual space that this same body occupies, or its real quantity of matter, is called its *mass*. When the body is such that equal parts of its volume always correspond to equal parts of its mass, we say that it has a uniform *density,* or that it is equally *dense* in all of its parts; and the relationship from mass to volume, or the quotient of one times the other, is called the *density of* this body. But if unequal masses correspond to equal volumes, we say that the density is variable and for each particle of matter, we call *density* the volume of this particle divided by its mass, or rather, the last reason of these two quantities. The empty parts or gaps lodged between the parts of the matter, and that make the volume or apparent space greater than the actual space are called *pores*.[9]
> [On the concept of force in the theory]. [...] in my opinion, no rigorous proof of the parallelogram of forces is possible: the mere existence of the *force* in the announcement of the proposition is able to make this demonstration impossible for the nature of things in itself. "It is extremely difficult", as Euler said, "to reason on primary principles of our knowledge [...]". This obscurity disappears in the second way [theory of motion] to conceive the mechanics, but another inconvenience appears; that is the fundamental principles that in the first way [theory of forces where cause produces motion] are established such as axioms in favor of the metaphysical expression [...] that is to say, [...] force, are, in this second case [theory of motion], nothing less than self-evident propositions, and in order to establish them, we need to include the recourse to the experience.[10]

At this point, what about the relationship between physics and mathematics? By considering the role played by physics–mathematics in a scientific theory (Pisano 2011a) – as previously mentioned – here we focus on the Newtonian paradigm until Laplace's physics (Fox 1974). Generally speaking, from a mathematical point of view, it interprets physical quantities, such as force and acceleration, by means of mathematical expressions[11] and by using differential equations a mathematical

---

[7]Carnot 1803a, p XVI, line 5.

[8]Carnot 1803a, p XVI, line 10.

[9]Carnot 1803a, pp 6–7, line 1. (Author's *italics*).

[10]Carnot 1803a, pp xiij–xiv, line 17.

[11]E.g, the second Newtonian law is not a strictly physical law. It is a second order differential equation that would interpret (physically) the law of motion. It does so by a mathematical–

result is obtained as well. In this sense, all of the potential physical effects and mathematical characteristics can be derived and discussed. Lazare Carnot's mechanics is an operative type of mechanics. To use Leibniz's words (Lazare Carnot was fashioned by the ideas of Leibniz[12]), theoretical physics must explain facts with facts. Therefore, the mathematics introduced is that which is absolutely necessary, adapted to represent a previously established physical argument.

In thermodynamics, Sadi Carnot should establish what the type of link between the new theory and mathematics would be. It should be remembered that, by beginning hydrodynamics, Leonhard Euler (1707–1783) had unveiled a new theoretical tendency. For non–reducible physical systems, such as the Newtonian system, which had a point or a discrete system of points, he had succeeded in maintaining the use of the very powerful infinitesimal analysis, conceiving the system as a fluid, to which he had extended mathematical Newtonian formalism. After this theorization, the scientists who followed him were naturally able to hypothesize that heat was a fluid (at an early stage: phlogiston, as a heavy fluid, then caloric, as a weightless fluid) to follow this type of physical–mathematical relationship. In the conception of heat as a caloric fluid, infinitesimal analysis constitutes the theoretical tool that is suitable for providing the solution to Sadi Carnot's problem. It suffices to express the differential Q, dQ through differentials of the variables on which it depends: t, V, etc.; that is to say: $dQ = AdV + Bdt + \dots$; from which the expression of function Q can be obtained and all of its applications can be easily deduced. This mathematical technique probably became very famous when in 1816 Laplace[13] pointed out that the speed of sound in air depended on the heat capacity ratio and corrected Newton's surprising error.

> Newton gave, in the second book of his *Mathematical Principles of natural Philosophy* the expression of the speed of sound: how he achieves this is one of the most remarkable features of his genius.[14]

---

physical equation, which, of course, one cannot establish experimentally, as instead can be done by dynamometer to measure magnitudes in a static equation (e.g., Hooke's law).

[12]In particular, one can see the concept of collision (adopted by Lazare Carnot) presented by Leibniz in his *Dynamica de Potentia et Legibus Naturae Corporeae* (Leibniz 1849–1863, II, sectio III, proposition 1–18, pp 488–507) and the early concept of potential energy (*Ivi*, II, sectio I, p 435). E.g., Lazare Carnot introduced an advancement of potential energy in his theory of motion applied to machines (Carnot 1803a, pp 36–38). On the Leibnizian background in Lazare Carnot, one can also see the famous correspondences in 1677 (*Ivi*, VI, pp 81–106) between Leibniz and Honoré Fabri (also Honoratus Fabrius, 1607–1688). For a panoramic view on Leibniz and his dynamics, see Pierre Costabel's (1912–1989) works (Costabel 1960); see also Drago 2003. For complete (works and letters) series of Leibniz's mathematical writings, see Eberhard Knobloch's VIII edition for "Berlin–Brandenburgische Akademie der Wissenschaften Leibniz–Edition, Reihe VIII" (Leibniz 2009).

[13]See also: Biot 1858, pp 1–9, 1802, pp 173–182.

[14]Laplace 1816, p 238, line 7. (Author's *italics* and capital letters).

> When the temperature of the air is raised, at constant pressure, only part of the heat is used to produce that effect [to raise the temperature]: the other part, which becomes latent, serves to increase its volume. This latter part of the heat is liberated when the air is reduced to its primitive volume by an increase in pressure. When two air molecules come close together in a vibration, the heat released raises their temperature and tends to radiate out into the nearby area; but if this happens very slowly relative to the speed of vibration, we can suppose that the amount of heat remains the same [for the two molecules]. Thus, as the two molecules approach, they meet a resistant force, first, because since their temperature being supposed constant, their [forces of] repulsions augment in inverse proportion to their distances; and second because the latent caloric which develops increase their temperature. Newton only considered the first of these causes of repulsion; but it is clear that the second cause must increase the speed of sound, since it increases the pressure. By entering it in the calculation, I come to the following theorem: "The real speed of sound is equal to the product of the Newtonian formula times the square root of the ratio of the specific heat of air at constant pressure of the atmosphere and at different temperatures, to its specific heat at constant volume".[15]

Newton's calculation gave 968 (920–1085) English feet per second (Newton 1687, pp 371–372), which is ca. 20% shorter than the value of speed of sound, and later 979 English feet per second appeared (Newton 1714, pp 343–344). It may have been convenient for the experimental data of the time, but it was undoubtedly too low a value.[16] In effect, the adiabatic compression of the air, which results in a local rise in temperature and pressure, has also been taken into account.

Laplace's investigations in practical physics were confined to those carried out jointly with Lavoisier on the specific heat of various bodies from 1782 to 1784 . It should also be noted that this is a similar technique that Émile Clapeyron would use in 1834 to reformulate Sadi Carnot's theory – but he would not succeed in doing so with his theorem (Clapeyron 1834, pp 153–190). Lazare Carnot, although he did not believe in caloric (Carnot 1990), considered the mathematical technique with the differential to be inaccurate. In fact, he considered infinitesimal analysis (Gillispie 1971, ft 1, p 12, Gillispie and Youschkevitch 1979, pp 251–298, § 13, p 256) to be a very clear mathematical apparatus in and of itself, which varies with continuity by means of concrete variables. However, for differentiated variables in the previous technique, the mathematical problem is the opposite: the aim is to determine the function Q by using an abstract calculation. Therefore, as Lazare Carnot explains in a footnote of *Principes fondamentaux de l'équilibre et du mouvement* (Carnot 1803a, p 11, ft 1), that infinitesimal analysis is not suitable in these cases. A different type of mathematics, in which geometry acquires a greater importance, is necessary. Lazare Carnot's mathematics selects geometric motions, which, by definition, admit their opposites.

---

[15] Laplace 1816, pp 238–239, line 24. (Author's quotation marks).

[16] Cfr.: Finn 1964, ft 19, p 8; Newton 1999, pp 772–778.

Lazare Carnot (1786, 1803a, 1813)	Sadi Carnot (1824)
LIX. When this first condition is fulfilled, there is nothing left to make a given Machine produce the greatest possible effect, to see that all of quantity$Q$ is used to produce this effect; [...] to make Machines produce the greatest possible effect, it is necessary that they never change motion, that by impervious degrees; [...][a] (1) Since they are heterogeneous, these two quantities, time and space, it is not exactly from the quotient of one to the other that we refer, but from the quotient of the relationships that these quantities have with respective rising unities conforming to the use admitted by geometry, when we refer to discharging, for example, a surface by a line, or multiplying a line by another [...] They are not [...] quantities that we express with algebraic characters; but abstract numbers that form the quotients of these quantities by their respective units[d] I am looking for the true sense of infinitesimal analysis.[e]50. Of the algorithm on differential calculus. [...] If the two systems are supposed to get as close to each other as we like, the difference between the two values of the same variable can be made as small as we like, and will become what we call a differential, and will only be the ordinary difference, simplified by the suppression of the quantities that, in their expression, could be found to be infinitely small, compared to the other terms of which it is composed. This is the general principle of differentiation[f].[...] that it to say, a method that unites the ease of a simple calculation of approximation with the exactitude of the most rigorous methods, [...][g]	Every change of temperature which is not due to a change of volume or to chemical action (an action that we provisionally suppose not to occur here) is necessarily due to the direct passage of the caloric from a more or less heated body to a colder body. This passage occurs mainly by the contact of bodies of different temperatures; hence such contact should be avoided as much as possible. It cannot probably be avoided entirely, but it should at least be so managed that the bodies brought in contact with each other differ as little as possible in temperature[b]. In reality the operation cannot proceed exactly as we have assumed. To determine the passage of caloric from one body to another, it is necessary that there should be an excess of temperature in the first, but this excess may be supposed as slight as we please. We can regard it [Δt] as insensible in theory, without thereby destroying the exactness of the arguments[c]

[a]Carnot (1786), pp 91–92, line 22
[b]Carnot (1978), p 24, line14
[c]Carnot (1978), p 25, line 6
[d]Carnot (1803a), p 11, ft 1
[e]Carnot (1813), p 1, line 1
[f]Carnot (1813), pp 65–66, line 24
[g]Carnot (1813), p 39, line 15

The weakness of differential techniques applied to heat machines before Sadi Carnot's book should also be noted (Lacroix1813). Moreover, as previously stated, differential technique was far from Sadi Carnot's interests in his *Réflexions sur la puissance motrice du feu*. In the unpublished manuscript *Recherche d'une formule propre à representer la puissance motrice de la vapeur d'eau*[17] (Drago and Vitiello

[17]In the first two decades of the nineteenth century, other essays, which are interesting for our book, are by Hachette (1811), Petit (1818), Navier (Navier in Belidor 1819) and Combes. These essays were similar in aim (Challey1971), but not in the approach and original synthetic method used (see above Chapters 7 and 9) for Sadi Carnot's project on heat engines.

1986, pp 391–402) found in 1966 (Gabbey and Herivel 1966), which should be placed temporally before the draft of *Réflexions sur la pusance motrice du feu,* Sadi Carnot indicated a mathematical expression for motive power. As in *Réflexions sur la pusance motrice du feu,* a general solution to cover all types of efficiency of heat machines was sought out. For the sake of brevity, we concisely reduce his search and determination by means of the following main operations:

- A quantity Q is considered as a function of the variables which involved transformations of a cycle.
- The motive power is intended as an integral of pdV.
- Formulas of the different transformations are necessary to know.
- Three basic different transformations–stages for a cycle[18]: an isothermal expansion, an adiabatic expansion, and an isothermal compression in the condenser.
- Employing Clément's law for saturated vapors.
- Set up an approximation function for Dalton's table of vapor pressures.
- The motive power as a function of the initial and final temperatures and pressures of the steam.
- Due to the importance of the *re–establishment of equilibrium of heat*, the calculation of the efficiency (see, Chapter 9) and Dalton's table relate pressure to temperature, the motive power can be expressed as a function of temperature alone (Table 11.3).

Table 11.3 On common use for non–sophisticated mathematics

Lazare Carnot (1780, 1786)	Sadi Carnot (1824)
No differential equations	No differential equations
No idealization in interpretation of physical phenomena	No idealization in interpretation of physical phenomena
Interaction motion–work	Interaction heat–work

Sadi Carnot abandons the path followed in the *Réflexions sur la puissance morice du feu* (e.g., see Carnot 1978, ft 1, pp 73–79) because he (like his father) thought a mathematical interpretation that, while still reasoning *en général* (e.g. see Carnot 1780 § 102; Gillispie 1971, Appendix C, p 301) followed the physical operations step by step. With this, a substantial detachment from the traditional physics of the 17th and 18th centuries (Fox 1992, 1995; Fox and Weisz 2009; Fox and Guagnini 1993) based on analysis was achieved. Additionally, a detachment from

[18]Even if Sadi Carnot argued the fact that the cycle was complete, we should remark that there it was possible only with respect to the motion of the piston and not, surely, with respect to the working substance, since it could not return to the temperature of the boiler.

the traditional theory of caloric took place since *Réflexions sur la puissance morice du feu* does not follow the common mathematical foundations, which are analytical and therefore aprioristic in regard to the physics of observing phenomena.

11.4 On the Constraints and Production of Work

We have just remarked that Lazare Carnot's mechanics does not depend on the metaphysical concept of force–cause since in his physics the nature of the cause is not strictly emphasized. In fact, while mechanics was concerned with the cause of motion, it was relegated to a theory of the *communication of motion*, measurable on any body by the quantity of motion (Table 11.4).

Table 11.4 On the common conception *en général* of the *motive power* and a theory

Lazare Carnot (1786, 1803a)	Sadi Carnot (1824)
Mechanical work: f_i and $\delta s (\neq 0)$	Heat work: Q and $\Delta t (\neq 0)$
Preface. Although the theory here presented is applicable to all issues concerning the communication of motions, *Essay on machines in general* was given as title of this pamphlet; first of all, because they are mainly the Machines that are considered the most important argument of mechanics; secondly, because no particular machine is dealt with but we only deal with properties which are common to all of them.[a] [...] XXXII. If a force *P* moves having velocity *u*, we call *z* the angle formed by *u* and *P*, the quantity *Pudtcosz*, where *dt* is the element of time, will be called *moment of activity* consummated by the force *P* during *dt*; that is the *moment of activity* consummated by a force *P* in an infinitesimally short time, is the product of this force, orientated such as its velocity, and the path that the point, [*fds*]where this force is applied, does in an infinitesimally short time. I will call *moment of activity*, consummated by this force in a given time, the sum of *moments of activity* consummated by it at every instant [...].[b] [...] we come back specifically to the second way [theory of motion] of looking at the problem, that is to say, that mechanics are nothing else than the theory of the laws of the communications of the motions.[c]	In order to consider in the most general way the principle of the production of motion by heat, it must be considered independently of any mechanism or any particular working substance. It is necessary to establish principles applicable not only to steam–engines[footnote] but to all imaginable heat–engines, [...].[e] [...] The production of motive power is then [since re–establishing of equilibrium in the caloric] due in steam engines not to an actual consumption of caloric, but to its transportation from a warm body to a cold body [...]. This condition is found to be fulfilled if, as we remarked above, there is produced in the body no other change of temperature than that due to change of volume, or, what is the same thing in other words, if there is no contact between bodies of sensibly different temperatures.[f] [...]. The steam is here only a means of transporting the caloric. It fills the same office as in the heating of baths by steam, except that in this case its motion is rendered useful. [...]. The production of motive power is then due in steam-engines not to an actual consumption of caloric, but *to its transportation from a warm body to a cold body*, that is, to its re–establishment of equilibrium

(continued)

Table 11.4 (continued)

Lazare Carnot (1786, 1803a)	Sadi Carnot (1824)
[...] The first method [theory of forces where cause produces motion] offers much more ease; so it is, as I mentioned here above, almost generally followed. Nevertheless, I adopted the second [theory of motion] as I already did in the first edition; because I wanted to avoid the notion of metaphysics of forces, to leave undistinguished the cause and the effect, in short, to bring everything to the only theory of communication of motions[d]	an equilibrium considered as destroyed by any cause whatever, by chemical action, such as combustion, or by any other. We shall see shortly that this principle is applicable to any machine set in motion by heat.[g] [...] [Production of engine power and importance of thermostats for different temperatures] [...] it is necessary that there should also be cold; without it, [the cold], the heat would be useless[h]

[a]Carnot (1786), p iij, line 1
[b]Carnot (1786), pp 65–66, line 21; see also pp 96–97
[c]Carnot (1803a), p xiij, line 4
[d]Carnot (1803a), pp xv–xvj, line 24
[e]Carnot (1978), p 8, line 1
[f]Carnot (1978), pp 9–10; see also p 38
[g]Carnot (1978), p 10, line 2
[h]Carnot (1978), p 11, line 8

Lazare Carnot considers the *production of work* (by mechanical machines) to be produced by mechanical machines. The f_i–forces are only important when they are linked to δs_i–displacements of bodies. In order to have *work* (by heat machine), in Sadi Carnot's thermodynamics, the heat (parallel to force) must move, passing from one thermostat to another. In his father's mechanical theory, the *production of mechanical work* occurs with the transference of motion from one body to another, both bodies being constrained. In thermodynamic theory, the *production of work* occurs with the transference of heat, transported from one body between two hard bodies, thermostats. It should be remembered that even later, William Thomson –Lord Kelvin (1824–1907) discussed the content in the second principle of thermodynamics and the necessity of a second thermostat with the aim of executing a passage of heat between a difference in temperature (Thomson 1848–1849, pp 541–574; see also *Id.*, 1882–1911, pp 113–155; *Id.*, 1852, pp 248–255; Thomson in Thurston 1943, pp 127–204). However, it should be noted that the analogy does not go any further. If Q is analogous to f, since neither are state functions f must be substituted by potential $\Delta V = f\Delta s$, while Q must be substituted by entropy, which however has a different formula $\Delta S = \Delta Q/t$. (Thomson 1851b, I, pp 175–183; see also Clausius 1850, vol 155, pp 368–397, pp 500–524). Moreover, it should be also noted that in the second case, it is not a special physical distance but it is temperature–range, $\Delta t \neq 0$. Sadi Carnot wrote this at the beginning of the discursive part of *Réflexions sur la puissance motrice du feu* and repeats it several times as well as at the end of the demonstration of his celebrated theorem (Carnot 1978, p 38): *work can be obtained every time there is a difference in temperature between which heat passes*. Thus, it is possible to note a common way of conceiving work in comparison with special and heat motions. To be more precise, according to Lazare Carnot, an action of every force with the weight force can be reproduced and, in the end, thanks to the *communication of motion* (Carnot 1803a, pp xiij–xvj), one can also theorize on the work it completes. In thermodynamics, Sadi Carnot maintains that the theory of mechanical machines (which at the time had his father's mechanical theory as a point of reference) is already complete (Carnot 1978, p 8) and that the theory of heat machines, just begun, must follow the example of the former. Sadi Carnot includes the *production of the motion of heat* with the aim of establishing a general principle. In order to theorize on the science of machines *en général* and produce a new physical situation of the conversion of heat to work, the two theories have the same methodological principles of virtual work, and the impossibility of perpetual motion. Let us also note that the former is expressed by different variables because of the two different fields of phenomena and the two different aims of the theories, *communication* of and *production*, whereas in both of these physical theories the role played by *motion* is fundamental.

Now let us see the work as produced by machines. In this regard, both mechanical and heat theories present the same type of effective abstraction. As with mechanical constraints, the relation $M \gg m$ between M–mass of the effective body and the m–masses (Carnot 1780; Gillispie 1971, Appendix C) of the bodies being studied is valid. In this way, in thermodynamics, for the thermal capacity of C of each of the two thermostats $C \gg c$ is valid as regards the thermal capacities c and the bodies in question. Therefore, each theory is based on a quantity that is obtained by the same type of abstraction. All of the bodies examined in this theory act between two bodies characterized by a physical quantity whose values are almost infinite. However, we should of course claim that, this quantity is specific for each theory because the field of phenomena is different.

It is well known that in the case of constrained motion, one vinculum or simply one constrain alone cannot produce work. Like Lazare Carnot (1786, pp 60–61, pp 85–86; 1780, § 108), Sadi does not think that heat, although emitted by a large reservoir (for example, the sea) provides a motive force on its own (Carnot 1978, pp 11–12, pp 40–41) (see in the following Table 11.5).

In both theories, it can be maintained that, otherwise, we would have, as Sadi Carnot notes further ahead (Carnot 1978, p 21, ft 1), a perpetual motion. This (*Ibidem*) is the first *argument ad absurdum* and is the first that can be said to be at the root of each theory.

In Lazare Carnot's theory, it is implicit that many bodies with infinite mass, that is to say the constraints, alone, do not form a machine (Carnot 1786, pp 58–59) and therefore never produce work. It can be asserted with the reasoning that once again, otherwise, it would confirm the possibility of perpetual motion. Following this analogy, we can clearly affirm, with the same reasoning as before, that it is impossible that, connecting *in a way only directed at different temperature of the thermostats*, that is to say letting heat pass without restrictions, they *produce work*. In other words, the reflection on the old experiment of the exchange between two bodies inside a calorimeter cannot show how work is produced. In fact, to produce work, other intermediary mechanisms are necessary in addition to thermostats in order to adequately utilize the transference of heat between the two temperatures.

Table 11.5 On the common way of conceiving vincula and production of work

Lazare Carnot (1780, 1786)	Sadi Carnot (1824)
The work as a *product* of a mechanical machine; vincula bodies	The work as a *product* of heat Machine; vincula bodies
Mechanical vincula: $M \gg m$	Heat vincula: $C \gg c$
(*Principle of virtual wok*). Systems of bodies, non–infinitesimal points, but global and with vincula	A physical complex system of machines + thermostats + vincula
More than one body having infinite masse cannot be a machine: no work from vincula, only	More than one body having infinite masse cannot be a machine: no work from vincula, only
It is impossible to link (in a direct way) different potentials systems to produce work freely. (*Impossibility of Perpetual Motion*)	It is impossible to link (in a direct way) thermostats by different temperature to produce work freely. (*Impossibility of Perpetual Motion*)

Lazare Carnot (1780)	Sadi Carnot (1824)
108. When a body acts on another one it is always directly or through some intermediary body. This intermediate body is in general what we call a machine. The motion that is lost at every moment in each of the bodies applied to this machine is partly absorbed by the machine itself and partly revised by the other bodies of the system but as it may happen that the subject of the matter is only to find the interplay of the bodies applied to the intermediate bodies without the need to known the effect on the intermediate bodies, we have imagined, in order to simplify the question, to ignore the mass of this body, however keeping all the other properties of matter. Hence the science of machines has become a sort of isolated branch of mechanics in which it is to be considered the mutual interplay of different parts of a system of bodies among which there are some that, lacking the inertia as common to all the parts of the matter as it exists in nature, withheld the names of machines. This abstraction might simplify in special cases where circumstances indicating those bodies for whom it was proper to neglect the mass to make it easier for the objective, but we easily know that the theory of machines in general has become much more complicated than before because then this theory was confined in the theory of motion of bodies as they are offered to us by nature, but now it is necessary to consider at the same time two kinds of bodies, one kind as actually existing, the other partially deprived of its natural properties. Now it is clear that the first	According to this principle, the production of heat alone is not sufficient to give birth to the impelling power: it is necessary that there should also be cold; without it, the heat would be useless. And in fact, if we should find about us only bodies as hot as our furnaces, how can we condense steam? What should we do with it if once produced? We should not presume that we might discharge it into the atmosphere, as is done in some machines; [* Footonote] atmosphere would not receive it. It does receive it under the actual condition of things, only because it fulfills the office of a vast condenser, because it is at a lower temperature; otherwise it would soon become fully charged, or rather would be already saturated[*Footonote]. Wherever there exists a difference of temperature, wherever it has been possible for the equilibrium of the caloric to be re-established, it is possible to have also the production of impelling power. [. . .] Then the motions of the piston being slight during the periods 3 and 5, these periods might have been suppressed without influencing sensibly the production of motive power. A very little change of volume should suffice in fact to produce a very slight change of temperature, and this slight change of volume may be neglected in presence of that of the periods 4 and 6, of which the extent is unlimited. If we suppress periods 3 and 5, in the series of operations above described, it is reduced to the following:

(continued)

Table 11.5 (continued)

Lazare Carnot (1780)	Sadi Carnot (1824)
problem is a special case, since it is more complicated than the other so that by similar hypotheses, we easily find the laws of the equilibrim and of motion in each particular machine such that the lever, the winch, the screw, resulting in a blend of knowledge whose binding can be hardly perceived and only by a kind of analogy; this must necessarily happen as we will resort to the particular figure of each machine to show the property which is common to it and to all the others. Since these properties are the ones we have mainly seen in this first section, it is clear that we will be able to find them only by putting aside the particular forms. So let us start by simplifying the state of the issue by ceasing to consider the system bodies of different natures; finally giving back to machines their inertia it will be easy afterwards to neglect the mass in the result, we will hold the possibility to consider it or not, and therefore the solution of the problem will be general and easier at the same time	(1) Contact of the gas confined in abcd (Fig. 2) with the body A, passage of the piston from *cd* to *ef*. (2) Removal of the body A, contact of the gas confined in abef with tile body B, return of the piston from *ef* to *cd*. (3) Removal of the body B, contact of the gas with the body A, passage of the piston from *cd* to *ef*, that is, repetition of the first period, and so on. The motive power resulting from the ensemble of operations 1 and 2 will evidently be the difference between that which is produced by the expansion of the gas while it is at the temperature of the body A, and that which is consumed to compress this gas while it is at the temperature of the body B. Let us suppose that operations 1 and 2 be performed on two gases of different chemical natures but under the same pressure under atmospheric pressure, for example.These two gases will behave exactly alike under the same circumstances, that is, their expansive forces, originally equal, will remain always equal, whatever may be the variations of volume and of temperature, provided these variations are the same in both[a]

[a]Carnot (1978), pp 11–12, line 3, pp 39–40, line 17. (Author's *italics*)

This is the *second argument ad absurdum* that unites the (implicit) development of the two theories, according to their common model of theory based on a problem.

11.5 On the *machine en général* and its Working Substance

When Lazare Carnot dealt with mechanical machines *en général* he reasoned independently from the working substance bodies and the particular mechanisms.

> But, I repeat, this Essai only concerns machines in general; each of them have their own particular properties [. . .][19]

In other words, he commonly conceived a machine independent from the working substance utilized. Therefore, the idea was to establish the general principles of physics in order to organize a general and global system–machine.

[19]Carnot 1786, p x, line 14.

Lazare Carnot (1780, 1786)	Sadi Carnot (1824)
109. The science of machines in general and all mechanics is thus reduced to the following question. Knowing the virtual motion of a system of bodies that is the one that it would be taken by each body if it was free to find the real motion that it will have the next moment because of the interplay of bodies assuming that each of them is endowed with inertia as common to all the parts of the matter. And since this problem is simpler if we would find, among the bodies, some that are deprived of this inertia it is clear that we cannot have a general theory of machines without having solved this problem in its full scope. That is what we will try to do[a]. But, I repeat, this Essai only concerns machines in general; each of them have their own particular properties [...][b]. [...] we compare these different efforts regarding the working substances that produce them, because the nature of the working substances cannot change the forces they must exert to fulfill the different objects for which the Machines are intended[c]	The phenomenon of the production of motion by heat has not been considered from a sufficiently general point of view[d]. [...].This is the character of a complete theory. A similar theory is evidently needed for heat–engines. We shall have it only when the laws of physics shall be extended enough, generalized enough, to make known beforehand all the effects of heat acting in a determined manner on any body[e]. [Footnote] (1) We assume here no chemical action between the bodies employed to realize the motive power of heat. The chemical action which takes place in the furnace is, in some sort, a preliminary action,–an operation destined not to produce immediately motive power, but to destroy the equilibrium of the caloric, to produce a difference of temperature which may finally give rise to motion[f]

[a]Carnot 1780, §§ 108–109; see also Gillispie 1971, Appendix C, §§ 108–109, pp 303–305; see also Carnot 1786, §§ XXIX—XXX, pp 60–61
b20
[c]Carnot 1786, p 62, line 2
[d]Carnot 1978, p 7, line 11
[e]Carnot 1978, pp 8–9, line 14
[f]Carnot 1978, p 23, ft 1

Therefore, they want to theorize "independently of any mechanism or any particular working substance" and to apply the reasoning "[...] to all imaginable heat–engines [...]".[21] In fact, later, Sadi Carnot's famous theorem on heat machines would achieve the same objectives expressed by Lazare Carnot, e.g., in his two works in *Mémoire sur la théorie des machines* (Carnot 1778, 1780; Gillispie 1971, Appendices B and C). The argument is also remarked in the footnote:

> We distinguish here the steam-engine from the heat–engine in general. The latter may make use of any working substance whatever, of the vapor of water or of any other, to develop the motive power of heat.[22]

[20]Carnot 1786, p x, line 14.

[21]Carnot 1978, p 8, ft 1.

[22]Carnot 1978, p 8, ft 1.

> The motive power of heat is independent of the working substances employed to realize it; its quantity is fired solely by the temperatures of the bodies between which is effected, finally, the transfer of the caloric.[23]

Therefore neither are physical quantities that concern points or local passages, but only global quantities.

11.6 On the Analogy between Hydraulic and Heat Machines

Unrelated to the concept of Newtonian paradigm concerning mathematical use of the force-cause concept, Lazare Carnot considers work as a fundamental quantity. Even Sadi Carnot uses it and gives it an analogous role. Moreover, in order to calculate the work of a gas, Sadi Carnot does not consider intermolecular forces like his very influential contemporaries did (e.g., *Laplace* and *Poisson*).

> Although the *Réflexions* was regarded by contemporaries as primarily an essay on steam engines, Carnot's most important innovations lay in a new approach to the study of heat. While he accepted, and in some theorems furthered, the theory of heat developed by Laplace and Poisson, Carnot also shifted the emphasis from the microscopic to the macroscopic. Rather than build upon the notion of gas particles surrounded by atmospheres of caloric, he began with the directly measurable entities of volume, pressure, temperature, and work. Of his concepts of an ideal engine, completeness, and reversibility, there were some vague anticipations. The notion of an abstract heat engine was approximated in the work of Hachette and was more clearly present in the studies by Cagniard de la Tour and Clément of the motive power produced by a bubble of gas rising adiabatically in water. Jacob Perkins's team engine, widely discussed in 1823, represented an attempt to design a closed, complete system, and engineers were aware that certain types of hydraulic engines were reversible. [...]. Although the exact reasons are impossible to determine, the *Réflexions* had almost no influence on contemporary science.[24]

Table 11.6 On the common conception of *work* for a machine and its working substance

Lazare Carnot (1786)	Sadi Carnot (1824)
Work as a primary magnitude	Work as a primary magnitude
No intermolecular forces for the calculation of the work	No intermolecular forces for the calculation of the work
W_{max} and $W=(F/S)S\Delta s$	W_{max} and $W=p\Delta V$

Instead, Sadi Carnot chooses those variables consider the physical system globally:

$$W_{\max}/Q = f\,(Q, V, t)\,.$$

By means of those variables, he obtains a very simple mathematical formula:

$$W = p\Delta V,$$

[23]Carnot 1978, p 38, line 4.

[24]Challey 1971, pp 82–83, line 53 (Author's *Italic*)

which, also has theoretical merit: in the exemplary thermal case, which is a gas in a cylinder, it formally belongs both to mechanical theory and heat theory.

In fact, generally speaking we have

$$W = F\Delta s = (F/S)\,S\Delta s = P\Delta V.$$

It should be noted that when Lazare Carnot extends his mechanical theory to hydraulic machines he uses the case of a fluid in a cylinder as an example.

Lazare Carnot (1803a)	Sadi Carnot (1824)
Corollary IV. 219. The above–stated theorem can be applied to the case of motion, since the motion with which each point tends to move breaks down in two, of which one rests and operates the subsequent motion and of which the other is destroyed. Yet this destroyed motion is liable (196,190) for the law in the above–stated theorem; that is to say, that whatever the state of rest or motion there is a system of various forces applied to a machine, if we suddenly make it subject to any geometric motion, without changing these forces; *the sum of the products of each of them times the speed that we will have in the first instant the point it is applied, estimated in the direction of this force, will be equal to zero.* [...] let us imagine, for example, a winch with a wheel and cylinder whose weights are suspended by vertical cords: if there is equilibrium, or if the motion is uniform, the weight attached to the wheel will be that of the cylinder, as the radius of the cylinder is the radius of the wheel. But it is a different matter altogether, since the machine accelerates or slows down: therefore it appears that the forces are not the direction of these forces; as the proposition should follow. The response to this is that in the case of non–uniform motion, the weights in question are not the only forces exerted in this system [...][a]	According to established principles at the present time [First announcement of his theorem on the independence of used working substance[b]] we can compare with sufficient accuracy the motive power of heat to that of a waterfall. Each has a maximum that we cannot exceed, whatever may be, on the one hand, the machine which is acted upon by the water, and whatever, on the other hand, the substance acted upon by the heat. The motive power of a waterfall depends on its height and on the quantity of the liquid; the motive. power of heat depends also on the quantity of caloric used, and on what may be termed, on what in fact we will call, the height of its fall[*] that is to say, the difference of temperature of the bodies between which the exchange of caloric is made. In the waterfall the motive power is exactly proportional to the difference of level between the higher and lower reservoirs. In the fall of caloric the motive power undoubtedly increases with the difference of temperature between the warm[c] [*] (1) The matter here dealt with being entirely new, we are obliged to employ expressions not in use as yet, and which perhaps are less clear than is desirable

[a]Carnot (1803a), Corollary IV, n. 219, pp 196–197, line 19. (Author's *italic*)
[b]Carnot (1978), pp 21–22
[c]Carnot (1978), pp 28–29, line 1

Therefore, the concept of work used by Sadi Carnot in thermodynamics has the same mathematical formula as mechanical work in a case that is particular to the old theory, but which is central to the new theory; in this way, an element of formal continuity is given to the two theories. Just after the first announcement of his theorem (Carnot 1978, pp 21–22) the analogy (Carnot 1978, pp 8–9, pp 28–29) between the falling of water on a hydraulic wheel and the falling of caloric in a heat machine is proposed. According to Charles Gillispie (1971, pp 96–98), this analogy between two apparently different physical systems also reveals the scientific sensibility and attitude of the young French scholar.

Lazare Carnot was also interested in finding out the criteria (Carnot 1986, ft 23, p 124) for obtaining the greatest possible effect from machines (Carnot 1780, §§ 149–152, §§ 155–157; Carnot 1786, pp 89–94; Gillispie 1971, pp 327–330, pp 332–333). His reference to the hydraulic engine (Carnot 1786, pp ix–x, pp 88–81; Carnot 1803a, pp xxi, p 149, pp 247–250) is interesting for our aim. As previously stated in the first paragraph, the product of *communication of motion* (Carnot 1780, ft *, § 148; Carnot 1786, p iij–iv, p 44; Carnot 1803a, pp xiij–xvj) and power should always be transmitted without percussion and all shocks should be avoided (Carnot 1786, pp (45–48 and) 91–95, pp 89–91, pp 93–99; Carnot 1780, §§ 146–147, § 152, § 157; Gillispie 1971, pp 323–325, pp 328–329, p 333). In particular,

> [Sadi] Carnot here remarked that the performance of work by heat is quite analogous to that by waterfall. By the fall of heat (*chute du calorique*) the performance of work is determined in quite similar manner to that performed by the fall of water (*chute d'eau*).[25]

Below is a summary of that ability (Fig. 11.6; see in the following Table 11.7).

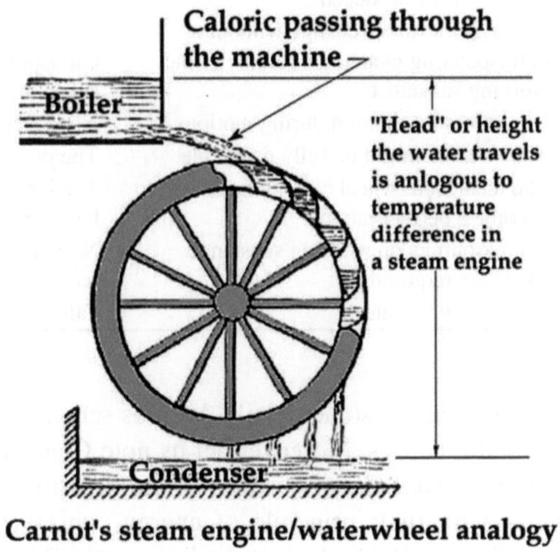

Fig. 11.6 The analogy ("The Engines of Our Ingenuity is Copyright © 1988–2004 by John H. Lienhard" (via: http://uh.edu/engines/epi1958.htm))

[25]Mach 1986, p 201, line 37. (Author's *italics* and "( )").

Table 11.7 On the analogy and magnitudes employed

Lazare Carnot (1780, 1786, 1803a)	Sadi Carnot (1824)
In a *waterwheel,* the power only depends on the difference in heights and water *falls* through them	In a heat machine, the power only depends on the difference in temperatures between boiler and condenser and caloric *falls* through them
Δh (heights values–range)	Δt (heat values–range)
Two systems–levels of water	Two systems–levels of heat
The water falls from higher to lower system	The heat falls from higher to lower system
Any motion escaping with the water leaving the *waterwheel* should be a minimum	Any motion escaping with the heat leaving the heat machine should be a minimum
Force expansive of *water* (and force–weight) on the system	*Force expansive de la chaleur* on the system
Efficiency: $\eta = f(h_1 - h_2, Q)$; $(h_1 > h_2)$	Efficiency: $\eta = f(t_1, t_2, x, Q)$; $(t_1 > t_2)$ (plus an *unknown variable during* Δt)
In an ideal waterwheel, no water's power would go to waste	In an ideal heat machine no heat's power would go to waste
The reasoning is based on the conservation of the mechanical energy, and the impossibility of perpetual motion	The reasoning is based on the conservation of the caloric, and the impossibility of perpetual motion
In a hydraulic machine the decrease in the potential energy of the water should be equal to the communication of motion–work produced	In a heat machine, the transference of heat without loss follows as well as *production* (of part of the heat) into work
No height's values change without a corresponding change of volume of the working substance	No temperature values change without a corresponding change of volume of the working substance
No useless dissipation during motions	No useless dissipation during motions
The *motions* might be fully reversible	The *productions* might be fully reversible
The ideal waterwheel can run backward to become a perfect pump	The ideal heat machine can run in reverse and become a perfect pump
No reasonings on working substance speed are required	No reasonings on working substance speed are required
Percussion of water	Fall of caloric

Let us examine some details. Various scholars revealed that this analogy presents some difficulties. Moreover, let us note Clausius and Thomson's perplexity when they read Sadi Carnot's work for the first time. Reconciling Sadi Carnot's caloric ideas with their principle of energy conservation was a problem after Joule's experiment (1844). In fact, Sadi Carnot's *Réflexions sur la puissance motrice du feu* was essentially based on (which, as previously mentioned, he stated with some doubt) caloric conservation ($Q_1 = Q_2$) and not on energy conservation. If heat corresponds to water then the fall should be conserved as occurs with water. Therefore, the analogy should link Sadi Carnot's book to the caloric hypothesis, not to the modern *equivalent* of heat and work (in which heat is consumed by the production of work). However, this difficulty would be overcome if – as Wilhelm Ostwald (1853–1932) asserted (Ostwald 1892) – the anticipation of the concept of entropy is attributed to Sadi Carnot. According to Ostwald, Sadi Carnot substantially defined that concept when he utilized heat at a constant[26] temperature, under certain conditions, *Q/t* (Carnot 1978, p 32). Nevertheless, that assumption is still weak. However, presuming that this assumption (or hypothesis) is valid, , within analogy, the *falling of water* should be considered analogous to the *falling of the quantity of heat* at constant temperatures. That is to say:

$$Q_1/t_1 - Q_2/t_2 = \Delta S \text{ between the two temperatures}$$

> [*chute du calorique* is quite similar *chute d'eau*]. But whilst for the water the performance of work is simply proportional to the height of fall, we may not put this performance in the case of heat proportional to the difference of temperature without a closer investigation.[27]

While in the hydraulic wheel W_{max} is proportional only at Δh, in the heat machine W_{max} can depend on unspecified variables. However, overall, as Sadi Carnot remarks (Carnot 1978, p 29) this work is not proportional to Δt since W_{max} seems to have greater experimental values at low temperatures (Carnot 1978, p 72). Therefore, this is a new type of function. Sadi Carnot's theoretical effort reaches a standstill at this last difficulty, although he even attempted a calculation in his previously discussed famous footnote (Carnot 1978, ft 1, pp 73–79) to determine the efficiency function. However, let us note that the analogy is more persuasive than it appears in modern times. Moreover, around the eighteenth century, it was common to consider machines by performing an abstraction from the masses of bodies (Carnot 1786, § XXX, p 60) and separate the wire of a pendulum from the mass. Therefore, the water that falls on the hydraulic wheel could have been thought of without mass, that is, as a weightless fluid, as caloric fluid was conceived. This way of envisioning mechanical machines allowed (still within the caloric hypothesis) for the consideration of the analogy as a true connection of heat machines to the theory of mechanical machines which includes the case study of falling water and the hydraulic wheel. Therefore, for Lazare Carnot, this is a fundamental analogy and for Sadi Carnot, who doubts caloric, it is merely striking; for us it plays a central, but not essential role.

11.7 On the Efficiency of a Machine

During Sadi Carnot's time, cannons had become very effective, having reached an astonishing power; and even ordinary thermal machines revived the chimera of obtaining the greatest effect possible and limitless results. In the case of mechanical machines, Lazare Carnot's idea refuted limitless power (Carnot 1786, 1803a). In fact, from the beginning, Sadi Carnot also clearly considers the problem of if, when

[26] Carnot 1978, p 32.

[27] Mach 1986, pp 201–202, line 40. (Author's *italics* and "( )").

utilizing a given quantity of heat, there is or is not a limit greater than the production of heat (Carnot 1978, pp 6–7) (Table 11.8).

Table 11.8 On the common way to consider efficiency

Lazare Carnot (1780, 1786, 1803a)	Sadi Carnot (1824)
What is the best way of utilizing the *greatest possible effect produced* by a mechanical machine in motion?[a]	Is the motive power of a heat machine bounded?[b]
The *effect produced* is always limited[c]	No creation of an "indéfinie" quantity of motive power[d]

[a]Carnot (1786), pp ix–x, pp 89–94; see also Carnot (1780), §§ 149–160; Gillispie (1971), Appendix C, §§ 149–160, pp 327–340; Carnot (1803a), p xxj, p 149, pp 247–250
[b]Carnot (1978), p 6; see also Carnot (1878a), folio 2rv(*Ia*); Carnot (1986), pp 183–185; Picard 1927, p 73
[c]Carnot (1786), pp vij–ix, pp 86–87; see also Carnot (1780), §§ 151–152; Gillispie (1971), Appendix C, §§ 151–152, pp 328–329
[d]Carnot (1978), pp 21–22; see also Carnot (1878a), folio 5rv; Carnot (1986), pp 189–190; Picard 1927, pp 78–79

He will also conclude with resolution that there is a limit and for this reason, the idea of obtaining unlimited work from heat is a chimera. *Therefore neither theory is based on axioms, but on the program of scientifically resolving a crucial problem that in the minds of the lay people of the time coincided with metaphysics.*

Overall, Sadi Carnot writes his book while reasoning according to the idea that in every thermal engine, in the end, efficiency only depends on the global Δt and on the condition of reversibility of the machine. In this sense, the variation on volume ΔV depends on the condition of reversibility. Recently some scholars (Guemeza, Fiolhaisb and Fiolhaisb 2002) worked on Sadi Carnot's theorem by using The International System of Units (SI) and compared the results with those obtained using modern data (see in the following Fig. 11.7):

11.8 On the Impossibility of Perpetual Motion

As mentioned in the previous paragraph, the idea of obtaining unlimited work from heat was a chimera; both Lazare and Sadi Carnot's theories obtain this limiting result because they are based on the fundamental affirmation that perpetual motion is impossible.

In Lazare Carnot's mechanics, the impossibility of perpetual motion (Carnot 1786, p ix, pp 94–96) has a special role in controlling and developing his special assumptions (Carnot 1786, pp v–xii, pp 88–95) and theses in the theory of mechanical machines (Carnot 1786, p 21; see Chapters 2 and 3). Sadi Carnot continues this tradition by dealing with that impossibility (Carnot 1978, pp 21–22) basing it also on the role played by science and the social progress (Carnot 1978, p 1, pp 5–6) of heat machines.

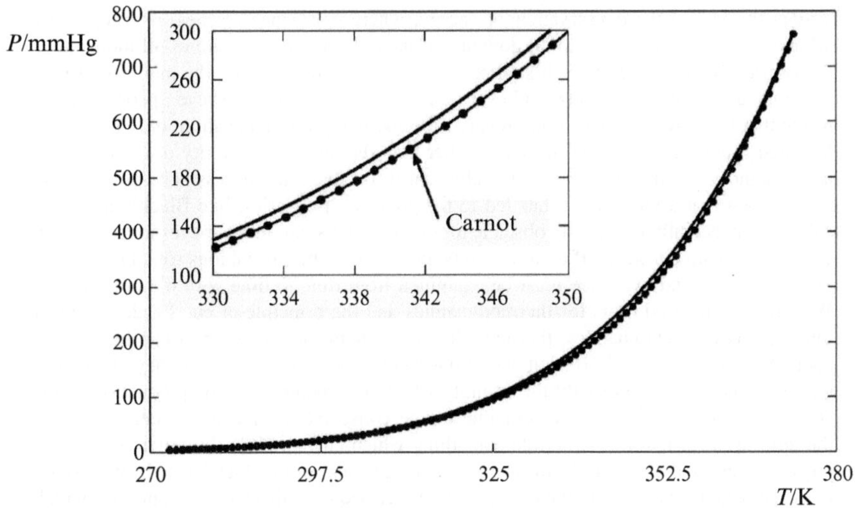

Fig. 11.7 Sadi Carnot's cycle in original and modern data (Guemeza, Fiolhaisb and Fiolhaisb 2002, p 46)

> Now if there existed any means of using heat preferable to [reversible] those which we have employed, that is, if it were possible by any method whatever to make the caloric produce a quantity of motive power greater than we have made it produce by our first series of operations, it would suffice to divert a portion of this power in order by the method just indicated to make the caloric of the body B return to the body A from the refrigerator to the furnace, to restore the initial conditions, and thus to be ready to commence again an operation precisely similar to the former, and so on: this would be not only perpetual motion, but an unlimited creation of motive power without consumption either of caloric or of any other working substance whatever. Such a creation is entirely contrary to ideas now accepted, to the laws of mechanics and of sound physics. It is inadmissible. We should then conclude that the maximum of motive power resulting from the *employment of steam is also the maximum of motive power realizable by any means whatever.* We will soon give a second more vigorous demonstration of this theory. This should be considered only as an approximation (see p 29).[28]

Let us remark that Sadi Carnot's emphasis was made here just before establishing his famous theorem (Carnot 1978, p 38), which is properly demonstrated by an *ad absurdum* proof in which the impossibility of perpetual motion is emphasized. Moreover, he dedicates a great deal of space to the matter, also inserting a lengthy note in which his point of view on the impossibility of perpetual motion both in thermodynamics and other areas of science such as then nascent electricity, emerges when he cites Volta's battery which had recently been shown to the King of France (1799). In this regard, we provide Mach's reasonings:

[28]Carnot 1978, pp 20–22, line 13. (Author's *italics* and capital letters).

To-day the law of the conservation of energy, wherever science reaches, is accepted by all and receives applications in all domains of natural science. The fate of all momentous discoveries is similar. On their first appearance, they are regarded by the majority of men as errors, as Mayer, Helmholtz, and even Joule found. Gradually, however, people are led to see that the new view was long prepared for and ready for enunciation, only that a few favoured minds had perceived it much earlier than the rest. The majority of the man who use it cannot enter into a deep-going analysis of it; for them, its success is its proof. It can thus happen that a view which has led to the greatest discoveries, like Black's theory of caloric, may actually become an obstacle to progress by its blinding our eyes to facts which do not fit in with our ideas. If a view is to be protected from this dubious rôle, the grounds of its evolution and existence must be examined from time to time with the utmost care. We will here try to do this for thermodynamics and the principle of energy. 2. The most multifarious physical changes, thermal, electrical, chemical and so forth can be brought about by mechanical work. If such alterations can be completely reversed, they yield anew the mechanical work in exactly the quantity which was required for the production of the change in question. This is the principle of the conservation of energy, "energy" being the term used for that indestructible something which characterizes the difference of two physical states and of which the measure is the mechanical work which has to be performed in the passage of one state to the other. How did we acquire this idea? The opinions which are held concerning the foundations of the law of energy diverge very widely from one another. To many physicists it now suddenly appears to be evident a priori. Others trace the principle to the impossibility of a *perpetuum mobile* which they regard as self–evident. Others start from the theory that all physical processes are purely mechanical processes and hence deduce the impossibility of a perpetuum mobile in the whole physical domain. Other inquirers, finally, are for accepting only purely experimental establishment of the law of energy. We will investigate these views, and it will appear from the discussion to follow that there is also a logical and purely formal source of the principle of energy which has hitherto been little considered. [...]. As far as the history of physics reaches, from the time of Democritus to the present day, there has been an unmistakable tendency to explain all physical processes mechanically. [...]. It is only from *experience that* we can know whether and how thermal processes are connected with mechanical ones. Technical interest and a need for clearness met in the brain of Sadi Carnot and drew his attention to this point. The great industrial importance of the steam-engine was very influential here, although it is only a historical accident that the development of science referred to was not connected with electrotechnics. Franz Neumann, indeed, followed exactly Carnot's way of thinking when establishing the laws of induced electric currents (1845). The peculiarity in Carnot's idea consists in the fact that he was the first to exclude the *perpetuum mobile* in a wider domain than that of pure mechanics, and assumed that even a use of thermal processes cannot give *perpetuum mobile.* However, the modem principle of energy 'was not held by Carnot, for he still kept Black's notion of caloric, which completely dominated Black for psychological reasons that we have already discussed.[29]

Sadi Carnot wrote a long and deeply interesting footnote in which all his attention and the tentative of to produce a mathematical interpretation on the matter emerge. In particular, and very important from an historical point of view, there is a declaration on the science and the role played by mechanical science and its laws at the time. In this regard, we provide this declaration in its entirety together with his father's assumptions on *perpetuum mobile*.

[29]Mach [1896] 1986, pp 295–298. (Author's *italics*).

Lazare Carnot (1786)	Sadi Carnot (1824)
[...] everyone repeats that in Machines in motion time or speed is always lost when force is gained [...][a]. LVII. What is finally the veritable purpose of moving machines? [...] *the machines in motion, always lose in time and velocity what is gained in force*[b]. The reflections I propose on this law [Ivi, p vi] lead me to say something about perpetual motion and I will show not only that every machine which is aborted must stop, but I will assign the very instant when this must occur[c]. LXII. We can conclude from that which we have just said regarding friction and other passive forces, that perpetual motion is absolutely impossible, using it to produce only bodies which are not sollicited by any motive forces and even heavy bodies [...][d]. It is therefore evident that we must absolutely give up the hope of producing that which we call perpetual motion if it is true that all of the motive forces that exist in nature [...][e].	Such a creation is entirely contrary to ideas now accepted, to the laws of mechanics and of sound physics. It is inadmissible[*][f].[*(1)...]. The objection may perhaps be raised, that perpetual motion, demonstrated to be impossible by mechanical action alone, may possible not be so if the power either of heat or electricity be exerted; but is it possible to conceive the phenomena of heat and electricity as due to anything else than some kind of motion of the body, and such as should they not be subjected to general law of mechanics? Do we know besides, *à posteriori*, that the all of the attempts made to produce perpetual motion by any means whatever have been fruitless? – that we have never succeeded in producing a motion veritably perpetual, that is, a motion which will continue forever without alteration in the bodies set to work to accomplish it? The electromotor apparatus (the pile of Volta) has sometimes been regarded as capable of producing perpetual motion; attempts have been made to realize this idea by constructing dry piles said to be unchangeable; but however it has been done, the apparatus has always exhibited sensible deteriorations when its action has been sustained for a time with any energy. The general and philosophic acceptation of the words *perpetual motion* should include not only a motion susceptible of indefinitely continuing itself after a first impulse received, but the action of an apparatus, of any construction whatever, capable of creating motive power in unlimited quantity, capable of starting from rest all the bodies of nature if they should be found in that condition, of overcoming their inertia; capable, finally, of finding in itself the forces necessary to move the whole universe, to prolong, to accelerate incessantly, its motion. Such would be a veritable creation of motive power. If this were a possibility, it would be useless to seek in currents of air and water or in combustible this motive power. We should have at our disposal an inexhaustible source upon which we could draw at will[g]

[a]Carnot (1786), p vi, line 14; see also p viii, line 20
[b]Carnot (1786), pp 88–89, line 24. (Author's *italics*)
[c]Carnot (1786), p ix, line 16
[d]Carnot (1786), p 94, line 16
[e]Carnot (1786), p 95, line 33
[f]Carnot (1978), p 21, line 5
[g]Carnot (1978), p 22, ft 1, line 11

Let us continue our historical comments.

In thermodynamics, by using thermostats only, a connection cannot produce positive work. In fact, in addition to the thermostats, heat, machines, and an appropriate way to link them by exploiting their differences in temperature is necessary. The heat machines basically decompose heat Q in more than one operation, as well as mechanical forces applied to a body (e.g., using d'Alembert's principle) can be decomposed into those absorbed by the vincula and the effectives, which – in the presence of δs – produce work. In fact, in his *Réflexions sur la puissance motrice du feu,* Sadi Carnot decomposed the heat from the reservoir in two ways: first by means of two isotherms and two isochors (note that the latter do not work) and then considering all of the absorbed heat on two isotherms only, followed by two adiabatic curves (without heat exchange) (Carnot 1978, pp 39–40).

Therefore, the two theories begin with the same *problem* and are based on the same *first fundamental principle* (it should be noted that this principle does not initiate analytical deductions, obtainable from its conceptual concepts, but begins a new method, capable of resolving the scientific problem proposed at the beginning).

11.9 On the Synthetic Method

In regard to the common choice for the two Carnots' use of the synthetic method, we refer the reader to Chapter 9. We fully reported the historical aspect of the synthetic method and the sections of *Réflexions sur la puissance motrice du feu* where Sadi Carnot used it both as an alternative method to using the sophisticated mathematics of the time, and an original method of reasoning for his cycle.

Lazare Carnot's synthetic method (e.g., Carnot 1813, pp 217–253) explains the nature and the frame of mind behind his research against the metaphysical conception – which prevailed at the time – of the infinitesimals. Sadi Carnot's theory states that this is possible when using infinitesimal calculus (e.g., Carnot 1978, p 18, ft 1). We saw that Sadi Carnot introduced the adiabatic as the synthetic method auxiliary variable to simplify the solution to the problem of how much the efficiency of a heat machine is; therefore (within an ideal cycle), the contributions of the work on the two infinitesimal adiabatic transformations mutually cancel each other out. Thus, the *suppression of the adiabatics* in Sadi Carnot's cycle may mean precisely that operation, typical in the synthetic method, of eliminating the auxiliary variable ε.

In Table 9.2 (see Chapter 9) bibliographical references (Drago and Pisano 2005) are presented in which Sadi Carnot applied synthetic method in his *Réflexions sur la puissance motrice du feu.* Particular attention is paid to Sadi Carnot's reasoning process while theorizing from a three–phase primitive cycle (or composed cycles), until he creates the idea of a four–alternate–phase cycle.

11.10 On the Principle of Virtual Work

Generally speaking, one can consider two main traditions for the formulation of the *principle of virtual work:*

1. By Aristotelian mechanics,[30] also called *principle of virtual velocities,* where a physical system (e.g., masses subjected to forces) is in an *equilibrium* state if and only if the (forces–)weights are inversely proportional to virtual velocities (to their points of applications).

> [On the lever, Problem 3]. Why is it that small forces can move great weights by means of a lever, as was said at the beginning of the treatise, seeing that one naturally adds the weight of the lever? For surely the smaller weight is easier to move, and it is smaller without the lever. Is the lever the reason, being equivalent to a beam with its cord attached below, and divided into two equal parts?[31] [. . .] now the ratio of the weight moved to the weight moving it is the inverse ratio of the distances from the centre. Now the greater the distance from the fulcrum, the more easily it will move. The reason has been given before that the point further from the centre describes the greater circle, so that by the use of the same force, when the motive force is farther from the lever, it will cause a greater motion[32].

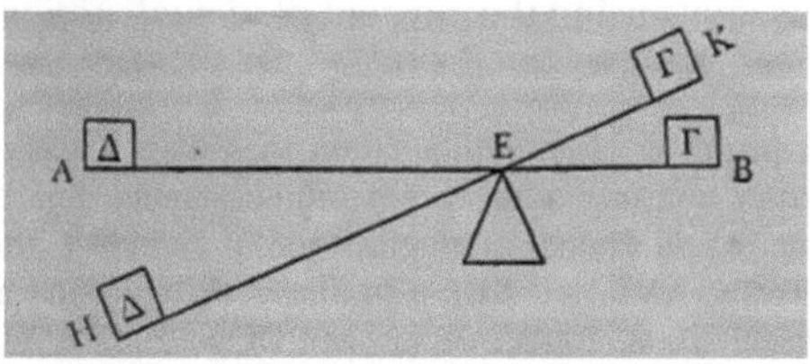

2. By Jordanus[33] de Nemore's mechanics (de Nemore 1533), also called principle of virtual displacement, where a physical system (e.g., masses subjected to

[30]Aristotle 1963, 1984; see also Baldi 1621; Aristotle 2000. The main Aristotelian works on mechanical arguments are in *Physics* (Aristotle 1999), *On the Heaven* (Aristotle 1984), and in *Problemata Mechanica* (Aristotle 1963). In particular, for the *principle of virtual velocity* one can also see *Physics*, $249^{b}30$–$250^{a}7$ and *de Caelo*, $301^{b}4$–11. In the Aristotelian or Aristotelian school, *Problemata Mechanica* seems to remain an argument which is still debated. In this regard, see Drake (Rose and Drake) and, recently, Winter (Winter). See also: Duhem 1905–1906, II, p 292, 1906–1913; Clagett ; Clagett and Moody; Brown; Lindberg; Truesdell 1968a, b. From an epistemological point of view, Aristotle dealt with the organization of science particularly in *The posterior analytics* (Aristotle 1853; see also *Id.*, 1949, 1963, 1996).

[31]*Problemata Mechanica* 850a 30 In: Aristotle 1963, 850a 30, p 353; see also Aristotle 1963, pp 347–349 and image, p 349.

[32]*Problemata Mechanica* 850b 5 In: Aristotle 1963, 850b 5, p 353. (*Idem* page for the image).

[33]See also Jordanus de Nemore in Tartaglia's edition (Tartaglia 1565; see also Clagett and Moody). Moreover, both *Elementa Jordani super demostrationem ponderum* (1229) and *Liber de ratione ponderis* (fl XIII c.) show an interesting proof of Archimedes' law of the lever, *Quaestio Sexta* (*Liber de ratione ponderis,* in Tartaglia's *Iordani Opvscvlvm de Ponderositate* edition: Tartaglia 1565, p 5(–6), line 13), by means of the application of the *principle of virtual work* and where the fall of geometric directions of *displacements* is considered vertical by Jordanus de Nemore. Moreover, one can also see Jordanus de Nemore's *Supposiotio Sexta* (Tartaglia 1565, p 1, line 13; see also *Liber de ratione ponderis* edited by Clagett and Moody, pp 174–175) where one can read that a body is able to raise another lighter body if a lever is utilized, that is like a embryonic engine.

forces) is in equilibrium state if only if the (forces–)weighs are in inversely proportional to their virtual displacements.

> If two weights descend along diversely inclined planes, then, if the inclinations are directly proportional to the weights, they will be of equal force in descending [idem force – equilibrium].[34]

Particularly, Lazare Carnot dealt with the *principle of virtual work* and then, by means of *geometric motion* (modernly, *virtual velocities*), canonically formulated the *principle of virtual velocities* starting from a *fundamental theorem* (Carnot 1786, § XXXIV, pp 68–69). In effect, since in his theory *geometric motions* coincide with *velocities* and not with displacements, this allowed Lazare Carnot to avoid, in the formulation of the *principle of virtual work* infinitesimal displacements, which could have produced some scientific embarrassment with respect to his assumptions (Carnot 1813). *Idem* situation for Sadi Carnot when he considered the small range of two reservoirs–temperatures. Furthermore, for the *principle of virtual velocity* related with any (general) mechanical machine, one can claim that the (forces–) weights that balance each other are reciprocal to their virtual velocities. Incidentally, the two conceptually different approaches/formulations can be mathematical equivalents using the concept of *virtual motion* as key reasoning.[35]

---

[34]"Quaestio decima. Si per diversarum obliquitatum vias duo pondera descendant, fueritque declinationum et ponderum una proportio eodem ordine sumpta, una erit utriusque virtus in discendendo [idem force – equilibrium]" (Tartaglia 1565, p 7v, line 1). See also Tartaglia 1554, Quesito XV–Def. XII; Quesito XVI–Def. XIII, p 84, line 7. (Pisano and Capecchi 2010; Capecchi and Pisano 2010a, b; Pisano 2008, 2009c; Pisano and Drago 2013; Pisano and Capecchi 2014 forthcoming).

[35]Explicit comments regarding the *principle of virtual work* are reported by Galilei in *Mecaniche* (Galileo 1890–1909, II, pp 155–191) and in *Discorsi intorno alle cose che stanno in sùl'acqua* (Galileo 1890–1909, IV, pp 3–141). Particularly, in this latter manuscript, Galileo clearly attributed the law of *virtual velocity* to Aristotle (1963, 847a 10–15, 847b 10, pp 329–332) also adding that the idea of the *principle of virtual work* was born thanks to the observation of the motion of points, which rotate along a circumference. Galileo also dealt with the law of the *virtual displacement* in more than one situation (Galileo 1890–1909 II, pp 240–242, IV, pp 68–69; VIII, pp 310–331, pp 329–330). We should wait for 1644, when Evangelista Torricelli (1608–1647), in his *Opera geometrica* (Torricelli 1644) claimed a rational criterion for the equilibrium, playing a fundamental role in mechanics and in the history of mechanics (Capecchi and Pisano 2007, 2010a, b). It can surely be considered the origin of the modern statement of the *principle of virtual work:* "Two heavy bodies linked together cannot move by themselves unless their common centre of gravity does not descend" (Torricelli 1644, *Liber primus de motu gravium naturaliter descendentium*, p 99, line 4).With regards to Torricelli's principle, one can also consider John Wallis's assumptions (Wallis 1693), and Pierre Varignon's (1654–1722) (Varignon 1725) essential and rigorous formulation as a scientific production which aimed at founding all statics upon an easily geometric principle: the composition of forces. In this sense, it is also alternative to the *principle of virtual work*. Let us remark that in his letter to *Johann* Bernoulli (1667–1748), Varignon also dealt with concept of *virtual velocities*, as components of *virtual infinitesimal displacements* towards the direction of the forces (Bernoulli J 1742, II). After Bernoulli, the most significant contribution to the development of the *principle of virtual work* is probably thanks to Vincenzo Riccati (1707–1775) who tried to establish it upon simple principles easily accepted by his contemporaries, introducing *Principles of actions* in *Dialogo di Vincenzo Riccati della compagnia di Gesù* (Riccati 1749) and in *De' principi della meccanica* (Riccati 1772). Recently, on mechanical science of 18th and 19th centuries see Pisano and Capecchi (2013).

A history[36] of *principle of virtual work* (Capecchi 2002, 2012) states that even if one could admit that the *principle of virtual work* was anterior to all of the laws of mechanics (but not everyone agreed with this), and could therefore be derived by the *principle of virtual work*, in the end, the fact that the *principle of virtual work* was self–evident could not be accepted. In other words, one could not accept it as a principle only. A proof was necessary; or a reduction to a theorem of another approach to the mechanics or an attempt to provide a more convincing version were necessary.[37] The *Principle of virtual work*, however, deals with extended systems of bodies that, differently from Euler's reasonings on fluids[38] (Euler 1757, p 286), include constraints in an essential way. These given forces are constraining reactions that are not included in the Newtonian scheme because they are unknown *a priori* (Lagrange 1788, pt II, IV). Therefore, when we follow Lazare Carnot's theoretical attitude, which is based on the *principle of virtual work, we consider the theory of mechanical machines to be theoretically self–sufficient* and consequently independent from Newtonian mechanics.

Lazare Carnot formulated the *principle of virtual work* by starting from his law of collisions (Carnot 1786, 1803a) and without (generally speaking) using classical Newtonian. forces[39] Particularly Lazare Carnot uses the *principle of virtual work* to discuss and define the conditions of equilibrium of the forces appied to the bodies.

> General principle equilibrium and of motion in machines
>
> XXXIV. *Whatever is the state of repose or of motion in which any given system of forces applied to a Machine, exists, if we take it all at once assume any given geometric motion, without changing these forces in any respect, the sum of the products each of them, by the velocity which the point at which it is applied will have in the first instant, estimated in the direction of this force, will be equal to zero.* That is to say, by calling F each of these forces (I), u the velocity which the point where it is applied will have at first instant, if we make the Machine assume a geometric motion, and z the angle comprehended between the directions

---

[36]In 1788, the principle of virtual work was referenced by Joseph–Louis Lagrange (1736–1813) as the fundamental principle for all of mechanical theory (Lagrange 1788). The Lagrangian principle of virtual work is usually associated with the first edition of Méchanique analytique and his comments (Lagrange 1788, p 11, B_2 § 18, p 21). However, their crucial elements were already given in *Recherches sur la libration de la Lune* (Lagrange 1764) in which he introduced his dynamical equations of motion by means of a new principle of Mechanics, or the principle of virtual work.

[37]The problem of the proof of the *principle of virtual work* sparked a heated debate, especially in France where Lazare Carnot (1786, 1803a), Vittorio Fossombroni (1754–1844, 1794), Fourier (1798; see also *Id.*, 1888–1890, pp 475–521), Ampère (1806) and Poinsot (1838; see also *Id.*, 1806) provided the main contributions. In effect, a specific difficulty was to connect the problem to Newtonian principles and to obtain its formal validity. In fact, initially this principle is independent from the Newtonian principles, which concerned an isolated particle (or the systems derived from it).

[38]The partial derivates in Euler's equations or Euler's fluids are applicable to compressible as well as to incompressible flow. It consists of an application of either an appropriate equation of state or assuming that the divergence of the flow velocity field is zero, respectively.

[39]In this regard, please see Lazare Carnot's very interesting assumptions on science, and forces (Carnot 1803a, p xj, p 47) and double–confuse use of the concept of space: Descartesian and Newtonian (Ivi, p 6).

> of F and of u, it must prove that we shall have for the whole system $[\Sigma]Fucosz = 0$. Now this equation is precisely the equation (AA) [$\Sigma FucosZ = 0$ (Carnot 1786, p 63, line 15)] found (XXX) [*Ivi*, p 60] which is nothing else in the end but the same [second] fundamental equation (F) [$\Sigma muUcosz = 0$ (*Ivi*, p 32, line 6)] presented under another form. It is easy to perceive that this general principle is, properly speaking, nothing else than that *Descartes*, to which a sufficient extension is to be given, in order that it may contain not only all the conditions of the equilibrium between two forces, but also all those of equilibrium and of motion, in a system composed of any number of powers: thus the first consequence of this theorem will be the principle of *Descartes*, rendered complete by the conditions which we have seen were waiting in it (V).[40]

Without discussing its historical epistemology,[41] below, we investigate whether Lazare Carnot's *principle of virtual work* could have coherently reordered Sadi Carnot's thermodynamics, comparing the theoretical developments of the two theories and their respective exemplary machines as well. Based on previous *excursus*, let us also note that

> Carnot's inference about the impermissibility of the *perpetuum mobile* was then repeated, and a reference was made to the similarity of Carnot's method to Lagrange's proof by pulleys of the principle of virtual displacements.[42]

On the other hand, like his father's mechanics, for the basis of his theory, Sadi Carnot did not use a single body (let alone infinitesimal) subject to thermal solicitation, but a complex system of bodies that includes thermostats (or thermal constraints) in an essential way. Therefore, Sadi Carnot also theorized in a way that is completely different from Newton's methods and is based on the *principle of virtual work*.

The principle stating that the total virtual work performed by all the forces acting on a system in static equilibrium is zero for a set of infinitesimal *virtual displacements* from equilibrium. The infinitesimal displacements are virtuals because they need not be obtained by a displacement that actually occurs in the physical system. The virtual work is the work performed by the virtual displacements, which can be arbitrary and are consistent with the constraints of the system. Its common mathematical expression is:

$$\delta W = \sum_i F_i^{(a)} \delta s_i = 0.$$

The theory of mechanical machines may be based on the *principle of virtual work*, and thought of as a consequence of the principle of the impossibility of perpetual

[40] Carnot 1786, § XXXIV, pp 68–69 and footnote "(I)". (Author's *italics* and capital letters).

[41] It can be argued that the role played by the *principle of virtual work* in modern classical mechanics is still not easily defined, although it is clear that it is important only for the historical part. In the theoretical treatises of rational mechanics, where one adopts a strongly axiomatic approach, the principle of virtual work is often not even mentioned, although the techniques of argument on which the mechanics, such as the Lagrange equations and Hamiltonian ones may be derived from it.

[42] Mach [1896] 1986, p 211, line 24.

motion, e.g., applied to machines and constraints: *it is impossible that the reactions of the constraints on the actions of the bodies, which make up the machine, produce positive work.* In others words it is impossible that constraints forces of bodies produce work:

$$\sum_i R_i d s_i \leqslant 0$$

Thus, in the end *what can the principle of virtual work suggest in both of the two Carnots' theories?* In previous paragraphs we discussed the famous analogy of the efficiency of a general theory of the heat machine (Carnot 1978, pp 7–8) independently from a working substance,[43] as Lazare Carnot demonstrated in *Essai sur le machines en général*, e.g., for any kind of working substance.[44] As we stated (see Chapter 9), starting from a function having three variables

$$W_{\max}/Q = f\,(Q, V, t)\,,$$

Sadi Carnot essentially reduced them to a Δt, as one variable only (Carnot 1978, pp 38–41). We remark

1. Since the machine that reversibly uses Δt runs cyclically, then, in order to obtain work, only the Δt of thermostats is important (and the amount of heat transferred; Carnot 1978, ft 1, pp 73–79).
2. Sadi Carnot used local variables for the working substance but, at the same time, in his thermodynamic theory they were also global variables.

Consequently, we can imagine the following physical Carnot system (Fig. 11.8):

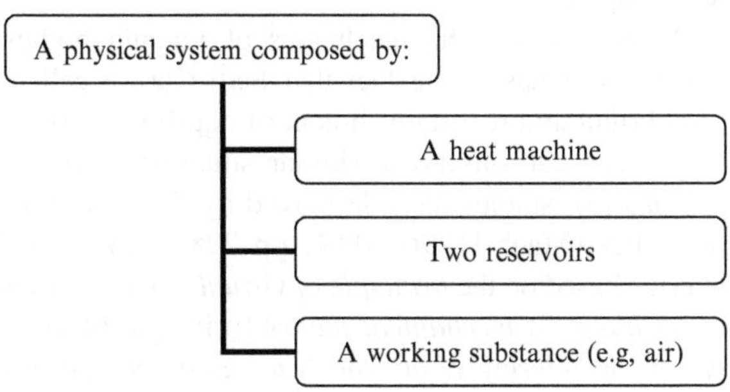

Fig. 11.8 A simplified performance model of a Carnot heat machine

[43]Carnot 1978, p 28, p 35, pp 37–39, p 112; see also Carnot 1878a, folio 2r(*Ia*); Carnot 1986, pp 183–184 folio 3r(*Ib*); Carnot 1986, pp 185–186; Carnot 1878a folio 5rv; Carnot 1986, pp 189–190; Carnot 1878a folio 6v; Carnot 1986, pp 188–189; Picard 1927, p 73, pp 76–79. (As ust mentioned, be careful attention to the page/folio numbers).

[44]Carnot 1786, pp 86–87, pp 89–93; see also Carnot 1780, §§ 155–156; Gillispie 1971, Appendix C, §§ 155–156, pp 332–333. Carnot 1878a, folio 5r; Carnot 1986, pp 189–190; Picard 1927, pp 77–78.

where from a global point of view, in the end, the magnitudes that produce work are W_{max}, t_1, and t_2. This assumption is crucial for the scientific advancement of his theory and it is based on the following consideration[45]:

In a general heat machine,
local variables are substituted by global variables.

The following chart is provided in regard to his analogy with the hydraulic wheel (Carnot 1978, pp 28–29) (Fig. 11.9):

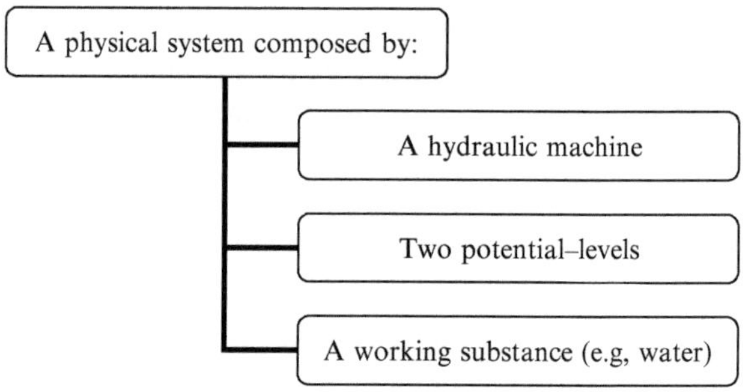

Fig. 11.9 A simplified performance model of a Carnot (analogy) hydraulic machine

The work produced only depends on the two levels of potential. Thus, generally speaking, based on previous analogy, for his heat machine we can write:

$$W_{\max}/Q = f\,(Q, t_1, t_2)\,.$$

This is the exact model of final function Sadi Carnot used to obtain thermal work (see Chapter 9).

By considering the mechanics of vincula–bodies for a global machine, we can also emphasize the fact that Sadi Carnot called attention to the concept of equilibrium and re–establishment of equilibrium (Carnot 1978, pp 9–12). In other words, one can observe a similar situation in mechanics when the *principle of virtual work* studies the role played by the equilibrium of forces that take action on bodies (Mach [1896] 1986, pp 228–230). In a theoretical hypothesis of an analogy based on the *principle of virtual work*, the following situation would occur: *in mechanics a mechanical theory* (principle of virtual work) *can be established without wondering about the true nature of the forces* (as Lazare Carnot did). *Therefore, in thermodynamics a heat machine theory can be established without questioning the validity of the nature of heat or caloric (as Sadi did;* see also in Appendix DNSs 61, 62). Moreover, Sadi Carnot also did not follow the *chimera* of large reservoirs producing enormous engine power, in mechanics as well, where one should not expect great results from great forces. This confutation is a crucial step in more than one of Lazare Carnot's works (Carnot 1778, 1780, 1786, 1803a)

[45]We should also note that Sadi Carnot did not give explicit details on it, even though he used it.

in which he proposed the crucial concept of work (Carnot 1786, pp 65–66, line 21; see also pp 96–97): mechanical work.[46]

Based on Lazare Carnot's discussions (Carnot 1786, pp 28–30), we can claim that a *geometric motion*[47] essentially expresses a non–mechanical interaction. The same situation occurs in thermodynamics where the dynamic of the process is based on the difference of the two temperatures without any interaction between work and heat: by an isochoric (without producing work) or adiabatically (without heat exchange), until the concept of reversibility. Lazare Carnot also defined these motions as *invertible*: *a motion assigned to a physical system of interacting bodies is geometric if the opposite motion is also possible*.[48] The result is a *possible motion*, but it is not always *invertible* (e.g., the motion of a sliding ring on a rotating rod). Therefore, one should add the hypothesis of *invertibility* for obtaining the concept of *geometric motion*. Conversely, a *geometric motion*, when integrated, gives an *invertible motion*. At this point, for contraints independent of time, a *geometric displacement* is equivalent to a *virtual invertible displacement* is valid (but not vice versa). On the contrary, a *possible displacement* only if it is *invertible*, produces, after its derivative, a *geometric motion*. In this sense, we note that initially, the geometric motion is a kind of uniform motion moving on the whole physical system when one considers the *equivalent* of the state of rest and the state of uniform motion. Consequently, by using the double negative sentences previously discussed (see Chapter 7) we can write:

$$\neg\,[(v = 0) \neq (v \neq 0)]\,.$$

In thermodynamics, one can also state $\neg\,(W \neq Q)$ where work W is a *static* situation since it only concerns thermodynamic phenomena. Heat Q concerns a *dynamic* situation because it involves the transfer of heat. Consequently, in thermodynamics, the analogous *geometric motion* is obtained when one adds Q to the physical system by implementing a process at a constant–temperature. In fact, Sadi Carnot considered a cycle composed of two isotherms, the first in one direction, and the other in the opposite direction (Carnot 1978, pp 29–38). Sadi Carnot also followed the same idea in his mathematical footnote (Carnot 1978,

[46]He instead referred to Archimedean equilibrium (e.g., Carnot 1780, § 152).

[47]Carnot 1786, pp 28–34, pp 41–45; see also Carnot 1780, § 113; Gillispie 1971, Appendix C, § 113, pp 308–309.

[48]Carnot 1978, pp 23–35 included footnotes. Let us also note a special passage where Sadi Carnot also discussed an irreversible isochoric (Carnot 1978, pp 25–26) just after the statement of his theorem (Carnot 1978, pp 21–22). For completeness, we add that in modern terms, to define the *principle of virtual work*, one can also specify that a *displacement* is possible if it is compatible with the fixed constraints. Moreover, it is *virtual* if it is compatible with the constraints even if they are moving. Limiting ourselves to the case of time–independent constraints, we can also derive in time a *possible displacement*. In this discussion, the term *displacement* may refer to a translation or a rotation (and the term force to a force or a momentum). When the virtual quantities are independent variables, they are also arbitrary.

ft 1, pp 73–79) in which at certain point the two isotherms essentially composed a unique process running in two directions, as well as Lazare Carnot's *geometric motions. What kind of similar results did the two Carnots obtain?* By means of his main equations (see Chapters 2 and 3), Lazare Carnot obtained his invariants of motion: conservation of quantity–of–motion and conservations of momentum of quantity–of–motion[49] (see above Chapters 2 and 3). Following another path, Sadi Carnot also obtained his invariants with regard to the efficiency and reversibility of a heat machine (Fig. 11.10).

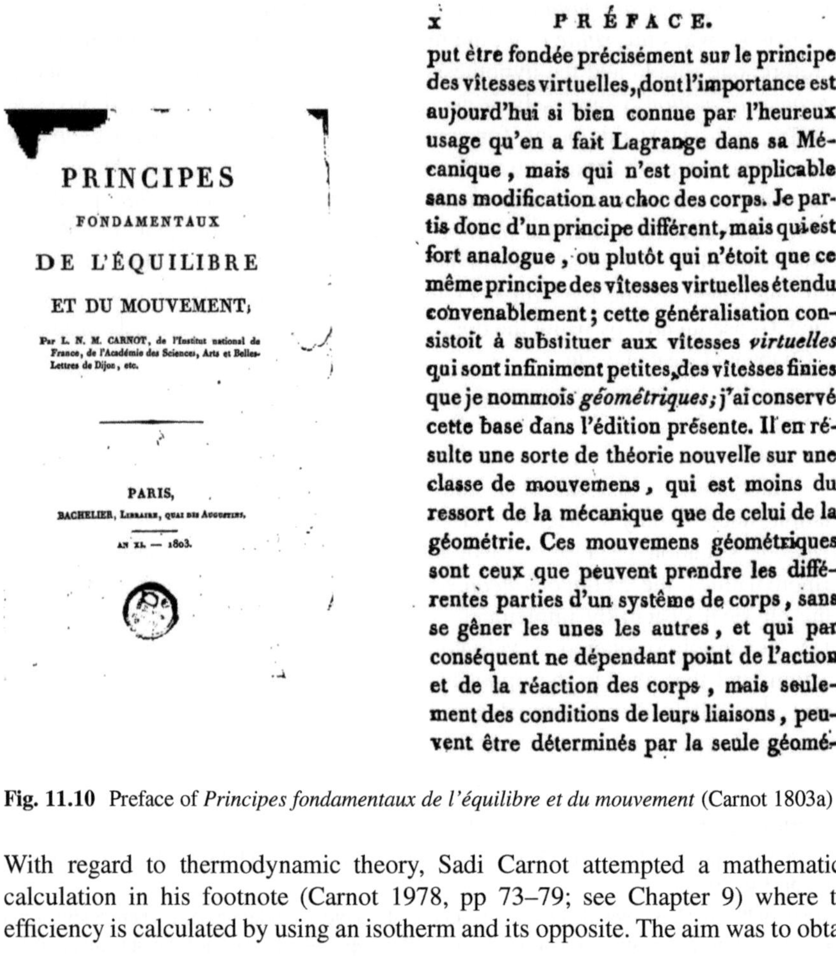

PRINCIPES
FONDAMENTAUX
DE L'ÉQUILIBRE
ET DU MOUVEMENT;
Par L. N. M. CARNOT, de l'Institut national de France, de l'Académie des Sciences, Arts et Belles-Lettres de Dijon, etc.

PARIS,
BACHELIER, Libraire, quai des Augustins,
AN XI. — 1803.

x PRÉFACE.

put être fondée précisément sur le principe des vîtesses virtuelles, dont l'importance est aujourd'hui si bien connue par l'heureux usage qu'en a fait Lagrange dans sa Mécanique, mais qui n'est point applicable sans modification au choc des corps. Je partis donc d'un principe différent, mais qui est fort analogue, ou plutôt qui n'étoit que ce même principe des vîtesses virtuelles étendu convenablement; cette généralisation consistoit à substituer aux vîtesses *virtuelles* qui sont infiniment petites, des vîtesses finies que je nommois *géométriques*; j'ai conservé cette base dans l'édition présente. Il en résulte une sorte de théorie nouvelle sur une classe de mouvemens, qui est moins du ressort de la mécanique que de celui de la géométrie. Ces mouvemens géométriques sont ceux que peuvent prendre les différentes parties d'un systême de corps, sans se gêner les unes les autres, et qui par conséquent ne dépendant point de l'action et de la réaction des corps, mais seulement des conditions de leurs liaisons, peuvent être déterminés par la seule géomé-

Fig. 11.10 Preface of *Principes fondamentaux de l'équilibre et du mouvement* (Carnot 1803a)

With regard to thermodynamic theory, Sadi Carnot attempted a mathematical calculation in his footnote (Carnot 1978, pp 73–79; see Chapter 9) where the efficiency is calculated by using an isotherm and its opposite. The aim was to obtain

[49]Lazare Carnot's reasonings upon his laws of conservation, are mainly reported in both *Essai sur les machines en général* (Carnot 1786) and *Principes fondamentaux de l'équilibre et du mouvement* (Carnot 1803a). The connections between the bodies constrain the *communication* of motion of the bodies. E.g., the theory of interaction–collisions by means of insensible degrees (Carnot 1803a, § 293, pp 261–262) as the result of a sequence of infinitesimally small percussions.

a mathematical expression of its invariant, or the efficiency of a heat machine with respect to all possible kinds of working substances. Therefore, in an analogy with Lazare Carnot's general *principle of virtual work* and his mechanical invariants, we can summarize (Fig. 11.11):

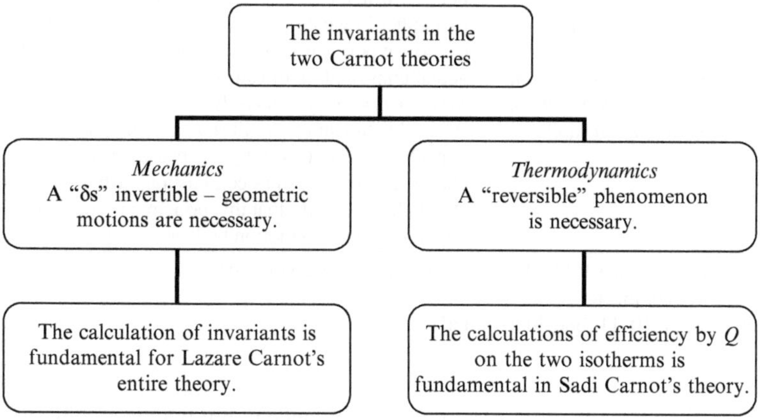

Fig. 11.11 The invariants in the two Carnot theories

Finally, we should also remark that the previous discussion on the *principle of virtual work* in Lazare and Sadi's work is surely suggestive but it has a limit in the two different mathematical approaches: *Lazare's approach is more mathematical than Sadi's*. However, this kind of analogy based on the two Carnots' *principle of virtual work* is not yet completed (Drago and Pisano, Pisano, forthcoming). In fact, we should clarify the common mathematical formalism which both theories evidently lack even though Lazare Carnot himself opened his program of research by explaining the foundation of his mechanics to the reader (Carnot 1803a, p x), that is to say, the *principle of virtual work*. In this sense, a new mechanics – with respect to Newtonian mechanics – would be reworked.

11.11 On the Principle of Virtual Work and the Interactions

We will now note that in the traditional mechanical theory of hard bodies, the *principle of virtual work* formally defines the condition of equilibrium of the forces that act on the bodies. Sadi Carnot's theory also strongly emphasizes the same situation: *equilibrium* reasoning on the production of work when it re–establishes heat equilibrium, which means that, in a cycle, it passes from a higher to a lower temperature.

Lazare Carnot (1786)	Sadi Carnot (1824)
Corollary II. General principle of equilibrium in weighing Machines. XXXVI. *When several weights applied to any given Machine, mutually form an equilibrium, if we make this Machine assume any geometric motion, the velocity of the centre of gravity of the system, estimated in the vertical direction, will be null at the first instant*[a]	The production of motion in steam–engines is always accompanied by a circumstance on which we should fix our attention. This circumstance is the re–establishing of equilibrium in the caloric; that is, its passage from a body in which the temperature is more or less elevated, to another in which it is lower. [. . .][b]. The production of motive power is then due in steam–engines not to an actual consumption of caloric, *but to its transportation from a warm body to a coldbody,* that is, to its re–establishment of equilibrium an equilibrium considered as destroyed by any cause whatever, by chemical action, such as combustion, or by any other. We shall see shortly that this principle is applicable to any machine set in motion by heat. [. . .][c]

[a]Carnot 1786, p 71, line 1. (Author's *italics*). See also Carnot 1803a
[b]Carnot 1978, p 9, line 8
[c]Carnot 1978, pp 10–11, line 20. (Author's *italics*)

Therefore, the two theories focus on the same concept, but with regard to different physical quantities. Moreover, Lazare Carnot, having the mathematical formula for the *principle of virtual work*, studies the theoretical conditions that translate the practical conditions of equilibrium. Sadi Carnot, instead, who does not have the *principle of virtual work* formulas for heat phenomena, tries to reason on that which, according to the conception of the *principle of virtual work* in anthropomorphic motions, causes equilibrium to be found in an analogy with δs motions in mechanics, and Δt in thermodynamics while in mechanics, in order to find equilibrium, a δs_i motion of a few bodies in the system is given. Consequently, it can be observed whether or not the system returns to its initial state. In thermodynamics, the generation of a quantity of caloric with fire unbalances the system, causing a change in temperature, Δt; in this instance heat also moves backwards to find the condition of equilibrium, or the initial temperature, then work W is obtained. Therefore, Sadi Carnot can conclude that wherever there is a Δt, *there is work to be gained.* Using the *principle of virtual work* as a reference confirmed the results obtained by reasoning only on the constraints by using the analogy between mechanical machines and heat machines.

In *Réflexions sur la puissance motrice du feu,* in order to clarify the problem of physical interaction with facts, the inside of a machine is examined as an example: a water vapor machine that worked according to particular mechanisms, which were common at the time. Two pages of *Réflexions sur la puissance motrice du feu* were dedicated to the description, in operative terms, of the work–heat interaction (Carnot 1978, pp 9–10) in a fairly general case, that is to say, the case of those single technical mechanisms that take advantage of the transformation of heat inside the water vapor machine (it should be noted that here there is no adiabatic). The result obtained is announced at the beginning of the description and then correctly repeated as a conclusion at the end

> Wherever there exists difference of temperature, motive–power can be produced.[50]

11.12 On the Principle of Virtual Work and State Equations

In the hydraulic wheel, the production of work depends on the difference in levels between waterfalls and makes the wheel turn (Carnot 1978, pp 28–29). In Sadi Carnot's theory as well, the production of motive power in a heat machine depends on a difference in value, that is to say, it is always associated with the transference of a quantity of heat from a higher temperature to a lower temperature (*Ivi*, pp 9–10). For example, in the case of a heat machine a quantity of heat Q_1 passes from the boiler, which is at temperature t_1, to the cylinder; and then, at the end of the operation, a quantity of heat Q_2 (which, in the caloric hypothesis, is equal to Q_1) passes to the condenser at temperature t_2. Here, once again, Lazare Carnot's mechanical theory on the *principle of virtual work* is helpful. In fact, the *principle of virtual work* is a global statement regarding the internal behaviour of the entire system and in order to apply it, a comment is necessary: e.g., entrance/exit (Carnot 1786, § XXXII, p 65), active forces/passive forces. In the first dualism, given any entrance, it acts as a first methodological principle, that is, it offers a path of reasoning to obtain the exit. In mechanical machines, Lazare Carnot only schematizes their operations by means of the entrance/exit relations (*Ibidem*). Sadi Carnot also considers the in–out efficiency of the physical system in question: his expression regarding the production of work W from a given quantity of heat Q involves the concept of the entrance/exit of the machine, efficiency η. It should be noted that this quantity links the two uncertain quantities into one (Q because it has an uncertain nature; W because it does not belong to a state), thereby eliminating part of the problem that they create by themselves. Therefore, any hypothesis on the nature of heat is valid. Q is no longer the central variable of which the transformation into work is comprised (as occurs with the calculation on $dQ = 0$) and there is nothing left to solve if heat Q is conserved or not in the passage between the two temperatures. The problem also does not interact with the work inside the machine; however it is necessary to take only the dependency of η on other variables of state into account.

The unpublished manuscript found in 1966 (Carnot S–EP; see also Gabbey and Herivel 1966) covers 21 folia in a sewn notebook and provided Sadi Carnot with an important advancement, most likely obtained by following the paternal concept of *state of the system*. In the theory of the equation of state of perfect gases, three variables p, t and V (in effect density d in place of V) are expressed. In this manuscript, his calculation of the efficiency is incorrect because it closes the cycle

[50] Carnot 1978, p 16, line 7.

by means of the V–variable only (Carnot S–EP, folia 1–2; see also Gabbey and Herivel 1966, pp 153–154). A comparable situation also occurs in *Réflexions sur la puissance motrice du feu* in which Sadi Carnot's incomplete cycle consisted of three stages only: production, expansion and condensation (Carnot 1978, pp 17–18, 82–86). As previously stated (see Chapters 6 and 9), the adiabatic phase and its suppression would be a central topic in his book.

In *Réflexions sur la puissance motrice du feu*, however, he presents the valid concept of state of the system since he clearly states that it is necessary to consider all of the variables that define the state. *Sadi Carnot reasons referring to something that is clear in all of the possible transformations: in modern times, the state of the system.* In fact, after having established that Q is connected to Δt when it produces *Work W* and that W is connected to ΔV, he also expresses his opinion on the *sufficient number* of variables. In Sadi Carnot's words:

> Wherever there exists a difference of temperature, wherever it has been possible for the equilibrium of the caloric to be re–established, it is possible to have also the production of impelling power. Steam is a means of realizing this power, but it is not the only one. All substances in nature can be employed for this purpose, all are susceptible of changes in volume, of successive contractions and dilatations, through the alternation of heat and cold. All are capable of overcoming in their changes of volume certain resistances, and of thus developing the impelling power. A solid body–a metallic bar for example–alternately heated and cooled increases and diminishes in length, and can move bodies fastened to its ends. A liquid alternately heated and cooled increases and diminishes in volume, and can overcome obstacles of greater or less size, opposed to its dilatation. An aeriform fluid is susceptible of considerable change of volume by variations of temperature. If it is enclosed in an expansible space, such as a cylinder provided with a piston, it will produce motions of great extent. Vapors of all substances capable of passing into a gaseous condition, as of alcohol, of mercury, of sulphur, etc., may fulfil the same office as vapor of water. The latter, alternately heated and cooled, would produce motive power in the shape of permanent gases, that is, without ever returning to a liquid state. Most of these substances have been proposed, many even have been tried, although up to this time perhaps without remarkable success. We have shown that in steam–engines the motive power is due to a re–establishment of equilibrium in the caloric; this takes place not only for steam–engines, but also for every heat engine, that is, for every machine of which caloric is the motor. Heat can evidently be a cause of motion only by virtue of the changes of volume or of form, which it produces in bodies.[51]

As previously mentioned, Sadi Carnot also proposed his equation of state for gases in his unpublished *Recherche d'une formule propre à représenter la puissance motrice de la vapeur d'eau* (Carnot S–EP, folio 5; see also Carnot 1978, pp 223–234; Carnot 1986, pp 167–180) and in the *Réflexions sur la puissance motrice du feu* (Carnot 1978, ft 1, p 75) (Fig. 11.12)

---

[51]Carnot 1978, pp 12–14, line 5.

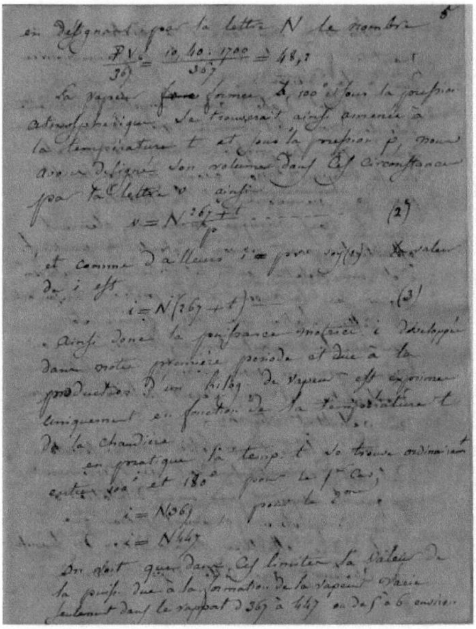

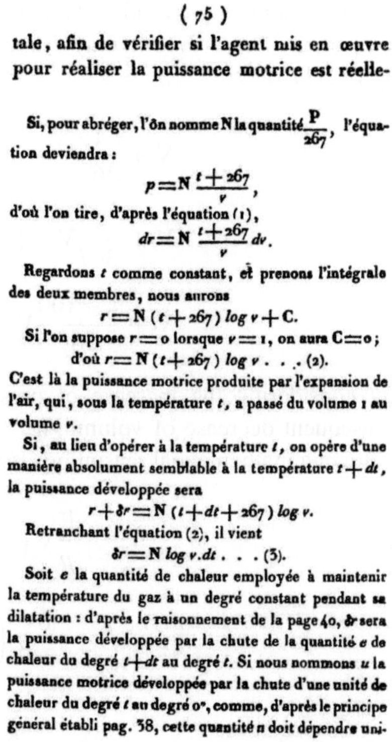

(75)

tale, afin de vérifier si l'agent mis en œuvre pour réaliser la puissance motrice est réelle-

Si, pour abréger, l'on nomme N la quantité $\frac{P}{267}$, l'équation deviendra :

$$p = N\frac{t+267}{v},$$

d'où l'on tire, d'après l'équation (1),

$$dr = N\frac{t+267}{v}dv.$$

Regardons t comme constant, et prenons l'intégrale des deux membres, nous aurons

$$r = N(t+267)\log v + C.$$

Si l'on suppose $r = 0$ lorsque $v = 1$, on aura $C = 0$;

d'où $r = N(t+267)\log v$ . . . (2).

C'est là la puissance motrice produite par l'expansion de l'air, qui, sous la température t, a passé du volume 1 au volume v.

Si, au lieu d'opérer à la température t, on opère d'une manière absolument semblable à la température $t+dt$, la puissance développée sera

$$r+\delta r = N(t+dt+267)\log v.$$

Retranchant l'équation (2), il vient

$\delta r = N \log v.dt$ . . . (3).

Soit e la quantité de chaleur employée à maintenir la température du gaz à un degré constant pendant sa dilatation : d'après le raisonnement de la page 40, δr sera la puissance développée par la chute de la quantité e de chaleur du degré $t+dt$ au degré t. Si nous nommons u la puissance motrice développée par la chute d'une unité de chaleur du degré t au degré 0°, comme, d'après le principe général établi pag. 38, cette quantité u doit dépendre uni-

Fig. 11.12 Sadi Carnot's state equation on gas theory in *Recherche d'une formule propre à représenter la puissance motrice de la vapeur d'eau* (*left*) and in *Réflexions sur la puissance motrice du feu*. (*On the left*: Carnot S–EP, folio 5, © *Collections archives de la bibliothèque de l'École polytechnique de Paris* (see also Gabbey and Herivel). *On the right*: Carnot 1953, ft 1, p 75)

As was common at the time, Sadi Carnot combined Mariotte's law and Gay–Lussac's law to obtain the following result for a state equation:

$$p = N\frac{t+267}{v} \quad \text{(Carnot 1953, ft.1, p 75)}, \qquad (11.1)$$

$$i = pv \quad \text{(Carnot S-EP folio 3rv)}. \qquad (11.2)$$

Here, it is interesting to note the calculation that he included in *Recherche d'une formule propre à représenter la puissance motrice de la vapeur d'eau* (CarnotS–EP; see also Gabbey and Herivel 1966; Carnot 1986, pp 167–180) to establish his state equation of gases (see Eq. 11.2).

Sadi Carnot mainly considered the increase of volume of vapor at various values of temperatures according to Gay–Lussac's law (Carnot S–EP, folio 4; Carnot 1986, pp 172–173):

$$\mathrm{V} + \frac{1}{367}\mathrm{V} \quad (\text{at } 101^\circ)\,;$$

$$\mathrm{V} + \frac{2}{367}\mathrm{V} \quad (\text{at } 102^\circ)\,;$$

$$\mathrm{V} + \frac{t-100}{367}\mathrm{V} \quad (\text{at a } t - \text{temperature}).$$

By considering the increase of pressure from P to p, Mariotte's law and the consequent decrease of volume, he considered the inverse ratio and obtained the following mathematical reasonings (*Ibidem*):

$$\frac{\mathrm{P}}{p}\mathrm{V}\left(1 + \frac{t-100}{367}\right),$$

$$\frac{\mathrm{PV}}{367}\left(\frac{267+t}{p}\right),$$

$$\mathrm{N}\frac{267+t}{p}$$

where N–value[52] (*Ivi*, folio 5; see also Carnot 1986, p 172) is:

$$\frac{\mathrm{PV}}{367} = \frac{10,40 \cdot 1700}{367} = 48,2.$$

In the end, for the vapor water transferred to t–temperature, p–pressure and thus v–volume, and substituting $i = \mathrm{N}\,(267 + t)$ in Eq. (11.2), he obtained the following final state equation[53] written in his unpublished *Recherche d'une formule propre à représenter la puissance motrice de la vapeur d'eau*:

$$v = \mathrm{N}\frac{267+t}{p}.$$

[52] At the beginning Sadi Carnot suggested the following initial conditions: P = 10,40 m. column of water and V = 1,700 L (that is the volume of 1 kg of steam). (Carnot S–EP, folio 4; see also Gabbey and Herivel 1966, p 154; Carnot 1986, p 171).

[53] Carnot S–EP, folio 20; see also Carnot 1986, p 179.

(In modern terms, we can also write[54] PV $= n$RT). The ratio 1/267 was the value belonging to Gay–Lussac's law (Gay–Lussac 1802). Currently, its value is 1/273. Thus the term "267 + t" is structurally comparable to absolute temperature T $= 273 + t$.

It is quite probable that Sadi Carnot, during his studies at Conservatoire national des arts et *métiers* in Paris, worked closely with his professor and friend Clément (Clément 1970) on the magnitude calorie, applied and industrial chemistry and other topics related to his coursework[55] (Fig. 11.13).

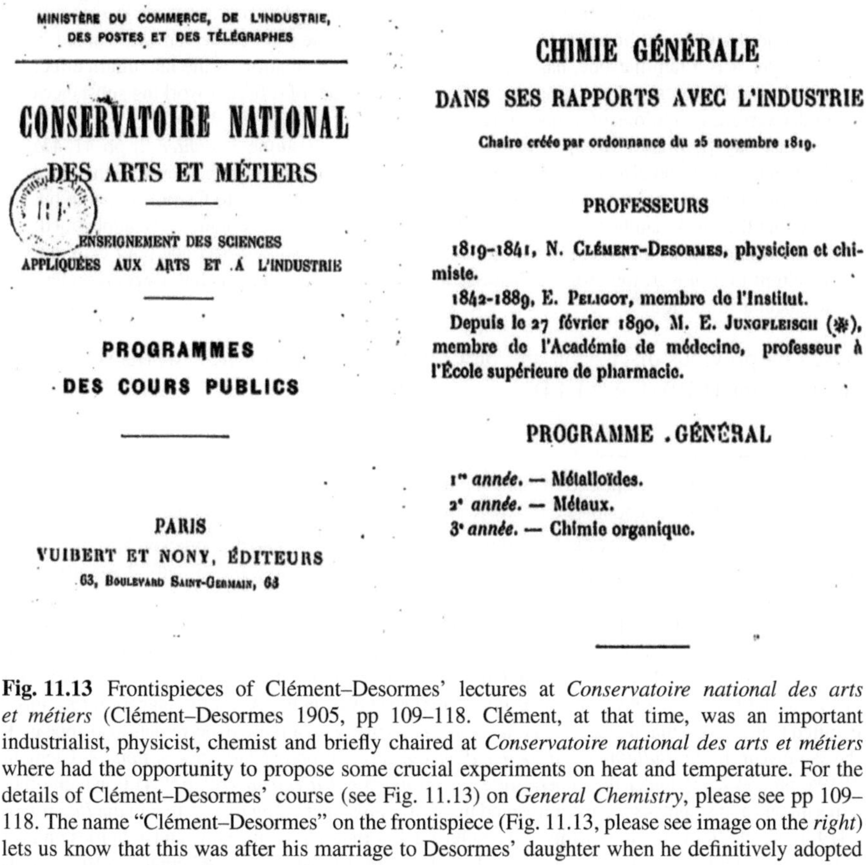

MINISTÈRE DU COMMERCE, DE L'INDUSTRIE, DES POSTES ET DES TÉLÉGRAPHES

CONSERVATOIRE NATIONAL DES ARTS ET MÉTIERS

ENSEIGNEMENT DES SCIENCES APPLIQUÉES AUX ARTS ET À L'INDUSTRIE

PROGRAMMES DES COURS PUBLICS

PARIS
VUIBERT ET NONY, ÉDITEURS
63, BOULEVARD SAINT-GERMAIN, 63

CHIMIE GÉNÉRALE
DANS SES RAPPORTS AVEC L'INDUSTRIE

Chaire créée par ordonnance du 25 novembre 1819.

PROFESSEURS

1819-1841, N. CLÉMENT-DESORMES, physicien et chimiste.
1842-1889, E. PELIGOT, membre de l'Institut.
Depuis le 27 février 1890, M. E. JUNGFLEISCH (✻), membre de l'Académie de médecine, professeur à l'École supérieure de pharmacie.

PROGRAMME GÉNÉRAL

1re *année*. — Métalloïdes.
2e *année*. — Métaux.
3e *année*. — Chimie organique.

Fig. 11.13 Frontispieces of Clément–Desormes' lectures at *Conservatoire national des arts et métiers* (Clément–Desormes 1905, pp 109–118. Clément, at that time, was an important industrialist, physicist, chemist and briefly chaired at *Conservatoire national des arts et métiers* where had the opportunity to propose some crucial experiments on heat and temperature. For the details of Clément–Desormes' course (see Fig. 11.13) on *General Chemistry*, please see pp 109–118. The name "Clément–Desormes" on the frontispiece (Fig. 11.13, please see image on the *right*) lets us know that this was after his marriage to Desormes' daughter when he definitively adopted the new composed family surname)

[54]The modern equation cited in the running text officially refers to the number of moles.

[55]On Sadi Carnot's debt (especially *influence* on the free expansion of the working substance after cut–off) to Clément and Desormes, see: Carnot 1978, pp 98–108 and footnote; Fox 1970, pp 235–238; Lervig 1985; Carnot 1986, pp 10–11, 19–21, En. 53, p 48, pp 166–170, p 180.

In fact, the calculation of N is one of the topics that Sadi Carnot dealt with in his unpublished manuscript between 1819 and 1827 while he was attending *Conservatoire national des arts et métiers.* While it is unnecessary to provide a detailed description, we wish to point out a fact that goes on to become historically significant for the birth and development of Carnot's science regarding the motive power of steam:

> Carnot had learnt of Clément's views, if not personally, then certainly through a joint paper by Clement and Desormes which had been read before the *Académie des sciences* in [23 and 30] August 1819 [Clément 1819b]. Although this paper was never published, an extract from it appeared almost simultaneously in the monthly bulletin of the *Société Philomathique* in Paris [Desormes and Clément 1819] and a copy of the complete paper was made available to Carnot by Clément himself [Carnot 1978, ft 1, p 98]. The problem which was tackled in the paper was a familiar one, namely the theoretical determination of the maximum effect which could be obtained in a heat engine from a given mass of different working substances under various conditions of temperature and pressure [Fox 1970, En. 31, p 249].[56]
> On that day [8 March 1827] (*Conservatoire* notebooks, volume 3, *cahier* 2, pp 41–43) Clément dictated precisely the result which Carnot arrived at in his paper, even to the point of using the same nomenclature. He did not acknowledge his debt to Carnot by name but said that the result had been given to him by "un mathématicien distinguée". Baudot added: "La formule algébrique [i.e. Carnot's] n'est ici que comme sujet d'exercice pour ceux qui voudront l'employer; toutefois, le Professeur avoue qu'il n'en a jamais fait usage; il préfère le calcul arithmétique".[57]

Based on previous discussion and on Clément[58] and Desormes' works (Desormes and Clément 1819) (Fig. 11.14):

[56]Fox 1970, pp 236–237, line 36. (Author's *italics*. The brackets "[...]" are ours. They explicitly report the content of the endnotes cited by Robert Fox. This first one only is ours).

[57]Fox 1970, p 247, line 5. (Author's *italics* and quotation marks. The first bracket "[...]" is ours, only). The formal announcement of Clément's experiments lectured on 23 and 30 August at the *Académie des sciences* (Clément 1819b), reported in a few lines in *Procès–verbaux des séances de l'Académie* (Clément 1819b, pp 480–481), and its next spread in the *Bulletin* des *Sciences par la Société Philomathique* de Paris (Desormes and Clément 1819, concern a previous presentation of these experiments, made by Clément at Christophe Oberkampf's factory at Jouy–en–Josas (April 1819), to the *Société d'Encouragement pour l'Industrie Nationale.* A brief explanation (but more detailed than the one that he lectured at *Académie des sciences*) of these experiments also appeared, the same year, in the *Bullettin de la société d'encouragement pour l'Industrie nationale* (Clément 1819a, pp 254–255) under the general title *sur les machines à vapeur* (*Ivi*, p 254) in which other experiments and scientists were commented upon. It is also interesting to note that in the *Bulletin* of 1819, more of Clement's comments and results are reported. Here, his name and activities on heat machines sometimes appeared with "[...] MM *Isnard*, *Olivier* [...]" (Clément 1819a, p 174, p 301, p 302, p 304) and, of course, Desormes: e.g., "[...] la théorie de Clément et Desormes [...]" upon "[...] la formation de l'acide sulfurique [...]" is also cited (*Ivi*, p 174, p 375, p 385; see also the *Bullettin de la Société d'Encouragement pour l'Industrie Nationale* of 1825, vol XXIV, pp 219–223). Sadi Carnot on his work published in 1824 is never cited.

[58]See also Clément–Desormes course on *Chimie industrielle* at *Conservatoire national des arts et métiers* (reported in *Conservatoire national des arts et métiers*. Journal des Cours de 1825 a 1830). A significance was noted by Robert Fox (1970, ft 21, p 248) who also cited Payen (1968; see also *Id.*, 1971) for a detailed discussion on Clément–Desormes' courses and his relationship with Sadi Carnot.

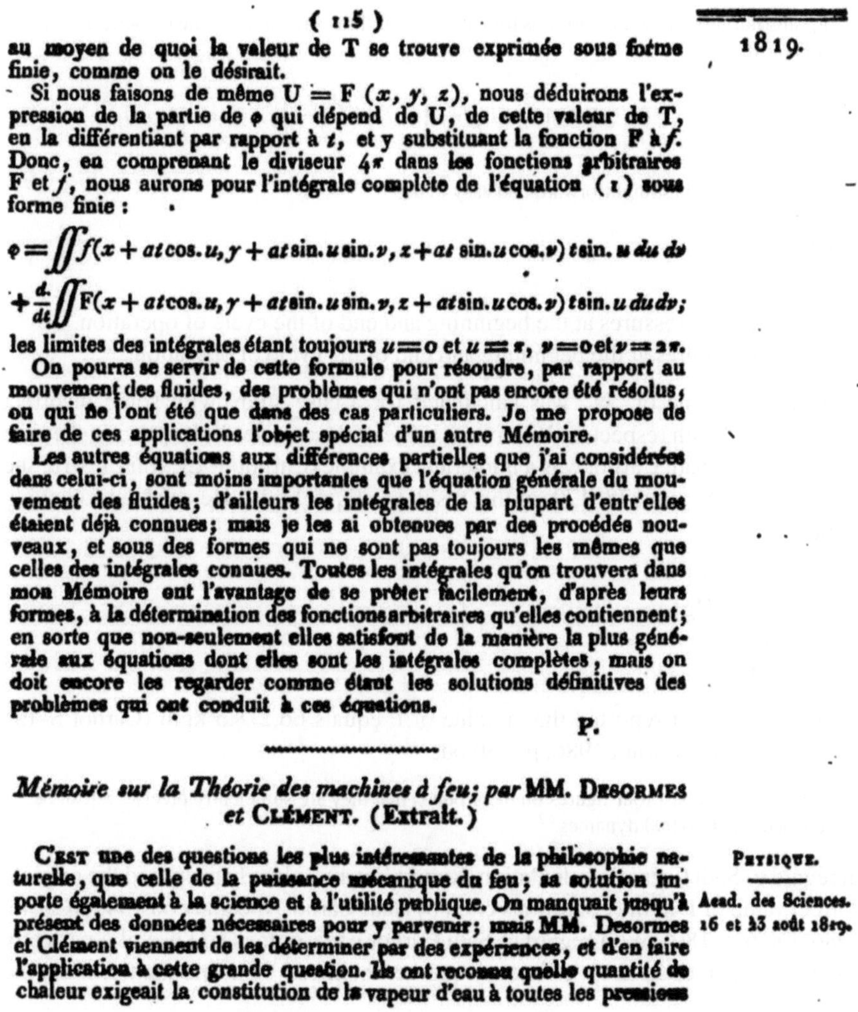

(115)

1819.

au moyen de quoi la valeur de T se trouve exprimée sous forme finie, comme on le désirait.

Si nous faisons de même U = F (x, y, z), nous déduirons l'expression de la partie de φ qui dépend de U, de cette valeur de T, en la différentiant par rapport à t, et y substituant la fonction F à f. Donc, en comprenant le diviseur 4π dans les fonctions arbitraires F et f, nous aurons pour l'intégrale complète de l'équation (1) sous forme finie :

$$\varphi=\iint f(x+at\cos.u, y+at\sin.u\sin.v, z+at\sin.u\cos.v)\, t\sin.u\, du\, dv$$

$$+\frac{d.}{dt}\iint F(x+at\cos.u, y+at\sin.u\sin.v, z+at\sin.u\cos.v)\, t\sin.u\, du\, dv;$$

les limites des intégrales étant toujours $u=0$ et $u=\pi$, $v=0$ et $v=2\pi$.

On pourra se servir de cette formule pour résoudre, par rapport au mouvement des fluides, des problèmes qui n'ont pas encore été résolus, ou qui ne l'ont été que dans des cas particuliers. Je me propose de faire de ces applications l'objet spécial d'un autre Mémoire.

Les autres équations aux différences partielles que j'ai considérées dans celui-ci, sont moins importantes que l'équation générale du mouvement des fluides; d'ailleurs les intégrales de la plupart d'entr'elles étaient déjà connues; mais je les ai obtenues par des procédés nouveaux, et sous des formes qui ne sont pas toujours les mêmes que celles des intégrales connues. Toutes les intégrales qu'on trouvera dans mon Mémoire ont l'avantage de se prêter facilement, d'après leurs formes, à la détermination des fonctions arbitraires qu'elles contiennent; en sorte que non-seulement elles satisfont de la manière la plus générale aux équations dont elles sont les intégrales complètes, mais on doit encore les regarder comme étant les solutions définitives des problèmes qui ont conduit à ces équations.

P.

Mémoire sur la Théorie des machines à feu; par MM. Desormes et Clément. (Extrait.)

Physique.

Acad. des Sciences. 16 et 23 août 1819.

C'est une des questions les plus intéressantes de la philosophie naturelle, que celle de la puissance mécanique du feu; sa solution importe également à la science et à l'utilité publique. On manquait jusqu'à présent des données nécessaires pour y parvenir; mais MM. Desormes et Clément viennent de les déterminer par des expériences, et d'en faire l'application à cette grande question. Ils ont reconnu quelle quantité de chaleur exigeait la constitution de la vapeur d'eau à toutes les pressions

Fig. 11.14 Desormes and Clément's *Mémoire sur la théorie des machines à feu* in *Bulletin* des *Sciences* par la *Société Philomathique* (Desormes and Clément 1819, pp 115–118. Please note that the volume is divided into three parts by date. The order page numbers restarts in each part (-years): 1817 (*Ivi*, pp 1–200), 1818 (*Ivi*, 1–192) and 1819 (*Ivi*, pp 1–192). In our case, the third part is considered)

In the following we comment *Recherche d'une formule propre à représenter la puissance motrice de la vapeur d'eau* where Sadi Carnot encountered the state equation:

> If we presume that $p'' = p'$, the equation [the total motive power produced] can be reduced to
>
> $$\mathrm{F} = \mathrm{N}\ln\frac{p}{p'}\left(267+\frac{t+t'}{2}\right) \qquad [(11.3)]$$

or, substituting common logarithms for natural logarithms and inserting the numerical value of N (=48,2), to

$$F = 110,8 \log \frac{p}{p'} \left(267 + \frac{t + t'}{2} \right).$$

This expression is straightforward and easy to use.[59]

Where

N = 48,2
p, p' = Vapors–pressures at the beginning and end of the cycle of operation.
t, t' = Temperatures at the beginning and end of the cycle of operation.

Thus, the value of "N" was calculated for a case study with some incorrect percentages[60] with respect to the modern value of the calorie. To be precise, the case study given by Sadi Carnot was a numerical example (Carnot S–EP, folia 17–18; see also Carnot 1986, pp 178–179) of that result see Eq. [(11.3)]. He used the following values[61]:

$p = 760$ [mmHg]
$p' = 9{,}47$ [mmHg]
$t = 100°$[C]
$t' = 10°$[C]

Then, Sadi Carnot reported that a value of F equals 66,278.5 kgm (Carnot S–EP, folio 18; see also Carnot 1986, p 179) but

> If we ignore the last four figures on the grounds that they are sufficiently precise, F [can be rounded to] = 66,000 dynames.[62]

In regard to Sadi Carnot's calculation, Robert Fox had already proposed the original correct value of F in the conditions cited by Sadi Carnot. It equals 66,734.8 (Carnot 1986, En. 6, p 180). With regard to Sadi Carnot's reasonings and use of a perfectly efficient machine, ca. 24% efficiency can be obtained.[63] On this matter, it can be

[59]Carnot S–EP, folio 17; see also Carnot 1986, p 178, line 16.

[60]1 cal = 4,186 J. Some texts use the thermochemical calorie (cal_{th}) equals 4,184 J (Cfr.: "Heat", Quantities and units, Part 4. ISO–*International Organization for Standardization* 31–4, 1992). The Kilocalorie, e.g., *Kcalorie* or *Calorie* is basically defined as the amount of heat required to raise the temperature of 1 kg of water from 0 °C to 1 °C at 1 atm of pressure. The *small calorie* or *g-calorie* is the amount of heat required to raise the temperature of 1 g of water by 1 °C with a temperature change from 14,5 °C to 15,5 °C. Currently, other equivalents used can be available with respect to specific experiments, SI metric and disciplines (*Ivi*).

[61]We added (in brackets) the units of measurement since in the unpublished manuscript they were implicitly lacking (Carnot S–EP, folio 18; see also Carnot 1986, pp 178–179).

[62]Carnot S–EP, folio 18; see also Carnot 1986, p 179, line 3, En. 79, pp 147–148; Fox 1970, En. 34, p 250.

[63]Let us note that the physical system should operate under particular conditions which are not cited.

historically relevant to report that Sadi Carnot calculated the value of the calorie (1,000/2.70 = 370 kg/cal) in another manuscript, *Notes sur les mathématiques, la physique et autres sujets* (Carnot 1878a) in which he claimed:

> According to certain ideas that I have conceived on the theory of heat, the production of one unit of motive power requires the destruction of 2.70 units of heat.[64]

Finally, in *Recherche d'une formule propre à représenter la puissance motrice de la vapeur d'eau*, Sadi Carnot (and Clément) succeeded in establishing a value of the mechanical equivalent more than 20 years before Mayer or Joule began work on this subject. Currently, the method for calculating mechanical power is also sometimes called the *Law of Clément–Desormes.*

It is also known that Poisson suggested, ca. at the same time, the formula $p = a\rho T$, but he gave neither the method to obtain it, nor the value of the constant. Moreover, he used density ρ and not volume V. This fact produced some difficulty: e.g., the comparison between several gas formulas was difficult due to V being confused with the number of moles.

On Sadi Carnot's end, in work on two variables (see Chapter 9), a third is obtainable from the equation of state. In fact, further ahead in *Réflexions sur la puissance motrice du feu* (Carnot 1978, p 40), he presents those state laws, which at the time were lacking, for every possible thermodynamic transformation (adiabatic, isochoric, etc.), so that he completes the knowledge of all of the laws between variables p, V and t. In other words, few variables are considered. Let us note pressure p is obtainable from t and V. Moreover, he knows that work$W$ is not a function of state (at his time, it was clear that the production of W depended on the path travelled; materially it depends on the ingenuity of the heat machine). Therefore, Sadi Carnot should consider W as a variable on a theoretical level which is different from that of the other variables that define the state of the gas. It also should be emphasized that these results appear to have realized exactly that program which Sadi Carnot himself had clearly stated at the beginning of his *Réflexions sur la puissance motrice du feu* (Carnot 1978, pp 8–9).

11.13 Scholars, Formulas, Experiments and Sources cited by Sadi Carnot in his *Works*

Below, we provide Table 11.9 to show experiments, laws and scholars quoted by Sadi Carnot in his works. All of these references could be considered the main bibliography used by the young French intellectual.

---

[64] Carnot 1878a, *Notes sur les mathématiques, la physique et autres sujets*, folio 7v; Carnot 1878b, p 95; Carnot 1986, p 191, line 21; see also: *Ivi*, En. 27, p 209; see also Hoyer 1976.

Table 11.9 Scholars, formulas and laws quoted by Sadi Carnot in his works

Experiment/Law/ machine/Memoir/device	Scholar/Engineer (in Sadi Carnot's order)	*Réflexions sur la puissance motrice du feu* (Carnot 1978)
Machines	Savery, Newcomen, Smeaton, Watt, Woolf, Trevithick	p 6
Battery ("Appareil")	Volta	p 21, ft 1
Device	Thermometer of Breguet	p 29, ft 1
Experiment	Laplace	p 30, ft 1
Experiment	Gay–Lussac and Welter	p 30, ft 1
Law	Mariotte, Gay–Lussac and Dalton	p 41
Calculus	Poisson	p 43, ft 1
Experiment	Clément and Desormes	p 43, ft 1
Experiment /results	Gay–Lussac and Welter	p 43, ft 1
Law	Gay–Lussac	p 44
Law	Gay–Lussac and Dalton	p 46
Experiment	Delaroche and Bérard	p 46
Law	Mariotte	p 51 (and *Ivi*, ft 1)
Law	Gay–Lussac and Dalton	p 51, ft 1
Experiment	Dulong and Petit	p 51, ft 1
Experiment /law	Davy and Faraday	p 51, ft 1
Experiment	Delaroche and Bérard	p 55
Law	Mariotte	p 58
Experiment	Gay–Lussac and Welter	p 59, ft 1
Experiment /memoir	Delaroche and Bérard	p 60; p 61
Experiment	Dulong and Petit	p 64
Law	Clément and Desormes	p 65
Experiment /memoir	Dulong and Petit	p 65, ft 1
Experiment	Dalton	p 66
Experiment	Dulong and Petit	p 66, ft 1
Table	Dalton	p 67, ft 1
Law	Mariotte, Gay–Lussac	p 67, ft 1
Traité	Biot	p 68, ft 1

(continued)

Table 11.9 (continued)

Experiment/Law/ machine/Memoir/device	Scholar/Engineer (in Sadi Carnot's order)	*Réflexions sur la puissance motrice du feu* (Carnot 1978)
Experiment	Delaroche and Bérard	p 72; p 73
Rule	Gay–Lussac	p 74, ft 1
Law	Mariotte	p 74, ft 1
Law	Gay–Lussac	p 80
Experiment	Delaroche and Bérard	p 81
Law	Clément and Desormes	p 85
Memoir	Petit	p 86, ft 1
Experiment /results	Delaroche and Bérard	p 87
Experiment	Bétancour	p 87
Book	Prony	p 87
Law	Dalton	p 87, ft 1
Memoir	Desptrets [Despretz]	p 87, ft 1
Experiment	Davy and Faraday	p 87, ft 1
Law	Clément and Desormes	p 88
Experiment	Ørsted	p 93, ft 1
Memoir	Clément[a]	p 98
Machine	Perkins, Watt, Robinson	pp 99–102, ft 1
Law	Mariotte	p 102, ft 1
Machine	Hornblower, Woolf	p 103
Machine	Watt	p 105*bis*, ft 1
Machine	Héron de Villefosse	p 105*bis*, ft 1
Machine	Trevithick, Vivian	p 106
Machine	Niepce	pp 110, ft 1
Machine	*Pyréolophore*[b]	p 110, ft 1
A Machine called	*Wheal Abraham,* Cornwall	p 116, ft 1
Machine	Watt	p 116
Machine	In Chaillot	p 117

Experiment/Law/ machine/Memoir/device	Scholar/Engineer (in Sadi Carnot's order)	*Recherche d'une formule propre à représenter la puissance motrice de la vapeur d'eau* (Carnot S–EP[c])
Table	Dalton	folio 3
Traité	Biot	folio 3
Law	Gay–Lussac	folio 4
Law	Mariotte	folio 4
Experience	Clément	folio 6
Table	Dalton	folio 7
Law	Clément	folio 8
Table	Dalton	folio 8

(continued)

Table 11.9 (continued)

Experiment/Law/ machine/Memoir/device	Scholar/Engineer (in Sadi Carnot's order)	*Recherche d'une formule propre à représenter la puissance motrice de la vapeur d'eau* (Carnot S–EP[c])
Law	Clément	folio 8
Experiments	Clément	folio 8, ft *
Table	Dalton	folio 8
Table	Dalton	folio 10
Table /Traite	Biot	folio 17
Table	Dalton	folio 20

Experiment/Law/ machine/Memoir/device	Scholar/Engineer (in Sadi Carnot's order)	*Notes sur les mathématiques, la physique et autres sujets* (Carnot 1878a[d])
Invention	Martin	folio 2v (*Ia*)
Experiment	Rumford	folio 3r (*Ib*)
Experiment	Gay–Lussac and Welter	folio 3v (*Ib*)
Experiment	Gay–Lussac	folio 3v (*Ib*)
Memoir	Navier	folio 6r[e]
Experiment	Gay–Lussac and Welter	folio 6r
Value quoted	*Mécanique céleste*	folio 6r
Memoir	Poisson	folio 6r
Experiment	Ørsted	folio 6r
Machine	Perkins	folio 6r
Value	Laplace	folio 6r
Experiment	Berthollet	folio 8r
Experiment	Rumford	folio 10r
Experiment	Rumford	folio 12r
Thermometer	Bréguet	folio 12v
Experiment	Dalton	folio 12v
Experiment	Gay–Lussac	folio 13r
Experiment	Davy	folio 14r
Experiment	Rumford	folio 14r
Traité	Scheele	folio 15v
Notes by	Kirwan	folio 15v
Introduction by	Bergmann	folio 15v
Citing	Rumford	folio 15v
Dictionary	Macquer	folio 15v
Memoir	Rumford	folio 15v
Work by Rumford on	Thomson	folio 15v
Works by	Landriani	folio 15v
Thermometer	Bréguet	folio 16r
Law	Mariotte	folio 19r

(continued)

Table 11.9 (continued)

Sources cited by Sadi Carnot in his *Works*

Réflexions sur la puissance motrice du feu (Carnot 1978)[f]	*Recherche d'une formule propre à représenter la puissance motrice de la vapeur d'eau* (Carnot S–EP)	*Notes sur les mathématiques, la physique et autres sujets* (Carnot 1878b)
Traits élémentaire de physique ou de chimie [e.g.: *Traité élémentaire de chimie* by Lavoisier] (*Ivi,* p 15, ft 1)	*Traité de physique [expérimentale et mathématique]* by Biot, p [530–]531 (*Ivi,* folio 3)	*Annales de chimie et de physique* 1821, p 357 (*Ivi,* folio 6r)
Annales de physique et de chimie (*Ivi,* p 30, ft 1)	*Conservatoire des Arts et Métiers* (*Ivi,* folio 6)	*Mécanique céleste,* t 12, p 97 (*Ibidem*)
Annales de physique et de chimie 1818, t 7, p 122 (*Ivi,* p 51, ft 1)	*Traité de physique [expérimentale et mathématique]* , p [530–]531 (*Ivi,* folio 17)	*Annales [de chimie et de physique]*[g] 1823a, b, p 344 (*Ibidem*)
Annales de chimie [et de physique 1813] t 85, p 72, p 224 (*Ivi,* p 55, ft 1)	Dalton's table (*Ivi,* folio 20)	*Annales [de chimie et de physique]* 1823, p 192 (*Ibidem*)
[Traité de] Mécanique céleste(*Ivi,* p 59, ft 1)		On Perkins [*Annales de chimie et de physique* 1821, t 16, pp 321–327] (*Ibidem*)
Annales de physique et de chimie 1822, p 267 (*Ivi,* p 59, ft 1)		"Small advertisements 17 March. Manufacture of ice, rue Michel–le Comte, 27 [. . .]" (*Ivi,* folio 15v)
Annales de chimie et de physique 1818 (*Ivi,* p 65, ft 1)		*Journal du commerce,* 16 and 17 March[h] (*Ibidem*)
Traité de physique [expérimentale et mathématique] by Biot, vol 1, p 272, p 531 (*Ivi,* p 68, ft 1)		*American colonization*[i] (*Ibidem*)
Annales de chimie et de physique 1818, p 294 (*Ivi,* p 86, ft 1)		On Scheele and Kirwan's notes [*Supplement au Traité chimique de l'air et du feu de M. Scheele,* Trans. by le Baron de Dietrich 1785, Paris], p 149.[j] (*Ibidem*)

(continued)

Table 11.9 (continued)

Réflexions sur la puissance motrice du feu (Carnot 1978)[f]	*Recherche d'une formule propre à représenter la puissance motrice de la vapeur d'eau* (Carnot S–EP)	*Notes sur les mathématiques, la physique et autres sujets* (Carnot 1878b)
Architecture hydraulique by Prony, p 180, p 195 (*Ivi,* p 87)		On Scheele and Bergmann's introduction [*Traité chimique de l'air et du feu de M. Scheele,* Intr. by Bergmann, Trans. by Baron de Dietrich 1781, Paris, p 169n]. (*Ibidem*)
Annales de physique et de chimie [1821] t 16, p 105, t 24 [1823], p 323 (*Ivi,* p 87, ft 1)		*Biblioth.*[èque] *britannique* (*Ibidem*)
Annales de chimie et de physique 1824, p 80 (*Ivi,* p 88, ft 1)		On "Feu" in Macquer's Dictionnaire [1788, vol 1, pp 499–500][k] (*Ibidem*)
Annales de physique et de chimie 1823, p 192 (*Ivi,* p 93, ft 1)		*Journal de physique* [*de chimie et d'histoire naturelle et des arts*] 1798, XLVII, p 24, p 228, p 253 (*Ibidem*)
Académie des sciences (*Ivi,* p 98, ft 1)		*Bib.*[liothèque] *britann.*[ique], I,V,VIII, XIII (*Ibidem*)
Annales de physique et de chimie 1823, p 429 (*Ivi,* p 99, ft 1)		*Annales de chimie* [1802], XLI, p 177 (*Ibidem*)
Encyclopédie britannique (*Ivi,* p 105*bis*, ft 1)		On [Marsiglio] Landriani's[l] article listed[m]: *Observations sur la physique, sur l'histoire naturelle et sur les arts* XXIX 1786, p 410 (*Ibidem*)

[a]Here Sadi Carnot ascribes the authorship of the paper to Clément without citing Desormes. (Cfr.: Carnot 1986, p 153, En 97)
[b]On invention see Redondi (1980), pp 34–39
[c]See also English edition by Fox (Carnot 1986, pp 166–180)
[d]See also English edition by Fox (Carnot 1986, pp 181–212)
[e]We adopted Raveau (1919) and Fox's (Fox 1986, p 187, p 208, ft 13) suggestions: folio 6 (r and v) should be considered before folio 5rv
[f]The names are reported as Sadi Carnot cited them. We have added some references in "[...]". Moreover, some "*Annales de physique et de chimie*" may be evidently *Annales de chimie et de physique*
[g]In this paper, Poisson (as previously discussed Clément and Desorrmes and Sadi Carnot) proposed his law for adiabatic changes
[h]Cfr.: Carnot 1986, p 210, ft 42
[i]*Ibidem*
[j]Page 149 is a part of the pages that include *notes* added by Irish chemist Richard Kirwan (1733–1812) to Scheele's previously cited *Traité*
[k]See also: *Dictionnaire de chimie* (1786) vol 1, pp 498–507; *Dictionnaire de chimie* (edition of 1788) vol 1, pp 481–500. The articles on "Feu" are very different from one another. (Cfr.: Carnot 1986, p 211, ft 45)

(continued)

Table 11.9 (continued)

[l]Marsiglio or Marsilio Landriani (1751–1815) was an Italian chemist. He published *Ricerche fisiche intorno alla salubrit dell'aria* (1775). In 1776, he was appointed *Regius Professor of Physics* at the Ginnasio of Brera (in Milan). Historians of chemistry consider *Ricerche fisiche intorno alla salubrit dell'aria* as one of the earliest examples of chemical–pneumatic–analysis of atmospheric air undertaken in Italy. (He also described the *eudiometer* to measure the quality of air). Nevertheless, his research is also considered – *a poteriori* – discussible. Generally speaking, he was mainly concerned with the nature of different gases, atmosphere, respiration, electric studies, important correspondences with Volta and heat. Particularly, studies on latent heat (Landriani 1785, pp 197–207) could have been an object of interest for Sadi Carnot. At the time, several contributions to heat and latent heat were circulating in France by scholars who had different scientific backgrounds and aims (e.g., the different approaches to heat between analytical theories and purely physical theories)

[m]Sadi Carnot quotes Landriani referring to a long index printed at the end of *Observations sur la physique, sur l'histoire naturelle et sur les arts* (Landriani 1786, pp 401–473). In this index, Landriani is quoted on pages 410, 411, 413, 436 (See also *Ivi*, p 40). Sadi Carnot certainly referred to a study entitled *Suite de la dissertation de M Landriani, sur la chaleur latente* (Landriani 1785, pp 197–207; see also p 88) which is quoted (twice) in in the above cited index (Landriani 1786, p 410). It is also possbile that Sadi Carnot's reference might be also addressed to another index (cfr. Carnot 1986, p 211, ft 48) where Landriani is quoted: *Table générale des articles contenus dans les vingt–six derniers volumes du Journal de physique, depuis 1787 jusqu'en 1802, pour faire suite à celle qui est imprimée à la fin du second volume de l'année 1786 par L Cotte*. In this *Table générale*, Landriani is quoted three times concering 3 letters from 1790 and 1791 (Landriani 1806, p 24, p 51, p 52). This *Table générale* was published in *Journal de physique de chimie, d'histoire naturelle et des arts* where several issues from 1806 were also published (Landriani 1806, pp 1–480). The *Table générale* appears at the end of the last issue of the above cited *Journal* in December 1806 (Landriani 1806, pp 1–106. Let us note that the page numbers ended with page 480 and restarted with page 2. On the frontispiece of *Table générale* the number "1" is not cited)

As we have cited in various parts of this research, although his scientific relationship with Lazare was strong, the lack of reference to his father's works is made evident by the previous tables. It is very interesting to note his interest in chemical studies. This should be considered if one thinks that thermodynamics, like chemistry, were far from the scientific paradigm produced by Newtonian mechanical approach to science. For that reason the scientific novelty of gas and heat should have appeared in other scientific environments.

In the end this consideration also confirms the fact that, since Lazare Carnot's references do not include Sadi's works, it it is very probable that the book was written by four hands, *père et fils*.

11.14 On the Equivalence of Work–Heat

According to the hypothesis of the *equivalent* of work–heat, the heat is not conserved during the transference between two thermostats at two different temperatures. As already stated, this goes against the hypothesis of the analogy with the hydraulic wheel discussed above, where, instead, the mass of water is conserved during the fall between the two differences. We also cannot follow the innovation of restoring the weight of the elements of the machines proposed by Lazare Carnot for which water is also considered a heavy fluid:

> Scholium XLIX. This scholium is directed at the development of the principle announced in corollary V; in fact, this proposition contains the principle which is part of the theory of Machines in motion because the majority of them create motion by working substances which can only exert dead force or pressure; like all animals, springs, weights, etc. which usually makes the Machine change state by imperceptible degrees.[65]

However, this is not possible in thermodynamics because then heat would be treated as a heavy fluid, which would take us back to the hypothesis of phlogiston, which had already been discredited in Lazare and Sadi Carnot's time. Therefore, the scientific foundations of the analogy with the hydraulic wheel proposed by Sadi Carnot should be improved and focused on the hypothesis of caloric. In a heat machine, the caloric, since it is a weightless fluid (in mechanics according to the seventeenth century simplified performance model of machines, the fluid–water is weightless) it merely produces work returning, without losses, to the initial state of equilibrium to the thermostat at a lower temperature. Clearly, the only variation is *t*. However, it is necessary to find a response to the problem regarding the interaction. At first glance, a response (Drago and Pisano's works) could have come from the analogy of gases with rigid bodies in elastic collision amongst themselves so that the gases were commonly defined "*fluides élastiques*" (see in the following Table 11.10).

According to Ernst Mach, Sadi Carnot, knowing that mechanical work by collision is transformed into heat, may have conceived that the opposite also occurs, that is, the *equivalent* hypothesis. Briefly, following his ideas on the role played by analogies in different fields of physics, e.g., as just mentioned,

> [...] the first great step in Carnot's discovery was the consideration of an analogy between water which, by falling, performed [mechanical] work; and heat which, by sinking in temperature performed [thermal] work.[66]

Generally speaking, following Mach's reasonings on the analogy between the transformations of a gas in thermodynamics and the mechanism of collision in mechanics, one might suppose that, e.g., the transformation of heat into work and vice–versa could became an object of discussion for Sadi Carnot by means of the phenomenon of collisions – which on the other hand – was already part of his father's mechanics. In particular, the formula $\Delta m_i U_i^2$ concerned lost kinetic energy (as Leibniz had already indicated) as the energy lost in the environment. Thus, he could have reasoned on the transference of these concepts to thermodynamics, e.g., on the compression of a gas in a given closed space. In this sense, generally speaking, he could have also reasoned on the opposite process of cooling by the adiabatic expansion of gas. However, some reflections are necessary. This analogy suggests a kind of interaction, but does not theoretically explain the *equivalent*.

[65]Carnot 1786, p 81, line 1. (Author's *italics*).

[66]Mach [1896], 1986, p 306, line 12.

Table 11.10 On the common way of conceiving elastic fluid

Lazare Carnot (1803a)	Sadi Carnot (1824)
The *Elasticity* is the quality that certain compressible bodies have of returning, since compression ceases, to their initial state. A perfectly elastic body is that in which compression and restitution operate by the same degrees in opposite directions. These bodies are called elastic bodies or spring bodies. Ivory, tempered steel and glass are solid elastic bodies; air and gas are elastic fluids[a]	Experiment has taught us nothing on this subject. It has only shown us that this caloric is developed in greater or less quantity by the compression of the elastic fluids[b] The elastic fluids, gases or vapors, are the means really adapted to the development of the motive power of heat. They combine all the conditions necessary to fulfill this office. They are easy to compress; they can be almost infinitely expanded; variations of volume occasion in them great changes of temperature; and, lastly, they are very mobile, easy to heat and cool, easy to transport from one place to another, which enables them to produce rapidly the desired effects. We can easily conceive a multitude of machines fitted to develop the motive power of heat through the use of elastic fluids; but in whatever way we look at it, we should not lose sight of the following principles: (1) The temperature of the fluid should be made as high as possible, in order to obtain a great fall of caloric, and consequently a large production of motive power. (2) For the same reason the cooling should be carried as far as possible. (3) It should be so arranged that the passage of the elastic fluid from the highest to the lowest temperature should be due to increase of volume; that is, it should be so arranged that the cooling of the gas should occur spontaneously as the effect of rarefaction. The limits of the temperature to which it is possible[c]

[a]Carnot (1803a), p 9, line 14. (Author's *italic*)
[b]Carnot (1978), p 32, line 9
[c]Carnot (1978), pp 93–94, line 14

In fact, the *equivalent* hypothesis process explains the situation only if another function can be introduced, $U = U(t)$ (that is, a *reserve* of heat or something of the sort). This is also applicable when $Q = 0$ can produce work $W > 0$. In fact, following Mach's intuition, the heat–work interaction is supposable as a collision between the caloric fluid of the upper thermostat and (the molecules of) the gas in the cylinder. In such a collision, (intended globally) the vapor (absorbing heat) becomes deformed. Consequently, it can be hypothesized that caloric produces work since, being conceived as a weightless fluid, passing into vapor, it produces a variation in volume and therefore thermal work, making it expand. Paraphrasing Sadi Carnot, in the initial premise of his text: it is "[...] the expansive force of vapor [...]"[67] (Carnot 1978, p 5, line 15) that produces work.

[67]The term *expansive force of vapor* or *expansive force of heat* or *expansive force of caloric* was used at the time (e. g., Payen 1967, pp 231–232; Betancourt 1792, p iiij; p 17, p 25, p 26, p 27, 37; Petrini 1808, p 69; Brocchi 1808, pp 229–230; Bouvier 1816–1817, p 132, p 134; Poli 1817, pp 198–200; Carpi 1836, pp 292–294; Reech 1853, ft *, p 363; Magrini 1861, I, p 259).

However, this explanation cannot take Watt's invention into account (which Carnot is well aware of); which achieves further work, detaching the cylinder from the thermostat and allowing the fluid to expand (that is, allowing for an adiabatic transformation). One question remains: *why in the isolated cylinder, does heat already captured by vapor still produce work with the reduction of the temperature?* The elasticity of gas could produce further work only if, previously, the isothermal fall had compressed the gas while the contrary is in fact true. Here, the caloric hypothesis has an *aporia* in the adiabatic transformation: *the physical mechanism of the interaction is mysterious.* Instead (maybe) Sadi Carnot could have followed (Carnot 1978, p 32) another simpler path: passing from physical quantity Q to function $Q(t)$. He may have thought of and presumably introduced another possible concept, Q/t, referring to *calorique*.

> *When a gas passes without change of temperature from one definite volume and pressure to another volume and another pressure equally definite, the quantity of caloric absorbed or relinquished is always the same, whatever may be the nature of the gas chosen as the subject of the experiment.*[68]

Of course, this is a possible explanation but we do not have enough historical facts and clear book–passages to support it. Thus, it appears as an epistemological and tentative interpretation.

It should be noted that in caloric theory, the reversibility, intended as an advancement "[. . .] for insensible degrees [. . .]"[69] does not make enough physical sense since caloric Q is conserved and is therefore a state function. Moreover, (1) there is no qualitative difference between infinitesimal and finite processes when Q is involved; (2) when a value of caloric is fixed, the magnitude W is the only non–state variable within a system that can have different values.[70] Let us note that the latter reflection has such important results that it cannot be explained by using the imprecision of an imperfect world in comparison with mathematical precision. Therefore, either the fact that W (physically and mathematically) depends on other variables in addition to state variables (including Q), or the fact that W is not a state function, contradicts the caloric hypothesis: so Q is also not a state function. Overall, since Sadi Carnot brings all of the variables characterizing the behaviour of the system into play and wants to base his theory on reversibility, (1) Q is not a state function and is therefore true only for the *equivalent* theory. For this reason, Sadi Carnot could have thought of the *equivalent* as a hypothesis of his theory. Currently, we know that in order to fully develop this hypothesis (as the physicists who reformulated thermodynamics did) (2) the relation J between the measurements of heat and work must be specified and both the (3) internal energy function $U(t)$ and the definition of the first principle must be invented. However, these three theoretical steps were too advanced for Sadi Carnot and more importantly, were outside of

---

[68] Carnot 1978, p 41, line 20. (Author's *italics*).

[69] Carnot 1786, p 92, line 8.

[70] Carnot 1978, p 23.

his scientific context, which was also technological. Therefore, he only used the calculation of J (later); but in this instance he was free to not believe that heat is conserved in the fall. After having presented the analogy of the wheel, Sadi Carnot announced the second demonstration of his famous theorem:

> We shall give here a second demonstration of the fundamental proposition enunciated on page 22, and present this proposition under a more general form than the one already given.[71]

and just after, he noted that in the compressions or dilations the constant temperature of gas can be maintained in order to respectively remove or give heat or gas as the transformation is executed.

> Similarly, if the gas is rarified, we can avoid lowering temperature [of the gas] by supplying it with a certain quantity of caloric.[72]

This Q at t–constant is defined as caloric at a constant temperature produced by a non–null variation in volume.

> [The change of temperature of a gas caused by a variation of volume can be seen as one of the most important phenomenon in whole physics . . .]. It seems [this phenomenon] in some respects singular anomalies.[73]

Therefore, following the analogy of the water wheel, he could have thought that work is produced by a physical interaction. That is to say *à la* two Carnots:

In mechanics, there is an interaction of water that, falling on vanes, produces work as an effect of its weight	In thermodynamics, there is an interaction of expansions and rapid compressions with heat exchanges which, maintaining a constant temperature, give work, that is, they produce a $\Delta h'$ in the piston which corresponds to the Δh of the vanes

In the case of the isochoric transformation, Sadi Carnot thought that passing to infinitesimal transformations brings reversibility. He applies this to heat Q as well, which in the *equivalent* hypothesis is not a state function. Thus, Sadi Carnot knew how heat best produces work: heat at a constant temperature (e.g., entropy) is that which gives the maximum ΔV. In other words, it is heat that produces work under conditions of reversibility. Overall the problem of the interaction could clarify all dynamical processes and his relationship with his father's culture: when t is constant or quasi–constant, a maximum W corresponds to every Q relative to that temperature and viceversa.

[71]Carnot 1978, p 29, line 6.

[72]Carnot 1978, p 31, line 5.

[73]Carnot 1978, p 31, ft 1, line 7.

11.15 On the Cycle and Reversibility of a Machine

Lazare Carnot's theory on machines clearly indicates that they can produce both the *maximum* efficiency and minor efficiencies, until reaching a null efficiency. It depends on a condition that was well established by Lazare Carnot: invertibility (generally speaking reversibility in thermodynamics). In order to obtain the *maximum*, the transformation should occur while avoiding collisions and abrupt changes in direction: "[...] insensible degrees [degrés insensibles] [...]" (Carnot 1786, p 92). In Lazare Carnot's words:

> *Corollary V. Particular law concerning the Machines whose motion changes by imperceptible degrees. XLI. In a Machine whose motion changes by imperceptible degrees,* the moment of activity *in a time given by soliciting forces, is equal to* the moment of activity, *exerted at the same time by resistant forces.*[74]

Sadi Carnot does not employ the concept of reversibility in his *posthumous manuscript*. We maintain that if he had completed his manuscript, he would have only obtained value W as a consequence of the variables characterizing a heat machine. However, with regard to natural phenomena, he states that these machines, even having the same values of variables, produce very different work. These are not mere approximations; they represent a preventative problem for the use of mathematics to interpret a phenomenon. Thus, Carnot is addressing the problem of the relationship between mathematics and physics in the theory. In the discursive part of *Réflexions sur la puissance motrice du feu* (Carnot 1978, pp 1–73), Sadi Carnot introduces his original concept of the cycle, both as a method of reasoning and a method of calculation alternative to infinitesimal analysis.[75] He dedicates several pages (Carnot 1978, pp 14–22) to presenting his reasonings regarding the first demonstration for a three–phase[76] cycle and goes on to discuss the cycle of inverse operations (Carnot 1978, p 19). He then presents the first expression of his theorem (Carnot 1978, pp 21–22). In the following section, we discuss his main reasoning:

> If we wish to produce motive power by carrying a certain quantity of heat from the body *A* to the body *B* we shall proceed as follows: (1) To borrow caloric from the body *A* to make steam with it-that is, to make this body fulfill the function of a furnace, or rather of the metal composing the boiler in ordinary engines-we here assume that the steam is produced at the same temperature as the body *A*. (2) The steam having been received in a space capable of expansion, such as a cylinder furnished with a piston, to increase the volume of this space, and consequently also that of the steam. Thus rarefied, the temperature

[74]Carnot 1786, pp 75–76, line 29. (Author's *italics*).

[75]After the second half of nineteenth century, the infinetisaml calculation was also widely developped for purely geometrical studies. For example, Jean–Gaston Darboux (1842–1917)'s contribution was very important for advanced studies between mathematics and geometry (Darboux 1887–1896). He was also a biographer of Poincaré and he edited the *Selected Works* of Fourier.

[76]He will then also discuss a cycle completed in four phases (Carnot 1978, pp 29–38).

will fall spontaneously, as occurs with all elastic fluids; admit that the rarefaction may be continued to the point where the temperature becomes precisely that of the body *B*. (3) To condense the steam by putting it in contact with the body *B*, and at the same time exerting on it a constant pressure until it is entirely liquefied. The body *B* fills here the place of the injection water in ordinary engines, with this difference, that it condenses the vapor without mingling with it, and with-out changing its own temperature.[Footnote]* [...] By our first operations there would have been at the same time production of motive power and transfer of caloric from the body *A* to the body *B*. By the inverse operations there is at the same time expenditure of motive power and return of caloric from the body *B* to the body *A*. But if we have acted in each case on the same quantity of vapor, if there is produced no loss either of motive power or caloric, the quantity of motive power produced in the first place will be equal to. that which would have been expended in the second, and the quantity of caloric passed in the first case from the body *A* to the body *B* would be equal to the quantity which passes back again in the second from the body *B* to the body A; so that an indefinite number of alternative operations of this sort could be carried on without in the end having either produced motive power or transferred caloric from one body to the other. Now if there existed any means of using heat preferable to those which we have employed, that is, if it were possible by any method whatever to make the caloric produce a quantity of motive power greater than we have made it produce by our first series of operations, it would suffice to divert a portion of this power in order by the method just indicated to make the caloric of the body *B* return to the body *A* from the refrigerator to the furnace, to restore the initial conditions, and thus to be ready to commence again an operation precisely similar to the former, and so on: this would be not only perpetual motion, but an unlimited creation of motive power without consumption either of caloric or of any other agent whatever. Such a creation is entirely contrary to ideas now accepted, to the laws of mechanics and of sound physics. It is inadmissible. We should then conclude that the *maximum of motive power resulting* from the *employment of steam is also the maximum of motive power realizable by any means whatever.* We will soon give a second more vigorous demonstration of this theory. This should be considered only as an approximation.[77]

Sadi Carnot's miscalcution (see Chapter 9) when he closed the cycle by means of a finite isochors transformation is noteworthy. The inaccuracy is related to the fact that the cycle reduced itself to be irreversible. We explained (Chapters 7 and 9) this plausible inexactness by the application of Lazare Carnot's synthetic method (Carnot 1978, pp 18–19) where the isochors were conceived as an auxiliary variable added whereas in Lazare' s mechanics, the auxiliary variable is a geometric motion. If we stress the common role played by the synthetic method in the theories of the two Carnots, we can hypothesize that since Lazare's Carnot considered the auxiliary variables indifferently as infinitesimal and finite, Sadi, following his father, could be persuaded to do the same for his thermodynamic cycle. All of the reasonings are made up of operations and transformations (now including the adiabatic) aimed at resolving the crucial problem previously presented (Chapters 8 and 9): *how to establish the maximum work, that is, how to optimally use Q to obtain maximum W*. (Carnot 1978, pp 21–22, p 38). It is in this part of *Réflexions sur la puissance motrice du feu* that Sadi Carnot introduces his crucial concept for thermodynamics, the reversibility for a thermodynamic process which he himself affirms to be fundamental for his theory:

---

[77]Carnot 1978, pp 17–22, line 12.

> The necessary condition of the maximum is, then, *that in the bodies employed to realize the motive power of heat there should not occur any change of temperature which may not be due to a change of volume.* Reciprocally, every time that this condition is fulfilled the maximum will be attained. This principle should never be lost sight of in the construction of heat-engines; it is its fundamental basis. If it cannot be strictly observed, it should at least be departed from as little as possible.[78]

It is obvious that, having $W = p\Delta V$, only then is work W, produced by a ΔV, *maximum.* Otherwise, the utilization of a Δt can fail; that is, $W = 0$ for $\Delta V = 0$. Sadi Carnot was interested in instances in which ΔV is different from *0* and if a possible *maximum* for every Δt possible. Let us examine the correlation between the two Carnots for this concept.

Lazare Carnot likens the concept of reversibility not to a transformation, but to a motion (Carnot 1786, pp 28–30). In Lazare Carnot's mechanics, reversibility is linked to "geometric motions [mouvements géométriques]" and is defined as the invertibility in a geometrical space: a motion is geometric if it is invertible. Let us note that according to Sadi Carnot's definition, a transformation is irreversible if the transformation of heat occurs with $\Delta V = 0$, that is to say, that it is the same when the thermal work produced is null. Therefore, in both cases, the criterion of reversibility is geometrically based. However, in this case, it is related to volume; it is still on a length but this time is thought of globally, in its three dimensions. It also should be noted that shortly after this theorem, Carnot states that:

> According to established principles at the present time, we can compare with sufficient accuracy the motive power of heat to that of a waterfall.[79]

This concerns the very intuitive and striking analogy of the hydraulic wheel which we previously examined. Here, we will attempt to examine its epistemological role. One may wonder *why does Sadi Carnot propose this analogy just after the first proposition of his theorem?* In fact, the position of this analogy at this point in the text, that is, at a rather advanced point in his discussion, is decisively the object of attention. This analogy cannot act as a conclusion or verification for the previous demonstration of the first proposition of his theorem on cycles. Therefore, we suggest that Sadi Carnot introduces the analogy with the hydraulic wheel as a consequence of "[...] established principles [the first announcement of his theorem]" (*Ibidem*) suggesting *a condition of reversibility* by means of a new sequence of very small thermal changes which today we would call infinitesimal. Therefore, here we could also hypothesize that even though W, and therefore efficiency η, are not state variables, Sadi Carnot succeeds in producing a basis of reasoning in infinitesimal terms. It should be noted, however, that his reasoning on the infinitesimal degrees are far from pure metaphysical entities belonging to the infinitesimal analysis of the time. In fact, in Sadi Carnot's theory they are not part of a mathematics–physics theory; they are justified by cautious physical reasoning

---

[78] Carnot 1978, pp 23–24, line 12. (Author's *italics*).

[79] Carnot 1978, p 28, line 1.

on phenomena such as the application of gas theory since they were not yet well interpreted by previous scientific theories, e.g., mechanics. In Sadi Carnot's words:

> We may perhaps wonder here that the body B being at the same temperature as the steam is able to condense it. Doubtless this is not strictly possible, but the slightest difference of temperature will determine the condensation, which suffices to establish the justice of our reasoning. It is thus that, in the differential calculus, it is sufficient that we can conceive the neglected quantities indefinitely reducible in proportion to the quantities retained in the equations, to make certain of the exact result. The body B condenses the steam without changing its own temperature this results from our supposition. We have admitted that this body may be maintained at a constant temperature. We take away the caloric as the steam furnishes it. This is the condition in which the metal of the condenser is found when the liquefaction of the steam is accomplished by applying cold water externally, as was formerly done in several engines. Similarly, the water of a reservoir can be maintained at a constant level if the liquid flows out at one side as it flows in at the other. One could even conceive the bodies A and B maintaining the same temperature, although they might lose or gain certain quantities of heat. If, for example, the body A were a mass of steam ready to become liquid, and the body B a mass of ice ready to melt, these bodies might, as we know, furnish or receive caloric without thermometric change.[80]

11.16 Final Remarks *a mò* of Conclusion

The exposition of *Réflexions sur la puissance motrice du feu* is clear only through the obtainment of the gas laws (based on the result acquired on the thermodynamic cycle, that is, reasoning in an unordinary way). A hypothesis may be advanced: Sadi Carnot returned from his stay in Magdeburg and continued to develop his idea of resolving the problem of the efficiency, using some foundations of analysis according to the method already learnt at *École polytechnique*. Later, he understood that this method did not work for his purposes, so he decided to follow another method, possibly discussed with his father in Magdeburg: the synthetic method. Of course we do not suggest that Sadi Carnot effectively reasoned according to the sequence previously indicated. However, no one can claim this with certainty.We do maintain that the aforementioned hypothesises seems plausible since Sadi Carnot, from a historical point of view, was in Magdeburg before the publication of his book. Thus, we logically think that he discussed his first important manuscript with his father. However, in any case, Sadi Carnot showed limits as a son of his time:

> He was hasty (ca. 2–3 years to publish) full of corrections,[81] with intentional archaisms, not very attentive to the prevailing academic styles
> He was not very familiar with his father's mechanical theory to which he wished to relate.
> He commited an error (finite isochors in a finite cycle).

In effect, based on what he strictly wrote, Sadi Carnot followed neither the *principle of virtual work* nor (e.g.) the analogy on caloric and entropy, e.g.,

---

[80]Carnot 1978, p 18–19, ft 1, line 1.

[81]Carnot 1978, p 27.

mechanical potential energies and entropy presented by Brønsted in the middle of the past century. Johannes Nicolaus Brønsted (1879–1947) was an important Danish physical chemist. He elaborated (ca. 1937–1947) a simple and original formulation of macroscopic thermodynamics. His reasonings also convinced him to consider Sadi Carnot's theory, offering very interesting epistemological points of reflection. Briefly, Brønsted mainly (Brønsted 1955) based his reasoning on the observation diffused during the eighteenth century: trying to reduce all physical works to mgh, force–weight of a body with respect to its height and thereby produce new energetic principles. In other words, all energetic processes are valid thanks to an energetic quantity K in motion between two states at different p–potentials, the so–called *Fundamental Principles of Energetics* (Brønsted 1940, 1941). A work δA is formulated for each physical, fundamental energetic process: $\delta A = (p_1 - p_2)\,\delta K$. In this way, energetic quantities and potential states appeared as conjugated magnitudes (first extensives and then intensives) and each of their couples of values should correspond to a particular physical work. E.g. Table 11.11:

Table 11.11 Examples of physical works proposed by Brønsted

Physical work	Potential	Physical magnitude
$(T_1 - T_2)\delta S$	Temperature T	Entropy
$(\varphi_1 - \varphi_2)\delta m$	Gravitational potential φ	Mass
$(V_1 - V_2)\delta q$	Electric potential V	Electric charge
$(F_1 - F_2)\delta d$	Force F	Distance
$(P_1 - P_2)\delta V$	Pressure P	Volume
$(\mu_1 - \mu_2)\delta\eta$	Chemical potential μ	Quantity of substance

When it comes to Sadi Carnot's thermodynamics, he avoided two fundamental Carnot aspects: the impossibility of perpetual motion and the role played by cycles. Moreover,

(a) he used a finite Δt as difference between two potentials ,
(b) deleting the synthetic method on $\Delta t \to 0$,
(c) a priori S–entropy in the theory by direct measures is introduced to obtain the famous – sought after – analogy (at the time) with *Q*–caloric at *T*–constant,
(d) thus Sadi Carnot's theorem become needless,
(e) and at the same time, the reasoning on cycles is lost.

In this sense, Brønsted's formulation is interesting and in part generally acceptable, but it is far from being considered an advancement of Sadi Carnot's theoretical approach.

Based on previous discussion and hypotheses in this book, in the following table we finally provide a summary that lists the main, common concepts adopted by the two Carnots. Table 11.13 shows the main historical studies thermodynamics conducted in recent years which more or less concern Lazare and Sadi Carnot.

Table 11.12 Summary of the main common concepts in Lazare and Sadi Carnot's scientific theories

	Lazare Carnot (1753–1823)	Sadi Carnot (1796–1832)
Main concepts	Mechanics	Thermodynamics
Space, Time	Limited and Relational (volume); *idem.*	Limited and Relational (volume); *idem*
Bodies	Global, machines	Global, machines
Inertia	Impossibility of perpetual motion	Impossibility of perpetual motion
Basic–concept	Transformations	Transformations
Synthetic method	Yes	Yes
Interaction	Work	Work [moment–of–activity]
Setting of theory	The laws in *Collision theory*	Integration of *dq/t*
Techniques	Geometric motion; vector calculus	Cycle
Solutions	Invariants; Geometric motions for mechanical machines	Maximum efficiency of heat machines
Analysis mathematics,	No	No
Lagrangean mathematics	No	No
Laplace intermolecular forces	No	No

See also Pisano (2010)

Table 11.13 A concise view of historical studies on thermodynamics, Lazare Carnot and Sadi Carnot

Years	Main scholars (alphabetic order)	Main Topics (alphabetic order)
1920s–1960s	Barnett	Sadi's biography and manuscript
	Brønsted	Classical thermodynamics
	Buchdahl	Efficiency of a machine
	Gabbey and Herivel	Energy and principles
	Gillispie	Lazare and Sadi
	Kelly	*Réflexion sur la puissance* …
	Kerker	Sadi and Cagnard
	Koenig	Sadi Carnot's sources
	Mendoza	The age of objectivity
	Payen	The birth of thermodynamics
	Picard	The second law
	Reinhard	Thermodynamics efficiency
	Renaud	Unpublished Sadi's manuscript
	Rosenfeld	
1970s–1980s	Birembaut	Analogy with wheel–hydraulic
	Callen	Caloric and entropy in Sadi
	Challey	Caloric theory of gases
	Costabel	Closing Sadi's cycle
	Fox	Concept of state in the theory
	Gillispie and Youschkevitch	Conservation laws

(continued)

Table 11.13 (continued)

Years	Main scholars (alphabetic order)	Main Topics (alphabetic order)
	Gillispie	Cycles theory in Sadi
	Hoyer	Equilibrium in the theory
	Klein	Impossibility of perpetual motion
	Lervig	Laplace's intermolecular forces
	Payen	Lazare's mechanical machines
	Redondi	Lazare's infinitesimal analyses
	Scott	Lazare's works
	Taton	Lazare–Sadi filiaton
	Truesdell	No Lagrangean approach
		No Newtonian approach
		Operative mathematics
		Sadi and Clapeyron
		Sadi and Clement–Desormes
		Sadi and *École polytechnique*
		Sadi and his time
		Sadi and new technology
		Sadi as engineer
		Sadi Carnot's scientific ideas and French technology
		Sadi Carnot's biography
		Sadi's book edition
		The environment of Sadi Carnot
		Thermodynamics and symmetry
		Unpublished Sadi's manuscript
		Vapor machines
		Work as basic concept
1980s–1990s	Drago and Vitiello	Analogy with wheel–hydraulic
	Drago	Essay reviews
	Gillispie	Heat and thermodynamics
	Gillispie and Youschkevitch	Impossibility of perpetual motion
	Hornix	Lazare's mathematics
	Lervig	Lazare's infinitesimal analyses
	Redondi	Lazare's mechanics
		Logical approach to history
		No Lagrangean approach
		No Newtonian approach
		Operative mathematics
		Science, old regime in France
		Work as basic concept
1990–2011s	Dhombres	A reversible and irreversible
	Drago and Pisano	Absorbed and produced energy
	Drago	Advancement of caloric theory
	Gillispie	Analogy wheel–hydraulic
	Koetsier	Caloric–heat–work *equivalent*
	Kostic	Cycle and hypothesis on its birth

(continued)

Table 11.13 (continued)

Years	Main scholars (alphabetic order)	Main Topics (alphabetic order)
	Lemons and Penner	Double negative sentences and non–classical logic
	Pisano	In–out as basic machine system
		Lazare and Ampere
		Lazare Carnot, man and scientist
		Lazare's infinitesimal analyses
		Logics and mathematics in Sadi
		On principles of Sadi Carnot
		Potential and kinetic energy
		Principles of virtual work
		Problematical organization
		Relationship mathematics physics in Carnot's theory
		Sadi and heat reversible engine
		Sadi and the second law
		Sadi Carnot and Volta's battery
		Science and Policy in France
		State function in Sadi Carnot
		Synthetic method in two Carnots
		The equilibrium in the theory
		The mathematics in Lazare
		The works in the theory
2013	Gillispie and Pisano	*Lazare and Sadi Carnot. A scientific and filial relationship* Lazare and Sadi's machines
	Pisano and Drago	Lazare's mechanics
	Pisano	On cycle in the theory
	...	Sadi and Lazare's science
		Sadi's thermodynamics
		Science before and after Sadi
		The birth of the Sadi's book

In Table 11.13, the content of the third column is a résumé of the arguments studied by the scholars cited. (We precise that the Table 11.13 should be only read per column since its formatted–arrangement in this page). By considering the importance of the authors included, to link each author with his argument is not necessary. The reader should accept our apology if we, in our limitations, omitted some authors. The complete references are listed in the main bibliography at the end of this book.

We would like to conclude this extensive work with the thought that Lazare and Sadi coexist in one book as *pére et fils* and as scientists as well.

Appendix to Chapter 7: DNSs Listed in *Réflexions sur la puissance motrice du feu*

In square parentheses we added explanations regarding the statement A corresponding to statement $\neg\neg A$ so that it is easier to notice the difference in the meanings of the two sentences. Due to epistemological investigation on $\neg\neg A$, we decided to use the original French version (Carnot 1978). Moreover, in brackets we attribute to each sentence (excluding the first two which do not have scientific relevance) its theoretical role played within scientific theory: PR, GP, TP, MP, *, OP as presented in Chapter 7.

(1.) Il n'existe peut–être pas en Angleterre un établissement d'industrie dont l'existence ne soit fondée [. . .] [≠ all of them are founded]. (p 4, line 5).

(2.) La destruction de sa marine [. . .] lui serait peut–être moins funeste [≠ more favourable]. (p 4, line 11).

3. L'on a souvent agité la question [= we do not know] de savoir si la puissance motrice de la chaleur est limitée ou si elle est sans bornes [≠ we are able to say that it has bounds] [. . .]. (p 6, line 18) [PR]
4. La phénomène de la production du mouvement par la chaleur n'a pas été considéré sous un point de vue assez général. [= not dependent from the details]. [≠ it has been considered in dependent manner from the details] (p 7, line 11) [PR]
5. Pour envisager dans toute sa généralité le principe de la production du mouvement par la chaleur, il faut le concevoir in dépendamment [≠ separately] d'aucun mécanisme, d'aucun agent particulier [. . .]. (p 8, line 1). [PR]
6. Une semblable théorie [a general theory for heat machines] manque évidemment pour les machines à feu. On ne la possédera que lorsque [≠ we will have this theory] les lois de la physique seront assez étendues, assez généralisées, pour faire à connaître à l'avance tous les effets de la chaleur agissant d'une manière déterminée sur un corps quelconque. (p 8, line 15) [PR]
7. La vapeur n'est ici qu'un [≠ it is] moyen de transporter le calorique (p 10, line 2). [PT]

C.C. Gillispie and R. Pisano, *Lazare and Sadi Carnot: A Scientific and Filial Relationship*, History of Mechanism and Machine Science 19, DOI 10.1007/978-94-017-8011-7,

8. La production de la puissance motrice est donc due, dans les machines à vapeur, non à une consommation réelle [≠ to conservation] du calorique, mais à son transport d'un corps chaud à un corps froid [...] [that is a formal manner of its conservation] (p 10, line 20). [MP]
9. [Production of engine power and importance of thermostats for different temperatures – see 8]. [...] sans lui [the cold], la chaleur serait inutile [≠ through it is always useful]. (p 11, line 8). [MP]

10*. [About vapor] Où la placerait–on une fois qu'elle aurait pris naissance? Il ne faudrait pas croire que l'on put [indefinitely] [≠ you can blow off it], ainsi que cela se pratique dans certains machines, la rejeter dans l'atmosphère: l'atmosphère ne la recevrait pas. (p 11, line 11) [TP]

11*. L'existence de l'eau à l'état liquide, admise nécessairement ici, puisque sans elle les machines à vapeur ne pourraient pas s'alimenter, [...] [≠ through it, it can stoke]. (p 12, ft 1, line 1). [TP]

12. La chaleur ne peut évidemment que [≠ it is] être une cause de mouvement qu'en vertu des changements de volume ou de forme qu'elle fait subir aux corps. (p 14, line 6). [MP]
13. [...] La puissance motrice de la chaleur est–elle immuable [≠ constant] en quantité, ou varie–t–elle avec l'agent dont on fait usage pour la réaliser avec la substance intermédiaire, choisie comme sujet d'action de la chaleur? (p 14, line 18). [PR]
14. [The motive power of heat is unchangeable...]. Il est clair que cette question ne peut être faite que [≠ it can be seen] pour une quantité de calorique donnée, la différence des températures étant également donnée. (p 15, line 2). [TP]
15. [Difference of temperature and heat motive power]. La percussion, le frottement des corps ne sont–ils pas en effet des moyens d'élever leur température, de faire arriver spontanément à un degré plus haut que celui des corps environnans, et par conséquent de produire une rupture d'équilibré dans le calorique, là où existait auparavant cet équilibre? [≠ they are methods to...] (p 16, line 12). [TP]
16. Le corps B remplit ici le rôle de l'eau d'injection dans les machines ordinaires, avec cette différence qu'il [B] condenses la vapeur sans se mêler [≠ staying isolated] [to the vapor] avec elle et sans changer lui–même de température [≠ maintaining]. (p 18, line 9). [TP]
17. [The body B has the same temperature of vapor]. [...] Sans doute [≠ certainly] cela n'est pas rigoureusement possible; mais la plus petite différence de température déterminera la condensation, ce qui suffit pour établir la justesse de notre raisonnement. (p 18, ft 1, line 2). [TP]
18. Le corps B condense la vapeur sans changer [≠ conservancy] lui–même de température [...] (p 18, ft 1, line 8). [TP]
19. [The operations we have just described could have been reserved and carried out in the opposite direction]. Rien n'empêchait [≠ we can ...] de former de la vapeur avec le calorique du corps B [...] (p 19, line 2). [PT]

20. Si, par exemple, le corps A était une masse de vapeur prête à se liquéfier, et le corps B une masse de glace prête à se fondre, ces corps pourraient, comme on sait, fournir ou recevoir du calorique sans changer [≠ conservancy] de degré thermométrique. (p 19, ft 1, line 9). [PT]
21. [Production and consumption of motive power]. Mais si l'on a agi de part et d'autre sur la même quantité de vapeur, s'il ne s'est fait aucune perte [≠ there has been constancy] ni de puissance motrice ni de calorique, la quantité de puissance motrice produite dans le premier cas, sera égale à celle qui aura été dépensée dans le second, et la quantité de calorique passée, dans le premier cas, du corps A au corps B sera égale à la quantité qui repasse, dans le second, du corps B au corps A [...](p 20, line 2). [PT]
22*. [Return of caloric from B body to A body]. [...] De recommencer une opération entièrement semblable à la première [caloric from A body to B body] et ainsi de suite: ce serait là, non seulement le mouvement perpétuel, mais une création indéfinie de force motrice sans consommation [≠ maintaining constant] ni de calorique ni de quelque autre agent que ce soit. (p 21, line 1). [GP]
23*. Une semblable création est tout–à–fait contraire [≠ motive power creation is according to] aux idées reçues jusqu'à présent, aux lois de la mécanique et de la saine physique; [but this creation] elle est inadmissible. [≠ the conservancy is admissible]. (p 21, line 5). [GP]
24. On objectera peut–être ici que le mouvement perpétuel [= without end], démontré impossible [≠ every motion has an end has shown] par les seules actions mécaniques, ne l'est peut–être pas lorsqu'on emploie l'influence soit de la chaleur, soit de l'électricité; [...]. (p 21, ft 1, line 1). [PR]
25. [It is impossible to obtain perpetual motion through electric actions]. [...] Mais peut–on concevoir les phénomènes de la chaleur et de l'électricité comme dus à autre chose qu'à [≠ to] des mouvements quelconques de corps, [...]. (p 21, ft 1, line 3). [GP]
26. [Carnot reasons about the possibility of obtaining perpetual motion through electric phenomena – see 25]. [...] Et comme tels ne doivent–ils pas être soumis aux lois générales de la mécanique? [≠ they are subjected by]. (p 21, ft 1, line 5). [GP]
27. [Perpetual motion and electricity – see 24] Ne sait–on pas d'ailleurs *à posteriori* que toutes les tentatives faites pour produire le mouvement perpétuel par quelque moyen que ce soit ont été infructueuses? [= perpetual motion was not realized ≠ every motion has had an end] [≠ we know it] (p 21, ft 1, line 6). [GP]
28. Que l'on n'est jamais parvenu à produire un mouvement véritablement perpétuel [≠ we were able to produce motion that has a sure end], c'est–à–dire un mouvement qui se continuât toujours sans altération [≠ constant] dans les corps mis en œuvre pour le réaliser? [≠ we are] (p 21, ft 1, line 8). [GP]

29*. Si elle était possible [motive power creation], il serait inutile [≠ conservation makes useful] de chercher dans les courans d'eau et d'air, dans les combustible, cette puissance motrice [. . .]. (p 22, ft 1, line 11). [GP]

30*. [see 29*] [If there is motive power creation]. [. . .] Nous eu aurions à notre disposition une source intarissables [≠ conservation leads exhaustion] où nous pourrions puiser à volonté. (p 22, ft 1, line 13) [GP]

31. [To re–establish equilibrium without motive power production is a real loss]. [. . .] Or, pour qu'on y réfléchisse, on s'apercevra que tout changement de température qui n'est pas dû à un changement de volume des corps ne peut être qu'un rétablissement inutile d'équilibre dans le calorique. [≠ every change due to ΔV is useful]. (p 23, line 9). [OP]

32. La condition nécessaire du maximum est donc qu'il ne se fasse dans les corps employés à réaliser la puissance motrice de la chaleur aucun changement de température qui ne [≠ there is change that] soit dû à changement de volume. (p 23, line 12). [OP]

33. Ce principe [to re–establish equilibrium – see 31] ne doit jamais être perdu de vue [≠ It must constantly be kept in mind] dans la construction des machines à feu; [. . .] (p 24, line 3). [OP]

34. Si [principle of equilibrium re–establishing] l'on ne peut pas l'observer rigoureusement il faut du moins s'en écarter le moins [≠ to be as near as] possible. (p 24, line 5). [OP]

35. Tout changement de température qui n'est pas dù à un changement de volume ou à une action chimique (action que provisoirement nous supposons ne pas se rencontrer ici) est nécessairement dù au passage direct du calorique d'un corps plus ou moins échauffé à un corps plus froid. Ce passage a lieu principalement au contact de corps de températures diverses: aussi un pareil contact [=bodies with different temperatures] doit–il être évité autant que possible. [≠ contact between bodies with $\Delta t = 0$]. (p 24, line 7) [OP]

36. [see 35] Il ne [≠ we can realize contact between bodies] peut pas être évité complètement, sans doute; [. . .] (p 24, line 14) [OP]

37. [see 36] [. . .] Mais il faut du moins faire en sorte que les corps mis en contact les uns avec autres diffèrent peu entre eux de température. [≠ about equal].
(p 24, line 14). [OP]

38. A la vérité les chose [maximum motive power production using bodies of peer temperature] ne peuvent pas se passer rigoureusement [. . .]. [≠ the things can happen with variations]. (p 25, line 6). [PT]

39. On peut le regarder [Δt] comme nul en théorie, sans que pour cela les raisonnements perdent rien de leur exactitude. [≠ they continue to be exact results] (p 25, line 10). [TP]

40. [Irreversible isochoric]. Si l'on veut recommencer une opération semblable à la première [three phases cycle], si l'on veut développer une nouvelle quantité de puissance motrice avec le même instrument, avec la même

vapeur, il faut d'abord rétablir les choses dans leur état primitif, il faut rendre à l'eau le degré de température qu'elle avait d'abord. Cela peut se faire sans doute en la remettant immédiatement en contact avec le corps A; mais il y a alors contact entre des corps de températures diverses et perte de puissance motrice: il deviendrait impossible d'exécuter [without work] [≠ you can do it through work] l'opérations inverse, c'est–à–dire de faire retourner au corps A le calorique employé à élever la température du liquide. (p 25, line 20). [OP]

41. La force produite par la petite machine n'eut couté aucune dépense de chaleur [≠ heat conservation], puisque toute celle qui été employée serait rentrée dans la chaudière avec l'eau de condensation. (p 26, ft 1, line 10). [OP]
42. [In Gay–Lussac and Welter's experiment]. [...] L'on n'a pas observé d'abaissement [≠ constancy has been observed] sensible dans le degré de température marqué par le thermomètre. (p 30, ft 1, 4°, line 15). [OP]
43. [Compression of a gas and consequent increase of its temperature]. [...] L'on peut éviter qu'il ne baisse de [of gas] température [≠ constant temperature] en lui fournissant une certaine quantité de calorique. (p 31, line 5) [OP]
44. [The change of temperature of a gas caused by a variation of volume can be seen as one of the most important phenomenona in all of whole physics...]. Il semble [this phenomenon] présenter dans [≠ it has] plusieurs circonstances des anomalies singulières. (p 31, ft 1, line 7) [OP]
45. [About air as an substance working]. L'air nous a donc servi de machine à feu; nous l'avons même employé de la manière la plus avantageuse possible, car il ne s'est fait aucun rétablissement inutile [= not useful] [≠ a useful re–establishment has been realized] d'équilibre dans le calorique. (p 35, line 4). [OP]
46. [see 45]. Nous supposons implicitement dans notre démonstration que, lorsqu'un [= it seems unexplainable] [≠ it is explainable] corps a éprouvé des changemens quelconques et qu'après un certain nombre de transformations il est ramené identiquement à son était primitif, c'est–à–dire à cet état considéré relativement à la densité, à la température, au mode d'agrégation nous supposerons, dis–je, ce corps se trouve contenir la même quantité de chaleur qu'il contenait d'abord [...]. (p 37, ft 1, line 1). [PR]
47. [Suppositions: quantities of caloric given and absorbed are equal]. Ce fait n'a jamais été révoqué en doute [≠ it is sure]; [...]. (p 37, ft 1, line 8). [MP]

48*. [$Q = 0$ in a cycle]. Le nier [to negate caloric theory], ce serait renverser [≠ to affirm it would mean to maintain = so theory is correct] toute la théorie de la chaleur, à laquelle il sert de base]. (p 37, ft 1, line 9). [MP]

49. La puissance motrice de la chaleur est in dépendante [≠ constant] des agens mis en œuvre pour la réaliser; [...]. (p 38, line 4). [OP]

50. [Motive power is independent from kind of gas]. Cette condition se trouvera remplie si, comme nous l'avons remarqué plus haut, il ne se fait dans les corps aucun changement de température que ne [$\neq$ every change of temperature] soit dû à un changement de volume, [...]. (p 38, line 10). [OP]
51. [Motive power is independent from kind of gas] [...] N'y a jamais de contact entre des corps de températures sensiblement différentes [$\neq$ there is contact between bodies of equal temperature]. (p 38, line 14). [OP]
52[1]. Lorsqu'un gaz passe, sans changer [$\neq$ staying constant] de temperature, d'un volume et d'une pression déterminés à un autre volume et à une pression également déterminés, la quantité de calorique absorbée ou abandonnée est toujours la même, quelle que soit la nature du gaz choisi comme sujet d'expérience. (p 41, line 20). [OP]
53. La quantité de chaleur que les fluides élastiques dégagent ou absorbent dans leurs changemens de volume n'a jamais été mesurée par aucune expérience direct, expérience qui offrirait sans doute [$\neq$ surely] de grandes difficultés; mais il existe une donnée qui est à peu près l'équivalent pour nous: [...]. (p 42, line 16). [OP]
54. Imaginons maintenant qu'au lieu d'échauffer de 1° [C] l'air soumis à une pression constant et pouvant se dilater librement, on le renferme dans une capacité inextensible [= not variable $\neq$ constant], et qu'en cet état, on lui fasse acquérir 1° de température. (*sic*). (p 44, line 11). [OP]
55. La capacité des gaz pour la chaleur change avec leur volume; il est très–possible [$\neq$ it is probable] qu'elle change aussi avec la température. (p 49, line 8). [OP]
56. Ces chaleurs spécifiques [c_p and c_V] augmentent toutes deux à mesure que la densité du gaz diminue, mais leur différence ne varie pas. [$\neq$ it is constant] (p 59, line 9). [OP]
57. [Constancy of specific heat for different temperatures]. Cette constance n'est [= it is not more than $\neq$ it is] qu'une hypothèse, admise pour les gaz par analogie, vérifiée passablement pour les corps solides et liquides dans une certaine étendue de l'échelle thermométrique, [...]. (p 64, line 14). [OP]
58. [see 57]. L'on ne voit pas de raison pour admettre *a priori* la constance de la chaleur spécifique des corps à diverses températures, c'est-à-dire pour admettre que des quantités égales de chaleur produiront des accroissemens égaux dans le degré thermométrique d'un corps, quand même ce corps ne changerait [$\neq$ they are constant] ni d'état ni de densité; quand ce serait, par exemple, un fluide élastique renfermé dans une capacité inextensible. (p 64, ft 1, line 1). [OP]

[1] From here forward, the part of *Réflexions sur la puissance motrice du feu* concerning gas theory begins.

59. Non seulement [$\neq$ they are only] les chaleurs spécifiques ne restent pas les mêmes aux diverses températures, mais en outre elles ne conservent pas entre elles les mêmes rapports; [...]. (p 65, ft 1, line 13). [OP]
60. D'après la théorie ci–dessus exposée [Davy–Faraday experiments], l'on peut prévoir que l'emploi de ces liquides ne présenterait pas d'avantages [$\neq$ it will be an inferior result] relativement a l'économie de la chaleur. (p 88, ft 1, line 3). [OP]
61. [Doubts about caloric theory]. La loi fondamentale que nous avions en vue de confirmer [*Q=0*] nous semblerait exiger cependant, pour être mise hors de doute [$\neq$ it is considered as a valid law], des vérifications nouvelles; [...].
(p 89, line 5). [PR]
62. [see 61]. [...] Elle [*Q=0*] est assise sur la théorie de la chaleur telle qu'on la concoit aujourd'hui, et, il faut l'avouer, cette base ne nous parait pas d'une solidité in ébranable [$\neq$ it is questionable]. (p 89, line 8). [PR]
63. [Heat of high pressure machines]. Elle [heat machine built by Perkins] est composée d'un seul cylindre, de dimensions fort petites, qui, chaque pulsation, se remplit entièrement de vapeur formée sous la pression de 35 atmosphères. La vapeur ne produit aucun effet par l'extension de son volume, car on ne [$\neq$ one has effect if there is space] lui présente aucune capacité où cette extension puisse avoir lieu; [...] (p 99, ft 1, line 7). [OP]
64. [Watt heat machine]. Watt supposait ici que la vapeur observe dans sa dilatation la loi de Mariotte: ce qu'il devait pas regarder comme exact, parce que, d'une part, le fluide élastique, en se dilatant, s'abaisse de température, et que, de l'autre, rien ne prouvait qu'il ne [$\neq$ has shown that a par...] se condense pas une partie de ce fluide par l'effet de son expansion. (p 102, ft 1, line 3). [OP]
65. [He quotes Niepce's attempt to produce motive power through *Pyréolophore*]. Au lieu d'opérer comme le faisaient MM. Niepce, il nous eut semblé préférable de comprimer l'air par des pompes [...]. Les principales difficultés que l'on rencontrées dans ce mode d'opérations eussent été de renfermer le foyer dans une enveloppe d'un solidité suffisante, d'entretenir cependant la combustion à un état convenable de maintenir les diverses parties de l'appareil à une température modérée, et d'empêcher les dégradations rapides du cylindre et du piston: nous ne croyons pas ces difficultés insurmontables [$\neq$ they are surmountable]. (pp 111–112, ft 1, line 14). [OP]

Notes on Manuscripts and Documents

In this section the main manuscripts & documents studied concerning Lazare Carnot and Sadi Carnot are listed.[1] In the previous Chapters special documents and related references were also classified, e.g, unknown[2] letters on Leibniz in Sadi Carnot's folder, Sadi Carnot's engineering design–projects during his study life at X.

Lazare Carnot's Manuscripts and Documents	
Lazare Carnot: three portraits on paper (20.5 × 28)cm; (19.5 × 27.5)cm; (25.5 × 34.5)cm	[We have plates from orginal by *Académie des sciences, Institut de France,* Paris]
Dissertation sur la théorie de l'infini mathématique, ouvrage destine it concourir au prix qu'a propose *l'Académie royale des sciences*, arts et belles–lettres de Berlin, pour l'année 1786. The manuscript is dated from Arras 8 September 1785. It is conserved in the Archives of the *Deutsche Akademie der Wissenschaftenzu Berlin,* and consists of 100 paragraphs in 90 folia. It is reproduced in its entirety In: Gillispie (1971), Appendix A, pp 171–267, *op. cit.*; and microfilm copies of the complete manuscript are also deposited in the Firestone Library of Princeton University.	Gillispie (1971)
Carnot (1778) *Mémoire sur la théorie des machines pour concourir au prix de 1779 proposé par l'Académie royale des sciences de Paris.* The manuscript is dated 28 March 1778. It is conserved in the *Archives de l'Académie des sciences, Institut de France.*	[We have plates from original by *Académie des sciences, Institut de France,* Paris]

(continued)

[1]We remind the reader to other lists by Charles Gillispie (1971) and by Robert Fox (Carnot 1986).

[2]Maybe, for the first time cited. One of us (RP) is working on.

C.C. Gillispie and R. Pisano, *Lazare and Sadi Carnot: A Scientific and Filial Relationship*, History of Mechanism and Machine Science 19, DOI 10.1007/978-94-017-8011-7,

(continued)

Ms. (18.5 × 23.5)cm consists of: 85 sections in 63 folia (31 folia r/v) Sections 27–60 are reproduced In: Gillispie (1971), Appendix B, pp 271–296, *op. cit.*; and microfilm copies of the complete manuscript are also deposited in the Firestone Library of Princeton University	
Carnot (1780) *Mémoire sur la théorie des machines pour concourir au prix que l'Académie royale des sciences de Paris doit adjuger en 1781.* The manuscript is dated from Béthune 15 July 1780. It is conserved in the *Archives de l'Académie des sciences, Institut de France.* Ms. (20.5 × 32)cm consists of 191 sections in 106 folia. (47 folia r/v + "2 planches de dessins de même dimension"). Sections 101–160 are reproduced In: Gillispie (1971), Appendix C, pp 299–343, *op. cit.*; and microfilm copies of the complete manuscript are also deposited in the Firestone Library of Princeton University	[We have plates from original by *Académie des sciences, Institut de France,* Paris]
Lazare Carnot's portraits	[We have many plates from the original kindly authorized by Monsieur Gaetan Carnot of *Carnot's family.* The original are conserved at *Académie François Bourdon*].
Sadi Carnot's Manuscripts and Documents Sadi Carnot: three portraits "d'après négatif (dont two diapos)". They are conserved at *Académie des sciences, Institut de France,* Paris	[We have plates from original by *Académie des sciences, Institut de France,* Paris. We also have several plates from the original kindly authorized by Monsieur Gaetan Carnot of *Carnot's family.* The original are conserved at *Académie François Bourdon*].
Recherche d'une formule propre à représenter la puissance motrice de la vapeur d'eau is conserved at *Bibliothèque de l'École polytechnique,* Paris. Ms. s.d., (17 × 21.5)cm, 21 folia. It was a "don du Professor Paul Carnot"	[We have plates from original by *Archives, bibliothèque de l'École polytechnique route de Saclay de Paris.* © Collections École polytechnique]
Carnot (1878a) *Réflexions sur la puissance motrice du feu* Ms. presented by Hippolyte Carnot to the *Académie des sciences* on 16 December 1878. It is conserved at *Académie des sciences, Institut de France,* Paris Ms. in 4 cahiers (16.5 × 21.5)cm. 22 folia, 24 folia, 20 folia, 26 folia; 92 folia r/v	[We have plates from orginal by *Académie des sciences, Institut de France,* Paris]

(continued)

(continued)

Machine à Vapeur de M. Watt Ms. s.d. (21.5 × 33)cm On the cover one can read "[1816]" It is a translation of Watt's paper by Sadi Carnot from English to French. It is conserved at *Académie des sciences, Institut de France,* Paris.	[We have plates from original by *Archives, bibliothèque de l'École polytechnique route de Saclay de Paris*. © Collections École polytechnique]
Dossier Carnot n° 12: various documents and letters[3] Various documents during his military life *Copie d'une correspondance d'Arnaul à Leibniz 1686, 1687, 1688* *Machine à vapeur de M Watt* (s.d. 1816X, 40 folia).	[We have plates from original by *Archives, bibliothèque de l'École polytechnique route de Saclay de Paris*. © Collections École polytechnique]
Recherche d'une formule propre à représenter la puissance motrice de la vapeur d'eau.	*Piece mise a la disposition de l'École polytechnique par M P–S Carnot.*

[3]Particularly, this dossier and correspondences are object of study by one of us (RP).

Special Images of Carnot's Family

C.C. Gillispie and R. Pisano, *Lazare and Sadi Carnot: A Scientific and Filial Relationship*, History of Mechanism and Machine Science 19, DOI 10.1007/978-94-017-8011-7,

Lazare–Nicolas–Marguérite Carnot (Nolay, May 13, 1753 – Magdeburg, August 2, 1823). *Plate from the original portrait conserved at Académie des sciences, Paris – France*

Lazare Carnot, portrait en uniforme de capitaine du génie en 1792, peinture de F. Bouchot (1836). Plate from the original portrait conserved at *Académie François Bourdon, Le Creusot – Archives Lazare Carnot*. A very kind authorization from Monsieur Gaetan Carnot, member of Carnot's family

Portrait en uniforme de Gouverneur d'Anvers en 1814, gravure d'après le tableau de Van Bree (s.d.). Plate from the original portrait conserved at *Académie François Bourdon, Le Creusot – Archives Lazare Carnot*. A very kind authorization from Monsieur Gaetan Carnot, member of Carnot's family.

Lazare Carnot à Wattignies, « L'Eventail patriotique » aquarelle dessinée par H. Pillé (s.d.). Plate from the original portrait conserved at *Académie François Bourdon, Le Creusot – Archives Lazare Carnot*. A very kind authorization from Monsieur Gaetan Carnot, member of Carnot's family.

Le général Carnot, gravure de H. Mayer, publiée par le Journal Illustré (1887). Plate from the original portrait conserved at *Académie François Bourdon, Le Creusot – Archives Lazare Carnot.* A very kind authorization from Monsieur Gaetan Carnot, member of Carnot's family

« Les hommes illustres de la Révolution française 1789-1793 », lithographie de Wentzel (s.d.). Plate from the original portrait conserved at *Académie François Bourdon, Le Creusot – Archives Lazare Carnot*. A very kind authorization from Monsieur Gaetan Carnot, member of Carnot's family

« La constitution de la république française » en 1795, Barras et Carnot, lithographie (s.d.). Plate from the original portrait conserved at *Académie François Bourdon*, *Le Creusot – Archives Lazare Carnot*. A very kind authorization from Monsieur Gaetan Carnot, member of Carnot's family

Nicolas Léonard Sadi Carnot (Paris, June 1, 1796 – Paris, August 24, 1832) 1813 at the age of 17 in his Polytechnique uniform by Louis Boilly. *Plate from the original portrait conserved at Académie des sciences, Paris – France*

Lazare et son fils Hippolyte à Magdebourg recevant des nouvelles de France (s.d.). Plate from the original portrait conserved at *Académie François Bourdon, Le Creusot – Archives Lazare Carnot*. A very kind authorization from Monsieur Gaetan Carnot, member of Carnot's family

Nicolas Léonard Sadi Carnot (Paris, June 1, 1796 – Paris, August 24, 1832) 1832 at the age of 36 years old. *Plate from the original portrait conserved at Académie des sciences, Paris – France*

Portrait attribué à Félie Carnot (s.d.). Plate from the original portrait conserved at *Académie François Bourdon, Le Creusot – Archives Lazare Carnot*. A very kind authorization from Monsieur Gaetan Carnot, member of Carnot's family

Portrait attribué à Félie Carnot (s.d.). Plate from the original portrait conserved at *Académie François Bourdon, Le Creusot – Archives Lazare Carnot*. A very kind authorization from Monsieur Gaetan Carnot, member of Carnot's family

Sadi en uniforme de polytechnicien en 1813, d'après une peinture de Boilly (1878). Plate from the original portrait conserved at *Académie François Bourdon, Le Creusot – Archives Lazare Carnot*. A very kind authorization from Monsieur Gaetan Carnot, member of Carnot's family

Sadi en uniforme de polytechnicien en 1813, copie d'après Boilly (XXe siècle). Plate from the original portrait conserved at *Académie François Bourdon, Le Creusot – Archives Lazare Carnot*. A very kind authorization from Monsieur Gaetan Carnot, member of Carnot's family

La famille Carnot (s.d.). Plate from the original portrait conserved at *Académie François Bourdon, Le Creusot – Archives Lazare Carnot*. A very kind authorization from Monsieur Gaetan Carnot, member of Carnot's family

References

Achard A (1866) Revue de Saint–Robert 1865. Bibliothèque universelle de Genève, Archives des sciences physiques et naturelles 27:77–81

Agazzi E (1980) Logic and methodology of empirical sciences. In: Agazzi E (ed) Modern logic – a survey. Reidel, Dordrecht, pp 255–282

Ampère AM (1806) Démonstration du principe des vitesses virtuelles, dégagée de toute considération des infiniment petits. Journal de l'Ècole Polytechnique 6(13):247–269

Ampère AM (1822a) Recueil d'observation électrodynamiques. Crochard, Paris

Ampère AM (1822b) Extrait d'une lettre d'Ampère au Professeur de la Rive sur des expériences électro–magnétiques et sur la formule qui représente l'action mutuelle de deux portions infiniment petites de courants électriques —12 juin 1822. Bibliothèque universelle des sciences, belles– lettres, et arts XX:185–192

Ampère AM (1826) Théorie des phénomènes électro–dynamiques uniquement déduite de l'expérience. Méquignon–Marvis, Paris

Ampère AM (1827) Théorie mathématiques des phénomènes électro–dynamiques uniquement déduite de l'expérience. Chez Firmin Didot, Paris

Arago FJD, Dupin C (1827) Rapport suivant sur un mémoire de M Poncelet intitulé Cours de mécanique appliquée aux machines. In: Institut de France. Académie des sciences. Procès–verbaux des séances de l'Académie. Imprimerie [publication 1918] de l'observatoire d'Abbadia, Hendaye Basses–Pyrénées, 7 Mai, Année 1824–1827, Tome VIII, pp 527–531

Aragao FJD, Gay-Lussac JL (eds) (1833) Annales de Chimie et de Physique. Tome LII. Crochard, Paris

Aries E (1921) L'Œuvre scientifique de Sadi Carnot. Introduction a l'étude de la thermodynamique. Payot, Paris

Aristotle (2000) Problemi meccanici. In: Bottecchia ME Dehò (ed) Rubbettino Editore, Cosenza

Aristotle (1999) Physics (trans: Waterfield R). Oxford University Press, Oxford

Aristotle (1996) In: Waterfield R, Waterfield R (eds) The principles of nature–physics, vol 1. Oxford University Press, Oxford

Aristotle (1984) In: Barnes J (ed) The complete works of Aristotle. The Princeton University Press, Princeton

Aristotle (1963) Minor works (trans: Hett WS). The Harvard University Press/Heinemann W LTD, Cambridge, MA/London

Aristotle (1949) In: Ross WD (ed) Aristotle's prior and posterior analytics. A revised text with introduction and commentary. Oxford University Press, Oxford

Aristotle (1853) On the definition and division of principles. In: Owen OF, Porphyry (eds) The Organon, or Logical treatises, of Aritotle, vol 1. Bohn H G, London, pp 263–266

C.C. Gillispie and R. Pisano, *Lazare and Sadi Carnot: A Scientific and Filial Relationship*, History of Mechanism and Machine Science 19, DOI 10.1007/978-94-017-8011-7,

Armstrong HL (1960) Statement of second law of thermodynamics. American Journal of Physics 28(6):564
Bailyn M (1985) Carnot and the universal heat death. The American Journal of Physics 53(11):1092–1099
Baldi B (1621) In mechanica Aristotelis problemata exercitationes. Ioannis Albini, Moguntiae
Barnett M (1958) Sadi Carnot and the second law of thermodynamics. Osiris 13:327–357
Bassa G (1888–1889) Commemorazione del conte p. B. dì St. Robert. In: Proceedings of R. Accademia delle scienze di Torino, vol XXIV, pp 235–244
Batens D, Meheus J (2006) A formal logic for abductive reasoning. Logic Journal of the IGPL 14:221–236
Belhoste B (2003) La formation d'une technocratie. L'École polytechnique et ses élèves de la Révolution au Second Empire. Belin, Paris
Bélidor BF (1729 et 1830) La science des ingénieurs dans la conduite des travaux de fortification et d'architecture civile. Nouvelle éditions, avec des notes, par Navier de l'académie des sciences. Didot F, Paris
Bélidor BF (1819) Architecture hydraulique, ou l'art de conduire, d'élever et de ménager les eaux pour les differens besoins de la vie. Nouvelle édition avec des notes et additions par M. Navier. Didot F, Paris
Bélidor BF (1813) La science des ingénieurs dans la conduite des travaux de fortification et d'architecture civile. Nouvelle éditions, avec des notes, par M. Navier, ingénieur ordinaire des Ponts et Chaussées. Didot F, Paris
Bélidor BF (1737–1739) Architecture hydraulique, ou l'Art de conduire, d'élever et de ménager les eaux pour les différents besoins de la vie. 2 vols, Jombert, Paris
Bellini L, Mauriello G, Drago A (2007) Ricostruzione della meccanica di Lazare Carnot come alternativa fondazionale alla meccanica newtoniana. In: L'eredità di Fermi, Majorana ed altri temi. Proceedings of XXIV SISFA. Bibliopolis, Napoli, pp 113–118
Berkeley G ([1734] 2002) The analyst, by Wilkins D.[1]
Berkeley G (1734) The analyst Dublin, Fuller S, Globe–in–Meath–Street, and J Leathly, Sect: XII–XIV
Bernoulli J (1742) Opera omnia, tam antea sparsim edita, quam hactenus inedita ... quibus continentur ea, quae ab A. 1690 usque ad A. 1727 prodierunt. Accedunt Lectiones Mathematicae de Calculo Integralium atque Anedokta, I–IV vols. Lausannae & Genevae, Bousquet [Facsimile 1968. Reprint Curavit Hofman JE. Georg Olms Verlag Hildeshe]
Bernoulli D (1738) Hydrodynamica sive de viribus et motibus fluidorum commentarii. Opus academicum, Sumptibus Johannis Reinholdi Dulseckeri, Typis Joh. Henr. Deckeri, Typographi Basiliensis, Strasbourg
Berthollet CL (1809) Sur le chaleur produite pour lo choc de compression. Mémoires de physique et de chimie de la Société d'Arcueil II:441–448
Berthollet CL (1803) Essai de statique chimique, 2 vols. Firmin Didot, Paris
Berthollet CL (1802) Extrait N°. 144 du Journal Britannique. Annales de chimie XLI/X:177–187
Betancourt MA (1792) Mémoire sur la force expansive de la vapeur de l'eau, Lu a l'académie royale des sciences par M. Betancourt. Chez Laurent, Paris (des registrer de Académie Royale des Sciences, septembre 1790)
Beth EW (1961) Semantic of physical theories. In: Freudenthal H (ed) The concept and the role of the model. In mathematics and natural and social sciences. Reidel, Dordrecht
Bevir M (1999) The logic of the history of ideas. The Cambridge University Press, Cambridge
Biot JB (1858) Une anecdote relative à Laplace, lue à l'académie française dans sa séance particulière du 5 février 1850. Mélanges scientifique et littéraires. Michel Lévy Frères, Paris, pp 1–9
Biot JB (1816) Traité de physique expérimentale et mathématique, vols 1–2. Deterville, Paris

[1][available via http://www.maths.tcd.ie/pub/HistMath/People/Berkeley/Analyst].

Biot JB (1802) Sur la théorie du son. Journal de Physique et de Chimie 55:173–182
Birembaut A (1974) A propos des notices biographiques sur Sadi Carnot: Quelques documents inédits. Revue d'histoire des sciences XXVII(44):355–370
Black J (1803) Lectures on the elements of chemistry delivered in the University of Edinburgh by the late Joseph Black, M.D, 2 vols. John Robison, Edinburgh
Borda JC (1770a) Mémoire sur les pompes. In: Histoire de l'Académie Royale des Sciences avec les Mémoires de Mathématique & de Physique, pour la même année 1768. Imprimerie Royale, Paris, pp 418–431
Borda JC (1770b) Mémoire sur les roues hydrauliques. In: Histoire de l'Académie Royale des Sciences avec les Mémoires de Mathématique & de Physique, pour la même année 1767. Imprimerie Royale, Paris, pp 270–287
Bossut C (1800a) Cours de mathématiques, géométrie et application de l'algèbre à la géométrie. Nouvelle édition revue et augmentée, Tome II. Chez Firmin Didot, Paris
Bossut C (1800b) Mécanique philosophique. Journal de l'École Polytechnique de Paris III, an. VIII, Cahiers 7&8:1–477
Bossut C (1775) Traité élémentaire de Mécanique et de Dynamique, appliqué principalement aux mouvements des Machines. Jombert, Paris
Bouvier AR (1816–1817) Mécanique appliquée. Description d'un projet de machine à vapeur, principalement applicable à faire mouvoir les bateaux. Annales de Mathématiques pures et appliqués 7:129–139
Briot C (1869) Théorie mécanique de la chaleur. Gauthier–Villars, Paris
Brocchi GB (1808) Trattato mineralogico e chimico sulle miniere di ferro. Bettoni, Brescia, I/25:229–230
Brønsted JN (1955) Principles and problems in energetics. [*Id.*, (1946) Principer og problemer i energetiken. Lunos Bogtryckeri, Copenhagen] (trans: Bell RP and Foreword by la Mer VK). Interscience, New York
Brønsted JN (1941) On the concept of heat. Kongelige Danske Videnskabernes Selskabs Skrifter. Matematisk–Fysiske Meddelelser XIX(8):1–79
Brønsted JN (1940) The fundamental principles of energetics. Philosophical Magazine 7(29):449–470
Brønsted JN (1939) The fundamental principles of energetics. Philosophical Magazine 29(7/196):449–470
Brown MT (1969) The electric current in early nineteenth–century French physics. Historical Studies in Physical Sciences 1:61–103
Brown EJ (1967–1968) The Scientia de Ponderibus in the later Middle Ages, Ph.D. dissertation. Tutor Clagett M. Madison: University of Winscosin [available via: UMI – Proquest Company].
Brown EJ (1967–1968) The Scientia de Ponderibus in the later Middle Ages. Ph.D. dissertation, Tutor Clagett M, The University of Winscosin Madison[2]
Brunold C (1930) L'entropie. Son rôle dans le développement historique de la thermodynamique. Masson, Paris
Buchdahl HA (1987) A variational principle in classical thermodynamics. The American Journal of Physics 55(1):81–83
Buchdahl HA (1966) The concepts of classical thermodynamics. The Cambridge University Press, Cambridge
Buchwald JZ (ed) (2012) A master of science history. Essays in honor of Charles Coulston Gillispie. Springer, Dordrecht
Bussotti P, Pisano R (2013) On the Conceptual and Civilization Frames in René Descartes' *Physical Works*. Advances in Historical Studies 2(3):106–125
Callen H (1974) Thermodynamics as a science of symmetry. Foundations of Physics 4:423–440
Callen HB (1960) Thermodynamics. Wiley, New York

[2][available via: UMI – Proquest Company].

Callendar HL (1910) Caloric theory of heat and Carnot's principle. Proceedings of the Physical Society–London 23(1):153–189

Callot JP (1980) Répertoire général des anciens élèves de l'École Polytechnique. 1794–1980. Tome I (A–B–C). Archives de la bibliothèque centrale de l'École polytechnique (Palaiseau). Société Amicale des Anciens élèves de l'Ecole polytechnique (AX), Paris

Caneva KL (1993) Robert Mayer and the conservation of energy. Princeton University Press, Princeton

Cantor GN (1975) William Robert Grove the correlation of forces, and the conservation of energy. Centaurus 19:273–290

Capecchi D, Pisano R (2010a) Reflections on Torricelli's principle in mechanics. Organon 42:81–98

Capecchi D, Pisano R (2010b) Scienza e Tecnica nell'Architettura del Rinascimento. Cisu, Roma

Capecchi D, Pisano R (2008) La meccanica in Italia nei primi anni del Cinquecento. Il contributo di Niccolò Tartaglia. In: Tucci P (ed) Proceedings of XXV SISFA Congress, Milano, pp C17.1–C17.6[3]

Capecchi D (2012) History of virtual work laws. A history of mechanics prospective. Springer Science-Networks-Historical Studies, Berlin

Capecchi D (2002) Storia del principio dei lavori virtuali. Hevelius, Benevento

Capecchi D, Pisano R (2007) La teoria dei baricentri di Torricelli come fondamento della static. Physis XLIV(1):1–29

Capecchi D, Drago A (2005) Lagrange e la storia della meccanica. Progedit, Bari

Cardwell DSL (1971) From Watt to Clausius. The rise of thermodynamics in the early industrial age. Heinemann, London

Cardwell DSL (1967) Some factors in the early development of the concepts of power, work, and energy. The British Journal for the History of Science 3:209–224

Cardwell DSL (1965) Power technologies and the advance of science 1700–1825. Technology and Culture 6(2):188–207

Carnap R (1943) Formalisation of logic. The Harvard University Press, Cambridge, MA

Carnot H (1861–1863) Mémoires sur Carnot par son fils, 2 vols. Pagnerre, Paris

Carnot L (1990) Rapport sur une nouvelle machine inventée par MM. Niepce, et nommée par eux Pyrélophore; par MM. Berthollet et Carnot. Lu le 15 décembre 1806, imprimé en 1807, 1er semestre. In: Charnay JP (ed) Lazare Carnot ou le savant–citoyen, La Herne, Paris, Presses de L'Université de Paris–Sorbonne, 1990 – Centre d'études et de recherches sur les stratégies et les conflits, Colloque en Sorbonne, 1988

Carnot L (1936) Razmyshleniia o metafizike ischisleniia beskonechno malykh, 2nd edn. Leningrad, Moscow

Carnot L (1933) Lazar Karno Razmyshlenia o metafizike ischislenia beskonechno malykh (trans: Soloviev NM, M.–Л. Гос). Технико–теоретическое издательство [ML Gos Technical and Theoretical Publishing] Moscow [see also: Id (1936) In: Idées sur le fondement de l'analyse mathématique au XVIII siècle. Rédaction et Suppléments de Youschkevitch AP, 2nd edn. Leningrad, Moscow]

Carnot L (1921) Réflexions sur la métaphysique du calcul infinitésimal. In: Les maitres de la pensée scientifique. Publication par les soins de Maurice Solovine, 2 Fascicules. Gauthier–Villars, Paris

Carnot L (1832) Reflexions on the methaphysical princiles of the infinitesimal analysis (trans: Browell WR, Parker et al). Oxford

Carnot L (1813) Réflexions sur la métaphysique du calcul infinitésimal. Courcier, Paris

Carnot L (1813) Note. In: Carnot L (1813), pp 213–257

Carnot L (1810) In: Schumacher HG (ed) Geometrie der Stellung oder über die Anwendung der Analysis auf Geometrie, 2 vols. Vorrede, Altona

[3][also available in .pdf via: http://www.brera.unimi.it/sisfa/atti/index.html]

Carnot L (1806a) Digression sur la nature des quantités dites négatives. In: Mémoire sur la Relation qui existe entre les distance respectives de cinq quelconques pris dans l'espace; suivi d'un essai sur la théorie des transversales. Courcier, Paris, pp 96–111

Carnot L (1806b) Mémoire sur la relation qui existe entre les distances respectives de cinq points quelconques pris dans l'espace, suivi d'un essai sur la théorie des transversales. Courcier, Paris

Carnot L (1806c) Rapport sur une nouvelle machine inventée par MM Niepce, et nommée par eux Pyrélophore; par MM Berthollet et Carnot. Lu le 15 décembre 1806, imprimé en 1807, 1er semestre. In: Charnay JP, Charnay JP (eds) Lazare Carnot ou le savant–citoyen. La Herne, Paris

Carnot L (1803a) Principes fondamentaux de l'équilibre et du mouvement. Deterville, Paris

Carnot L (1803b) Géométrie de position. Duprat, Paris

Carnot L (1801) De la corrélation des figures de géométrie. Duprat, Paris

Carnot L (1800a) Lettre du citoyen Carnot au citoyen Bossut, contenant quelques vues nouvelles sur la trigonométrie. In: Bossut (1800a), vol II, pp 401–421

Carnot L (1800b) Betrachtungen über die Theorie der Infinitesimalrechnung. Von dem Burger Carnot. Aus dem Französischen übersetzt und mit Anmerkungen und Zusätzen begleitet von Johann Karl Friedrich Hauff. Jager, Frankfurt am Main

Carnot L (1808a) Essai sur les machines en général (I part). In: Philosophical Magazine: comprehending the various branches of science, the liberal and fine arts, agriculture, manufactures, and commerce. Tilloch A (ed). Vol XXX. Murray J, London, pp 8–14; pp 154–158; pp 208–221; pp 310–320. [very interesting is the avant-titre page of the book where a portrait of Lazare Carnot is published]

Carnot L (1808b) Essai sur les machines en général (II Part). In: Philosophical Magazine: comprehending the various branches of science, the liberal and fine arts, agriculture, manufactures, and commerce. Tilloch A (ed). Vol XXXI. Murray J, London, pp 136–145; pp 220–228; 295–305

Carnot L (1797) Réflexions sur la métaphysique du calcul infinitésimal. Duprat, Paris

Carnot L (1786) Essai sur les machines en général. Defay, Dijon

Carnot L (1785) Dissertation sur la théorie de l'infini mathématique, ouvrage destine it concourir au prix qu'a propose l'Académie Royale des Sciences, arts et belles–lettres de Berlin, pour l'année 1786 [The manuscript is dated by Arras 8 Sept 1785]

Carnot L (1784) Lettre sur les aérostats. A manuscript memoir addressed to the Académie Royale des Sciences on 17 Jan 1784 [document lost]

Carnot L (1780) Mémoire sur la théorie des machines pour concourir au prix que l'Académie Royale des Sciences de Paris doit adjuger en 1781. The manuscript is dated from Béthune 15 July 1780. It is conserved in the Archives de l'Académie des sciences, Institut de France, and consists of 191 sections in 106 folia. Sections 101–160 are reproduced. In: Gillispie (1971), Appendix C, pp 299–343

Carnot L (1778) Mémoire sur la théorie des machines pour concourir au prix de 1779 propose par l'Académie Royale des Sciences de Paris. The manuscript is conserved in the Archives de l'Academie des sciences, Institut de France, and consists of 85 sections in 63 folia. Sections 27–60 are reproduced. In: Gillispie (1971), Appendix B, pp 271–296

Carnot S–EP (s.d.) Recherche d'une formule propre à représenter la puissance motrice de la vapeur d'eau. © Collections Archives de la Bibliothèque de l'École polytechnique, Paris

Carnot S (1986) Reflexions on the motive power of fire: a critical edition with the surviving scientific manuscripts translated and edited by Robert Fox. The Manchester University Press, Manchester

Carnot S (1980) L'Accueil Des Idées de Sadi Carnot Et la Technologie Française de 1820 à 1860 par Pietro Redondi. Vrin J, Paris

Carnot S (1978) Réflexions sur la puissance motrice du feu sur les machinés propre à développer cette puissance, édition critique par Fox Robert. Vrin J, Paris

———— Recherche d'une formule propre à représenter la puissance motrice de la vapeur d'eau. In: Carnot S (1978), pp 223–234

Carnot S (1953) Réflexions sur la puissance motrice du feu et sur les machines propres a développer cette puissance. Blanchard, Paris

Carnot S (1943) Reflections on the motive power of heat and on machines fitted to develop this power. In: Thurston RH (ed), Waverly Press, Inc., Baltimore
Carnot S (1878a) Réflexions sur la puissance motrice du feu sur les machinés propre à développer cette puissance.[4] Archives of the Académie des Science–Institut de France, Paris
Carnot S (1878b) Réflexions sur la puissance motrice du feu sur les machinés propre à développer cette puissance. Gauthier–Villars, Paris
Carnot S Notes sur les mathématiques, la physique et autres sujets. In: Carnot S (1878b), pp 89–102 [The manuscript is conserved In: Carnot S (1878a)]
Carnot S (1996) Riflessioni sulla potenza motrice del fuoco. Cuen, Napoli
Carnot S (1872) Réflexions sur la puissance motrice du feu sur les machinés propre à développer cette puissance. Annales scientifiques de École Normale Supérieure XII(1):393–457
Carnot S (1824) Réflexions sur la puissance motrice du feu sur les machinés propre à développer cette puissance. Bachelier, Paris
Carpi P (1836) Lezioni di fisica sperimentale di Saverio Barlocci professore di detta facoltà nell'università roma della Sapienza e del collegio filosofico, ad uso degli studenti della medesima università. Tom. I parte I e II, Roma tipografica di Pietro Aureli. Giornale arcadico di scienze, lettere, ed arti, Tomo LXIX. Accademia Letteraria Italiana, Boulzaler, Roma, pp 292–294
Cauchy AL (1829) Mémoire sur un nouveau principe de mécanique, par M. Aug. Louis Cauchy. In: Sturm et Gaultier de Claurbry (eds) Bulletin universel des sciences mathématiques, physique et chimiques, Tome XII. Firmin A Didot, Paris, pp 116–122
Cauchy AL (1828) Mémoire sur un Nouveau principe de mécanique rationnelle. In: Procès–verbaux des séances de l'Académie, 21 Juillet, Institut de France. Académie des Sciences, Imprimerie [publication 1921] de l'observatoire d'Abbadia, Hendaye Basses–Pyrénées, Année 1828–1831, Tome IX, p 94
Cavalieri B (1635) Geometria Indivisibilius continuo rum nuova quadam ratione promota. Ferrone, Bologna
Cerruti L (1999) Dante's bones. In: Gavroglu K (ed) The sciences in the European periphery during the enlightenment. Archimedes. Kluwer, Dordrecht
Challey F (1971) Sadi Carnot. In: Gillispie (1970–1980), vol III, pp 79–84
Chasles M (1875) Aperçu historique sur l'origine et le développement des méthodes en géométrie, 2nd edn. Gauthier–Villars, Paris
Chasles M (1852) Traité de géométrie supérieur. Bachelier, Paris
Chasles M (1839) Énoncé de deux théorèmes généraux sur l'attraction des corps et la théorie de la chaleur. Comptes rendus de l'Académie des sciences 8:209–211
Chatzis K (2009) Charles Dupin, Jean–Victor Poncelet et leurs mécaniques pour "artistes" et ouvriers. In: Christen C, Vatin F (eds) Charles Dupin (1784–1873): ingénieur, savant, économiste, pédagogue et parlementaire du Premier au Second Empire. PUR, Rennes, pp 99–113
Cheng KC (1998) Some observations on Carnot cycle as the genesis of the heat pipe and thermosyphon. International Journal of Mechanical Engineering Education 28(1):69–87
Clagett M, Moody E (1960) The medieval science of weights. Scentia de ponderibus. The University of Wisconsin Press, Madison
Clagett M (1959) The science of mechanics in the Middle Ages. The University of Wisconsin press, Madison
Clagett M, Murdoch J (1958–59) Medieval mathematics, physics, and philosophy: a revised catalogue of photographic reproductions. Manuscripta 2:131–154, Manuscripta 3:19–37

[4]The manuscript *Notes sur les mathématiques, la physique et autres sujets* is also conserved at *Archives of the Académie des Science–Institut de France*, Paris. Thus, we used "Carnot S 1878a" to cite both of the two original manuscripts studied. The difference in the running text is presented by the titles of the two manuscripts. We let note that the manuscripts edited by Gauthier–Villars (Carnot S 1878b) are not always integrally reproduced. For any complete consulting of the Sadi Carnot's manuscripts, please see Fox 1986.

Clagett M, Moody EA (1952) The medieval science of weights. The University of Wisconsin, Madison
Clapeyron ÉBP (1834) Mémoire sur la puissance motrice du feu. Journal de l'École Royal Polytechnique XXIII(XIV):153–190
Clausius RJE (1850) Über die bewegende Kraft der Wärme und die Gesetze, welche sich daraus für die Wärmelehre selbst ableiten lassen. Annalen der Physik und Chemie 79:368–397, 500–524. (English trans: Id., On the Motive Power of Heat and on the Laws which can be deduced from it for the theory of heat itself. In: Mendoza (1960), pp 73–74, 109–152)
Clausius RJE (1868–1869) Théorie Mécanique de la Chaleur, II vols, French édition (trans: Folie). Lacroix éditeur, Paris
Clausius R (1865a) Sur la détermination de la disgregation d'un corp et de la vraie capacité calorifique. Bibliothèque universelle de Genève, Archives des sciences physiques et naturelles XXV:117–124
Clausius RJE (1865b) Über die bewegende Kraft der Wärme und die Gesetze
Clausius RJE (1864) Die Mechanische Wärmetheorie. Vieweg, Braunschwei–Poggendorff's. Annalen der Physik und Chemie 125:352–400
Clément N (1970) Nicolas Clément. In: Gillispie (ed) (1970–1980), III:315–317
Clément–Desormes N (1905) Chimie générale dans ses rapports avec l'industrie Chaire créée par ordonnance du 25 novembre 1819. In: Conservatoire national des arts et métiers. Enseignement des sciences appliquées aux arts et à l'industrie. Programmes des cours publics. Vuibert et Nony Éditeurs, Paris, pp 109–118
Clément N (1819a) Sur les machines à vapeur. Bulletin de la Société d'Encouragement pour l'Industrie Nationale. Imprimerie de Madame Huzard, Paris, XVII:254–255
Clément N (1819b) Un mémoire sur les machines a vapeur. In: Procès–verbaux des séances de l'Académie, 23 et 30 Aout, Institut de France, Tome VI. Académie des sciences. Imprimerie de l'observatoire d'Abbadia, Hendaye (Basses Pyrénées), pp 480–481
Cohen B, Taton R (eds) (1964) Mélanges Alexandre Koyré, 2 vols. Hermann, Paris
Combes C (1863) Exposé des principes de la théorie mécanique de la chaleur et ses applications principales, Bulletin de la Société d'Encouragement pour l'Industrie Nationale. Janvier, Paris
Combes C (1867) Exposé des principes de la théorie mécanique de la chaleur et ses applications principales. Bouchard–Huzard, Paris
Commandino F (1565) Federici Commandini Vrbinatis Liber de centro gravitatis solido rum. Ex Officina Alexandri Benacii, Bononiae
Condillac EB (1821) La logique par Condillac. Verdier Quai Des Augustins, Paris
Coriolis GG (1844) Traité de la mécanique des corps solides et du calcul de l'effet des machines, 2nd edn. Carilian–Goeury et Vor Dalmot, Paris
Coriolis GG (1830–1837) École centrale des arts et manufactures. Cours de M Coriolis. Leçons sur la mécanique, 1830–1831, 1836–1837 (slnd), 2 vols[5]
Coriolis GG (1829) Du calcul de l'effet des machines ou considérations sur l'emploi des moteurs et sur leur évaluation. Servir d'introduction a l'étude spéciale des machines. Carilian–Goeury, Paris
Costabel P (1976) L'Œuvre de Sadi Carnot, son contexte, ses suites. In: Taton (1976), pp 122–124
Costabel P (1960) Leibniz et la dynamique: les textes de 1692. Histoire de la pensée–Hermann, Paris [English trans: Herman 1973]
Coulomb CA (1799) Résultat de plusieurs expériences destinées a déterminer la quantité d'action que les hommes peuvent fournir par leur travail journalier suivant les différentes manières dont ils emploient leurs forces, vol II. Mémoires de l'Institut National des sciences et arts, Paris, pp 340–428

[5][also available in .pdf, vols. 14276–14277 at *Bibliothèque Nationale de France* via: http://www.bnf.fr]

Coulomb CA (1785) Théorie des machines simples en ayant égard au frottement de leurs parties et à la roideur du cordage. Mémoire de mathématique et de physique présentes à l'Académie Royal des Sciences par divers savants. Bachelier, Paris

Coulomb CA (1784) Observations théoriques et expérimentales sur l'effet des moulins à vent, & sur la figure de leurs ailes. In: Histoire de l'Académie Royale des Sciences avec les Mémoires de Mathématique & de Physique, pour la même année 1781. Imprimerie Royale, Paris, pp 65–81

Crombie AC (1994) Styles of scientific thinking in the European tradition, 3 vols. Duckworth, London

Crombie AC (ed) (1963) Scientific change: historical studies in the intellectual, social and technical conditions for scientific discovery and technical Invention, from antiquity to the present – Symposium on the history of science, University of Oxford, 9–15 July 1961. Heinemann education books, London

Crombie AC (1959) Medieval and early modern science, 2 vols. Doubleday, Garden City

Cropper HW (1987) Carnot's function: origins of the thermodynamic concept of temperature. The American Journal of Physics 55(2):120–122

D'Alembert JLR ([1743] 1758) Traité de dynamique. Nouvelle édition, revue et fort augmentée par l'Auteur Chez David Libraire, Paris

d'Arcy P (1763) Théorèmes de dynamique. Histoire de l'Académie Royale des Science: année I758. Imprimerie Royal, Paris, pp 1–8

d'Arcy P (1754) Suite d'un mémoire de dynamique. Histoire de l'Académie Royale des Science: année 1750. Imprimerie Royal, Paris, pp 107–108

d'Arcy P (1752) Problème de dynamique. Histoire de l'Académie Royale des Science: année 1747, vol II. Imprimerie Royal, Paris, pp 344–356, with an addendum, Ivi, pp 356–361

Da Costa N, French S (2003) Science and partial truth: a unitary approach to models and reasoning in science. Oxford University Press, New York

Dalla Chiara ML, Toraldo di Francia G (1999) Introduzione alla filosofia della scienza. Laterza, Bari, pp 87–88

Dalla Chiara ML (1985) Some foundational problems in mathematics suggest by physics. Synthése 62:303–315

Darboux G (1887–1896) Leçons sur la théorie générale des surfaces et les applications géométriques du calcul infinitésimal, 4 vols. Gauthier–Villars, Paris

Darrigol O (2012) A History of Optics: From Greek Antiquity to the Nineteenth Century. The Oxford University Press, Oxford

Dartnall WJ, Reizes J (2005) A novel approach to the teaching of thermodynamic cycles and the laws of thermodynamics. In: Proceedings of international mechanical engineering congress and exposition 2005. American Society of Mechanical Engineers, New York, pp 1–9

Daub EE (1970) Entropy and dissipation. Historical Studies in Physocal Sciences 2:321–354

Décombe L (1919) Sadi Carnot et le Principe de l'Équivalence. Comptes rendus 168:268–271

Delaroche F, Bérard JE (1813) Mémoire sur la détermination de la chaleur spécifique des différens gaz. Annales de chimie et de physique 85(72–110):113–182

Derry TK, Williams TI (1960) A short history of technology, 2 vols. The Clarendon Press, Oxford

Desaguliers JT (1751) Cours de physique expérimentale, Traduit de l'anglais par le RP Pezenas de la Compagnie de Jésus, Professeur Royal d'Hydrographie a Marseille, 2 vols [trans: Desaguliers 1734). Rollin J–Jombert, Paris

Desaguliers JT (1734) A course of experimental philosophy, 2 vols. Innys W–Senex M–Longman T, London

Descartes R (1897–1913) Œuvres de Descartes. Par Charles Adams et Paul Tannery, Paris, 12 vols; also see Discours de la méthode et Essais, Specimina philosophiae, vol VI; Physico–mathematica, vol X, Le Monde ou Traité de la lumière, vol XI (Id, 1964–1974 by Rochot B, Costabel P, Beaude J et Gabbery A, Paris)

Desormes CB, Clément N (1819) Mémoire sur la Théorie des machines à feu (Extrait). In: 1817–1819. Bulletin des Sciences par la Société Philomathique de Paris. Imprimerie de Plassan, Paris, pp 115–118

Despretz MC (1823) Note extraite d'un Mémoire sur les chaleurs latentes de diverses vapeurs [It was read at Académie des Sciences on 3 October 1818]. Annales de chimie et de physique XXIV:323–331

Despretz MC (1821) Note extraite d'un Mémoire lu à l'Institut le 29 novembre 1819, par M.C. Despretz. Annales de chimie et de physique XVI:105–109

Destouches JL (1966) Logique et Théorie Physique. Synthèse 16:66–73

Destouches JL (1959) Physico–logical problems. In: Brouwer L (ed) Studies in logic and the foundations of mathematics. The axiomatic method with special references to geometry and physics. North–Holland Publishing Co, Amsterdam, pp 390–405

Destouches JL (1951) Sur la Mécanique Classique et l'Intuitionnisme. Indagations Mathematicae 13:74–79

Destouches JL (1948) Physico–logical problems. Synthése 411–417

Dhombres J, Dhombres N (1997) Lazare Carnot. Fayard, Paris

Dias PMC, Pinto SP, Cassiano DH (1995) The conceptual import of Carnot's theorem to discovery of the entropy. Archive for History of Exact Sciences 49(2):142–143

Dickinson HW (1939) A short history of the Steam Engine. The University of Cambridge Press, Cambridge

Dijksterhuis EJ (1955) The principal works of Simon Stevin–Mechanics, vol I. NV Swets & Zeitlinger, Amsterdam

Dirichlet JPGL (1837) Über die Darstellung ganz willkürlicher Funktionen durch Sinus–und Cosinus–reihen. Repertorium der Physik 1:152–174

Drago A (2005) Kolmogoroff and the relevance of the double negation law in science. In: Sica G (ed) Essays on the foundations of mathematics and logic–advanced studies in mathematics and logic. Polimetrica International Scientific Publisher, Monza, pp 57–81

Drago A (2000) Il principio di d'Alembert non è un principio. Sua relazione con il principio dei lavori virtuali. In: Tucci P (ed) Proceedings of XIX SISFA congress. Istituto Fisica Università di Milano, Como, pp 185–210[6]

Drago A (1999) Incommensurability as bound of hermeneutic science. In: Fehér M, Kiss O, Ropolyi L (eds) Hermeneutics and science. Kluwer Academic Publishers, Dordrecht, pp 135–155

Drago A (1991) Le due Opzioni. La Meridiana, Molfetta

Drago A (1992) Thermodynamics vs mechanics: a new look and a new didactics. In: Giaquinta P et al (eds) Proceedings of Taormina conference on thermodynamics. Accademia Pericolanti, Messina, pp 311–329

Drago A (1989a) La rivoluzione francese ha realizzato il programma giacobino di rifondazione di tutta la scienza. In: Proceedings of XL Riunione Società Italiana Progresso Scientifico. L'età della rivoluzione e Il progresso delle scienze, Bologna, pp 335–342

Drago A (1989b) The birth of symmetry in theoretical physics: Lazare Carnot's mechanics. In: Darvas G, Nagy D (eds) Symmetry in structure. Hungarian Academy of Science, Budapest, pp 98–101

Drago A (1988) A characterization of Newtonian paradigm. In: Scheurer PB, Debrock G (eds) Newton's scientific and philosophical legacy. Kluwer Academy Press, Dordrecht/Boston, pp 239–252

Drago A (1995) The choice of the kind of mathematics in historiography: the case–study of S Carnot. In: Kovàks L (ed) Studia Physica Savariensia I, History of science in teaching physics, pp 102–111

Drago A (1993a) Analisi critica dei principi della dinamica. In: Proceedings of Società dei Naturalisti di Napoli, pp 117–135

Drago A (1993b) Forza ed energia: l'analisi critica di Saint–Robert. Quaderni PRISTEM 4:91–113

[6][also available in .pdf via: http://www.brera.unimi.it/sisfa/atti/index.html]

Drago A (1993c) The principle of virtual works a source of two traditions in 18th century of mechanics. In: History of physics in Europe in 19th and 20th century. SIF, Bologna, pp 69–80

Drago A (2003) La riforma della dinamica secondo GW Leibniz. Hevelius, Benevento

Drago A (2007a) The genius of infinitesimal analysis: Berkeley's criticism and its generalization by Lazare Carnot. Speech In: International conference George Berkeley: religion and science in the age of the enlightenment. International Berkeley Society and University of Cassino, Gaeta, 26th–26th, Sept

Drago A (2007b) There exist two models of organization of a scientific theory. Atti della della Fondazione Ronchi 62(6):839–856

Drago A, Pisano R (2008) Note storiche sul superamento della teoria del calorico in Sadi Carnot. In: Tucci P (ed) Proceedings of XXV SISFA congress, Milano, pp C22.1–C22.6[7]

Drago A, Pisano R (2007) La novità del rapporto fisica–matematica nelle Réflexions di Sadi Carnot. Atti della Fondazione Giorgio Ronchi 62(4):497–525

Drago A, Pisano R (2005) La nota matematica nelle Réflexions sur la puissance motrice du feu di Sadi Carnot: interpretazione del calcolo col metodo sintetico. Quaderni di Storia della Fisica–Giornale di Fisica 13:37–57

Drago A, Perno A (2004) La teoria geometrica delle parallele impostata coerentemente su un problema (part I). Periodico di Matematiche VIII(4):41–52

Drago A, Cerreta P (2003) Il programma storiografico di Kuhn caratterizzato secondo due programmi di ricerca sui fondamenti della scienza. In: Garuccio A (ed) Proceedings of XXIII congress SISFA. Progedit, Bari, pp 120–130

Drago A, Boiano L (2002) La geometria di Lazare Carnot: Rinascita e potenziamento del metodo sintetico. Atti della Fondazione Giorgio Ronchi 57:9–122

Drago A, Pisano R (2002) S Carnot's Réflexions: a theory based on non–classical Logic. The Bulletin Symbolic Logic 8:130–131

Drago A, Pisano R (2000) Interpretazione e ricostruzione delle Réflexions di Sadi Carnot mediante la logica non–classica. Giornale di Fisica 40:195–217

Drago A, Manno SD, Mauriello A (2001) Una presentazione concettuale della meccanica di Lazare Carnot. Giornale di Fisica 42:131–156

Drago A, Capriglione M (1993) Lo status logico del principio dei lavori virtuali. In: Cellucci C (ed) Logica e filosofia della scienza. ETS, Pisa, pp 35–44

Drago A, Vitiello O (1991) Nuovi risultati su Sadi Carnot e le leggi dei gas. In: Calascibetta F (ed) Proceedings of del II Congresso Nazionale di Storia e Fondamenti della Chimica XLV/12: 391–402

Drago A, Cerreta P (1989) La struttura delle rivoluzioni scientifiche di T Kuhn rilevate dall'analisi dei suoi esempi storici. In: Proceedings of X congress SISFA, pp 49–61[8]

Drago A, Manno SD (1989) Le ipotesi della meccanica secondo L Carnot. Epistemologia 12:305–330

Drago A, Vitiello O (1986) Sulla datazione del manoscritto inedito di Sadi Carnot. Proceedings of VII SISFA congress, Padova, pp 149–154[9]

Drake S (2000) Two new sciences. A history of free fall, Aristotle to Galileo. Wall and Emerson, Toronto

Drake S (1985) Galileo's accuracy in measuring horizontal projections. Annali dell'Istituto e Museo di Storia dlle Scienze di Firenze 10(1):3–14

Drake S (1973) Galileo's experimental confirmation of horizonal inertia. Isis 64:291–305

Drake S, Drabkin IE (1969) Mechanics in sixteenth century Italy. The University of Wisconsin Press, Madison

[7] [also available in .pdf via: http://www.brera.unimi.it/sisfa/atti/index.html]

[8] [also via: http://albinoni.brera.unimi.it/Atti/index.html]

[9] [also available in .pdf via: http://www.brera.unimi.it/sisfa/atti/index.html]

Dugas R (1955) Histoire de la Mécanique (trans. Maddox JR) [1950: Editions du Griffon, Neuchâtel]. Dover, New York
Duhem PM (1977) Aim and structure of physical theory. Princeton University Press, Princeton
Duhem PM (1906–1913) Études sur Leonard de Vinci. Hermann, Paris
Duhem PM (1905–1906) Les origines de la statique, 2 vols. Hermann, Paris
Duhem PM (1905–06) Les origines de la Statique, Tome I–II. Hermann, Paris
Dulong PL (1829) Recherches sur la chaleur spécifique des fluides élastiques. Lu à *Académie des sciences* le 18 mai 1828. Annales de Chimie et de Physique XLI:113–159
Dulong PL, Petit AT (1816) Recherches sur les lois de la dilatation des solides, des liquides et des fluides élastiques, et sur la mesure exacte des températures. Annales de Chimie et de Physique II:240–263
Dummett M (1975) Principles of intuitionism. The Clarendon Press, Oxford
Dupré MA (1869) Théorie mécanique de la chaleur. Gauthier–Villars, Paris
Duthuron G (1940) Un Éloge de Vauban. Annales historiques de la révolution française, 17 May–June:152–165
Edmunds CK (1902) The motive power of heat. Popular Science Monthly, pp 117–128
Einstein A (1949 [1970]) Autobiographical notes. In: Schilpp PA (ed) Albert Einstein: philosopher–scientist, library of living philosophers, vol VII. The Cambridge University Press, London
Erlichson H (1999) Sadi Carnot, founder of the second law of thermodynamics. The European Journal of Physics 20:183–192
Euler L (1757) Principes généraux du mouvement des fluides. Mémoires de l'Academie des Sciences de Berlin 11:274–315
Euler L (1751) De machinis in genere Novi Commentarii Academiae Scientarum Imperialis Petro–Politanae. Petropoli Typis Academiae Scientiarum
Falk G (1985) Entropy, a resurrection of caloric – a look at the history of thermodynamics. The European Journal of Physics 6:108–115
Falk G, Hermann F, Schmid GB (1983) Energy forms or energy carriers? The American Journal of Physics 51(12):1074–1077
Faujas de Saint Fond B (1784) Raccolta di esperienze, osservazioni, riflessioni sui palloni aerostatici. In: Giusti E (ed) Il Giardino di Archimede, Gesiniana, Genova S III/23
Fermi E (1937) Thermodynamics. Dover (ed), New York
Feyerabend P (1989) Against method. Humanities Press, New York
Finn BS (1964) Laplace and the speed of sound. Isis 55(1):7–19
Fossombroni V (1794) Memoria sul princpio delle velocità virtuali. Gaetano Cambiagi, Firenze
Fourier JBJ (1888–1890) Œuvres de Fourier par les soins de M. Gaston Darboux, 2 vols. Gauthier–Villars, Paris
Fourier JBF (1878) Analytical theory of heat (trans: Freeman A). The Cambridge University Press, London
Fourier JBJ (1829) Mémoire su la théorie analytique de la chaleur dans les solides. Mémoires de l'Académie royale des sciences de l'Institut de France pour l'année 1825. Firmin Didot, Paris VIII:581–622
Fourier JBJ (1808) Théorie de la propagation de la chaleur dans les solides. Nouveau Bulletin des Sciences par la Société philomathique de Paris I(6):112–116
Fourier JBJ (1807) Théorie de la propagation de la chaleur dans les solides. Nouveau Bulletin des Sciences par la Société philomathique de Paris, Tome I(6):112–116. (n. 6, mars 1808), Bernard, Paris
Fourier JBJ (1822) Théorie analytique de la chaleur. Firmin Didot, Paris
Fourier JBJ (1798) Mémoire sur la statique, contenant la démonstration du principe de vitesses virtuelles, et la théorie des momens. Journal de l'École Polytechnique 2(5):20–60 [see also In: Fourier (1888–1890), pp 475–521]
Fox R (2012) Sadi Carnot on Political Economy. Science, Morals, and Public Policy in Restoration France. In: Buchwald (ed) 2012, pp 417–427

Fox R, Weisz G (2009) The organization of science and technology in France 1808–1914. The Cambridge University Press, Cambridge
Fox R (1995) Science, industry and the social order in post–revolutionary France, Variorum collected studies series 489. Ashgate Variorum, Aldershot
Fox R, Guagnini A (eds) (1993) Education, technology and industrial performance in Europe, 1850–1939. The Cambridge University Press, Cambridge
Fox R (1992) The culture of science in France, 1700–1900, Variorum collected studies series 381. Ashgate Variorum, Aldershot
Fox R (1988) Les Réflexions sur la puissance motrice du feu de Sadi Carnot et la leçon se leur éditions critique. La vie des sciences – Comptes rendus, série générale 5(4):283–301
Fox R (1976) The challenge of new technology: theorists and the high–pressure steam engine before 1824. In: Taton (ed) (1976), pp 149–168
Fox R (1974) The rise and fall of Laplacian physics. Historical Studies in the Physical Sciences 4:89–136
Fox R (1971a) The intellectual environment of Sadi Carnot: a new look. In: Proceedings of XII Congres International d'Histoire des sciences, vol IV. Blanchard, Paris, pp 67–72
Fox R (1971b) The caloric theory of gases: from Lavoisier to Regnault. The Clarendon Press, Oxford
Fox R (1970) Watt's expansive principle in the work of Sadi Carnot and Nicolas Clément. Notes and records of the Royal Society of London 24:233–253
Fox R (1969) James Prescott Joule, 1818–1889. In: North J (ed) Mid–nineteenth–century scientists. Pergamon Press, New York, pp 72–103
Fuchs HU (1987) Entropy in the teaching of introductory thermodynamics. The American Journal of Physics 55(3):215–216
Fuchs HU (1986) A surrealistic tale of electricity. The American Journal of Physics 54(10):907–909
Fuchs HU (1996) The dynamics of heat. Springer, New York
Gabbey AW, Herivel JW (1966) Un manuscrit inédit de Sadi Carnot. Revue d'histoire des sciences et de leurs applications 19(2):151–166
Galileo G (2002) Galileo Galilei. Le Mecaniche. Introduction and Notes by Gatto R. Olschki, Firenze
Galileo G (1954) Dialogues concerning two new sciences (Trans: Crew H, de Salvio A, an introduction by Favaro A). Dover Publications INC, Mineola
Galileo G (1890–1909) Le Opere di Galileo Galilei: Edizione nazionale sotto gli auspici di sua maestà il re d'Italia. Introductions and Notes by Favaro A. Barbèra G, Firenze
Galluzzi P (1979) Momento Studi Galileiani. Ateneo & Bizzarri, Roma
Galvani L ([1841] 1998). In: Gherardi S (ed) Opere edite ed inedite di Luigi Galvani. Tipografia di Emidio Dall'Olmo, Bologna
Galvani L (1794a) Dell'uso e dell'attività dell'arco conduttore nelle contrazioni de' muscoli. S Tommaso d'Aquino. Bologna. In Galvani L ([1841] 1988), pp 155–278
Galvani L (1794b) Supplemento al trattato dell'arco conduttore. S Tommaso d'Aquino. Bologna. In Galvani L ([1841] 1988), pp 279–299
Galvani L (1791) De viribus electricitatis in motu musculari commentarius. De Bononiensis Scientiarum et Artium atque Accademia Commentarii VII, pp 363–418 [*Id.*, (1953) Commentary on the Effects of Electricity on Muscolar Motion. Translation by Foley M, Burndy Library, Norwalk]
Galvani L (1797) Memorie sulla elettricità animale al celebre abate Lazzaro Spallanzani Professore nella Università di Pavia. In: Galvani L (1841), pp 299–434
Gaspard M (1799) Géométrie descriptive. Leçons données aux école normales l'an de la République, an 7, Badouin, Paris
Gauthey EM (1832) Traité de la construction des ponts. In: Navier CLMH (ed) 2 vols [next: Liège, Leduc, 1843]. Didot F, Paris
Gauthier Y (1991) De la logique interne. Vrin J, Paris

Gay–Lussac JL (1802) Recherche sur la dilatation des gaz et des vapeurs. Annales de Chimie 43:137–175

Ghisalberti A (1963) Dizionario Biografico degli Italiani. Istituto Enciclopedia Italiana, Roma, 5, pp 564–565

Gianelli G (1888) Cenni necrologicì del conte Paolo Ballada conte di St. Robert. Rivista mensile del Club Alpino. Novembre, Tipografia editrice Candeletti G, Torino, pp 387–389

Giannetto E (1989) Fisica Quantistica e verità logica. Epistemologia XII:261–276

Giles R (1979) The concept of a proposition in classical and quantum physics. Studia Logica 38:345–353

Giles R (1970) Foundations for quantum mechanics. Journal of Mathematical Physics 11:2139–2151

Gille B (1964) Les ingénieurs de la Renaissance. Hermann, Paris

Gillispie CC (2004) Science and policy in France: the revolutionary and napoleonic years. Princeton University Press, Princeton

Gillispie CC (1983) The Montgolfier brothers and the invention of aviation. Princeton University Press, Princeton

Gillispie CC, Youschkevitch AP (1982) Lazare Carnot Savant et sa contribution à la théorie de l'infini mathématique. Revue d'histoire des sciences XXXV(1):79–82

Gillispie CC (1980) Science and polity in France at the end of the old regime. Princeton University Press, Princeton

Gillispie CC (ed) (1970–1980) Dictionary of scientific biography. Charles Scribner's Sons, New York

Gillispie CC, Youschkevitch AP (1979) Lazare Carnot savant et sa contribution à la théorie de l'infini mathématique. Vrin, Paris [Collection des Travaux de l'Académie Internationale d'Histoire des Sciences, 20]

Gillispie CC (1970–1979) Carnot Lazare–Nicolas–Marguerite. In: Gillispie (ed) (1970–1980), III:70–79

Gillispie CC (1976) The scientific work of Lazare Carnot, and its influence on that of his son. In: Taton (ed) (1976), pp 23–33

Gillispie CC (1971) Lazare Carnot Savant. A monograph treating Carnot's scientific work, with facsimile reproduction of his unpublished writings on mechanics and on the calculus, and an essay concerning the latter by Youschkevitch AP. Princeton University Press, Princeton

Gillispie CC (1960) The age of objectivity. The Princeton University Press, Princeton

Gillmor CS (1971) Coulomb and the evolution of physics and engineering in eighteenth–century France. Princeton University Press, Princeton

Girard PS (1824) Réflexions sur la puissance motrice du feu sur les machinés propre à développer cette puissance, par S. Carnot, ancien élève de l'école polytechnique [...] prix 3 fr. Revue Encyclopédique XXIII–Juillet:414

Gondinet AD (1833) Necrologie Sadi Carnot. Association Polytecnique, Compte-Rendu Trimestriel, Janvier. Imprimerie Guiraudet [BNF, V 30915], p 46

Grant E (1971) Physical science in the Middle Ages. The Cambridge University Press, Cambridge

Grattan–Guinness I (1990a) Convolutions in French mathematics. Birklausen, Basel

Grattan–Guinness I (1990b) Work for the workers: engineering mechanics and its instruction. In: Grattan–Guinness (ed), pp 1046–1121

Grattan–Guinness I (1972) Joseph Fourier (1768–1830): a survey of his life and work. MIT Press, Cambridge MA

Grattan-Guinness I (1970) Berkeley's criticism of the calculus as a study in the theory of limits. Janus 56:215–227

Grattan-Guinness I (1969) Joseph Fourier and the revolution in mathematical physics. Journal of the Institute of Mathematics and its Applications 2(5):230–253

Grattan–Guinness I, Ravetz J (1980–1990) Fourier Jean Baptiste Joseph. In: Gillispie (1970–1990), V:93–99

Grinevald J (1976) Présentation d'un manuscript inédit de Sadi Carnot. In: Taton (ed) (1976), pp 383–395

Grove WR (1846) The correlation of physical forces. Longmans Green & Co., London
Guemeza J, Fiolhaisb C, Fiolhais M (2002) Sadi Carnot on Carnot's theorem. The American Jornal of Physics 70(1):42–47
Guenyveau A (1810) Essai sur la science des machines. Des Moteurs. Des Roues hydrauliques, des Machines à colonne d'eau, du Bélier hydraulique, des Machines à vapeur, des Hommes et des Animaux. Kindelem JB, Lyon
Guerlac H (1970–1990) Black Joseph. In: Gillispie (ed) (1970–1980), II:173–183
Guerlac H (1963) Some historical assumptions of the history of science. In: Crombie (1963), pp 797–817
Gustav Z (1869) Théorie mécanique de la chaleur: avec ses applications aux machines. Gauthier–Villars, Paris
Haack S (1974) Deviant logics. The Cambridge University Press, Cambridge
Hachette JNP (1811) Traité élémentaire des machines. Klostermann fils, Paris/Saint Pétersbourg
Hallyn F (2004) Les structures rhétoriques de la science. De Kepler à Maxwell. Éditions du Seuil–Collection des Travaux, Paris
Harned HS, Owen BB (1950) The physical chemistry of elecrtolytic solutions. Reinhold Publishing Corporation, New York, pp 297–325
Helm G (1887) Die Lehre von der Energie. Felix, Leipzig
Helmholtz H (1853) H. von Helmholtz, Scientific Memoirs (trans: Taylor R). Tyndall & Francis, London
Helmholtz H (1847) Über Die Erhaltung der Kraft, eine physikalische Abhandlung, Reimer, Berlin. In: Helmholtz H (1853)
Henrion MRA (1833) Sadi Carnot. Annuaire biographique année 1830–1834, Tome I/1833. Méquignon P (ed), Paris
Hirn GA (1887) La thermodynamique et le travail chez les êtres vivants. Revue scientifique 22:673–684 [May, Paris, Bureau des revues]
Hirn GA (1868) Conséquences philosophiques et métaphysiques de la Thermodynamique, Analyse élémentaire de l'univers. Gauthier–Villars, Paris
Hirn GA (1867) Mémoire sur la Thermodynamique. Gauthier–Villars, Paris
Hirn GA (1864a) Esquisse élémentaire de la théorie mécanique de la chaleur et ses conséquences philosophiques. Colmar, Paris
Hirn GA (1864b) Théorie mécanique de la chaleur. Exposition analytique et expérimental. Gauthier–Villars, Paris
Hirn GA (1862a) Exposition analytique et expérimental de la théorie mécanique de la chaleur. Mallet-Bachelier, Paris
Hirn GA (1862b) Théorie mécanique de la chaleur. Tramblay, Leiber
Hirn GA (1863) Galilei G (1954) Dialogues concerning two new sciences. Translated by Crew H and de Salvio A and an introduction by Favaro A. Dover Publications INC, Mineola, NY. Tramblay, Leiber
Hodges W (1998) The laws of distribution for syllogism. Notre Dame Journal of Formal Logic 39:221–230
Hodges W (1997) A shorter model theory. The Cambridge University Press, Cambridge
Hodges W (1983) Elementary predicate logic. In: Gabbay DM, Guenthner F (eds) Handbook of philosophical logic – elements of classical logic, vol I. Reidel, Dordrecht, pp 1–131
Hornix WJ (1982) Essay review of *Réflexions sur la puissance mortrice de feu* by S. Carnot. Annals of Science XXXIX:399–406
Hoyer U (1976) La théorie de Carnot première et seconde approximations de la thermodynamique. In Taton (ed) (1976), pp 221–228
Hoyer U (1975) How did Carnot calculate the mechanical equivalent of heat? Centaurus 19:207–219
Hoyer U (1974) Über den Zusammenhang der Carnotshen Theorie mit der Thermodynamik. The Archive for History of Exact Sciences 13:359–375
Jacquier E (1867) Exposé élémentaire de théorie mécanique de la chaleur appliquée aux machines. Gauthier–Villars, Paris

Jamin MJ (1886) Cours de Physique, Tome I–IV . Gauthier–Villars, Paris, pp I–II
Jamin MJ (1868) Cours de Physique à l'École Polytechnique, Tome I–IV. Gauthier–Villars, Paris, pp I–II
Jammer M (1961) Concepts of mass in classical and modern physics. The Harvard University Press, Cambridge, MA
Jammer M (1957) Concepts of force: a study in the foundations of dynamics. The Harvard University Press, Cambridge, MA
Jouguet E (1924), Lectures de mécanique. La mécanique enseignée par les auteurs originaux, 2 vols. Gauthier–Villars, Paris
Joule JP (1965) The scientific papers of James Prescott Joule, vol I. Dawson's, London, pp 277–281
Joule JP (1847) On the mechanical equivalent of heat, as determined by the heat evolved by the friction of fluids. Philosophical Magazine, 3rd series, XXXI(207):173–176
Joule JP (1845) On the mechanical equivalent of heat. British Association meeting, Cambridge [read before the British Association at Cambridge, June]
Joule JP (1844) On the changes of temperature produced by the rarefaction and condensation of air. Philosophical Magazine, 3rd series, XXVI:369–383
Kelly EM (1964) Simple treatment of thermodynamic efficiency. The American Journal of Physics 32(8):643
Kerker M (1957) Sadi Carnot. The Scientific Monthly 85:143–149
Keston D (1988) Joseph Fourier, Politician & Scientist[10]
Kieseppä IA (2000) Rationalism, naturalism, and methodological principles. Erkenntnis 53(3):337–352
Klein F (1967) Development of mathematics in the nineteenth century (trans: Ackerman M). Math SciPress, Brokline
Klein F (1926) Vorlesungen über die Entwicklung der Mathematik im 19. Jahrhundert, Berlin (English Translation in Klein 1967)
Klein MJ (1976) Closing the Carnot cycle. In: Taton (ed) (1976), pp 213–219
Klein MJ (1969) Gibbs on Clausius. Historical Studies in the Physical Sciences 1:127–149
Knobloch E (2005) Géométrie pratique, Géométrie savante. Albertiana 8:27–56
Knobloch E, Vasoli C, Siraisi N (2001) Il Rinascimento. In: Medioevo, Rinascimento. Storia della scienza. Istituto della Enciclopedia Italiana, Roma, 4, pp 605–1044
Koenig F (1959) On the history of science and the second law of thermodynamics. In: Evans HM (ed) Men and moments in the history of science. The Washington University Press, Seattle, pp 57–111
Koetsier T (2010) Simon Stevin and the rise of Archimedean Mechanics. In: Ceccarelli M, Paipetis S (eds) Proceedings of the genius of archimedes. Springer, Dordrecht, pp 85–111
Koetsier T (2007) Lazare Carnot's theory of machines and the background of Ampère's introduction of the word kinematics. In: Ceccarelli M (ed) Proceedings of 13th national conference on mechanisms and machines. Springer, Dordrecht, pp 29–35
Kolmogorov AN (1991–1993) Selected works of A.N. Kolmogorov. In: Mathematics and mechanics, vol I. Kluwer, Dordrecht
Kolmogorov AN (1932) Zur Deutung Der Intuitionistischen Logik. Matematicheskij Zeitschr 35:58–65 [English trans In: Kolmogorov (1991–1993), I:151–158]
Kolmogorov AN (1925) On the principle tertium non datur. Matematicheskij Sbornik 32:646–667 [English translation In: Kolmogorov (1991–1993), I:40–68]
Kostic MM (2008) Sadi Carnot's ingenious reasoning of ideal heat engine reversible cycles In: Proceedings of IASME/WSEAS international conference on Energy, Environment, Ecosystems and Sustainable Development (EEESD'08), Algarve, Portugal, pp 159–166

[10][also available at: http://www.todayinsci.com /F/ Fourier_JBJ/FourierPoliticianScientistBio.htm]

Koyré A (1986) De la mystique à la science. Cours, conférences et documents 1922–1962. In: Redondi P (ed) Cours, conférences et documents 1922–1962. EHESS Editions, Paris
Koyré A (1973) Études d'histoire de la pensée scientifique. Gallimard, Paris
Koyré A (1971) Études d'Histoire de la pensée philosophique. Gallimard, Paris
Koyré A (1966) Études galiléennes. Hermann, Paris
Koyré A (1965) Newtonian studies. The Harvard University Press, Cambridge, MA
Koyré A (1963) Scientific Change. In: Crombie (ed) (1963), pp 847–857
Koyré A (1961a) Du monde de « à–peu–près » à l'univers de là précision. M Leclerc et Cie – Armand Colin Librairie, Paris (Id, les philosophes et la machine. Du monde de l'«à–peu–près» à l'univers de la précision. Études d'histoire de la pensée philosophique)
Koyré A (1961b) Message d'Alexandre Koyré à l'occasion du centenaire de la naissance d'Emile Meyerson. Bulletin de la Société française de philosophie 53:115–s
Koyré A (1957) From the closed world to the infinite universe. The Johns Hopkins University Press, Baltimore
Kragh H (1987) An introduction to the historiography of science. The Cambridge University Press, Cambridge
Kuhn TS (1955) Carnot's version of cycle. The American Journal of Physics 23:91–94
Kuhn TS ([1961] 1980) Engineering precedent for the work of Sadi Carnot. In: Proceedings of IX Congrès international d'Histoire des sciences (Barcelona – Madrid 1959). Asociatión para la historia de la ciencia Española, Barcelona/Paris, pp 530–535
Kuhn TS (1974) Second thoughts on Paradigms. In Suppe F (ed) The structure of scientific theories. The Illinois University Press, Urbana, pp 459–482
Kuhn TS (1970) Reflections on my Critics. In: Lakatos I, Musgrave A (eds) Criticism and growth of knowledge. Proceedings of international colloquium in the philosophy of science. The Cambridge University Press, Cambridge, pp 231–278
Kuhn TS (1963) The function of Dogma in Scientific Research. In: Crombie (1963), pp 347–369
Kuhn TS (1962) The structure of scientific revolutions. The Chicago University Press, Chicago
Kuhn TS (1961) Sadi Carnot and the Cagnard Engine. Isis 52(4):567–574
Kuhn TS (1960) Engineering precedent for the work of Sadi Carnot. Archives Internationales d'Histoire des Sciences 13:251–255
Kuhn TS (1959) Energy conservation as example of simultaneous discovery. In: Clagett M (ed) Critical problems in the history of science. Wisconsin University Press, Madison, pp 321–356
Kvasz L (1999) On classification of scientific revolutions. Journal for General Philosophy of Science 30(2):201–232
La Mer VK (1954) Some current misinterpretation of NL Sadi Carnot's memoir and cycle. The American Journal of Physics 22:20–27
Laboulaye C (1860) Essai sur l'équivalent mécanique de la chaleur. Lacroix et baudry, Paris
Lacroix SF (1813) Traite elementaire de trigonometrie rectiligne et spheriue, et d'application de l'algebra a la geometrie par Lacroix SF Ve Courcier, Paris
Lagrange JL (1973a) Œuvres de Lagrange. Seconde édition, Courcier (ed), I–XIV vols, (in X) Paris. Gauthier–Villars, Paris
Lagrange JL (1973b) Théorie des fonctions analytiques. In: Lagrange (1973), vol IX
Lagrange JL (1881) Théorie des fonctions analytiques contenant les principes du calcul différentiel, dégagés de toute considération d'infiniment petits, de limites et de fluxions, et réduits à l'analyse algébrique des quantités finies. In: Serret JA (ed) Œuvres de Lagrange. Cinquième édition, vol IX, Chapter VII. Gauthier–Villars, Paris, pp 409–410
Lagrange JL (1813) Théorie des fonctions analytiques. Courcier, Paris
Lagrange JL (1806) Leçons sur le calcul des fonctions, Nouvelle edn. Courcier, Paris
Lagrange JL (1797) Théorie des fonctions analytiques contenant les principes du calcul différentiel, dégagés de toute considération d'infiniment petits, de limites et de fluxions, et réduits à l'analyse algébrique des quantités finies. Journal de l'École Polytechnique III(9):91–93
Lagrange JL (1788) Mécanique Analytique. Desaint, Paris

Lagrange JL (1764) Recherches sur la libration de la Lune dans lesquelles on tache de résoudre la question proposes par l'Académie Royale des Sciences, pour le Prix de l'année 1764. Prix de l'académie Royale des Sciences de Paris, IX (1777), pp 1–50. See also In: Lagrange (1973), Tome VI, pp 5–61
Lamé G (1861a) Discours préliminaire. In: Leçons sur la théorie analytique de la chaleur. Mallet–Bachelier, Paris
Lamé G (1861b) Leçons sur la théorie analytique de la chaleur. Mallet–Bachelier, Paris
Lamé G (1852) Leçons sur la théorie mathématique de l'élasticité des corps solides. Bachelier, Paris
Lamé G (1836) Cours de physique de l'école polytechnique, vol I. Bachelier, Paris (Id, Cours de physique de l'école polytechnique, 2nd edn. Bruxelles)
Landriani M (1806) [he is quoted in] Table générale des articles contenus dans les vingt–six derniers volumes du Journal de physique, depuis 1787 jusqu'en 1802, pour faire suite à celle qui est imprimée à la fin du second volume de l'année 1786 par L Cotte. Journal de physique de Chimie, d'histoire naturelle et des arts LXIII:24, 51, 52 [1–106[11]]
Landriani M (1786) [he is quoted in] Table générale des articles. In: Observations sur la physique, sur l'histoire naturelle et sur les arts XXIX:410, 411, 413, 436 [401–473]
Landriani M (1785) Suite de la dissertation de M Landriani, sur la chaleur latente. Observations sur la physique, sur l'histoire naturelle et sur les arts XXVI:197–207
Laplace PS (1836 [1984]) Exposition du système du monde. Fayard, Paris
Laplace PS (1822) Notes sur la vitesse du son. Annales de chimie et de physique XX:266–268
Laplace PS (1816) Sur la vitesse du son dans l'air et dans l'eau. Annales de chimie et de physique III:238–241
Laplace PS (1805) Traité de mécanique céleste. Courcier, Paris
Lavoisier AL (1789 [1937]) Traité élémentaire de Chimie. Gauthier–Villars, Paris
Lavoisier AL (1862–1893) Œuvres de Lavoisier publiées par le soin de Son Excellence le Ministre de l'instruction publique et des cultes, 6 vols. Imprimerie Impériale, Paris
Lavoisier AL, Laplace PS ([1780] 1784) Mémoires sur la Chaleur. In: Histoire de l'Académie Royale des Sciences. (First reading on 18th June 1783). Imprimerie Royale, Paris, pp 355–408
Leibniz GW (2009–) Gottfried Wilhelm Leibniz Naturwissenschaftlich–medizinisch–technische Schriften Reihe VIII der Akademieausgabe. Berlin–Brandenburgische Akademie der Wissenschaften, Российская Академия Наук, in Zusammenarbeit mit dem Centre National de la Recherche Scientifique, der Gottfried Wilhelm Leibniz Bibliothek Hannover und der Herzog August Bibliothek Wolffenbüttel. Knobloch E (project leader editor). There is a printed version of the volume(s) and there is an online–version available via: http://leibnizviii.bbaw.de/
Leibniz GW (1849–1863) G.W. Leibniz Die mathematische Schriften, 7 vols. Gerhardt CI, Berlin–Halle [many reprints by Hildesheim]
Lemons DS, Penner MK (2008) Sadi Carnot's contributions to the second law of thermodynamics. The American Journal of Physics 76(1):21–25
Lervig P (1985) Sadi Carnot and the steam engine: Nicolas Clément's lectures on industrial chemistry, 1823–28. The British Journal for the History of Science 18(59):147–196
Lervig P (1982) What is heat? C. Truesdell's view of thermodynamics. A critical discussion. Centaurus 26(2):85–122
Lervig P (1976) The existence of work function in Carnot's theory. In: Taton (ed) (1976), pp 199–212

[11]Here the *Table générale des articles contenus dans les vingt–six derniers volumes du Journal de physique, depuis 1787 jusqu'en 1802* is published in *Journal de physique de Chimie, d'histoire naturelle et des arts* related to several issues of the year 1806, pp 1–480. This *Table générale* is printed in the end of last issue on December 1806 of the up mentioned *Journal.* The numbers of pages started with new number "2". The number "1" on the frontispiece is not cited.

Lervig P (1972) On the structure of Carnot's theory of heat. Archive for History of Exact Sciences 9(2):222–239
Lindberg DC (1976) Theories of vision from Al–Kindi to Kepler. The Chicago University Press, Chicago
Liouville J (1836) Thèses de mécanique et d'astronomie. Bachelier, Paris
Lucreti CT (1922) De rerum natura. Oxford University Press, Oxford
Mach E ([1883] 1996) The science of mechanics – a critical and historical account of its development, 4th edn. Open Court–Merchant Books, La Salle
Mach E (1986) Principles of the Theory of Heat, Historically and Critically Elucidated, McGuinness B (ed), vol 17. Reidel D Publishing Co, Boston
Mach E (1896) Die Prinzipien der Wärmelehre. Historisch–kritisch entwickelt. Barth JA, Brockhaus, Leipzig
Magrini L (1861) Notizie biografiche e scientifiche su Alessandro Volta desunte da' suoi autografi recentemente rinvenuti comunicate da prof. Luigi Magrini, M. E. nelle adunenze del 21 febbrajo e del 7 marzo 1861. Atti del Reale Istituto lombardo di scienze, lettere ed arti, Reale Istituto lombardo di scienze e lettere, vol I
Magrini L (1859) Notizie biografiche e scientifiche su Alessandro Volta desunte da' suoi autografi recentemente rinvenuti comunicate da prof. Luigi Magrini ME nelle adunenze del 21 febbrajo e del 7 marzo 1861. In: Proceedings of Reale Istituto lombardo di scienze, lettere ed arti, Reale Istituto lombardo di scienze e lettere, vol 2, 21/1:259
Mahoney MS (1994) The mathematical career of Pierre de Fermat, 1601–1665. Princeton University Press, Princeton
Maquer PJ (ed) ([1786] 1788) Dictionnaire de chimie contenant la théorie et la pratique de cette science. Son application à la Physique, à l'Histoire Naturelle, à la Médecine, & aux Arts dépendans de la Chymie. Théophile Barrois, Paris.
Marrou HI ([1954] 1975) De la connaissance historique. Éditions du Seuil, Paris
Maxwell JC (1871) Theory of heat. Longmann Green, Roberts & Green, London
Mayer RJ (1842a) Bemerkungen Über die Kräfte der unbelebten Natur. Annalen der Chemie und Pharmazie 42:233–240
Mayer RJ (1842b) Bemerkungen Über die Kräfte der unbelebten Natur. In: Weyrauch JJ (ed) Die Mechanik Warme, Stuttgart
Mayer RJ (1840) Über Die Quantitative und Qualitative Bestimmung der Kräfte [sent to Poggendorff's Annalen der Physik and Chemie on June 16]
Mendoza E (1963) Sadi Carnot and the Cagnard engine. Isis 54:262–263
Mendoza E (ed) (1960) Reflections on the motive power of fire and other papers on the second law of thermodynamics by E Clapeyron and R Clausius. Dover, New York
Mendoza E (1959) Contributions to the study of Sadi Carnot and his work. Archive International d'Historie des Sciences XII:377–396
Meusnier de la Place J–B–M–CH (1783) Mémoire sur l'équilibre des machines aérostatiques, sur les différents moyens de les faire monter et descendre, et spécialement sur celui d'exécuter ces manœuvres, sans jeter de lest, et sans perdre d'air inflammable, en ménageant dans le ballon une capacité particulière, destinée à renfermer de l'air atmosphérique, présenté à l'Académie, le 3 décembre 1783, avec une addition contenant une application de cette théorie au cas particulier du ballon que MM. Robert construisent à St–Cloud et dans lequel ce moyen doit être employé pour la première fois. Extrait du Journal de physique, Juliet: 1784
Meyer M (1979) Découverte et justification en science. Klincksieck, Paris
Meyer M (1989) De la problématologie. Philosophie, science et language. Mardaga, Bruxelles
Moigno A (ed) (1865) Les mondes Revue hebdomadaire des sciences et de leurs applications aux arts et l'industrie. Tome VIII. Giraud E, Paris
Monge G (1799) Geometrie descriptive. Leçons données aux écoles normales, L'an 3 de la République ; par Gaspard Monge, de l'Institut national. Baudouin, Imprimeur du Corps législatif et de l'Institut national, an VII, Paris
Montbrial AL (1976) Interprétation économiques suggérées par le théorème de Carnot. In: Taton (ed) (1976), pp 333–335

Morveau LBG, Lavoisier AL, Berthollet CL, de Fourcroy AF (1787) Méthode de nomenclature chimique. Cuchet, Paris

Mottelay PF (1922) Bibliographical history of electricity and magnetism. Charles Griffin and Company Limited Publication, London

Muller I (2007) A history of thermodynamics. The doctrine of energy and entropy. Springer, New York, pp 13–21

Nagel E (1961) The structure of science: Problems in the logic of scientific explanation. Harcourt–Brace & World Inc, New York

Navier CLMH (1864) De la résistance des corps solides 3e ed. avec des notes et des appendices par M. Barre de Saint–Venant, Dunod, Paris

Navier CLMH (1856) Résumé des Leçons d'analyse données à l'École Polytechnique, Suivi de notes par M Ernest Liouville. Deuxième édition revue par M Ernest Liouville, 2 vols. Dalmot, Paris

Navier CLMH (1841) Résumé des leçons de mécanique donnée à l'École Polytechnique. Carilian–Goeury et Vor Dalmot, Paris

Navier CLMH (1826) Résumé des leçons d'analyse l'École des Ponts et Chaussées, sur l'application de la mécanique a l'établissement des constructions et des machines. Didot F père et fils, Paris (Id., (1833 et 1838), Deuxième édition. Carilian–Goeury, Paris; Id., (1839), Société Belge de Librairie, Bruxelles)

Navier CLMH (1821) Notes sur l'action mécanique des combustibles. Annales de chimie et de physique XVII:357–372

Navier CLMH (1819) Notes. In: de Bélidor (1819), pp 111–113

Navier CLMH (1818) Détails historiques sur l'emploi du principe des forces vives dans la théorie des machines, et sur diverses roues hydrauliques. Annales de Chimie et de Physique par Gay–Lussac et Arago (eds) 9:146–159

Navier CLMH, de Prony G, Girard S (1829) Du calcul de l'effet des machines, ou Considérations sur l'emploi des moteurs et sur leur évaluation, pour servir d'introduction à l'école spéciale des machines; par M Coriolis, ingénieur des Ponts et Chausses. Rapport fait le 8 Juin 1829 à l'Académie des sciences par une commission composée de MM. De Prony, Girard, et Navier rapporteur. Bulletin des sciences mathématiques, Firmin A Didot, Paris, XII:103–116

Nemore J (1533) Liber Iordani Nemorarii viri clarissimi, de ponderibus propositiones XIII & earundem demonstrationes, multarumque rerum rationes sanè pulcherrimas complectens, nunc in lucem editus.Cum gratia & priuilegio Imperiali, Petro Apiano Mathematico Ingolstadiano ad xxx. annos concesso. M. D. XXXIII

Newton I (1999) In: Cohen IB, Whitman A (eds) The principia: mathematical principles of natural philosophy. The University of California Press, London

Newton I (1803) The mathematical principles of natural philosophy, by Sir Isaac Newton (trans: English by Motte A), vol I. Symonds, London

Newton I (1730) Queries. In: Innys W (ed) Optiks, 4th English edition corrected, London

Newton I (1714) Philosophiae Naturalis Principia Mathematica. Reprints of 2nd English edition [1713, Cotes R (ed), Cantabrigiae], Rogerus Cotes Editoris Praefatio, Sumptibus Societatis, Amstaelodami

Newton I (1687) Philosophiae naturalis Principia mathematica. Londini, jussi Societatus Regiae ac typis Josephi Streater, prostat apud plures bibliopoles [two variant title pages]

Nickles T (2001) The logic and methodology of science. Early modern thought: seven studies by Fred Wilson. Isis 92(4):775–776

Oresme N (1966) De proportionibus proportionum and Ad pauca respicientes. In: Grant E (ed), The University of Wisconsin Press, Madison

Ørsted HC (1822) Instruments pour mesurer la compression de l'eau. Annales de chimie et de physique XXI:99–100

Ørsted HC (1823) Sur la compressibilité de l'eau. Annales de chimie et de physique XXII:192–198

Ostwald W (1892) Klassiker der exakten Wissenschaften. Engelmann 37, Lipsia

Palmer R (1941) Twelve who ruled: the committee of public safety during the terror. Princeton University Press, Princeton

Pancaldi G (2005) Volta: science and culture in the age of enlightenment. Princeton University Press, Princeton

Panza M (2003) The origins of analytic mechanics in the 18th century. In: Jahnke HN (ed). A history of analysis. Proceedings of the American Mathematical and London Mathematical Society, S.L., pp 137–153

Pappus from Alexandria (1588) Mathematicae collectiones, by Federico Commandino. Pesaro. See also: Pappus of Alexandria, Matematical collections

Parent A (1745) [1706]) Sur la plus grande perfection possible des machines. In: Histoire et Mémoires de l'Académie Royale des Sciences. Année [29 November] MDCCIV. Avec les Mémoires de Mathématique & de Physique, pour la même année. Coignard G JB & Guerin HL, Paris, pp 323–338

Parigi S (ed) (2010) George Berkeley. Religion and science in the age of enlightenment. Springer, Dordrecht

Partington JR (1960) Joseph Black's "Lectures on the elements of chemistry". Chymia VI:27–67

Payen J (1971) Deux nouveaux documents sur Nicolas Clément. Revue d'histoire des sciences 24(1):45–60

Payen J (1968) Une source de la pensée de Sadi Carnot. Archive international d'Historie des Sciences XXI:18–30

Payen J (1967) Betancourt et l'introduction en France de la machine à vapeur à double effet (1789). Revue d'histoire des sciences et de leurs applications XX(2):187–198

Perkins J (1820) On the compressibility of Water. Read on 29 June 1820. Philosophical Transaction of the Royal Society of London 110:324–330

Perkins J (1821) Sur la compressibilité de l'eau. Annales de chimie et de physique XVI:321–327

Petit AT (1818) Sur l'emploi du principe des forces–vives dans le calcul de l'effet des machines. Annales de Chimie et de Physique par Gay–Lussac et Arago (eds) 8:287–305

Petrini P (1808) Chimica e Fisica. Atti dell'imperiale Accademia Pistojese di scienze e lettere, vol I. Bracali e Figlio, Pistoja, I/10

Picard E (1927) Sadi Carnot, Biographie et Manuscrit. Gauthier–Villars, Paris

Pisano R (2011a) Physics–mathematics relationship. Historical and epistemological notes. In: Barbin E, Kronfellner M, Tzanakis C (eds) Proceedings of the ESU 6 European Summer University history and epistemology in mathematics. Verlag Holzhausen GmbH–Holzhausen Publishing Ltd, Vienna, pp 457–472

Pisano R (2012a) On Lazare and Sadi Carnot. A synthetic view of a historical epistemological research program. In: Matovani R (ed), Proceedings of XXX SISFA Congress. Argalia Editore, Urbino, pp 147–153

Pisano R (2012b) Historical reflections on the physics mathematics relationship in electromagnetic theory. In: Barbin E, Pisano R (eds) The dialectic relation between physics and mathematics in the XIXth century. Springer, Dordrecht, pp 31–58

Pisano R (2013) Reflections on the Scientific Conceptual Streams in Leonardo da Vinci and his Relationship with Luca Pacioli. Advances in Historical Studies 2(2):32–45

Pisano R, Bussotti P (2012) Galileo and Kepler. On Theoremata Circa Centrum Gravitatis Solidorum and Mysterium Cosmographicum. History Research 2(2):110–145

Pisano R, Casolaro F (2011b) An historical inquiry on geometry in relativity. Reflections on early relationship geometry–physics (Part One). History Research 1(1):47–60 [see also part Two: History Research 2(1):56–64]

Pisano R (2010) On principles in Sadi Carnot's thermodyamics (1824). Epistemological reflections. Almagest International Interdisciplinary Journal 2(2010):128–179

Pisano R (2009a) On method in Galileo Galilei' mechanics. In: Hunger H (ed) Proceedings of ESHS 3rd conférence. Austrian Academy of Science, Vienna, pp 147–186

Pisano R (2009b) Continuity and discontinuity. On method in Leonardo da Vinci' mechanics. Organon 41:165–182

Pisano R (2009c) Il ruolo della scienza archimedea nei lavori di meccanica di Galilei e di Torricelli. In: Giannetto E, Giannini G, Capecchi D, Pisano R (eds) Da Archimede a Majorana: La fisica nel suo divenire. Proceedings of XXVI SISFA congress. Guaraldi Editore, Rimini, pp 65–74

Pisano R (2009d) Towards high qualification for science education. The loss of certainty. International Journal of Baltic Science Education 8(2):64–68

Pisano R (2008) Il ruolo della scienza meccanica nella progettazione degli architetti e degli ingegneri del Rinascimento (The role played by mechanical science in the architects and engineers in the Renaissance), 2 vols. University of Roma la Sapienza, Roma. Ph.D. dissertations,[12] Roma

Pisano R (2007a) Note sui Principes de Thermodynamique di P de Saint Robert. In: Proceedings of XXIV SISFA congress. Biblipolis, Napoli–Avellino, pp 129–134

Pisano R (2007b) A history of chemistry à la Koyré? Introduction and setting of an epistemological problem. Khimiya 17(2):143–161

Pisano R (2007c) Évariste Galois' algebraic theory, epistemological reflections and educational elements. In: Barbin E, Stehlikova N, Tzanakis C (eds) Proceedings of the ESU 6 European Summer University history and epistemology in mathematics, pp 147–148

Pisano R (2006a) Mathematics of logic and logic of mathematics. Critical problems in history of science. The Bulletin of Symbolic Logic 12(2):358

Pisano R (2006b) La Science et la Hypothèse di JH Poincaré. Note epistemologiche Rivista di Epistemologia didattica 1:279–300

Pisano R (2005) Si può insegnare la pluralità delle logiche? Periodico di Matematiche 1:41–58

Pisano R (2004) Quanti sono i principi della termodinamica. In: Proceedings of XLIII AIF congress–LFNS, pp 203–211

Pisano R (2003) Il ciclo di S Carnot e la pila di A Volta. In: Garuccio A (ed) Proceedings of del XXIII SISFA congress. Progedit, Bari, pp 327–348

Pisano R (2001) Interpretazione della nota matematica nelle Réflexions sur la puissance motrice du feu di Sadi Carnot. In: Proceedings of XX SISFA congress. Osservatiorio Astronomico di Capodimonte, Napoli, pp 205–230[13]

Pisano R, Capecchi D (2013) Conceptual and Mathematical Structures of Mechanical Science in the Western Civilization around 18th Century. Almagest 4(2):86–121

Pisano R, Capecchi D (2014) Tartaglia's Science of Weights and Mechanics in Sixteenth–Century. Selections from Quesiti et inventioni diverse: Books VII–VIII, forthcoming

Pisano R, Capecchi D (2010) On Archimedean roots in Torricelli's mechanics. In: Ceccarelli M, Paipetis S (eds) Proceedings of the genius of archimedes. Springer, Dordrecht, pp 17–27

Pisano R, Capecchi D, Lukešová A (eds) (2013) Physics, Astronomy and Engineering. Critical Problems in the History of Science. In: Proceedings of international 32nd congress for The SISFA–Italian Society of Historians of Physics and Astronomy. The Scientia Socialis UAB & Scientific Methodical Centre Scientia Educologica Press, Šiauliai University, Lithuania

Pisano R, Drago A (2013) The modern thermodynamics as based on the principle of virtual laws. In: Pisano R, Capecchi D, Lukešová A, pp 345–352

Pisano R, Gaudiello I (2009a) Continuity and discontinuity. An epistemological inquiry based on the use of categories in history of science. Organon 41:245–265

Pisano R, Gaudiello I (2009b) On categories and scientific approach in historical discourse. In: Hunger H (ed) Proceedings of ESHS 3rd conference. Austrian Academy of Science, Vienna, pp 187–197

Pisano R, Capecchi D (2009) La Théorie Analytique de la Chaleur. Notes on Fourier and Lamé. In: Barbin E (ed) Proceedings of Gabriel Lamé, les pérégrinations d'un ingénieur du XIXe siècle. Bulletin de la Sabix 44:83–90

Pisano R, Capecchi D (2008) Leonardo da Vinci. Recenti riflessioni storico–epistemologiche sulla deformabilità dei corpi. In: Proceedings of XLVI AIF congress–LFNS, pp 120–129

Pisano R, di Pietrantonio F (2007) Meccanica ed architettura nel Rinascimento. Note su Leonardo da Vinci. Rivista di Epistemologia didattica 2:97–219

[12][available in .pdf via: *International Galilean Bibliography,* Istituto e Museo di Storia delle Scienze. Firenze: http://biblioteca.imss.fi.it/].

[13][also available in .pdf via: http://www.brera.unimi.it/sisfa/atti/index.html]

Poinsot L (1975) In: Bailache P (ed) La Théorie générale de l'équilibre et du mouvement. Vrin, Paris

Poinsot L (1838) Note sur une certaine démonstration du principe des vitesses virtuelles, qu'on trouve au chapitre III de livre 1er de la "Mécanique céleste". Journal de mathématiques pures et appliquées 1(3):244–248

Poinsot L (1806) La Théorie générale de l'équilibre et du mouvement des systèmes. Journal de l'Ècole Polytechnique 6(13):206–241

Poisson SD (1833) Traité de Mécanique. Bachelier, Paris

Poisson SD (1829) Mémoire sur l'équilibre et le mouvement des corps élastiques. Mémoires de l'Académie des sciences de l'Institut de France. [Read at Académie, 14 April 828]. Paris, vol 8, pp 357–570

Poisson SD (1823a) Sur la chaleur de gaz et des vapeurs. Annales de Chimie et de Physique XXXIII:337–352

Poisson SD (1823b) Sur la vitesse du son. Annales de Chimie et de Physique XXIII:5–16

Poli SG (1817) Elementi di fisica sperimentale, Tomo 4. Andrea Santini e Figlio, Venezia

Poncelet JV (1827–1829) Cours de mécanique industrielle, professé de 1828 à 1829, par M. Poncelet, 2e partie. Leçons rédigées par M. le capitaine du génie Gosselin. Lithographie de Clouet, Paris, (s.d.)[14]

Poncelet JV (1829) Note sur quelques principes de mécanique relatifs à la science des machines; par M. Poncelet, capitaine du Génie. Bulletin des sciences mathématiques, physique et chimiques. 1er section du Bulletin universel des sciences. Sturm et Gaultier de Claurbry (eds), Firmin A Didot, Paris, Tome XII: 323–335[15]

Poncelet JV (1874) In: Kretz MX (ed) Cours de mécanique appliquée aux machines. Gauthier–Villars, Paris

Poncelet JV (1845) Traité de mécanique appliquée aux machines. Deuxième partie, Leroux, Liège

Poncelet JV (s.d.) Notice analytique sur les travaux de M Poncelet. Bibliothèque Nationale de France, 4°–Ln–27, vol. 16475[16]

Popper K (1963) Conjectures and refutations: the growth of scientific knowledge. Routledge, London

Popper K (1959) The logic of scientific discovery. Hutchinson, London

Prawitz D (1977) Meaning and proof. The conflict between the classical and intuitionistic logic. Theoria 43:6–39

Prawitz D (1965) Natural deduction. A proof–theoretical study. Almqvist & Wilksell, Uppsala

Prawitz D, Melmnaas PE (1968) A survey of some connections between classical, intuitionistic and minimal logic. In: Schmidt A, Schuette H (eds) Contributions to mathematical logic. North–Holland, Amsterdam, pp 215–229

Prony G (1837) Notice biographique sur M. Navier. Annales des ponts et chaussées Mémoires et documents 7(1):1–19

Prony G, Girard S, Navier CLMH (1829) Rapport suivant sur le Mémoire de M. Coriolis, intitulé *du calcul de l'effet des machines*. In: Procès–verbaux des séances de l'Académie, 22 Jun, Institut de France. Académie des sciences. Imprimerie de l'observatoire d'Abbadia, Hendaye Basses Pyrénées, [publication 1921] Année 1828–1831, Tome IX, pp 266–272

Prony G (1796) Nouvelle architecture hydraulique. Contenant la description détaillée des machines à feu. Seconde Partie. Firmin Didot, Paris

[14] [see also in .pdf, Section 1, vol 5562 at *Bibliothèque Nationale de France* via: http://www.bnf.fr]

[15] [see also .pdf, vols 5866 and 5562 at *Bibliothèque Nationale de France* via: http://www.bnf.fr]

[16] [see also .pdf (for registered members only) at *Bibliothèque Nationale de France* via: http://www.bnf.fr and *Id.*, 1834? 1835? Notice analytique sur les travaux de M. Poncelet, chef de bataillon du Génie, Professeur de Mécanique appliquée aux machines, actuellement adjoint au Comité des fortifications à Paris. Bachelier, Paris]

Prony G (1790) Nouvelle architecture hydraulique. Contenant un traité de mécanique à l'usage de ceux qui se destinent aux constructions de tous les genres, et des Artistes en général. Première Partie. Firmin Didot, Paris
Przelecki M (1969) The logic of empirical theories. Routledge & Kegan Paul, London
Ramsey W (1918) Life and letters of Joseph Black, M.D. Constable, London
Raveau C (1919) Comment Carnot a calculé l'équivalent mécanique de la Chaleur. Un document Inédit. Comptes Rendus des Sciences de l'Académie des sciences 168:549–552
Redondi P (1980) L'accueil des idées de Sadi Carnot et la technologie française de 1820 à 1860. De la légende al l'histoire. Vrin, Paris
Redondi P (1976) Sadi Carnot et la recherche technologique en France de 1825 à 1850. Revue d'histoire des sciences XXIX(3):243–259
Redondi P (1974–75) Contributi alla conoscenza di Sadi Carnot e dei principi della termodinamica nell'opera di Paolo Saint–Robert. Atti della Accademia delle scienze di Torino, Classi Scienze Morali 109:281–318
Reech F (1869) Théorie des machines motrices et des effets mécaniques de la chaleur. Lacroix éditeur, Paris
Reech F (1854) Théorie générale des effets dynamiques de la chaleur. Journal de Mathématiques pures et appliquées. Bachelier, Paris
Reech F (1853) Théorie générale des effets dynamiques de la chaleur. Journal de Mathématiques pures et appliquées XVIII:357–378
Reech F (1852) Deuxième Note sur la théorie des effets dynamiques de la chaleur. Comptes rendus des séances de l'Académie des sciences XXXIV:21 [see also the first quotation of idem *Note* at Séance du lundi 1er Décembre in: Reech (1851), p 602]
Reech F (1851) Notice sur un Mémoir intitulé : Théorie de la force motrice du calorique. Comptes rendus des séances de l'Académie des sciences. Séance du lundi 24 Novembre. XXXIII:567–571
Reinhard M (1950–52) Le grand Carnot, 2 vols. Hachette (ed), Paris
Renaud P (1937) Analogies entre le principes de Carnot, Mayer et Curie. Hermann, Paris
Renn J (2000) Challenges from the past. Innovative structures for science and the contribution of the history of science. Innovative structures in basic research. In: Ringberg–Symposium Max–Planck–Gesellschaft zur Förderung der Wissenschaften, pp 25–36
Riccati V (1772) De' Principi della Meccanica. Stamperia Coleti, Venezia
Riccati V (1749) Dialogo di Vincenzo Riccati della Compagnia di Gesù dove ne' congressi di più giornate delle forze vive e dell'azioni delle forze morte si tien discorso. Stamperia di Lelio dalla Volpe, Bologna
Richmann GW (1750) Formulae pro gradu excessus caloris supra gradum caloris, mixti ex nivee et sale amoniaco. Novi commentarii Academiae Scientiarum Imperialis Petropolitanae I:171–172
Riemann GF ([1861] 1868) Memoir on the conduction of heat. In: Weber (1892), pp 391–404
Robelin CP (1832) Comment Carnot a calculé l'équivalent mécanique de la Chaleur. Un document Inédit. Revue Encyclopédique 55:528–530
Rose PL, Drake S (1971) The Pseudo–Aristotelian questions of mechanics in renaissance culture. Studies in the Renaissance 18:65–104
Rosenberg F (1895) Isaac Newton und seine physikalischen Prinzipien. Barth, Leipzig
Rosenfeld L (1941) La genèse des principes de la thermodynamique. Bulletin de la Société Royale de Liège 10:197–212
Rothstein J (1956) Information, logic, and physics. Philosophy of Science 23(1):31–35
Rumford[17] B (1798) Recherces sur la sources de la chaleur qu'excite le frottement. Journal de physique, de chimie et d'histoire naturelle et des arts XLVII:24–39

[17]On the first page of the paper on can read "Benjamin Rumford". In *Notes sur les mathématiques, la physique et autres sujets* Sadi Carnot cite the author with "Benjamin Thompson" (*Ivi*, f. 15v; see also Carnot S 1878b; Carnot S 1986).

S N (1785) Articolo VI. Histoire de l'Académie Royale des Sciences année 1780 avec les Mémoires & A Paris. Giornale de' Letterati, Tomo LVII. Jocopo Grazioli, Pisa, pp 231–232

S N (1796) Articolo I Recherches sur les causes des principaux faits physiques & c. par J. B. Lamark Professuer de Zoologìe au Muséum National d'Histoire Naturelle. A Paris T.I. & II. Seconde année de la République. Giornale de' Letterati, Tomo CI. Landi e Compagni, Pisa

Saint–Robert P (1883–84) Perché i ghiacciai si vadano ritirando. Atti della Reale Accademia nazionale dei Lincei, Classe Scienze Fisiche e Matematiche, s3, III, pp 22, 56–62

Saint–Robert P (1879) Sul moto sferico del pendolo avuto riguardo alla resistenza dell'aria e alla rotazione della terra. Memorie della Società Italiana delle Scienze III(48)

Saint–Robert P (1878) Poche parole intorno ad una memoria del Capitano Francesco Siacci sul pendolo di Leone Foucault. Atti della Reale Accademia delle scienze di Torino

Saint–Robert P (1876) Intorno al calore che deve prodursi nell'esperienza immaginata da Galileo per misurare la forza di percossa. Rendiconti dello Reale Instituto Lombardo di Scienze e Lettere. Milan IX:174–179

Saint–Robert P (1875) Del calore attuale contenuto ne' corpi. Rendiconti dello Reale Instituto Lombardo di Scienze e Lettere. Milan VIII:876–879

Saint–Robert P (1872–1874) Mémoires scientifiques reunis et mis en ordre. Balistique, vol I–1872; Artillerie, vol II–1873; Mécanique et hypsométrie, vol III–1874. Bona, Torino

Saint–Robert P et al (1873) Una salita alla Torre d'Orvada. Narrazione della salita. Appunti zoologici. Fratelli Bocca editore, Torino

Saint–Robert P (1872) Qu'est–ce que la force. La revue scientifique de la France et de l'étranger. 2s., Tome II. Germer Baillière, Paris, pp 985–993

Saint–Robert P (1870–1871) Altezza sul livello del mare di alcuni punti dell'alto Piemonte determinata col barometro. Atti della Reale Accademia delle scienze di Torino VI:272–285

Saint–Robert P (1870a) Note II. Jules–Robert Mayer. In: Saint–Robert 1870c, pp 451–470

Saint–Robert P (1870b) Note I. Sadi Carnot. In: Saint–Robert 1870c, pp 431–450

Saint–Robert P (1870c) Principes de Thermodynamique. 2nd edn. Loecher, Turin et Florence [see also idem edition printed by Gauthier – Villars, Paris]

Saint–Robert P, Berruti G (1871) Gita al Gran Sasso d'Italia. Bollettino del Clup Alpino Italiano, Torino

Saint–Robert P (1868–1869a) Teorema sul declinatore orario del prof Foscolo. Atti della Reale Accademia delle scienze di Torino IV:331–333

Saint–Robert P (1868–1869b) Notice biographique sur Sadi Carnot. Atti della Reale Accademia delle scienze di Torino, Classe Scienze Fisiche e Matematiche IV:151–170

Saint–Robert P (1869) Sadi Carnot. Notice biographique par le Comte Paul de Saint–Robert. This title is only cited in: Pubblicazione del Regno. Bibliografia d'Italia. Anno III. N°. 1, Gennaio 1869. Loesher, Torino e Firenze. [It would refer to Saint–Robert 1868–1869b; someone else also cited Classe Scienze Fisiche e Matematiche VI, but it is not verified]

Saint–Robert P (1867–1868a) Sopra un'opera del prof. Cavallero intitolata: sulle macchine motrici. Atti della Reale Accademia delle Scienze di Torino, Classe Scienze Fisiche e Matematiche III:16–20

Saint–Robert P (1867–1868b) Sul vero significato di una terzina di Dante. Atti della Reale Accademia delle scienze di Torino III:588–600

Saint–Robert P (1867–1868c) Sopra un'opera del prof. Cavallero intitolata: sulle macchine motrici. Atti della Reale Accademia delle scienze di Torino III:16–20

Saint–Robert P (1865–1866a) Aggiunta alla Memoria del prof. Govi sul metodo per determinare la lunghezza del pendolo. Atti della Reale Accademia delle scienze di Torino I:513–515

Saint–Robert P (1865–1866b) Note sur le travail mécanique dépensé dans la compression et sur le travail restitué par la détente d'un gaz permanent, par le Cte Paul de S^{t}–Robert. Atti della Reale Accademia delle scienze di Torino I:283–296

Saint–Robert P (1865–1866c) Intorno alla formula barometrica ed alla rifrazione atmosferica. Atti della Reale Accademia delle scienze di Torino I:193–203

Saint–Robert P (1865–1866d) Sulla Saxifraga fiorulenta Moretti. Atti della Reale Accademia delle scienze di Torino I:203–205

Saint–Robert P (1866) Remarques à l'occasion d'une note de M. Clausius. Archives des sciences physiques et naturelles 25:34–41

Saint–Robert P (1865) Principes de Thérmodynamique. Cassone (ed) 1st edition, Torino

Saint–Robert P (1863) Lettre concernant la théorie mécanique de la chaleur. Cosmos XXII:200

Sarton G (1952) A guide to the history of science. The Ronald Press, New York

Scheele CW (1785) Supplement au Traité chimique de l'air et du feu de M. Scheele, contenant un Tableau abrégé des nouvelles découvertes sur les diverses espèces d'air, par Jean–Godefroi Léonhardy; Des notes de M. Richard Kirwan, & une lettre du Docteur Priestley à ce chimiste anglois, sur l'Ouvrage de M. Scheele; Traduit et augmenté de Notes & du complement du Tableau abrégé de ce qui a éte publié jusqu'aujourd'hui sur les differentes espéces d'Air: Par M. le Baron de Dietrich. Rue et Hotel Serpente, Paris

Scheele CW (1781) Traité chimique de l'air et du feu de M. Scheele. Avec une introduction de Torbern Bergman, Professeur de Chimie & Pharmacie, Écuyer, Membre de plusierus Académies: Ouvrage traduit de l'Allemand, par le Baron de Dieterich. Rue et Hotel Serpente, Paris

Schmid GB, Herrmann F (1985) Analogy between mechanics and electricity. The European Journal of Physics 6(1):16–21

Schmid GB (1984) An up – to – date approach to physics. The American Journal of Physics 52(9):794–799

Schubring G (2005) Conflicts between generalization, rigor and intuition. Number Concepts Underlyng the development of Analysis in 17–19th Century France and Germany. Springer, New York

Scott WL (1971) The conflict between atomism and conservation laws 1640–1860. Elsevier Publishing Company, New York

Scott WL (1959) The significance of "hard bodies" in the history of scientific thought. Isis 50:199–210

Secchi A (1864) L'unità delle forze fisiche. Tipografia Forense, Roma

Singer C ([1954–58] 1993) A history of technology, 5 vols. The Clarendon Press, Oxford

Skolem T ([1920] 1967) Logico–combinatorial investigations in the satisfiability or provability of mathematical propositions. In: van Heijenoort J (ed) From Frege to G&oumldel: a source book in mathematical logic, 1879–1931. The Harvard University Press, Cambridge MA, pp 252–263

Spivak M (1979) A comprehensive introduction to differential geometry, vol II. Perish Inc., Berkeley, pp 179–182

Stevin S ([1605] 1608) Liber primus Staticae. De staticae elmentis. In: Tomus quartus mathematicorum hypomnematum de statica, Lugodini Batavoru

Steward B (1866) An elementary treatise on heat. The Clarendon Press, Oxford

Tait PG (1887) Conférences sur quelquesuna des pogrés récent de la physique. Trad. de la 3e éd. angl. par Krouchkoll M, Paris

Tait PG (1868) A sketch of history of thermodynamics. Edmoston and Douglas, Edinburgh

Tartaglia N (1565) Iordani Opvscvlvm de Ponderositate, Nicolai Tartaleae Stvdio Correctvm Novisqve Figvrisavctvm. Cvm Privilegio Traiano Cvrtio, Venetiis, Apvd Curtivm Troianvm. M D Lxv

Tartaglia N ([1554] 1959) La nuova edizione dell'opera Quesiti et inventioni diverse de Nicolo Tartaglia brisciano Riproduzione in facsimile dell'edizione del 1554. In: Masotti A (ed) Commentari dell'Ateneo di Brescia, Tipografia la Nuova cartografica, Brescia

Taton A (ed) (1976) Sadi Carnot et l'essor de la thermodynamique, Table Ronde du Centre National de la Recherche Scientifique. École Polytechnique, 11–13 Juin 1974. Éditions du Centre National de la Recherche Scientifique, Paris

Taton R (1964a) L'École Rouale du Génie de Mézières In: Taton R (ed) (1964), pp 559–615

Taton R (ed) (1964b) Symposium: Enseignement et diffusion des sciences en France au XVIII[è] siècle. Hermann, Paris

Taton R (ed) (1951) L'œuvre scientifique de Monge. Presses Universitaires de France, Paris

Tega W (1984) Arbor Scientiarum. Il Mulino, Bologna

Thackray A (1970) Atoms and powers. An essay on newtonian matter and the development of chemistry. The Harvard University Press, Cambridge, MA
Thomson W ([1890] 1943) Carnot's theory of the motive power of heat with numerical results deduced from Regnault's experiments on steam. In: Thurston (1943), pp 127–204
Thomson W (1882–1911) An account of Carnot's theory of the motive power of heat with numerical results deduced from Regnault's experiments on steam. Mathematical and physical papers by Sir William Thomson, vol I. The Cambridge University Press, Cambridge, pp 113–155
Thomson W (1852) An account of Carnot's theory of the motive power of heat with numerical results deduced from Regnault's experiments on steam. Annales de Chime et de Physique XXXV:248–255
Thomson W (1851a) On an absolute thermometric scale founded on Carnot's theory of the motive power of heat, and calculated from Regnault's observations. Mathematical and physical papers, vol I. The Cambridge University Press, Cambridge, pp 100–106
Thomson W (1851b) On the dynamical theory of heat, with numerical results deduced from Mr Joule's equivalent of a thermal unit, and M. Regnault's observations on steam. Mathematical and physical papers, vol I. The Cambridge University Press, Cambridge, pp 175–183
Thomson W (1848–1849) An account of Carnot's theory of the motive power of heat with numerical results deduced from Regnault's experiments on steam. Transactions of the Edinburgh Royal Society XVI(5):541–574
Thomson W (1848) On an absolute thermometric scale founded on Carnot's theory of the motive power of heat, and calculated from Regnault's observations. Proceedings of The Cambridge Philosophical Society 1(5):66–71
Thurston RH (1943a) Reflections on the motive power of heat and on machines fitted to develop this power. Waverly Press Inc, Baltimore
Thurston RH (1943b) The work of Sadi Carnot. In: Thurston (1943), pp 1–19
Torricelli E (1644) Opera geometrica. Massa–Landi, Firenze
Tredgold T (1838) The steam engine, its invention and progressive improvements, an investigation of its principles, and its application to navigation, manufactures, and railways. vol 2, a new edition, enlarged by the contributions op eminent scientific men, and extended to the science oprevised and edited by Woolhouse WSB, Weale, London
Truesdell CA (1980) The tragicomical history of thermodynamics. 1822–1854. Springer, Berlin
Truesdell CA (1976) Termodinamica razionale. Accademia Nazionale dei Lincei, Contributi del Centro Linceo Interdisciplinare di Scienze Matematiche e loro Applicazioni. Anno CCCLXXXIII, N. 20, Roma, pp 208–235
Truesdell CA (1970) The tragicomedy of classical thermodynamics. Springer, Berlin
Truesdell CA Bharatha S (1977) The concepts and logic of classical thermodynamics as a theory of heat engines: rigorously constructed upon the foundation laid by S Carnot and F Reech (Theoretical and mathematical physics). Springer, Berlin/Heidelberg/New York
Truesdell CA (1968a) Essay in the history of mechanics. Springer, New York
Truesdell CA (1968b) Whence the law of the moment of momentum? [Lecture 202, Whence the law of moment of momentum? Colloquium in Engineering Science, Mechanical Engineering Department, Columbia University, 8 Jan 1963] In: Essays in the history of mechanics, *op. cit.*, pp 239–271
Tyndall J (1863) Heat considered as a Kind of Motion being a course of twelve lectures delivered at the Royal institution of Great Britain in the season of 1862
Vailati G (1896–1897) Il principio dei lavori virtuali da Aristotele a Erone d'Alessandria. Atti della R. Accademia delle Scienze di Torino XXXII. Clausen, Torino
Varignon P (1725) La nouvelle Mécanique ou statique. Jobert C, Paris
Verdet MÉ (1868–1869) Cours de Physique a l'Ecole polytechnique, 2 vols. Imprimerie Impériale, Paris
Verdet MÉ (1868–1872) Œuvres de E. Verdet. Masson et Fils, Paris
Verdet MÉ (1868) Théorie mécanique de la chaleur. [Imprimerie Impériale, Paris]. In: Œuvres de E. Verdet (1868–1872), Tome VII

Verdet MÉ (1862) Exposé de la théorie mécanique de la chaleur. Leçon su la theorie mecanique de la chaleur professes en 1862 devant la Societé chimique de Paris le 7 et le 21 février. In: Œuvres de E. Verdet (1868–1872), Tome VII

Volta A ([1968] 1998) In: Gliozzi M (ed) Opere scelte di Alessandro Volta. Reprint with new bibliograhy. UTET, Torino

Volta A (1987) In: Recenti M (ed) L'elettromozione. Theoria, Roma

von Weizsacker CF (1973) Classical and quantum descriptions. In: Mehra J (ed) The physicist's conception of nature. Reidel, Dordrecht, pp 635–667

Volta A (1966) Aggiunte alle opere e all'epistolario di Alessandro Volta, Edizione Nazionale, vol I. Nicola Zanichelli, Bologna

Volta A (1949–1955) Epistolario di Alessandro Volta, Edizione Nazionale, vols I–V. Nicola Zanichelli, Bologna

Volta A (1918–1929) Le Opere di Alessandro Volta Edizione Nazionale, vols I–VII. Ulrico Hoepli, Milano

Volta A (1816) Lettera Settima [and Addizione alla Lettera Settima]. In: Collezione dell'Opere del Cavaliere Conte Alessandro Volta Patrizio Comasco. Membro dell'Istituto Reale del Regno Lombardo Veneto, Professore Emerito dell'Università di Pavia, e Socio delle più Illustri Accademie d'Europa. Tomo I. Parte II. Nella stamperia di Guglielmo Piatti, Firenze, pp 238–283

Volta A (1802) Memoria del prof. Alessandro Volta sull'identità del fluido elettrico col fluido galvanico. Annali di chimica e storia naturale 19:38–74

Volta A (1800a) Sull'elettricità eccitata dal semplice contatto di sostanze conduttrici di diversa natura in una lettera di Alessandro Volta a Sir Joseph Banks, da Como nel Milanese, 20 marzo 1800. In: Opere scelte di Alessandro Volta, pp 514–534

Volta A (1800b) On the electricity excited by the mere contact of conducting substances of different kinds. In a letter from Mr. Alexander Volta ... to the Rt. Hon. Sir Joseph Banks .... Philosophical Transactions of the Royal Society II:403–431

Volta A (1799) L'elettroscopio condensatore in una lettera a Lorenzo Mascheroni. In: Opere scelte di Alessandro Volta, p 511

Volta A (1796) Lettera Terza del Sig Cav Don Alessandro Volta P Professore, Al Sig Ab Anton Maria Vassalli Professore di Fisica a Torino, 24 ottobre 1795. Annali di chimica e storia naturale 11:84–128

Volta A (1782) XVI. Del modo di render sensibilissima la più debole Elettricità sia Naturale, sia Artificiale. By Mr. Alexander Volta, Professor of Experimental Philosophy in Como, &; communicated by Right Hon. George Earl Cowper, FRS. Read on March 14, 1782. Davis L, Elmsly Printers to the Royal Society. Philosophical Transaction of the Royal Society of London LXXII(I):237–280

Wallis J (1693) Geometriae professoris Saviliani, in celeberrima Academia Oxoniensi Opera mathematica by Wallis, John, 1616–1703. Oxoniae: E Theatro Sheldoniano

Wallis J (1668) A summary account of the general laws of motion. Philosophical Transactions III(26):864–866

Weber H (ed) (1892) Bernhard Riemann's Gesammelte Matheamtische Werk. Collected mathematical works, 2nd edn. Leipzig, Teubner

Winter TN (2007) The mechanical problems in the corpus of Aristotle. DigitalCommons.The University of Nebraska–Lincoln, pp 1–43[18]

Zenzen MJ (1990) Thermodynamics: paradigms, pedagogy and epistemology. Communication to European Conference on History Teaching Physics, Cambridge

Zeuner G ([1860] 1866) Grundzüge der mechanischen Wärmetheorie mit besonderer Rücksicht auf das Verhalten des Wasserdampfes, 2nd edn. Freiberg, Engelhardt

Zeuner G (1869) Théorie mécanique de la chaleur, avec des applications aux machines, 2nd édn (trans: Germany by Arnthal Maurice ad Cazin Achille). Gauthier–Villars, Paris

[18] [Retrieved March 16, 2008, avilable via http://digitalcommons.unl.edu/classicsfacpub/68]

Analytical Index: Names and Subjects

C.C. Gillispie and R. Pisano, *Lazare and Sadi Carnot: A Scientific and Filial Relationship*, History of Mechanism and Machine Science 19, DOI 10.1007/978-94-017-8011-7,

D

F

G

P

Q

MIX
Papier aus verantwortungsvollen Quellen
Paper from responsible sources
FSC® C105338

If you have any concerns about our products,
you can contact us on
ProductSafety@springernature.com

In case Publisher is established outside the EU,
the EU authorized representative is:
Springer Nature Customer Service Center GmbH
Europaplatz 3, 69115 Heidelberg, Germany

Printed by Libri Plureos GmbH
in Hamburg, Germany